D1728613

Contributions to Statistics

V. Fedorov / W. G. Müller / I. N. Vuchkov (Eds.) Model-Oriented Data Analysis, XII/248 pages. 1992

J. Antoch (Ed.) Computational Aspects of Model Choice, VII / 285 pages, 1993

W. G. Müller / H. P. Wynn / A. A. Zhigljavsky (Eds.) Model-Oriented Data Analysis XIII/287 pages, 1993

P. Mandl / M. Hušková (Eds.) Asymptotic Statistics X/474 pages, 1994

Peter Dirschedl
Rüdiger Ostermann (Eds.)

Computational Statistics

Papers Collected on the Occasion of the 25th Conference on Statistical Computing at Schloß Reisensburg

With 82 Figures

Physica-Verlag
A Springer-Verlag Company

Series Editors
Werner A. Müller
Peter Schuster

Editors
Dipl.-Inf. Peter Dirschedl
University of Munich
Dept. of Medical Information Processing,
Biometry and Epidemiology
D-81377 Munich, Germany

Dr. Rüdiger Ostermann
University of Siegen
Computer Center
D-57068 Siegen, Germany

ISBN 3-7908-0813-X Physica-Verlag Heidelberg

CIP-Titelaufnahme der Deutschen Bibliothek
Computational statistics: papers collected on the occasion of the 25th Conference on Statistical Computing at Schloss Reisensburg / Peter Dirschedl; Rüdiger Ostermann (ed.). - Heidelberg: Physica-Verl., 1994
(Contributions to statistics)
ISBN 3-7908-0813-X
NE: Dirschedl, Peter [Hrsg.]

88/2202-543210 - Printed on acid-free paper

Contents

Editorial
by P.Dirschedl & R.Ostermann 1

The Roots of Computational Statistics in Germany
by N.Victor 3

Twenty-Five Working Conferences on Statistical Computing — Reflections on Twenty Years of Reisensburg Meetings
by A.Hörmann 17

Remarks on the History of Computational Statistics in Europe
by P.P.Sint 37

Languages for Statistics and Data Analysis
by P.J.Huber 53

A Brief History of S
by R.A.Becker 81

Practical Guidelines for Testing Statistical Software
by L.Wilkinson 111

On the Choice and Implementation of Pseudorandom Number Generators
by J.Lehn & S.Rettig 125

Seven Stages of Bootstrap
by R.Beran 143

Special Resampling Techniques in Categorical Data Analysis
by I.Pigeot 159

Statistical Problems in Planning, Conduct and Analysis of Epidemiological Studies
by K.-H.Jöckel 177

Computer Aided Design of Experiments
by D.Rasch & P.Darius 187

Knowledge-Based Systems in Statistics: A Tutorial Overview with Examples
by U.Haag 211

Diagnostic Plots for One-Dimensional Data
by G.Sawitzki 237

Graphical Data Analysis Using LISP-STAT
by A.Benner 259

Multivariate Graphics: Current Use and Implementations in the Social Sciences
by R.Schnell & H.Matschinger 275

Interactive Analysis of Spatial Data
by M.Nagel 295

REGARDing Geographic Data
by A.Unwin 315

Applied Nonparametric Smoothing Techniques
by W.Härdle, S.Klinke & M.Müller 327

Missing Values: Statistical Theory and Computational Practice
by W.Vach 345

A Permutation Approach to Configural Frequency Analysis (CFA) and the Iterated Hypergeometric Distribution
by J.Röhmel, B.Streitberg & C.Tismer 355

Dynamic Modelling of Discrete Data
by L.Fahrmeir 379

Evaluating the Significance Level of Goodness-of-Fit Statistics for Large Discrete Data
by G.Osius 395

A Multiple Test Procedure for Nested Systems of Hypotheses
by G.Hommel & G.Bernhard 419

Kernel Estimation in the Proportional Hazards Model
by J.Kübler 435

Interval Censored Observations in Clinical Trials
by A.Koch 451

Covariates in Clinical Trials: Effects of Adjustment in Regression Models
by G.Antes & C.Schmoor 469

Classification and Regression Trees (CART) Used for the Exploration of Prognostic Factors Measured on Different Scales
by B.Lausen, W.Sauerbrei & M.Schumacher 483

Clustering Algorithms and Cluster Validation
by A.D.Gordon 497

Learning Statistics: Beyond Authoring Systems
by R.Schulmeister 513

Remarks on Protecting Patient Data Against Misuse and on its Consequences Concerning their Statistical Data Analysis
by R.Haux 533

Index 539

Alphabetical List of Authors, Reviewers and Editors 549

Acknowledgements 553

Editorial

Peter Dirschedl[1], Rüdiger Ostermann[2]
[1]Institut für Medizinische Informationsverarbeitung, Biometrie und Epidemiologie, Ludwig-Maximilians-Universität München, D - 81377 München
[2] Hochschulrechenzentrum, Universität-GH Siegen, D - 57068 Siegen

There are other books on *Computational Statistics*, e.g. vol. 9 of the Handbook of Statistics, edited by C.R.Rao. But the present volume is different from all others, because it does not try to define *Computational Statistics.* We believe that liveliness of any scientific area is not determined by interesting but sometimes fruitless definitions. These are liable to change quite rapidly in our developing field, hence the topic is defined by the contributions.

Our editorial effort to produce this book is closely linked to an anniversary, the 25th conference on *Statistical Computing* at Schloß Reisensburg, Bavaria, in June 1993.

Dozens of years ago, these conferences were founded by Prof. Norbert Victor, by now past president of the IASC - we all owe him so much! He provided an organizational basis for the annual meetings of the German working group *Statistische Auswertungssysteme* (Statistical Analysis Systems) of the GMDS (German Association of Medical Informatics, Biometry and Epidemiology), and for many years the conferences have been jointly organized with the working group *Computational Statistics* of the International Biometric Society, German Region.

The conference site was well chosen! Everybody who had the opportunity to join the meetings was pleased with the current relevance of topics, the quality of presentations, the possibility to discuss anything until late at night – and especially with the friendly and informal atmosphere of the meetings (obviously strongly influenced by some benevolent ghost residing on the pinnacles of the castle).

Thus, this volume is a collection of invited papers, covering at least the more important parts of a wide spectrum, with the intention of supplying tutorial and in depth overviews of those topics which were typically dealt with or aroused continuous interest at the Reisensburg meetings.

We start with contributions reflecting our historical background, while others discuss different approaches to developing something like a *language for statistics*. Two further papers draw on basic features of software, i.e. on testing the numerical reliability of packages and on construction principles for better pseudorandom number generators. Obviously, the following papers on bootstrapping and resampling assume that the numbers produced are of sufficient randomness.

In the sequel, a wide range of contributions covers areas like design of experiments, graphical techniques, modelling and testing problems, a review of clustering algorithms, classification and regression trees, and a concise discussion of cognitive aspects of computer aided learning. None of the authors missed the opportunity to provide a comprehensive overview of a specialized area, including a rich bibliography for more intense studies. Many of the papers never would have been written without this occasion.

To edit a volume like this is synonymous with keeping control of the heavy communications traffic between editors, authors and reviewers (more than 70 people altogether). Anything which might have happened eventually did and we observed dropouts and outliers in both space and time. But in the end we – hopefully – succeeded in keeping everything together and in remaining true to the style and contents of the *Reisensburg* meetings.

One may rate this collection to be heavily biased. This is true, but nevertheless, in our opinion, we have produced a volume that contains some *missing* papers, which may serve as a textbook and which is a valuable guide towards further reading as well.

Cordially we would like to thank all authors for joining us in the project, the reviewers for their valuable help (each paper was reviewed twice, and anonymously), and the sponsoring companies who supplied the necessary financial backing (see list enclosed).

Special thanks are due to everybody who did not hesitate a second to support us, in particular our colleagues in the working groups, Uwe Haag, Armin Koch, Matthias Nagel and Günther Sawitzki. The preparation of this volume would not have been possible without their help.

Munich and Siegen , March 1994

Peter Dirschedl
WG "Statistical Analysis Systems"
GMDS , German Association of
Medical Informatics,
Biometry and Epidemiology

Rüdiger Ostermann
WG "Computational Statistics"
International Biometric Society,
German Region

The Roots of Computational Statistics in Germany [1]

Norbert Victor
University of Heidelberg, Institute for Medical Biometry and Informatics,
Im Neuenheimer Feld 305, D-69120 Heidelberg, Germany

Key Words:
history of computational statistics in Germany

Abstract
The history of "Computational Statistics" in Germany is described from a subjective point of view. The importance of the "Reisensburg Symposia" for this development is shown.

Introduction

The development of electronic computers (the hardware) proceeded temporarily parallel in North America (Aiken, von Neumann and others → ENIAC) and Germany (Zuse). The formation of the discipline named "Informatik" in German (in English mostly "computer sciences" but increasingly "informatics", too) also took place almost simultaneously in North America and several European countries. In this development the German group around F.L. Bauer and K. Samelson played an important role. The mentioned names show that the beginnings of informatics were determined by numerical mathematicians. Computers and software were dominantly seen as a device for solving numerical problems.

The first attempts to use computers and appropriate software as tools for statistics can all be localized in Anglo-Saxon countries; I can refer to a ten year old survey paper about these early activities (Victor (1984)). Hence,

[1] In memoriam Liselotte Eder, who coworked in editing the SSN, for 15 years from its first to its last issue. Even though she was not a statistician, she committed herself to this task for she felt that we intended to contribute to a transformation of statistics.

the first phase of the development of "computational statistics" (for a definition see Victor (1984)), took place without German participation. The explanation for this fact is the languish of the mathematical and applied statistics in Germany after WW II. The bloodletting by the emigration and expulsion of German-speaking Jewish scientists under the Nazi regime was especially crucial to the just developing field of statistics; sufficient evidence is mentioning the emigration of Abraham Wald. In Germany there were no statisticians, who needed suitable tools for the solution of their problems.

Taking over American developments by applied statisticians for analyses of large studies and surveys was the beginning of the computational statistics ("CS") in Germany: The German "CS" is therefore rooted in North America. With the flourishing of mathematical and applied statistics in our country (the latter especially in the fields of economics and medicine) CS-activities developed quickly on the basis of these experiences. To give a report on these activities is the task given to me because I was involved in these developments.

The Subjective View

The presentation of a development by someone directly involved in it will always have a subjective imprint. The reason of this meeting — the 20th anniversary of the Reisensburg Symposia about computational statistics — however justifies a subjective view. I will allow myself to write in the first person, which is unusual in scientific papers. These symposia have emerged from a project which I was the initiator of; this project (grant no.: DVM-107) will play an important role in the following text. The DVM-107-activities and the Reisensburg Symposia belong undoubtedly to the strongest roots out of which the German "CS" have gained their strength to grow. They had inspiring influence on many developments and many young statisticians have become computer statisticians at the "Reisensburg".

Today there is no more backlog of "CS" in Germany. In the "International Association for Statistical Computing" (IASC) the Germans make up the second largest group next to the USA; German participation in scientific activities in "CS" (journals, publications, congresses etc.) is overproportionally large in size. The supposedly largest statistical software fair (SoftStat) is organized in Heidelberg; statistical software houses prefer Germany as the site for their European branches. The "Reisensburg" has contributed essentially to this success.

I myself have studied mathematics in Mainz (orientation numerics) under the leadership of Bauer and Samelson, and so I was lucky to directly experience the beginnings of informatics. I was one of the first students, who had to learn ALGOL in their practical course (in 1960). The experiences with these courses have been reported in the same issue of "Elektronische Datenverarbeitung" as the German version of the ALGOL-report Baumann & Paul (1961). Later I concentrated my interests on statistics and applied for the above mentioned DVM-107 project in 1972. The ambitious aims of this project were not reached. The youthful spirit with which I pushed developments wore out in the course of time. Nevertheless, I consider many of today's CS-activities as belated fruits of these efforts. I ask you to understand if I put too much emphasis on activities which I have been involved in and consequently neglect other activities.

Motivating Powers and Phases of Development

The striving for two aims has pushed the development of "CS":

- The feasibility and facilitation of calculation intensive (multivariate) statistical analyses and
- the handling and mastery of very large data sets like those occurring in big studies.

Later the extension of the spectrum of statistical methods became an important aim. Finally, improving the presentation of results by graphical methods and facilitating the method choice and improvement of statistical consulting through knowledge-based systems have been added as aims.

In his famous lecture on classification problems to the Royal Statistical Society, Rao (1948) presented discriminant functions for the separation of three Indian castes using four random variables. Anderson (1951) used Rao's example in his foundation-laying paper on discriminant analysis and corrected a series of Rao's calculating errors. In his standard monograph on multivariate analysis (Anderson (1958)), published seven years later, he again used this example and had to correct some of his own additional mistakes. Such troubles in practical application of multivariate procedures can not be imagined by today's users of statistical systems, who with a single statement can request complete canonical analyses for 25-dimensional random vectors.

The work-expensive data handling was the main obstacle for the analysis of large studies like the Framingham study or the DFG-study "Pregnancy and child development". Today we can manage data sets of this size with database systems without any problems on workstations (possibly even with statistical systems).

If we put the pre-computer era out of consideration (dealt with in Victor (1984) too), one can divide the CS-activities into three phases guided by the mentioned aims. I will relate these to the last three decades:

The Sixties

- Writing individual programs which make statistical analysis possible or facilitate them
- Extension of programs by routines for data handling to master large data sets
- Development of statistical program packages (usable also by non-statisticians) to promote the wider use of statistical methods

The Seventies

- Information services to improve the access to statistical software and increase the efficiency by shared use
- Enhancement of portability to avoid unnecessary programming work and double developments
- Evaluation of statistical programs to avoid defective analyses because of deficient software
- Definition of quality standards for software to secure the quality of statistical analyses
- Striving for user friendliness to decrease the efforts in analyses and to relieve the specialists

The Eighties

- Availability of stastistical software on everybody's desk with systems running on PCs and workstations
- Extension of the method spectra (robust methods, exact procedures with combinatorial problems, resampling methods etc.) through development and use of better algorithms and faster computers
- Shifting the main emphasis from inferential methods to descriptive procedures and methods of exploratory data analysis, especially graphical procedures

- Support of method choice and of statistical consulting by knowledge-based systems

As I should report on the roots of "CS" I will only deal with the first two decades. Some of these early developments are outdated today; yet, these activities created the basis for today's wide use of statistical methods and they are the basis on which the now booming "CS"-field has developed. The activities of the eighties do not belong to today's theme. I would like to mention only that I am not sure about the allocation of knowledge-based systems. They were intensively discussed in the eighties and some prototypes have been presented; however, no product has attained the maturity to be distributed within the eighties. Hopefully the nineties will become "the decade of the knowledge-based statistical systems" for CS.

The Sixties

Already before 1960 statistical programs were written, especially for calculation intensive multivariate procedures (e.g. Kaiser (1959)). The first collection of statistical programs (the "package" BMD) was ready for distribution in 1961. Such programs were used here and there in Germany by scientists, who had gotten in touch with these valuable tools for analyses during a stay in the USA. These scientists brought statistical programs back to Germany, installed them in the computer centers of universities and sometimes made them available to the general public. BMD, for example, was available at the "Deutsches Rechenzentrum" (German Computer Center) in Darmstadt since 1964.

With the spreading of the problem-oriented languages FORTRAN and ALGOL the doors were then open for individual programming activities; in the sixties a lot of statistical programs were produced in Germany, too. To give an overview over these first independent German CS-activities is impossible. I can only list a few names in my scientific environment during that period: K.Überla (Mainz) in whose textbook (Überla (1968)) programs for factor analysis are listed, E.Weber (Kiel) who mainly developed programs for analysis of variance and U.Feldmann (UKE, Hamburg) who together with colleagues, developed a program package called ADAM, for the analysis of medical data.

I would also like to refer to my own activities of this type. At that time I was an assistant at the Institute for Medical Statistics, Mainz University. Being educated in numerics it was obvious that I would become responsible for the "CS" there. As the BMD-programs could not be used on our CDC-computer and the workload for adaption could not be justified (portability was an unknown term at this time), I developed (supported only by a single programmer) the STATSYS-package in 1967-68. This BMD-oriented system was an improvement of BMD in some aspects, mainly extensions of the method spectrum; it was transferred to Siemens computers in 1970-71 and achieved a certain degree of spreading (even internationally); some of the programs were still used until about 1980 (e.g. Victor (1971)).

Besides institutes of applied statistics some university computer centers were also breeding grounds for "CS" in this decade. Statistical packages were installed there, single persons or small groups were put in charge of their maintenance and of the consulting of its users and a lot of own supplementary programs were also written to extend the packages. Many participants of the Reisensburg Symposia and many of today's computer statisticians originate from this group of persons in charge of the statistical systems.

Memories of pioneer years easily stimulate to tell anecdotes. Therefore, I will stop my report of the sixties and go over to the next decade, which was of great importance for the "Reisensburg".

The Seventies and the Project DVM-107

My report for this decade has to remain incomplete, too, but the quoted developments are suitable examples for typical activities, which had priority at this time.

Program-making was eagerly continued in Germany throughout the seventies; numerous statistical programs and systems were created. I would like to mention the medical information system MINDIUS by D.Hölzel (Munich) and the analysis system for clinical studies, ASPECT, which was developed by Schering in Berlin and sponsored by the German ministry of research (BMFT) from 1974 to 1980. A remarkable distribution was attained by Späth's cluster-analysis programs, which were also introduced to the international scene (cf. Späth (1975)).

The CS-activities in university computer centers increased in number and size over this decade. I would especially like to mention the URZ in Heidelberg which made numerous statistical systems available simultaneously. By the linkage of these systems P.Beutel kept a real statistical software monster alive for several years (cf. Beutel (1979)).

Of a different kind were the activities of the group created by J.Gordesch in Berlin. As the German part of the COMPSTAT-society they contributed to the distribution of CS-ideas by the organization of congresses and publications, before the COMPSTAT activities obtained a continuous sponsor by the integration into the IASC in 1978.

Activities mentioned until now were almost solely initiated by statisticians; but in the mid seventies German informaticians working in universities or in software houses began to show interest in "CS", too. I want to mention the KARAMBA-project of the Institute of Informatics at the University of Karlsruhe (Hüber (1977)) and the Management System for Method Banks, which was developed by mbp-Dortmund (Hauer (1977)).

In the seventies, however, the efforts for better information about statistical software and the striving for the improvement of the software itself were dominating. From 1973 to 1980 so-called 'signal information' concerning statistical programs was collected in the SIZSOZ-project. This information was made available in software catalogues (three editions: 1975, 1976 and 1980); W.Langenheder was the promoter of this project. Similar aims were pursued by the project DVM-107; besides a software information service, however, a standardization of program development was aspired to, too. A short description of this origin of the Reisensburg Symposia seems appropriate.

DVM-107 was applied for as a project for "coordination and standardization of statistical programs" in September 1972 (Victor (1972)). From July 1, 1973 to December 31, 1975 it was sponsored by the BMFT under the condition that the continuation of the project by a 'permanent institution' would be ensured. The medis-institute in Munich gave its consent, took over the project activities step by step from 1975 and continued them until the eighties. For that reason the medis-institute also has its merits in the Reisensburg Symposia and in the development of "CS" in Germany.

This project's impulse were the experiences with the insufficient portability of the systems available at that time (including the own STATSYS), and discontent with the gappy spectra of methods of the then proliferating statistical systems (BMDP, SPSS, P-STAT, OSIRIS etc.). Getting over the often horrible numerics of these packages was also an aim for me. Being educated in numerics, the results of multivariate methods produced with such packages were not acceptable for me, especially, if simple IBM-accuracy was used. Realizing that it would be impossible for a small team to master the whole method spectrum competently, it was also striven for to induce as many groups of statisticians as possible to participate so that in each field one could fall back on an expert. I consider it appropriate to name the most active groups and their representatives: Mrs.J.Berger (DKFZ, Heidelberg), H.J.Christl (DKD, Wiesbaden), U.Feldmann (MH, Hannover), H.J.Friedrich (IMSD, Gießen), E.Hartmann (Schering, Berlin), E.Hultsch (Münster), H.Nowak (Aachen), E.Scheidt (Mainz), H.K.Selbmann (Ulm) and E.Weber (Kiel). In addition, the group of the medis-institute (A.Hörmann, L.Eder and M.Sund) as well as my own collaborators P.Schaumann, H.Strudthoff, F.Tanzer, H.J.Trampisch and R.Zentgraf (Biomathematics,

Gießen) should be named. The list of DVM-107 workshops on Table 1 gives an impression of the intensive project work, the discussed topics and the acquired results.

1.	Oct. 10-12, '73:	'Hattsteiner Hof', Münzenberg near Gießen
	Topics:	Software Information Service (SIS), software evaluation by experts, standardization
	Resolution:	SSN-plan
2.	Jan. 24-26, '74:	Schloß Reisensburg
	Topics:	SIS-forms, subroutine pool, standards for FORTRAN and I/O
	Resolution:	Syntax checker plan
3.	Apr. 22-24, '74:	Schloß Reisensburg
	Topics:	Subroutine pool, self-describing data interfaces
	Resolutions:	SSN-start, application for a GMDS-WG
4.	Aug. 21-24, '74:	'Berghotel Hoher Knochen', Winterberg
	Topics:	Evaluation criteria, SIS-data base, structured FORTRAN
	Results:	First version of the syntax checker
5.	Apr. 2-4, '75:	Schloß Reisensburg ($\cup$ session of the GMDS-WG)
	Topics:	Software evaluation, existing subroutine pools, missing values
	Results:	Manual for syntax checker
6.	Sept. 18-19, '75:	'Schloßberg-Hotel', Wasserlos/Aschaffenburg
	Topics:	Segmentation of statistical methods, interfaces
	Results:	Lists of segments for statistical procedures
7.	Dec. 8-10, '75:	'Dünsbergheim', Gießen ($\cup$ session of GMDS-WG)
	Topics:	Structured FORTRAN, preprocessors, portability, monitoring systems for method bases, data handling in statistical systems
	Results:	Draft of a list of requirements for stat. systems

Table 1: DVM-107 Working Sessions

The aims were ambitious and they included almost everything that was striven for by computer statisticians in this decade. In retrospect it is obvious that even in an entire decade these goals could not be reached. In the application the following aims were explicitly named:

- Software information service with evaluation of the programs by experts

- Standardization for the purpose of better portability and a more efficient use

- Improvement of the quality by segmentation of the procedures and the development of the segments by specialists and furthermore by the integration of special software (e.g. numerical subroutines).

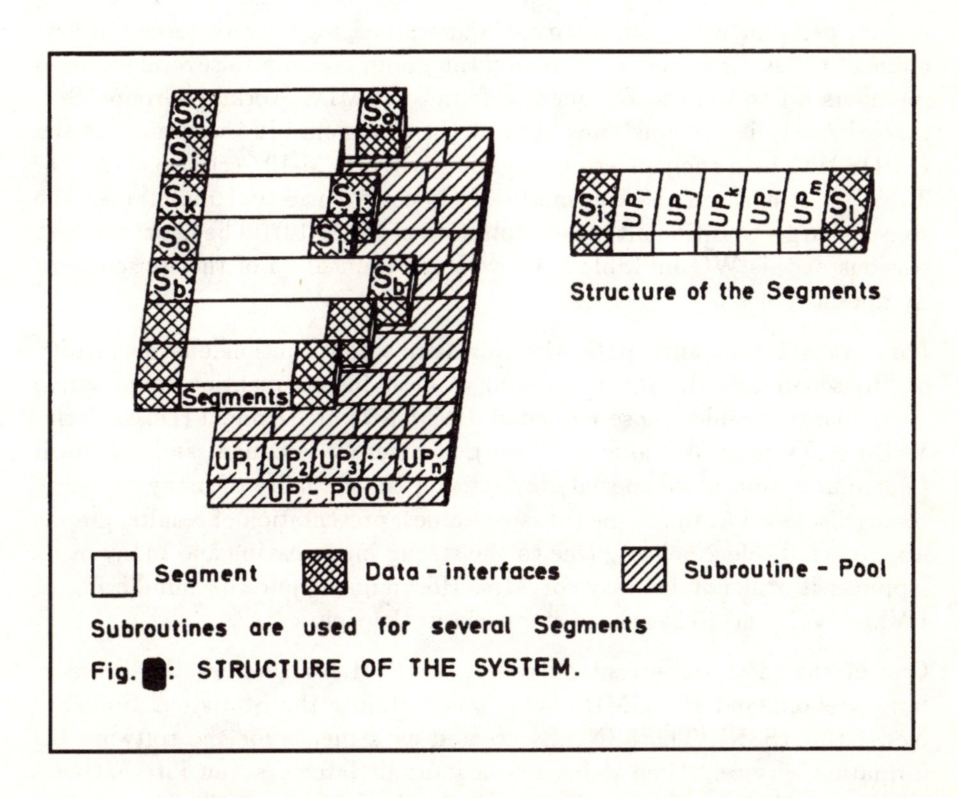

Figure 1: A figure taken from Victor et al. (1974)

The last aim resulted in the construction of a tool kit for statistical analyses. The concept of this tool kit is pictured in Fig.1, which is taken from the COMPSTAT-proceedings of 1974 (Victor et al. (1974)). The proceeding of an analysis is composed stepwise of segments exchanging data and provisional results over standardized interfaces and sharing a common subroutine-pool. Just as reasonable as this concept was, so was the hope that this tool kit could be carried out by so many groups of statisticians unrealistic.

Impulses Originated from DVM-107

Even though not all of the project's goals were reached the development of "CS" in Germany was considerably influenced by the activities induced by it and some of them still have some impact today. I will mention the most important of these impulses and their results.

The DVM-107 working sessions were first platforms for the discussion of CS-problems in Germany. A group of people interested in "CS" got together, debated with committment and worked together to form the first circle of CS-statisticians in Germany; the group grew up to several hundred members up to today. The idea to found a GMDS-Working-Group "Statistical Analysis Systems" was born at these sessions. In the beginning the GMDS-WG held their meetings together with DVM-107 (cf. No. 5 & 7 of Table 1), it ensured the continuity of CS-conferences and maintained the Reisensburg's tradition after the expiration of DVM-107. The list of the first sessions of this WG in Table 2 shows this ramification of the Reisensburg Symposia root stock.

I do not want to anticipate Mrs.Hörmann's contribution on the history of "Reisensburg". However, I would at least like to mention some active participants (besides those named at DVM-107): K.Dannehl (Düsseldorf), W.Dorda (Wien), W.Haufe (Einbeck), N.Krier (Wiesbaden) and E.Schach (Dortmund) and place special emphasis on the actuality of many mingling topics discussed at that time (missing values, presentation of results, graphics etc., cf. Table 2 below). Due to the strong interweaving and many overlappings it will not be easy for Mrs. Hörmann to put the numbering of DVM-, WG- and Reisensburg-symposia in the right order.

One of the most important outcomes of DVM-107, besides the Reisensburg sessions and the GMDS-WG, was certainly the Statistical Software Newsletter (SSN). The SSN was created as a means for the software information service. Then it became a journal (later even an international one) for statistical software papers, partly also for general CS-themes. The SSN became the official organ of the IASC, reached 1500 subscriptions and is still existing today as a section of "Computational Statistics and Data Analysis".

The list of requirements for statistical systems (Hultsch et al. (1980)) has to be mentioned as another product of DVM-107, in regard to the topic of quality improvement. The publication of this list, which prompted a lot of reactions from users and program developers, was preceded by numerous sessions and seemingly endless debates. The comparative software evaluations (method specific) which were published in the SSN, intended also to serve the quality improvement. The reports shocked the trust of many users

1. Oct. 2, '74	Mainz: Foundation Assembly (42 participants)
2. Apr. 2-4, '75	Reisensburg: see Table 1, No. 5
3. Oct. 1, '75	GMDS Annual Congress in Heidelberg: Business meeting
4. Dec. 8, '75	'Dünsbergheim', Gießen: see Table 1, No. 7
5. Sep. 13-15, '76	Reisensburg (with MEDIS-Institut, Munich) Topics: Flexibility, portability, data interfaces
6. Sep. 29, '76	GMDS Annual Congress in Hannover: Business meeting
....	....
....	....
Since then this pattern has been maintained.	
FURTHER IMPORTANT DATES:	
June 3, '80:	Handing over of the chairmanship to A.Hörmann
1984:	Foundation of the WG "Computational Statistics" of the Biometric Society (since then cooperation of both WG)
TOPICS OF THE SESSIONS UNTIL 1980:	
	Program evaluation, software information service Program developers and users Standardization, portability, flexibility Numerics of statistical programs, missing values Data manipulation, Linkage of statist. systems and data bases, Presentation of results, output-editor (?), graphics, Extensibility of program systems (method spectrum)

Table 2: Sessions of the GMDS-WG "Statistical Analysis Systems"

in the results of commercial software and initiated many improvements of these products.

To have made portability of statistical software a point of discussion, was another merit of DVM-107. A syntax checker for standard FORTRAN has been developed as a tool for the improvement of portability. It was used in many places and often made the domination of foreign programs easier.

Finally, the integration of the German CS-circle into the international community of computer statisticians was also a merit of the DVM-107-project. First contacts developed through the search for suitable numerical subroutines, especially with the NAG-group in Oxford, who sent the first international participants to the Reisensburg Symposia. Over the activities for program evaluation contacts were developed to the 'Centraal Rekeeninstituut' of the 'Rijksuniversiteit Leiden' (1976) and to I.Francis. Francis was (co)-author of the first list of evaluation criteria for statistical packages (Francis et al. (1975)) and became the protagonist of statistical software evaluation because of his reviews which appeared in 1975 and 1981. Further contacts were established through the SIZSOZ-project to the European Community activities for improvement of the information flow concerning software; the "European Association for Software Access and Information Transfer" (EASIT, in the beginning ECSIR) had similar aims like DVM-107 but on the European level (cf. Langenheder & Züllighoven (1976)). The facts, that I participated at the foundation of IASC as respresentative of the GMDS-WG in 1977 and that the COMPSTAT-activities, which were developed by cooperating groups in Wien and Berlin, were integrated into the IASC in 1978, also promoted the internationalization of the German CS-activities. This in turn was the prerequisite for the great German representation in the membership and in the steering committees of international associations.

Final Remarks

Developments in the field of "CS" were often pushed forward by outsiders and in the beginning these activities have often been judged depreciatorily by theoretical oriented statisticians. Today – with a big time delay – most statisticians have come to accept that practical application of statistics without the tools of "CS" is almost beyond thought and that "CS" also contributes to the development of statistics as a science. Nota bene, I am speaking about statisticians, who do not confine themselves to the axiomatic framework of mathematical statistics, but consider it as the task of our discipline to support empirical research in gaining knowledge.

The colleagues, who for the first time met at the Reisensburg about 20 years ago, with the aim to advance "CS" in Germany were mostly scientists of the marginal areas of statistics, too: In my retrospective – which is unavoidably biased – I tried to show that German "CS" is rooted in Reisensburg, because a lot of activities have originated from here. The continuity and the vitality of the "Reisensburg" demonstrate the significance of this initiative.

Comparing the titles of papers in this volume with those mentioned in Table 2, besides new and innovative themes you can discover a lot of well-known topics. The repeated discussion of old topics shows that many CS-problems in spite of all efforts have not yet been solved, or may even never be definitely solved. The pursuit of "CS"-problems has two sides, a very frustrating and a very enticing one. "CS" is on the one hand a collection of tools which must be improved by troublesome work over and over again, and on the other hand these tools permit the penetration into new and unknown fields of statistics. Let us not only look at the few highlights of big innovations but also be thankful for the numerous small increments in the improvement of the tool-kit which has contributed to the fact that statistics in between the empirical sciences could grow to its importance and spreading of today.

Acknowledgements

I thank my daughter Anja for her assistance with the English translation of my German manuscript as well as Mr.R.Schauwienold for the care taken in typing it.

References

Anderson, T.W. (1951), Classification by Multivariate Analysis. Psychometrika 16, 31-50

Anderson, T.W. (1958), An Introduction to Multivariate Statistical Analysis. Wiley, New York

Baumann, R. & Paul, M. (1961), Praktische Erfahrung im ALGOL-Betrieb. Elektron. Datenverarb. Beiheft 2 (1961) 51-56

Beutel, P. (1979), Kopplung von statistischen Auswertungssystemen – Vorschläge und Realisierung. Statist.Softw.Newsl. 5, 49-59

Centraal Rekeeninstituut (1976), Rijksuniversiteit Leiden: Enquete Statist. Programatuur. Techn. Report

Francis, I. (Ed.) (1977), A Comparative Review of Statistical Software. IASC, Voorburg

Francis, I. (1981), Statistical Software - A Comparative Review. North-Holland, New York

Francis, I., Heiberger, R.M. & Velleman, P.F. (1975), Criteria and Considerations in the Evaluation of Statistical Packages. Amer.Statist. 29, 52-56

Hauer, K. (1977), Realisierung eines portablen Softwaresystems zur Steuerung von FORTRAN-Moduln im Rahmen einer ökonometrischen Methodenbank. Statist. Softw. Newsl. 3, 46-53

Hüber, R. (1977), KARAMBA – Karlsruher Rahmensystem für Methodenbanken. Statist.Softw.Newsl. 3, 41-45

Hultsch, E., Jannasch, H., Krier, N., Sund, M. & Victor, N. (1980), Requirements for Program Systems used for Statistical Data Analysis. Stat. Softw. Newsl. 4, 3-30

Kaiser, H.F. (1959), Computer Program for Varimax Rotation in Factor Analysis. Educ.Psychol.Meas. 19, 413-420

Langenheder, W. & Züllighoven, H. (eds.) (1976), Proceedings of the 1.ECSIR-Workshop. GMD, Bonn

Rao, C.R. (1948), The Utilization of Multiple Measurements in Problems of Biological Classification (with discussion). J.Roy.Statist.Soc. (B) 10, 159-203

Späth, H. (1975), Cluster-Analyse-Algorithmen. Oldenbourg, München

Überla, K. (1968), Faktorenanalyse. Springer, Berlin

Victor, N. (1971), A Nonlinear Discriminant Analysis. Comp.Progr.Biomed. 2, 36-50

Victor, N. (1972), Ein Projekt zur Standardisierung und Koordinierung statistischer Auswertungssoftware. Unpublished Application, 2nd DP-Program of the BMFT (grant no.: DVM-107), Gießen

Victor, N. (1984), Statistique informatique - Science ou outil? Bull. ASU 9, 76-101. Engl. Translation: Statist.Sofw.Newsl. 10 (1984), 105-124

Victor, N., Tanzer, F., Trampisch, H.J., Zentgraf, R., Schaumann, P. & Strudthoff, W. (1974), A Project for Standardization and Coordination of Statistical Programs in the BRD. In: Bruckmann, G. et al. (eds.): COMPSTAT 1974. Physica, Wien, 526-535

Twenty-Five Working Conferences on Statistical Computing — Reflections on Twenty Years of Reisensburg Meetings

Allmut Hörmann
GSF – Forschungszentrum für Umwelt und Gesundheit,
Institut für Medizinische Informatik und Systemforschung,
Ingolstädter Landstr. 1, D-85758 Oberschleißheim, Germany

Key Words:
chronic of a series of meetings concerned with Statistical Computing in Germany, bibliography of related papers published in the Statistical Software Newsletter

Abstract
This contribution is intended to provide a personal outlook to the annals of a series of meetings known as Reisensburg meetings, originated by a working group on statistical computing within a German society (GMDS) aimed at discussing methods of statistics, informatics and epidemiology in the broad area of medicine with emphasis to data collection, data analysis, and data documentation. Founded in 1974 and called "Statistische Auswertungssyteme" (statistical analysis systems) this working group pursues its objectives in propagating methods of statistical computing in Germany. In 1985, within the German Region of the International Biometric Society, a working group with similar aims "Computational Statistics" was founded - henceforth joining, supporting, and coorganizing all Reisensburg meetings. An overview on some areas of activity during the 25 meetings is given.

Introduction

In the seventies computer science and statistics did evolve quite rapidly, a continuing process which nowadays has led to nearly everybody's access to computer facilities including powerful integrated statistical software. In this context, by reviewing the Reisensburg meetings, the typical work and results of the Working Group "Statistical Analysis Systems" formed the baseline and thus have to be reflected. Despite the the foundation of a related working group in the German Region of the International Biometric Society the direction was defined by the WG "Statistical Analysis Systems".

If one considers the topics of interest during this period, the first years were influenced by the non-availability of sufficient computer resources as well as the lack of on-line terminals in the seventies. The user had to defeat bulky problems related to portability, interfaces and reliability. In the eighties we were fascinated by the velocity of the computer hardware revolution and sometimes impressed by the number of new statistical software offered to the marked. Nowadays, by the widespread and cheap disposal of workstation-like PCs, features which should have already been settled are again of high interest, e.g. accuracy of calculations.

Foundation of the WG Statistical Analysis Systems

In the late sixties and the early seventies mainframe computers were installed at companies and universities and became available for statistical analyses. One of the germ cells in Germany can be found in the Mathematical Department of thc Univcrsity of Mainz, with need for calculations in the application area of medical statistics. In the late sixties Norbert Victor, one of the scholars of Professor Koller, developed the first German general purpose statistical analysis package (STATSYS), which could be enlarged in the early seventies at the Medis Institute of GSF (Institut für Medizinische Informatik und Systemforschung - Institute for Medical Informatics and System Science, Forschungszentrum für Umwelt und Gesundheit - Research Centre for Environment and Health, Munich/Neuherberg). In 1973 at the University of Gießen, Victor got funds from the German Ministry of Research and Technology (BMFT) for a multi-center project on "Standardizing Statistical Software in the Area of Medical Statistics - grant No. DVM107". The working conferences within this project turned out to become the germ cell of all statistical computing activities within this group. Three of the seven working conferences during the grant period already were realized at Reisensburg while the others were held at different places in Germany. Twenty-one of twenty-five meetings could be arranged at Reisensburg. To guarantee the continuity of the project's work it was decided to form the working group "Statistische Auswertungssysteme" (Statistical Analysis Systems), henceforth abbreviated with WG SA, which could be embedded in the multi-disciplinary scientific society GMDS (Deutsche Gesellschaft für Medizinische Informatik, Biometrie und Epidemiologie - German Society for Computer Science, Biometry and Epidemiology). The official act of foundation is dated October 2, 1974 and happened at the venue of the 20th GMDS Annual Meeting in Mainz. The principles are set to balance the gap between members who are emphasizing more the statistical methodology and those who are primarily computer scientists, thus providing a forum for discussion of the tools and their developments supplied by computer scientists as well as the methods contributed by statisticians. [1]

Officers of WG SA

Period	Chairman	Vice-Chairman	Secretary
1974-1976	Victor (Gießen)	-	Failing (Gießen)
1976-1978	Victor (Gießen)	-	Failing (Gießen)
1978-1980	Victor (Heidelberg)	Hörmann (Munich)	Failing (Gießen)
1980-1981	Hörmann (Munich)	Victor (Heidelberg)	-
1981-1984	Hörmann (Neuherberg)	Hultsch (Münster)	Haux (Aachen)
1984-1987	Hörmann (Neuherberg)	Haux (Aachen)	Haux (Aachen)
1987-1990	Haux (Tübingen)	Hörmann (Neuherberg)	Hörmann (Neuherberg)
1990-1992	Haux (Heidelberg)	Dirschedl (Munich)	Haag (Heidelberg)
1992-1995	Dirschedl (Munich)	Koch (Heidelberg)	Haag (Heidelberg)

Table 1: Officers of the WG "Statistische Auswertungssysteme"

Norbert Victor, as originator of DVM107 as well as of the WG, served as chairman in the very beginning. His main concern was recruiting of coworkers and convincing people of the necessity as well as the importance of the discipline statistical computing. He laid the topical foundation-stone and formed the style of the official sessions. When he changed to the University of Heidelberg, his new responsibilities did not allow him to serve in the way he was used to, so he decided to pass the duties to me, his deputy, one year before his official election period was finished. For two election periods afterwards, I chaired the group until 1987 when I decided to serve as scientific secretary for the IASC. Following the law of the series my deputy, Reinhold Haux (Universities of Tübingen and Heidelberg) followed for the coming two election periods, and since 1992, Haux's deputy, Peter Dirschedl (University of Munich) is responsible for the WG SA. Details are summarized in Table 1.

[1]WG SA's Objectives and Scope by Norbert Victor, Heidelberg, Founder

Foundation of the WG Computational Statistics

In the German Region of the International Biometric Society demands were expressed to form a method oriented working group aiming to pass the knowledge of a growing market of statistical software to all members of the society. Especially Heinz Hochadel (Ludwigshafen, Secretary of the German Region of the Biometric Society) initiated this foundation. In 1984, the Working Group "Computational Statistics" of the German Region of the Biometric Society was established in close cooperation with the WG SA. Henceforth it was decided to run the Reisensburg conferences as one joint task and to support mutually all other meetings and activities of the partner working group. This joint venture is running keeping in mind: "Together we are strong", and helps to enlarge the community of people interested in computational statistics in Germany.

Officers of WG CS

The society nominated Reinhold Haux (Aachen), and Karl-Heinz Jöckel (Bremen) as Haux's deputy, two members experienced in this fields of computational statistics for years and sustaining close connections to the GMDS working group. At the very first stage after the foundation it was decided not to follow separated paths but to cooperate as close as possible, especially to arrange the Reisensburg meetings jointly and such to profit from the different aspects of the two big German societies. Since 1988 the chairs of the WG were Karl-Heinz Jöckel and Rüdiger Ostermann, today Günther Sawitzki. Details are summarized in Table 2.

Period	Chairman	Vice-Chairman	Secretary
1985-1988	Haux (Aachen)	Jöckel (Bremen)	—
1988-1991	Jöckel (Bremen	Ostermann (Siegen)	—
1991-1994	Ostermann (Siegen)	Sawitzki (Heidelberg) Nagel (Zwickau)	—
1994-1997	Sawitzki (Heidelberg)	Nagel (Zwickau)	Ortseifen (Heidelberg)

Table 2: Officers of the WG "Computational Statistics"

Membership

The determination of the exact number of members of the WG SA at any time resulted to be a difficult task. Even if there were no restrictions for joining GMDS, however, it was not necessary to be a member of GMDS if one wishes to join the WG SA. As there were no membership fees the membership list was set up on an informal basis only and renewed by time to time. That is why we cannot speak of regularly registered members. A lot of the participants at the annual workshops (or other meetings) contributed to the progress without being forced to take part in the business sessions (where the names have to be registered in addition). The membership list, or better the list of people interested in the progress of the WG SA, showed 85 members in 1980 and 146 registered members in 1993.

Similar circumstances can be notified concerning the WG CS. Annual address checks and corrections are keeping the membership list of this WG to about 80 to 120 members who likewise need not be member of the Biometric Society. The intersection of both groups is approximately by 20 members.

What's About "the Reisensburg"

The facilities to be found at the Castle of "Reisensburg" ascertain everybody in having a stimulating meeting by being comfortably accommodated in one's own apartment, exquisitely maintained, with drinks even during night-time, and by using the opportunity to talk to every other participant during meals, coffee breaks and in special corner-rooms, where discussions may last until dawn. (Nowadays, facilities like these seem not to be noteworthy, but in the beginning seventies traveling funds for scientists from research centers or universities seldom allowed any hotel accomodation of that quality). The Castle of Reisensburg lodges the Centre for International Collaboration belonging to the University of Ulm and provides 33 rooms for up to 42 visitors for conferences of universities and scientific organisations. In addition, the local hotel facilities of the near-by town Günzburg can be used. The flair of the old medieval tower contrasts harmonically to the modern rooms erected along the castle-moat. The green of the Danube near-by and the picturesque bavarian-suebian baroque in the area invite participants to walks and excursions. In 1976 the WG SA decided to take possession of this unique place and any invitation to hold the annual conference at other location was denied until today. Meanwhile, the meetings of the WGs are well-known in Germany (Geidel & Lorenz (1993)).[2]

[2]Geidel, H. & Lorenz, R.: 40 Jahre Deutsche Region der Internationalen Biometrischen Gesellschaft, Biometrische Berichte, 1, 13-26, Münster-Hiltrup: Landwirtschaftsverlag GmbH (p.20).

Highlights Originated, Followed up or Finished During Reisensburg Meetings

Statistical Software Newsletter (1975 - 1993ff)

The Statistical Software Newsletter was founded in 1975 as a medium for information transfer in the project DVM107. From 1973 to 1976, the grant was given on condition that the project must be followed up by a continuously financed institution, such as GSF. GSF substantially supported the SSN from the very beginning until it joined CSDA in 1991. In the very first years, members of the WG SA were asked to use the SSN as their favorite medium for publishing papers in statistical computing. They were encouraged to continue when in 1983 the IASC - International Association for Statistical Computing, a section of the International Statistical Institute, elected the SSN to become its official journal. The impressive result is summarized in Table 3 and documented with respect to Reisensburg meetings in the bibliography.

Authors	1975-78 project period	1979-82 continuing in Europe	1983-86 start with IASC	1987-90 continuing world-wide	1991-93 SSN in CSDA
Members					
papers in German	44	42	5	1	0
papers in English	1	8	19	17	8
shorter comm.	2	1	2	2	7
Non-Members					
papers in German	0	2	1	0	0
papers in English	0	8	28	21	12
shorter comm.	0	0	0	5	4
Summary					
papers	45	60	53	39	20
shorter comm.	2	1	2	7	11

Table 3: Number of publications of members compared to non-members in the SSN during its different periods of development

List of Requirements (1975-1978)

The first challenging task of the working group completed after the project DVM107 has expired was the list of requirements for program systems used for statistical data analysis, which was first discussed in 1975 and started by Feldmann/ Hultsch/ Sund/ Busse/ Haarmann/ Grzesik/ Jannasch/ Victor in 1976 and was published in the SSN by HULTSCH et al. (1978) in German and English. This list considered the properties and performance of a statistical analysis system from the user's, statistician's and computer scientist's point of view.

Nonparametric Approach (1986-1988)

Users cannot expect to exploit all statistical methods ever developed in one unique system. However, broad agreement exists concerning the minimal set of methods in a specific field. A thorough investigation in nonparametrics was carried out in collaboration with both WGs and the Working Group on Nonparametrics of the German Region of the Biometric Society and published in the SSN by Bernhard et al. (1988).

Criteria for Software Comparison (1986-1990)

Twelve years after the presentation of the WG SA's List of Requirements an updated list of criteria for the comparison of statistical analysis systems was developed in close cooperation with representatives from software producers. Nearly everything had changed with computer environments and statistical software packages, for example by the advent of the PC, the on desk availability of computing power and graphics. Hence, a new attempt of evaluating software was started by E.-J. Zimmermann during the meeting in 1985. The new idea resulted in asking the software suppliers - with products being or intended to be comparable to some extent - to produce a comparison of their statistical software in form of a standardized description in cooperation with members of the working groups. The first task was to establish as measuring instrument a list of items worked out to be likewise reasonably restricted and fair, as well as sufficiently wide to meet this new approach to the old task of software comparability. This task took about two years and software producers were involved, too. The list was applied to compare BMDP, P-STAT, SAS and SPSS. The final comparison had to be done within a short period to be up-to-date. It was published 1990 in the final issue of the independent Statistical Software Newsletter. The WGs are continuing to observe this rapidly increasing market and checking the reliability of products.

Statistical Software Guide (1991-1993ff)

The latest task prepared and already published by members of the WGs (Koch & Haag (1993)) is the Statistical Software Guide, a standardized catalogue of statistical software products collected by means of electronic mail. This catalogue will be accessible for everybody interested in this topic by public mail box servers. As this medium will become more and more accepted timely information about statistical software can be collected and distributed at low costs. Further it must be investigated if a link between this pure information service and the pretentious evaluation approach can be established.

Numerical Reliability (1990-1993ff)

Based on L. Wilkinson's "Statistical Quiz" a series of tests on numerical reliability of data analysis systems had been carried out. Without the obstinacy of G. Sawitzki the partly frightening results could not have been compiled. The visible fact is: improvement is still needed. The publication is submitted to CSDA.

Cooperation with Other Groups (1974-1993ff):

Determined by the work within DVM107, we sustained close connections to a project of providing software information in the social sciences (SIZ-SOZ) in Germany, which tried cover Europe (ECSIR, EASIT), but failed as the funding ceased. Since the first of the SoftStat conferences, the WGs held informal contact to the organisers at ZUMA, Mannheim. In 1985, the conference was held jointly with GSF MEDIS in Munich/Neuherberg, where it was agreed that the chairmen of the WGs should be members of the programme committee ex officio. Up to now, WG members are invited to comment on this conference and asked for proposals in addition. The connections to international activities in statistical computing, e.g. to the COMPSTAT conferences and to the IASC are described in detail in the contribution of N. Victor. Contacts to similar groups in German speaking countries, e.g. to members of the ROES (the Austro-Swiss region of the International Biometric Society) are not formally established, but realized by presentation of papers at the Reisensburg meetings. Connections to activities outside Europe were not formally established. However, since the very beginning it was considered to be essential discuss the proceedings of the Statistical Computing Section of the American Statistical Association and those of the Computer Science and Statistics Sessions of the Interface.

Chronic of Reisensburg "Working Conferences"

In the first years we had at least one full day for discussions and preparation of draft papers, and in addition some tutorial presentations providing inexpensive continuing education. Until today the conferences comprise paper presentations which are discussed in detail as well as working sessions in the late afternoons and in the evenings. For coffee breaks and the evenings software demonstrations are scheduled.

Number of visits ever	Year 76	77	78	79	80	81	82	83	84	85
1 time	6	16	39	13	15	26	7	5	3	7
2-5 times	16	25	42	31	22	26	16	19	8	20
> 5 times	5	6	6	12	13	13	14	16	11	19
Summary	27	47	87	56	50	65	37	40	22	46
Number of visits ever	**Year 86**	**87**	**88**	**89**	**90**	**91**	**92**	**93**		**persons**
1 time	18	9	19	5	8	9	15	19		239
2-5 times	27	22	22	11	17	18	19	16		377
> 5 times	24	19	19	21	13	13	10	14		248
Summary	69	50	60	37	38	40	44	49		864

Table 4: 864 Reisensburg guests from 1976 to 1993. Single visits are rare events.

Main Topics of Interest Within 20 Years of Statistical Computing in Germany (1974-1993)

- Problems with the availability of hard- and software
- Problems with transportation, adaptation, programming of statistical software
- Disputes with statistical software producers
- Discussion on requirements for statistical software
- Presentations of statistical software packages available for many platforms
- Interactive use of statistical software systems
- General purpose statistical software packages: BMDP, SAS, SPSS, P-STAT, GENSTAT
- Use of PCs and workstations for statistical analyses

- Windows and graphical user interfaces (GUIs)
- Quality of software on different platforms

Dateline Chronic

Reisensburg meetings must be seen in close connections to the project DVM107. The Working Conferences were started as project meetings. That is why the 5th Working Conference in general at Reisensburg turns out to be the third meeting at the location Reisensburg and is simultaneously the first Reisensburg Meeting of the WG SA. The seventeenth conference (1985) is the first in the series with WG CS. Topics are marked with * .

Roots Within Project DVM107

10.-12.10.1973	1st Working Conference (DVM) at Münzenberg
24.-26.01.1974	2nd Working Conference (DVM) at Reisensburg
22.-24.04.1974	3rd Working Conference (DVM) at Reisensburg
21.-24.08.1974	4th Working Conference at Winterberg
02.-04.04.1975	5th Working Conference at Reisensburg (DVM) * Software Evaluation Service, Subroutine Pool, Standardization, Missing Values
18.-19.09.1975	6th Working Conference at Wasserlos (DVM) * Problem Oriented Structuring, Interfaces
08.-10.12.1975	7th Working Conference at Dünsbergheim/Gießen (DVM) * Standards, Structured Programming, Subroutine Pool, Layout of the List of Requirements for Statistical Data Analysis Systems

Cooperation with GSF (1976-1980)

13.-15.09.1976	8th Working Conference at Reisensburg in cooperation with GSF WG Systems for Statistical Analysis * Flexibility, Portability, Data Interfaces; List of Requirements
23.-25.05.1977	9th Working Conference at Reisensburg in cooperation with GSF WG Systems for Statistical Analysis * Integration of Database and Statistical Analysis Systems; List of Requirements

03.-05.07.1978 10th Working Conference at Reisensburg in cooperation with GSF Medis WG Systems for Statistical Data Analysis
* Statistical Computing: in the Point of View - Numerics and Presentation of Results

25.-27.06.1979 11th Working Conference at Reisensburg in cooperation with the GMDS WG Systems and System Development and GSF WG Systems for Statistical Data Analysis
* Data Management for Statistical Analysis

30.6.-2.7.1980 12th Working Conference at Reisensburg, last cooperation with GSF WG Systems for Statistical Analysis
* Programs in Addition to Statistical Program Packages

Statistical Computing Conferences at Reisensburg

06.-08.07.1981 13th Working Conference at Reisensburg
* Statistical Computing '81: Graphics in Statistical Packages, Software for Clinical Trials, Data Structures, List of Requirements

04.-07.07.1982 14th Working Conference at Reisensburg
* Statistical Computing '82: Graphics in Statistical packages, Languages for Data Analysis Software; Statistical Quality of Programs

26.-29.06.1983 15th Working Conference at Reisensburg
* Statistical Computing '83: Communications within Systems; Statistical Quality of Programs

01.-04.07.1984 16th Working Conference at Reisensburg
* Statistical Computing '84: Methods and Programs for Epidemiology; Aspects of Computer Science in Statistical Analysis Systems

Joining Computational Statistics Working Group

24.-27.06.1985 17th Working Conference at Reisensburg,
First Joint Session with the Working Group Computational Statistics (WG CS) of the German Region of the International Biometrics Society
* Statistical Computing '85: Statistical Analysis Systems and Statistical Programs for PCs, Algorithms in statistics especially time complexity

23.-26.06.1986	18th Working Conference at Reisensburg * Statistical Computing '86: Software for PCs, Expert Systems
05.-08.07.1987	19th Working Conference at Reisensburg in cooperation with WG Nonparametric Methods * Statistical Computing '87: Quality of Statistical Algorithms, Gene Research, Statistical Expert Systems, Graphics
12.-15.06.1988	20th Working Conference at Reisensburg * Statistical Computing '88: Graphics and Presentation of results, Software validation, Algorithms for Bootstrapping and Jack-knifing
18.-21.06.1989	21st Working Conference at Reisensburg in cooperation with WG Systems and System Development * Statistical Computing '89: Management and Analysis of Large Databases, Workstations within Statistics, Algorithms for Gene Technology, Statistical Expert Systems
05.-08.07.1990	22nd Working Conference at Reisensburg * Statistical Computing '90: Algorithms in Statistics, Workstations in Statistics, Numerical Aspects of Statistical Algorithms, Semantic Data Models, Statistical Expert Systems, Statistics as Decision Aid, Sample Size Algorithms and Systems, Comparison of Data Analysis Systems
05.-08.07.1991	23rd Working Conference at Reisensburg * Statistical Computing '91: Algorithms in Statistics, Algorithms for Bayes Methods, Analysis Systems for Epidemiological studies, Problems of Missing Values in Statistical Data Analysis; Statistical Data Analysis Systems, Comparison of Statistical Data Analysis Systems
21.-24.6.1992	24th Working Conference at Reisensburg * Statistical Computing '92: Graphical Methods for Data Analysis, Methods in Data Analysis, Statistical Knowledge Representation, Symbolic Calculation, Planning of Experiments for Data Analysis
05.-08.07.1993	25th Working Conference at Reisensburg in cooperation with WG Data Analysis and Numerical Classification of the German Society for Classification

* Statistical Computing '93: Analysis of Interactive Geographical Data, Data Analysis and Numerical Classification, Data Privacy, Methods of Data Analysis, Planning of Experiments for Data Analysis, Tools for Data analysis, Random Number Generators

Final Remarks

If one considers the main goals of Nobert Victor's project DVM107, it can be seen that an institution like the Medis Institute of GSF was not able to guarantee continuity in general. While the software information service could be established via the SSN for 15 years by GSF, and exists nowadays as a section of CSDA, the software evaluation project was substantially followed up and realized in first instance by the efforts of the WG SA and now by both WGs. The task of standardization with regard to portability became less important for GSF when general purpose statistical software systems and special software products became more widely offered commercially. The old interface problem may be handled satisfactorily by DBMS/Copy and general features available by methods of dynamic data exchange (DDE). The improvement of software quality remains a persisting task for users as well as for producers. While GSF is no longer involved in general program development for statistical applications, the contributions of the WGs to this field are impressive. I expect suggestions from the WGs on a modern approach of the old fashioned tool-kit concept in the near future. May be, the idea of a simple statistical program package as it is discussed now, will be the "grandfather" of its final realization.

My best wishes for the future are with the members of the WGs and especially with those being responsible in the coming challenging years: Peter Dirschedl, Uwe Haag and Armin Koch (for WG SA) as well as Rüdiger Ostermann, Günther Sawitzki and Matthias Nagel (for WG CS). My warmest thanks are directed to the management of the Reisensburg team, especially to Ursula Reck. May the enlargement of the hotel capacities and the modernization which are just on the way continue to support the work in computational statistics at "the Reisensburg". The 26th meeting will take place at the Reisensburg, June 19.-22., 1994.

Bibliography

Papers discussed and presented at working conferences on statistical computing especially Reisensburg meetings and published in the Statistical Software Newsletter

First Phase (1975-1978): In General Papers in German on Portability, Interface Problems and Standardization Tied Inseparably with DVM107

Bausch, B., Morgenstern, W., Scheidt, R., Kurz, E., Buchholz, L.: Practical experiences with data entry, data, data analysis and documentation in clinical epidemiology (in German). SSN 4, 98-101 (1978)

Bock, H.H.: Programs for cluster analysis (in German). SSN 2,72-84 (1976)

Christl, H.L.: Overview on some MANOVA programs (in German). SSN 1,10-12 (1975)

Döring, A.: NAG - a subroutine library (in German). SSN 2, 14-16 (1976)

Döring, A.: Portability of numerical software (in German). SSN 3, 22-24 (1977)

Dorda, W. Kogler, W.: WAMAS - statistical analysis of a patient oriented database system (in German), 3, 54-58 (1977)

Drechsler, K.H.: STRUCTRAN - structured programming in FORTRAN (in German). SSN 2, 23-25 (1976)

Edler, L.: Programs file (in German). SSN 1, 32-35 (1975)

Feldmann, U.: Several program systems for interactive data analysis (in German). SSN 2, 39-44 (1976)

Friedrich, H.J.: Data management as a variation of data structures (in German). SSN 2, 68-71 (1976)

Giere, W., Heger, J.P., Krier, N.: "Transportability" of FORTRAN programs to small computers (in German). SSN 3, 20-21 (1977)

Hague, S.: Implementation of the NAG Library - Principles and Practice. SSN 4, 84-90 (1978)

Hauer, K.: Realization of a portable software system to run FORTRAN modules in the context of a econometric method based system (in German). SSN 3, 46-53 (1977)

Haufe, W.: Data checking in planned trials (in German). SSN 4, 69-76 (1978)

Hölzl, D.: Problems by connecting data, methods, and model based systems (in German). SSN 3, 83-93 (1977)

Hörmann, A.: Design of a pool for subroutines (in German). SSN 2, 17-22 (1976)

Hörmann, A., Kelava, R.: STADAB - statistical database and analysis system (II): interactive analysis (in German). SSN 2, 50-56 (1976)

Hörmann, A., Victor, N.: Continuing the project DVM-107 within the IMD (in German). SSN 2, 37-38 (1976)

Hörmann, A., Victor, N.: Problems with the portability of statistical software (in German). SSN 4, 102-113 (1978)

Hueber, R.: KARAMBA - a frame for method based systems (in German). SSN 3, 41-45 (1977)

Hultsch, E.: Methods and use of interactive program systems in data analysis (in German). SSN 2, 45-49 (1976)

Hultsch, E.: The importance of data models for the use of statistical software systems (in German). SSN 3, 25-28 (1977)

Hultsch E.: Problems of statistical data analysis of retrospective and prospective questioning in database systems (in German). SSN 3, 76-78 (1977)

Jannasch, H., Naumann, K.: First experiences with an interface concept (in German). SSN 3, 79-82 (1977)

Jesdinsky, H.J.: Statistical analysis of large data files - a technical problem only? (in German). SSN 3, 68-75 (1977)

Koeppe, P.: Problems of portability in the user's point of view (in German). SSN 3, 3-12 (1977)

Langenheder, W.: EASIT - a new European software user's organisation (in German). SSN 3, 59-62 (1977)

Nickel, L.: A program for generalized checking of input data (in German). SSN 3, 94-95 (1977)

Nickel, L.: A program for preparing multi-factorial data for output of the standardized interface S0 (OPROS) (in German). SSN 4, 59-64 (1978)

Nowak, H.: General programs for linear models (in German). SSN 3, 42-47 (1975)

Precht, M.: On the numerical solution and condition of regression problems (in German). SSN 4, 36-41 (1978)

Prestele, H.: POLYP - a program for plotting time dependent data (in German). SSN 4, 51-58 (1978)

Raab, A., Selbmann, H.K.: SAVOD-Q: An interactive statistical analysis system for large medical data files (in German). SSN 2, 65-67 (1976)

Schaumann, Ph.: A general standardized interface for connecting data management and data analysis in statistical software (in German). SSN 2, 29-31 (1975)

Schaumann, Ph., Strudthoff, W.: A general file management system in standardized FORTRAN (in German). SSN 2, 3-12 (1976)

Selbmann, H.K.: Programs for factor analysis - attempt of an overview (in German). SSN 2, 20-25 (1975)

Sund, M.: The BMDP-System - an overview of the latest statistical software package of UCLA (in German). SSN 2, 26-28 (1975)

Sund, M.: We need standardizing— (in German). SSN 2, 26-27 (1976)

Sund, M.: Impressions of a trip to the US in portability matters (in German). SSN 2, 85-87, 1976

Tanzer, F.: Software information service: program systems (in German). SSN 1, 59-62 (1975)

Victor, N. (Editorial): Aims of the Newsletter (in German). SSN 1,1 (1975)

Victor, N.: Portability ? (in German). SSN 3, 13-19 (1977)

Victor, N., Trampisch, H.J.: Programs for discriminant analysis (in German). SSN 3, 48-58 (1975)

Weingarten, W., Sauter, K., Klonk, J.: Stepwise connection of a database with statistical analysis systems (in German). SSN 3, 34-40 (1977)

Wolf, G.K.: Methods and algorithms for robust regression and comparison of samples (in German). SSN 4, 65-68 (1978)

Wolf, H.H.: STADAB – statistical database and analysis system (I): database and data transformation (in German). SSN 1, 63-66 (1975)

Second Phase (1979-1985): Achieving More Importance in the Area of Statistical Computing

Abel, U., Berger, J.: a dialogue system for the analysis of clinical trend data (in German). SSN 9, 66-70 (1983)

Bechtold, H., Hörmann, A.: Information and contacts between producers of statistical software systems and their users (in German). SSN 5, 87-98 (1979)

Berger, J., Rittgen, W., Weber, E.: Graphical presentation of statistical data (in German). SSN 7, 110-117 (1981)

Beutel, P.: Connecting different statistical data analysis systems - proposals and realization (in German). SSN 5, 49-59 (1979)

Bloedhorn, H., Driever, R., Godehard, E.: Analysis of variance with missing data (in German). SSN 8, 61-72 (1982)

Blomer, R.J.: A database management system based on the formal description of questionnaires for clinical trials (in German). SSN 5, 104-107 (1979)

Blomer, S.: Evaluation of database concepts and their realization with respect to the usage in the area of medicine (in German). SSN Supplement 4, 1-24 (1980)

Bongartz, D., Lopez-Frank, U., Züllighoven, H.: The social science software catalogue 1980. SSN 6, 17-21 (1980)

Broszio, E.P.: A project for the development of interactivee programs for the analysis of medical data (in German). SSN 5, 30-34 (1979)

Budde, M., Wargenau, M.: Analysis of survival data in SAS and BMDP (in German). SSN 10, 126-135 (1984)

Dietlein, G.: Schematic plots - an alternative to the presentation of time dependent data (in German). SSN 7, 100-103 (1981)

Dietlein, G., Gammel, G.: Graphical presentation of medical data (in German). SSN 8, 130-136 (1982)

Dirschedl, P.: Ways to graphical presentations by use of small computers (in German). SSN 7, 89-99 (1981)

Doering, A.: Experiences with the transportation of the NAG Library to a Siemens 4004 - BS 2000 (in German). SSN 4, 94-96 (1978)

Dolejsky, W., Weiss, D.: Problems concerning the implementation of numerical software - experiences with the transportation of the NAG Library to a CDC 3300 (in German). SSN 4, 91-93 (1978)

Edler, L., Wahrendorf, J., Berger, J.: SURVIVAL - a program package for the statistical analysis of censored survival times. SSN 6, 44-53 (1980)

Exner, H.: An overview and evaluation on basic random generators in statistical program packages (in German). SSN 9, 71-77 (1983)

Failing K.: Missing methods within statistical analysis packages (in German). SSN 7, 3-9 (1981)

Feldmann, U., Pralle, H.: Proposal for a project of user friendly statistical method based systems (in German). SSN 5, 99-103 (1979)

Greis, J.: An interface between the database system RAMIS and the statistical analysis system SAS (in German). SSN 6, 27-31 (1980)

Hansert, E., Federkiel, H., Dierlich, G.: Statistical program or programmed statistics? (in German). SSN 5, 3-7 (1979)

Haufe, W., Nickel, L.: A program for randomized trials (in German). SSN 6, 59-63 (1980)

Haux, R.: Criteria for the development of programs for statistical data analysis (in German). SSN 6, 85-96 (1980)

Haux, R.: Some comments to the question of storage of hierarchical data structures in a database system by use of a hierarchical data model (in German). SSN 7, 58-63 (1981)

Haux, R.: Statistical analysis systems - construction and aspects of method design (part 1). SSN 9, 106-115 (1983)

Haux, R.: Statistical analysis systems - construction and aspects of method design (part 2). SSN 10, 14-27 (1984)

Held, G., Villot, G.: SAS/FSP: to start with data analysis (in German). SSN 9, 17-25 (1983)

Hitz, M., Hudec, M., Müllner, W.: A software package for the analysis of censored survival data. SSN 11, 43-54, (1985)

Hitz, M., Hudec, M., Takacs, H.: AMINO - An interactive system for the analysis of mixture of normal distributions. SSN 12, 111-115 (1986)

Holle, R., Leibbrand, D.: Data design in clinical trials. SSN 11, 9-14 (1985)

Hörmann, A., Victor, N.: Projects on statistical software information services. SSN 6, 3-4 (1980)

Holzreiter, P., Bausch, B.: Qualitative comparison of three database systems (in German). SSN 5, 23-29 (1979)

Hüber, R.: Software technologies in the area of user systems (in German). SSN 5, 79-86 (1979)

Hultsch, E. Jannasch, H., Krier, N., Sund, M., Victor, N.: Requirements for program systems used for statistical data analysis (in English and German). SSN 4, 3-30 (1978)

Jannasch, H., Naumann, K.: SIR: Evaluation and practical experiences (in German). SSN 7, 19-24 (1981)

König, A.: Interactive factor analysis with graphical support (in German). SSN 4, 42-50 (1978)

Kubale, R., Schmidt, F.W.: Supplemental programs for documentation and data analysis of patient records (in German). SSN 8, 21-26 (1982)

Kuhnert, A., Blomer, R.J., Haux, R., Zeh, Th.: On realizing an interface for database and statistical analysis systems (in German). SSN 6, 22-26 (1980)

Kunz, E., Neiss, A.: A program for determination of the sample size for clinical trials when the dependant variable has an exponential distributed lifetime (in German). SSN 7, 53-57 (1981)

Lambert, T.W.: The development of a graphical supplement for the NAG library. SSN 6, 107-110 (1980)

Mangstl, A., Bergermeier, J.: SPSS-Graph - a program for graphical presentations of analyses from the SPSS package (in German). SSN 6, 111-115 (1980)

Martin, W.: Methods for time series analysis with restriction to their availability (in German). SSN 6, 97-106 (1980)

Mau, J.: Some graphical methods for data monitoring and interactive analysis in randomized clinical trials. SSN 8, 120-129 (1982)

Meinzer, H.P.: Graphics in statistical program packages. Statistics and an interpretative graphics language (in German). SSN 7, 104-109 (1981)

Neiss, A., Jöckel, K.-H., Knappen, F., Nowak, H.: Does the result of statistical evaluation depend on the choice of the statistical consultant? SSN 11, 69-75 (1985)

Pahnke, K.: Programs for the analysis of qualitative dose-effect relations (in German). SSN 5, 8-22 (1979)

Pahnke, K.: WIRKAN – a program for the analysis of qualitative dose-effect relations if the relation is dichotomous (in German). SSN 6, 54-58 (1980)

Pahnke, K.: Repeated measurement models - analyses using SAS and BMDP (in German). SSN 10, 33-36 (1984)

Schach, E., Jöckel, K.-H.: A new topic: statistical data analysis and communication between user and statistician (in German). SSN 7, 37 (1981)

Schafroth, M.: The program library STATPLOT (in German). SSN 5, 108-112 (1979)

Schafroth, M., Flury, B., Kläy, M.: First experiences with the system STATPLOT (in German). SSN 6, 116-121 (1980)

Sund: M.: Findings of a statistician concerning the importance of database systems for scientific data analysis (in German). SSN 7, 10-13 (1981)

Tolxdorff, T.: Pictures on microfilm (in German). SSN 7, 122- 123 (1981)

Ulm, K., Weck, N., Schimper, K.: PHMOD - a program for the analysis of censored data. SSN 9, 61-65 (1983)

Victor, N.: Computational statistics - tool or science? Werkzeug oder Wissenschaft?: with discussion by Escoufier, Y., Havranek, T., Lauro, N., Nelder, J., Schach, S., Sint, P. SSN 10, 105-125 (1984)

Warncke, W., Selbmann, H.K.: Data preparation for epidemiological studies (in German). SSN 5, 60-68 (1979)

Weck, N.: Checking of distribution assumptions if the data are censored (in German). SSN 7, 47-52 (1981)

Wilke, H.: The forgotten 'user' in statistical package evaluation. SSN 6, 64-70 (1980)

Wilke, H.: ISSUE '80 - report on the SPSS user Conference in Alexandria (in German). SSN 6, 122-124 (1980)

Wolf, G.K.: Requirements concerning data preparation in the view of epidemiology (in German). SSN 6, 39-42 (1980)

Third Phase (1986-1993): Cooperation of Both WGs

Abel, V., Wahl, C.: A structogram for Hoadley's quality measurement plan (QMP). SSN 15, 20-22 (1989)

Bernhard, G., Alle, M., Herbold, M., Meyers, W.: Investigation on the reliability of some elementary nonparametric methods in statistical analysis systems with comments from BMDP, SAS and SPSS and a rejoinder. SSN 14, 19-30 (1988)

Blettner, M., Becher, H.: The analysis of cohort studies - evaluation of statistical software. SSN 16, 37-42 (1990)

Brunner, E., Compagnone, D.: Two sample rank tests for repeated observations - the distribution for small sample sizes. SSN 14, 36-42 (1988)

DeBoer, F., Schumacher, M.: Some algorithms for the exact distributions of nonparametric two-sample tests - considerations on complexity and empirical investigations. Part 2: Rank tests. SSN 12, 28-33 (1986), together with: Streitberg, B., Tritchler, D.: Comments on the papers of DeBoer/Schumacher and König/Giani/Haux. SSN 12, 34 (1986)

Gefeller, O., Woltering, F.: A general method of estimating measures of association and their asymptotic variances under the multinomial model using standard statistical software. SSNinCSDA 16, 127-130 (1993)

Grüger, J., Ostermann, R.: Construction and integration of a statistical expert system for binomial experiments. SSN 12, 124-128 (1986)

Haag, U., Koch, A., Vach, W.: Why and how to build a simple statistical computing system. SSNinCSDA 16, 237-244 (1993)

Härdle, W.: XPLORE - a computing environment for exploratory regression and density smoothing. SSN 14, 113-119 (1988)

Haux, R., Hilgers, R., Hörmann, A., Jöckel, K.-H., Lehmacher, W.: Recommendations on software for nonparametric statistical methods. SSN 14, 27 (1988)

Held, G.: Advances in graphical data analysis from SAS Institute. SSN 15, 85-90 (1989)

Hilgers, R.A.: Elementary nonparametric test: a review. SSN 14, 4-18 (1988)

Hörmann, A., Dirschedl, P., Haag, U., Koch, A., Himmelmann, W. (eds.): Comparing statistical analysis systems - a proposal for a list of comparison criteria with descriptions of BMDP (Fitzgerald, G.), P-STAT (Zimmermann, E.-J.), SAS (Held, G.), SPSS (Schmidtner, Chr.) and comments by members of the Working Group "Statistical Analysis Systems" of the German Society for Medical Documentation, Informatics and Statistics. SSN 16, 90- 127 (1990)

Holtbrügge, W., Glocke, M.H.: Data management in the pharmaceutical industry. SSN 16, 85-89 (1990)

Jöckel, K.-H.: Monte Carlo tests. SSN 12, 35-39 (1986)

Jöckel, K.-H.: Interrelation between the bootstrap and Monte Carlo test procedures. SSN 14, 35 (1988)

Koch, A., Haag, U.: The Statistical Software Guide '92/93. SSNinCSDA 15, 241-264 (1993)

Koch, R.: Hierarchical changes of the structure in a temporal statistical scientific data model. SSN 16, 49-53 (1990)

König, J., Giani, G., Haux, R.: Some algorithms for the exact distributions of nonparametric two-sample tests - considerations on complexity and empirical investigations. Part 1: Permutation tests. SSN 12, 20-27 (1986); together with: Streitberg, B., Tritchler, D.: Comments on the papers of DeBoer/Schumacher and König/Giani/ Haux. SSN 12, 34 (1986)

Küffner, H.: Teaching the use of statistical packages at a distance. SSN 14, 108-112, (1988)

Möhring, M., Biedassek, Th.: SAMOS - a system for applied matrix-oriented statistics based on a functional language. SSN 12, 116-123 (1986)

Neumann, N.: Some procedures for calculating the distributions of elementary nonparametric test statistics. SSN 14, 120-126 (1987)

Payer, M.: A tutorial introduction into the theory of complexity. SSN 12, 4-9 (1986)

Rahlfs, V.W.: TESTIMATE a special package with special reference to nonparametric methods. SSNinCSDA 14, 125-126 (1992)

Ritter, H., Züll, C.: Data exchange between statistical packages using the PC program DBMS/COPY Version 3.0. SSNinCSDA 14, 539-544 (1992)

Rothe, G.: Bootstrap: estimating and testing. SSN 14, 31-34 (1988)

Sayn, H., Merkle, W.: Statistical software for sample size estimation: Power, DESIGN-POWER, and IFNS. SSN 15, 56-59 (1989)

Sayn, H., Budde, M., Schach, S.: Small sample properties of the estimators of the regression coefficients in the Cox model. SSN 12, 76-79 (1986)

Schach, S.: On the impact of electronic data management on research and teaching in the area of statistics in the Federal Republic of Germany (in German). SSN 13, 62-65 (1987)

Slepica, J.M., Wolf, G.K.: A new algorithm and program for robust linear regression. SSN 12, 80-83 (1986)

Streitberg, B.: The permutation test problem is NP-hard. SSN 12, 18-19 (1986)

Streitberg, B.: On the non-existence of expert systems - critical remarks on artificial intelligence in statistics. With comments by Chambers, J.M., Gale, A., Pregibon, D., Hajek, P., Haux, R., Havranek, T., Nelder, J. and a rejoinder. SSN 14, 55- 74 (1988)

Streitberg, B., Röhmel, J.: Exact distributions for the permutation and rank test: An introduction to some recently published algorithms. SSN 12, 10-17 (1986)

Unkelbach, H.-D., Passing, H.: Quality assurance of statistical software. SSN 15, 49-55 (1989)

Wittkowski, K.M.: Statistical analysis of unbalanced and incomplete designs – experiences with BMDP and SAS. SSN in CSDA 14, 119-124 (1992)

Remarks on the History of Computational Statistics in Europe

Peter Paul Sint
Forschungsstelle für Sozioökonomie (Unit of Socioeconomics),
Österreichische Akademie der Wissenschaften
(Austrian Academy of Sciences), Kegelgasse 27,
A-1030 Wien, Austria

Key Words:
computational statistics, statistical computing, computers, history, statistics, technological development, statistical analysis systems, statistical software, Monte Carlo, data analysis

Abstract
Early statisticians and their impact on computers and computing. The first uses of the computer in statistics. Some statistical methods inherently computer oriented. Some organisational development.

The Forerunners

Before I go into the main topic of computational statistics or statistical computing on electronic computers I will mention some curious incidents which connect the history of statistics and the history of computing. This part draws on my 1982 paper on the Roots of Computational Statistics which describes the early developments in more detail. The environment of the history of computing as visible in the general history of statistics may be learned from a number of recent books on the history of statistics. For an overview see e.g. Fienberg (1992).

After Wilhelm Schickard (1592 to 1635) had built the first mechanical device for calculation in 1623 following an idea of the famous astronomer Kepler the next and most influential calculating machine was built by Blaise Pascal (1623-1662) whose purpose was to help his father in accounting problems. Pascal was also studying problems in the theory of games of chance, which we consider today a root of the modern theory of probability in itself an important ingredient of the modern theory of statistics.

The machine built 1642 to 1644 could handle only addition and subtraction. This was improved by Gottfried Wilhelm Leibniz (1646-1716) whose machine additionally handled multiplication and division. More or less his principles (especially the Leibniz wheel) governed the construction of mechanical calculating machines until their demise.

Leibniz is also interesting because he combined three traditions of statistics: The mathematical – computational aspect , the old "statistics" in its original meaning, and the political arithmetic.

The word statistics derives from the Italian "ragione di stato" and described originally the knowledge of statesmanship and of state affairs. Shakespeare uses the word statist in Hamlet and Cymbline in this way. Gabriel Naude used the adjective statisticus for the first time in print in 1639 (Pearson (1978, p.4)). The older comtemporary of Leibniz, Hermann Conring founded – based on Italian, French, Dutch and German forerunners - "statistics" as a continental school studying the art of collecting data relevant for decision making by government. He warned Leibniz who worked as a diplomat for the same patron – Baron Boineburg – not to waste his time on mathematics. Leibniz contributed to the "old" statistics e.g. by his "Entwurf gewisser Staatstafeln" (design of certain state tables). He used the language of Conring and his contemporaries but his concepts were related to his "Arte Combinatoria" (1666). This Arte was to do on the one hand with combinatorics and probability following Pascal and on the other hand with the art of logical inference and calculation.

As for political arithmetic the foundations of which were laid by John Graunt (1620-1674) and William Petty (1623-1687) Leibniz never wrote on the subject. But he brought the data based on registration figures of christenings and funerals in Breslau to the attention of the Royal Society of which he was a Fellow since 1673. These tables were handed over to the astronomer Halley who produced with their help in 1693 his famous paper on mortality which contained the first real life table (John (1884), Pearson (1978)).

The word statisticus had fallen into disrespect in Conring's time. It was used for a criminal or at least a "pseudopoliticus" and was only revived by the professors of jurisprudence Martin Schmelzer (†1747) and Gottfried Achenwall (1717-1772). The latter was considered as a founding father of "statistics" by the mainly German school predominant until the late 19th century. Achenwall's pupils - most notably August L.v.Schlözer fought the "table statisticians", the followers of political arithmetic. At that time the Scotsman John Sinclair had usurped the term statistics for the collection of numeric data (1790) and this usage quickly replaced "political arithmetic" in Britain.

The "Arithmetical" and the "Statistical" Machines

The next famous innovator was Charles Babbage. As is well known he planned his first calculating machine in 1812 and finished its first model in 1822. His report on this difference engine appeared 1822. While he finished a number of working parts he had difficulties with others. Though he got considerable funding, the work on the machine was stopped in 1833. But – and this is less well known – because of the technological difficulties involved Babbage studied carefully the technological problems and economic conditions under which a technology becomes viable. This was related to the "statistics" of his time and was probably the reason why Babbage became the driving force for the establishment of section F – Statistics – in the British Association for the Advancement of Science (of which he was a co-founder) and of the Statistical Society of London. William Farr (1807-1883) stated in a presidential address to the Statistical Society of London 1871 that "Babbage was, in reality more than any other man its founder". During the congress period of statistics (1853-78) he was an active participant and speaker in those congresses.

Babbage did not finish the difference engine but continued to develop the concept of an analytical machine, a real programmable mechanical computer with 1000 mechanical registers and a "mill", a kind of CPU, to do calculations guided by punched cards (developed to drive the loom invented by Jacquard in 1805). Peter Georg Scheutz (1785-1873) and Edward Scheutz from Sweden built a difference engine during Babbage's lifetime and displayed it in London with Babbage's active help. Later it was installed at the Registrar General's department and used for statistical calculations. The life tables which insurance companies used for several years were calculated on this machine. Babbage's analytical engine was displayed for the first time in 1991 on the occasion of his 200th birthday (it may be seen at the Kensington museum of Science and Technology, London).

The following period saw the introduction of statistics into the natural sciences: Francis Galton (1822-1911) a cousin of Charles Darwin introduced it into biology. This laid the foundation for British biometry and the British predominance which followed. The foundation of Biometrika in 1901 was a milestone in this development. James Clerk Maxwell (1831-1876) and Ludwig Boltzmann (1844-1911) founded statistical mechanics in physics. Their work was based on Laplace and French mathematicians from Condorcet to

Poisson and was largely independent of the work done in biometry. This development supported by the Russian school of probabilists led to the establishment of statistics as an abstract science remote from its roots in political arithmetic and state sciences.

The next step in the invention of statistical counting machines was initiated by John Shaw Billings (1839-1892), surgeon in charge of the work on vital statistics in the US censuses of 1880 and 1890, who gave Herman Hollerith (1860-1929) the idea of constructing a machine to do the hand tallying of statistical data in a mechanical way using "something on the principle of the Jacquard loom". The success in the 1880 US-census and the success of the improved machine of Otto Schäffler in the census of the Austro-Hungarian monarchy – which was programmable by a telephone switch board while Hollerith's machines had to be rewired for each process – in the same year established them firmly in statistical offices and in business.

Starting 1928 the Hollerith tabulating machines – improved during their use in the National Bureau of standards – were used for scientific purposes using mainly a method called progressive digiting (Hartley in Owen (1976), short note in Sint (1984)).

Back to Europe: Karl Pearson (1857-1936) the famous statistician calculated many important statistical tables and introduced the method of finite differences into ballistic computation in England (Goldstine (1972, p.76)). He edited a short lived series "Tracts for Computers" where the computers were humans doing computational work. The first published list of random numbers appeared in this series. While the first programmable electromechanical computers were built by Zuse in Germany the main development of the electronic calculator took place in the United States. In Europe Alan Turing (1912-1954), who had described "Computable Numbers" and the abstract notion of the yet unsurpassed "Turing machine", cooperated in the construction of COLOSSUS, an electronic computer to decipher the cryptographic military code produced by the German Enigma machines.

The Electronic Early Birds

One of the first applications of the electronic computer was the use of a statistical procedure, the Monte Carlo method. The "statistical experiment" as it was called by one of the fathers of the electronic computer, John v.Neumann, served to determine the behaviour of the atomic fission bomb. The very first paper on electronic computing in the Journal of the American Statistical Association stated that the problems around the Boltzmann equations governing fission gave rise to the Monte Carlo method (Metropolis & Ulam (1949)). The original document in Los Alamos from 1944 was by D.Hawkins and S.Ulam (Puri in Owen (1976)). The further development in Monte Carlo methods is visible in a number of conferences in the 1940's and in a monograph by Hammersley and Handscomb which appeared in London, in 1964 (a report by the National Bureau of Standards was already published in 1951). The Polish researcher S.Ulam had earlier found the association between product measures and the concept of independence in probability theory, 1932, one year before the appearance of Kolmogoroff's book (only published in 1934).

An important step in the development of computational statistics was the development of simulation systems, a more general form of statistical experiments. This was started by Geoffrey Gordon in 1954 at the Telecommunication research Establishment in England. Later he developed GPSS, the General Purpose Simulation System which made simulation accessible for a large variety of users. It became available in 1961.

Kristen Nygaard was working in the Norwegian Defence Research Establishment where Monte Carlo methods were used to make manual estimations related to the first nuclear reactor in Norway (1949/50). Ole-Johan Dahl joined the institution 1952 and they worked together on Operations Research using Monte Carlo methods. Nygaard started the development of a simulation Language SIMULA in the Norwegian Computing Center in 1961. In the first document dated 1962 he states: "The status of the Simulation Language (Monte Carlo Compiler) is that I have rather clear ideas on how to describe the queueing systems, ... I believe that the results have some interest even isolated from the compiler". The Algol based language had far reaching consequences on the development of computer languages. Today's Object Oriented Programming is heavily influenced by this early computer language, based on Algol. The practical requirements of developing a simulation language for stochastic systems served as a model which Nygaard and Dahl provided for program developers at large. This is another example of applied statistical work influencing the development of computing.

At the same time a number of other simulation languages and packages emerged which today form an important tool for many enterprises. Flows of telecommunication messages or of products in the production processes are studied with fully commercial packages, often tailored to the needs of a specific industry.

In elementary particle physics – the domain where it originated – the Monte Carlo method has become an indispensable tool for studying the fundamental behaviour of matter. This intense usage sometimes blocks the construction of useful analytic formulations, which give better insight into the underlying processes. Modern highly parallel computers are especially well suited for such simulations in high dimensional spaces.

The Monte Carlo method also led to important advances in pure mathematics: it was used as the only practical method for calculating high dimensional integrals (e.g. by Gibbs sampling). Statistically the sequences calculated converge with probability one to the correct value of the integral. Monte Carlo methods use pseudo-random numbers calculated by an appropriate algorithm. However mathematicians showed that the selection of specific series of "not so random" numbers could improve on this and guarantee faster convergence (for an overview see Hlawka et al. (1981)). For a more recent treatment of computer experiments, with a different emphasis, stressing the relationship of deterministic models and their run time behavior with "randomly" generated input, compare Sacks et al. (1989).

Monte Carlo was certainly the earliest and most spectacular usage of the computer in statistics. But most other branches of statistics were also influenced deeply. Practically all underwent considerable changes of perspective during the introduction of electronic computing. But there are a few applications which seem to be fundamentally related to this introduction. One group of them are resampling methods. It was Quenouille who first proposed the method but John Tukey introduced the method into the computing environment and gave it a name (which he did with many statistical and computing concepts e.g. the bit). Being a more natural way to look at statistical data than most traditional methods it had to wait for the computer for more widespread use. Today's evangelist is Bradley Efron (Efron & Tibshirani (1986)).

A more European development was the robust estimation of location parameters. Starting with a seminal paper by Peter J. Huber (1964), the development of "Robust Methods" in statistics needed considerable computing power. Similar to resampling methods it gives a sound basis to common sense principles and uses the computer to do what it can do best: calculate a wide variety of models which need a lot of computational effort. Huber (1981), Hampel et al. (1986), Rasch (1982,1984), Rousseeuw & Leroy (1987), Stahel (1991) give overviews on different varieties of this branch. All kinds of Bayesian methods fall into a similar category.

Another computer intensive application is automatic classification. Classical reviews by Sokal & Sneath (1963) and by Jardine & Sibson (1971) set the stage. David Wishart provided a standard with his Clustan package for a while, incorporating a number of his own ideas. Today many of the most frequently used cluster methods are available in the standard packages. Special efforts by Diday try to classify forms and logically characterised symbolic objects (1992).

There are also a number of fringe activities, inherently computer oriented, which use statistical concepts or concepts related to statistical reasoning.

Fuzzy data sets and fuzzy inference have statistical connotations and may be used in environments where statistical reasoning is important to the "fuzziness" of concepts. Similarly Genetic Algorithms (e.g. Schaffer (1989)), related to stochastic approximation. Neural Networks, Catastrophe theory and Chaos Theory are relevant for statistics. Chaos models are especially interesting as they show how seemingly chaotic behaviour may be governed by simple rules and equations. They generalize ideas which appeared in the generation of random numbers and are relevant to the study of "planned" forms of randomness.

Before the advent of the computer a lot of computational effort went into the production of tables which reduced tedious calculations in practical applications. The production of such tables inspired both Babbage and the developers of the first electronic computers. The statistical tables of Ronald A. Fisher and Yates contain electronically computed additions in the fifth edition (finished 1956). The work was done by Frank Yates mostly at Rothamsted, one of the first locations of statistical computing worldwide. In this sense Yates was probably the first statistician in Europe to use the computer extensively in scientific statistical work.

The overlapping of traditional and new methods of computation may be seen by the fact that Healy and Dyke from the same institution described in 1953 a Hollerith technique for the solution of normal equations using only a sorter and a tabulator following an idea of Yates.

In the same volume of the Journal of the American Statistical Association Brown, Houthakker and Prais wrote a paper on Electronic Computation in Economic Statistics which tried to promote the use of the new tool outside the community of electrical egineers and mathematicians which was involved in the development of the EDVAC. The authors had access to this machine, finished by Wilkes and Hartree for the Cavendish Laboratory in Cambridge, UK in 1949. They saw clearly the importance of this landmark machine to input-output analysis, the potential of Monte Carlo methods and the possibility of solving "full maximum likelihood" models in econometrics. Yet this was at a time when no programmes of that kind were even operational.

The development in Europe did not get to speed in the same way as the development in the USA. In 1972 the second volume of the Biometrika tables, edited by E.S.Pearson and H.O.Hartley stated: it is noteworthy that about three quarters of the 69 tables were computed in North America, sometimes at Biometrica's suggestion, a fact which shows the much greater extent, to which computer facilities have been made available in recent years in this continent compared with what has been the case in the United Kingdom". The volume is dedicated to Karl Pearson and L.J.Comrie who, without the aid of electronic computers contributed so much to the production of mathematical tables. Increasingly the production of tables became less relevant. For today's computer literate it is generally easier to calculate values anew each time they are needed. An extreme case is StatXact from Cytel which gives exact values for tests concerning contingency tables and nonparametric data where until recently asymptotic estimates had to be used. Powerful new algorithmic tricks make this possible on simple 386 or 486 PCs.

Packages Arrive

The algorithms needed are often buried in general purpose statistical packages and the results are often presented in a way where the less initiated hardly know which functions have been used to determine significance values. Nevertheless the introduction of the packages made learning the practical use of statistical methods much easier. The reduced turn-around times and the ease of use of alternate methods can improve the quality of the output.

Most of the early large packages were developed in the USA, which still provides most practically used packages Notable exceptions are the developments in Britain. GENSTAT was an early attempt to formulate a general matrix operating language well suited for a wide class of statistical procedures. Genstat was based on work by Nelder, Wilkinson and James on general methods for specification and computation in the analyis of variance. It was started by John Nelder in 1965 in Adelaide in South Australia and ran there on a CDC 3300. Later it was ported to Rothamsted, developed there by twelve members of the Statistics Department. This version became operational in 1971. To implement Generalized Linear Models and iteratively reweighted least squares, developed by Nelder and Wedderburn, a special package GLIM was developed in the seventies by R.J.Baker, M.R.B. Clarke and John A.Nelder under the patronage of the Royal Statistical Society. These packages are now marketed by NAG (Numerical Algorithm Group) Ltd.in Oxford which itself had developed mathematical subroutines for a number of computer languages and environments. The paradigm of Generalized Linear Models has become much more widespread and is an important ingredient for most packages in the field. Another European development, related to latent factor analysis by Lazarsfeld and Henry were the LISREL program and Partial Least Squares estimation. The first is based on covariance structures using Maximum Likelihood methods, the second analyses latent path models with data-structure oriented methods developed by Herman Wold (Jöreskog (1973), Jöreskog & Sörbom (1978, 1988), Wold (1978), Lohmöller (1982)).

The major packages SPSS, SAS, the now European BMDP (further development in Ireland) and SYSTAT are overshadowed by the S-language environment (Becker et al. (1988)). This provides in its current S-PLUS version a flexible base for graphical data analysis, statistics and computational programming. While it is an American product, many add-ons were developed in Europe, e.g. a whole set of algorithms for robust statistics by Marazzi (1993). This openness for extension and extensive user control is one of its major benefits. While a commercial institution may be necessary to guarantee continuity and further development, the possibility of incorporate one's own algorithms and of trying out new methods is an invaluable asset for the researcher. LISP-Stat does similar things in a public X-LISP or Common Lisp environment and draws on the experiences of S. Much of the material and implementation of Lisp-Stat code are freely available on the statlib archive (Internet: statlib@lib.stat.cmu.edu). This is only one example of the growing importance of international electronic networks, especially the Internet! The usenet newsgroups on statistics (sci.stat, sci.stat.consult ...) provide information on the availability of software and the solution of computational problems in statistics and are being used more and more. Other packages, like Data Desk, ISP, XPLORE etc. deliver other valuable resources and control over them to the user. Published overviews show the increasing number of available products (e.g. Koch & Haag (1993)).

The developments in Artificial Intelligence and Expert Systems triggered interest in formalizing the metaknowledge of statistics. Lisp-Stat is partially motivated by this development. However a lot of the development in this direction stagnates. It may be just too ambitious for the time being. But this development has still brought some advances for statistical databases: The development of statistical databases – especially in the Eurostat office – allows today the representation of general "statistical objects". The entity-relationship paradigm allows the simultaneous representation of numeric statistical data and the description of those data. During an analysis, the description of the data is transformed simultaneously with the data. Using concepts of expert systems has brought an important advance in the representation of statistical data. Some paradigms of AI have crept into the help systems of all major statistical packages. Especially fruitful was the impact in simulation tasks, which are well suited for some kinds of qualitative control. In combination with fuzzy methods, results are achieved by approximate reasoning which are hardly available from the analytic and parametric models of reality, typical of many statistical studies.

The use of dynamic graphics is one of the most "dynamic" developments in computational statistics. To look at clouds of data, the selection of subsets in appropriate low dimensional representations or projections and their immediate analysis gives insights in large data sets which were until recently unimaginable. The work started by Tukey with Exploratory Data Analysis diffused in a larger number of packages. The developers of the SAS package support JMP an interactive graphic analysis tool to do this on a Macintosh i.e. in a windows environment and in an added module SAS/INSIGHT. Other packages (S-Plus, ISP) move in the same direction. The use of icons in the computer environment is related to the implementation of ideas Otto Neurath developed in the 30's and 40's (1936). He used his "Isotypes" to produce picture statistics. The Macintosh and Windows environment is already creating an international picture language in Neurath's sense: it is understood by the masses.

An early effort to visualize statistical data by computer were Chernoff faces, where each face gives information on a multivariate item (more funny than useful). Two splendid books by Edward Tufte take up a tradition starting at least at the beginning of the last century and show statisticians how to present data. He describes historical forerunners like Minard's graphic representation of the decreasing number of soldiers during Napoleon's invasion of Russia. The appearance of efficient, graphically powerful workstations brings the graphical features of the packages to the forefront. Multivariate data exploration has changed in an irreversible way for the better.

More and more of the non-statistical packages contain functions which qualify them for statistical work: Mathematica, Derive, Maple and other algebraic systems are used to easily compute distributions, and extensive packages for econometric and statistical work are available in these environments. In a way they are the more general version of S: even more open, handling an even larger class of problems, but less efficient in some statistical problems. One possible future is that S and similar systems get more general and include symbolic manipulation, another one is the possibility of including efficient statistical functions in the symbolic manipulation packages. This is probably a signal for future developments. Programs have increasingly open interfaces wich may import data and export the results of their processing. Finally: even the spreadsheet Excel contains practically all elements needed for undergraduate statistics. For many casual users – even in a professional business environment – this may be enough.

Besides and complementary to the large and comprehensive packages there are many smaller ones, often serving specific need, e.g. on clustering from H.H.Bock (1974) in Germany or by Leonard Kauffman and Peter Rousseeuw (1990) in Brussels, or on survey processing (Blaise from the Netherlands Bureau of Statistics). The French have their own tradition, e.g. on correspondence analysis, and produce commercial software for training purposes (Scribe Tuteur from Aix-en-Provence). The countries in transition from eastern Europe had their own systems – often because of the lack of access to western software (although this problem was reduced by liberal copying practices). Two – rather arbitrarily chosen – examples of such programs available in the west are Mesosaur from the Russian Academy of Sciences, a time series analysis system, also distributed by Systat, USA, and Adstat from Trilobyte Software in Prague, advanced numerical Statistics for the PC, strong on nonlinear regression calibration and smoothing.

Specific packages for econometrics led to advances in multivariate models and models for time series analysis starting from Box-Jenkins approaches (RATS – Regression Analysis of Time Series) triggered again theoretical work for instance on co-integration of variables. The impact of sociological methodology on statistical methodology is emphasized by Clifford C.Clogg (1992). The extensive discussion of this article shows a number of additional influences, e.g. from econometrics. The discussions show also that comprehensive sciences like statistics are fed by different contributions. The statistical and computational subcultures often remain separated from each other for long periods. If the different cultures become aware of the work done in another application field very fruitful exchanges often emerge. Extensive statistical applications are used and handled in modern particle physics. Quantum mechanics and its derivatives are inherently statistical theories. The basic training of physicists comprises the calculation of reaction mechanisms of the interacting particles in time and space. Nonetheless most physicists get hardly any training in the methods taught in statistics departments. The feedback from Physics to Monte Carlo methods is comparatively small, considering the intensive use of some statistical methods in this area.

Some Organizational Developments

The lack of knowledge in the field where the computer and the statistician meet was felt more in Vienna than elsewhere. That was the reason why we, from the Institute of Statistics in Vienna, organized in 1974 a conference on "Computational Statistics" in Vienna (the expression was choosen in analogy with "theoretical" and "applied" statistics). It became the first of the biannually held series of COMPSTAT conferences. It was dedicated to the methodology of statistical computing and featured: Computational Probability, Automatic Classification, Numerical Aspects of Statistical Methods, Simulation and Stochastic Processes, and Software Packages. The conference wandered around in Europe, occasionally reminding participants of the changes in Europe: The conference in Prague was originally planned for Bratislava in Slovakia. A "central" decision brought along a change of location. Nonetheless the conference contributed to the relatively strong development of computational statistics in Czechoslovakia (or better in Prague?) compared to other eastern European countries. The conference in Dubrovnik was held just a year before the war started.

The contents of the conferences changed over the years : while some topics

– multivariate analysis, classification (discrimination and clustering) – remained strong, others – e.g. analysis of variance – disappeared. ANOVA has become a standard part of the general packages (and the specific target for some e.g. EASYSTAT, developed at the university of Cologne) and is handled more in texts concerning teaching and education. Some interests – teaching, optimization, expert systems – appeared, reappeared and disappeared. Symbolic manipulation, parallel processing and recursive techniques have not featured strongly in the conferences. Related hardware specific problems, even those concerning graphics are not an important topic, hardware is hardly mentioned.

During its Delhi meeting in 1977 the International Statistical Institute formed an International Association for Statistical Computing (IASC). In 1978 the former organizers of the COMPSTAT symposia together with its publisher Physica agreed to hand the organization of the symposia to a newly established European Regional Section of this Association.

The German Gesellschaft für Strahlen- und Umweltforschung and the Institut für Medizinische Informatik und Systemforschung have edited a Statistical Software Newsletter since 1975. In 1980 it was made available to the international community – at first free of charge – and became in the following years the periodical of the IASC.

The European scientific journal devoted to computing questions in statistics was for some years the Computational Statistics quarterly. In 1982 The Statistical Software Newsletter under the guidance of the indefatigable Allmut Hörmann was incorporated in the journal Computational Statistics and Data Analysis which became the official publication of the IASC and got a non-European publisher (Elsevier; but it is at least printed by North-Holland). The Newsletter gives a comprehensive picture of the relevant movements in the statistical software environment. The Statistical Software Guide 92/93 gives e.g. an overview of most relevant products (Koch & Haag (1993), available also by e-mail: v20@vm.urz.uni-heidelberg.de). As the article you are just reading is biased and reflects the perception of an individual – even if after some discussion with colleagues – compilations like those in the Newsletter may be a good starting point to look into areas which are more relevant to your interests and to correct a partial picture.

Naturally COMPSTAT did not remain alone. A variety of conferences were dedicated to specific topics: Pattern Recognition and Classification, Simulation, Multivariate Methods, workshops on nearly every single aspect. The applied sciences especially biometrics, physics and ecology lead partially independent lifes. The German Region of the International Biometric Society and the GMDS have had working groups for Statistical Data Analysis since 1974 and for Computational Statistics since 1985. Softstat has similar intentions to Compstat. Originating as a local German event, coorganized by two working groups, it has grown increasingly international. The traditional conferences of statistics, be they official statistics or even the theoretical mathematical conferences of the Bernoulli Society of the ISI, also treat computational aspects. A curiosity was the "First European Conference on Statistical Computing" held in Turkey at the time of the sixth COMPSTAT Symposium. Special conferences like those on analyse des données, classification or pattern recognition complement and substitute some efforts of COMPSTAT.

The widespread use of the computer, especially the PC, in statistical work – often by the untrained – has sharpened our view of the calculating machine as a tool to reflect reality and to understand social, economic, natural and biological systems. Similar to an X-ray machine it makes facts visible which are not discernible by the unaided eye. It makes the analysis of huge amounts of data in a short time possible and condenses them into a few facts relevant for decisions. At the same time it allows us to influence these systems with incredible speed, distributing the effects of our decisions via networks over wide areas, to many persons and to many decision-makers. Both, in the number crunching of large data sets and in information reduction in inference generation, the contribution of statistics is crucial. The inferences drawn are highly relevant for many societal problems: The first use of computational statistics in the development of the atomic bomb by the use of Monte Carlo methods shows that it is not an ivory tower business. The use in medicine, in assessing social behaviour and effects of political measures, and in the use in environmental monitoring brings it to the fore where non-scientific factors are clearly relevant. It should be an aim of methodologists to improve methods in such a way to make the results of their enquiries as "objective" as possible. We all know how many a priori assumptions influence the final outcome. To minimize their impact and make the remaining impact visible should be our task for the future. In all this a certain view of the whole shall be maintained, very befitting to the heirs of the "homo statisticus" whose purpose was the knowledge of statesmanship and of state affairs.

References

Becker, R. A.; Chambers, J. A.; Wilks, A. R. (1988) The New S Language. Pacific Grove, Calif., USA, Wadsworth;

Bock, H. H. (1974) Automatische Klassifikation. Göttingen: Vandenhoek und Ruprecht; (Studia Mathematica).

Brown, J.C., Houthakker, H.S. & Prais, S.J. (1953) Electronic Computation in Economic Statistics, JASA, 43, 411-428

Clogg, C.C. (1992), The Impact of Sociological Methodology on Statistical Methodology, Statistical Sciences, 7, 183-207

Comrie, J.L.; Hey, G.B. & Hudson, H. G. (1937) Application of Hollerith equipment to an agricultural investigation. J.Roy.Statist.Soc.Suppl. 4: 210-224.

Diday, E. (1992) Computational Statistics; Neucátel. Neuchâtel. Heidelberg/New York, Physica; 1: 193-213.

Efron, B. & Tibshirani, R. (1986) Bootstrap Methods for Standard Errors, Confidence Intervals, and other Measures of Statistical Accuracy. Statistical Science. 1(1): 54-77.

Everett, C.J. & Ulam, S. (1948) Multiplicative Systems. I. Proc. Nat. Acad. Sci. (USA). 34: 403-405.

Everett, C.J. & Ulam, S. (1948) Multiplicative systems in several variables III. LADC, Los Angeles Declassified Document. 533 and 534.

Fienberg, S.E. (1992) A Brief History of Statistics in Three and One-Half Chapters: A Review Essay. Statistical Science. 7(2): 208-225.

Goldstine, H.H. (1972) The Computer from Pascal to von Neumann. Princeton, New Jersey

Hammersley, J.M. & Handscomb, D.C. (1964) Monte Carlo Methods. London, Methuen

Hampel, F.R.; Ronchetti, E.M.; Rousseeuw, P.J. & Stahel W.A. (1986) Robust Statistics. The Approach Based on Influence Functions. John Wiley and Sons.

Hawkins, D. & Ulam S. (1944) Theory of multiplicative processes. Los Alamos, LADC; 265.

Hlawka, E. (1981) Numerische Analysis. Wien, Oldenbourg; (Schriftenreihe der Österreichischen Computergesellschaft).

Hlawka, E.; Firneis, F. & Zinterhof, P. (1981) Zahlentheoretische Methoden in der numerischen Mathematik. Wien, Oldenbourg; (Schriftenreihe der Österreichischen Computergesellschaft).

Huber, P.J. (1964) Robust Estimation of a Location Parameter. Ann. Math. Stat. 35: 73-101.

Huber, P.J. (1981) Robust Statistics. Wiley, New York.

Jardine, N. & Sibson, R. (1971) Mathematical Taxonomy, Wiley, London New York

John, V. (1884) Geschichte der Statistik. Stuttgart

Jöreskog, K.G. (1973) A General Method for Estimating a Linear Structural Equation System. In: Goldberger, A.S. & Goldberger, O. D. (eds.) Structural Equation Models in the Social Sciences. New York, Academic Press;

Jöreskog, K.G. & Sörbom, D. (1988) Advances in Factor Analysis and Structural Equation Models. Cambridge, Mass., USA, Abt Books;

Jöreskog, K.G. & Sörbom, D. (1988) LISREL 7: A Guide to the Program and Applications. Chicago, SPSS

Kaufman, L. & Rousseeuw, P.J. (1990) Finding Groups in Data, An Introduction to Cluster Analysis. New York, Wiley; 1990.

Koch, A. & Haag, U (1993) The Statistical Software Guide, Computational Statistics and Data Analysis, 15, 241-262

Lazarsfeld, P. (1977) Notes on the History of Quantification in Sociology. In: Kendall M.G & Plackett R.L (eds.) Studies in the History of Statistics and Probability.; 2: 277-333.

Lohmöller, J.-B. (1982) Path models with latent variables and Partial Least Squares (PLS) estimation. München: Hochschule der Bundeswehr München, Diss. Pädagogik

Marazzi, A. (1993) Algorithms, Routines, and S Functions for Robust Statistics. Pacific Grove, California, Wadsworth & Brooks/Cole. Advanced Books & Software; 1993.

Metropolis, N. & Ulam S. (1949) The Monte Carlo Method. JASA. 44: 335.

Neurath, O. (1936) International Picture Language. London, K.P.Trench, Trubner and Co.

Owen D.B. (1976) On the History of Statistics and Probability. New York

Pearson, E.S. & Kendall M.G. (eds.) (1970) Studies in the History of Statistics and Probability. London; 1.

Pearson, K. (1978) The History of Statistics in the 17th and 18th Centuries against the changing background of intellectual, scientific and religious thought. Pearson, E. S. (ed.) London

Rasch D. & Tiku M.L, (eds.) (1984) Robustness of Statistical Methods and Nonparametric Statistics. Berlin, VEB Berlin

Rasch, D. & Herrendörfer, G. (1982) Probleme der angewandten Statistik. Robustheit III, Arbeitsmaterial zum Forschungsthema Robustheit. Heft 7. Forschungszentrum für Tierproduktion Dummerstorf-Rostock der AdL der DDR.

Rousseeuw, P.J. & Leroy, A.M. (1987) Robust Regression and Outlier Detection. John Wiley and Sons

Sacks, J.; Welch, W.J.; Mitchell, T.J. & Wynn, H.P. (1989) Design and Analysis of Computer Experiments. Statistical Science. 4(4): 409-435.

Schaffer, J.D. (Ed.) (1989) Proceedings of the Third Conference on Genetic Algorithms, 3, Morgan Kaufman Publishers, Arlington, VA, USA

Sint, P.P. (1984) Roots of Computational Statistics. In: Havránek, T.; Sidák, Z. & Novak, M., (eds.) COMPSTAT 1984, Proceedings in Computational Statistics, 6th Symposium. Wien, Physica-Verlag

Sokal, R.R. & Sneath, P.H.A. (1963), Principles of Numerical Taxonomy, Freeman, San Francisco London

Stahel, W.; & Weisberg, S. Directions in Robust Statistics and Diagnostics. Part I und II. Springer Verlag; 1991.

Tufte, E. (1990) The Visual Display of Quantitative Information. Cheshire, Connecticut: Graphics Press

Tufte, E. (1992) Envisioning Information. Cheshire, Connecticut: Graphics Press

Tukey, J.W. (1964) Bias and confidence in not-quite large sample. Ann. Math. Stat. 29: 614.

Tukey, J.W. (1977) Exploratory Data Analysis. Reading, Mass., Addison-Wesley

Wilkes, M. V. (1956) Automatic Digital Computers. London

Wishart, D. (1978) CLUSTAN User Manual. Edinburgh, Program Library Unit, University of Edinburgh

Wold, H. (1975) Soft Modelling by latent variables: The Non-Linear Iterative Partial Least Squares (NIPALS) approach. In: Gani, J. Perspectives in probability and statistics: Papers in honour of M.S.Bartlett on the occasion of his sixty fifth birthday. London: Applied Probability Trust; 1975.

Languages for Statistics and Data Analysis

Peter J. Huber, University of Bayreuth, D-95440 Bayreuth

Key Words:
languages, data analysis

Abstract
Languages for data analysis and statistics must be able to cover the entire spectrum from improvisation and fast prototyping to the implementation of streamlined, specialized systems for routine analyses. Such languages must not only be interactive but also programmable, and the distinctions between language, operating system and user interface get blurred. The issues are discussed in the context of natural and computer languages, and of the different types of user interfaces (menu, command language, batch). It is argued that while such languages must have a completely general computing language kernel, they will contain surprisingly few items specific to data analysis – the latter items more properly belong to the 'literature' (i.e. the programs) written in the language.

Diverse lingue, orribili favelle...
(Dante, Divina Commedia, Inf. 3,25)

Introduction

In this paper, 'statistics' and 'data analysis' will be used synonymously. It has been said that statistics is the art of collecting and interpreting data. If we accept this, statistics logically covers all of data analysis (and some more). On the other hand, most data analysis is done by non-statisticians, and much of it is not statistical in the narrow sense of that word. Statisticians have had an unfortunate tendency to box themselves into the corner corresponding to the narrowest possible interpretation, that is, they would not consider an activity 'statistical', unless it explicitly involved chance models and formally optimized procedures. The warnings of J. W. Tukey (1968, p.20) against such tendencies are still fully valid after 25 years and should be heeded. I have come to feel that 'data analysis' is a more neutral term, so I shall give it preference.

Over the past 15 years or so, data analysis has emerged as the single most demanding application of interactive computing. Part of the challenge is that data analysis covers a spectrum ranging from research to repetitive routine. Data exploration is interactive by its nature. One needs a system flexible enough to deal with most diverse, unpredictable applied problems, and to improvise non-standard analyses. Often, this stage is followed by one or more rounds of fast prototyping, ultimately resulting in a streamlined routine version of a specialized data analysis program, so that similar data can be analyzed by less qualified personnel. Ideally, a data analysis system should be fast and stable enough to handle not only the research and the prototyping stages, but also the implementation of the final product.

This implies that a system suitable for data analysis must be interactive and must have the capabilities of a general purpose programming language as well as those customarily associated with an operating system.

We have argued elsewhere (Huber (1986a), Huber & Huber-Buser (1988)) that for all practical purposes this goal can be achieved only with a system built around a programmable command language. In the meantime the arguments have gained rather than lost strength, despite the still fashionable Windows-Icons-Menus-Pointing bandwagon.

It is necessary to stress that there is a fundamental distinction between an interactive language and an interactive program. An interactive language is not a language suitable for writing interactive programs, or for writing batch programs interactively, but a language suitable for interactive or immediate execution: when the user enters a line, it is immediately interpreted and executed. Since the terms 'interactive' and 'interpretative' are over-used and ambigous, I shall prefer to talk about 'immediate' languages. Among the two best known languages of this class, APL goes back to Iverson (1962), and BASIC arrived only two years later in 1964. What is needed for data analysis is not merely an interactive system, but an *immediate language.*

None of the classical programming languages are suitable for the spectrum of tasks encountered in data analysis. The batch languages (FORTRAN, PASCAL, C, ...) do not permit immediate use. Those languages that do (APL, BASIC, and shells of various operating systems) fall short in other respects.

APL is write-only: it is easy to write, but it cannot be read, only deciphered. BASIC is not array-oriented, and as a consequence, even very simple loops necessitate the writing of a small throw-away program; with many implementations of BASIC (in particular the standard versions of BASIC on the PC), this causes the loss of all previous results obtained in immediate mode.

Among the languages specifically designed for doing statistics and data analysis, S primarily is a high-level programming language. While it permits immediate use, its syntax is poorly suited for it. ISP was designed for immediate use; it was soon extended to include high-level programming features. I believe that among the two it comes closer to the ideal.

Though, talking about Platonic ideals may be out of place here: in order that a language be suitable for data analysis it must compromise its ideals by making concessions. It is more important that it does everything moderately well, than that it does something very well.

With immediate languages the customary distinctions between operating systems, programming languages and user interfaces disappear. There have been several computers whose operating system coincided with their principal programming language, from the tiny Timex-Sinclair using BASIC, to the Symbolics LISP Machine. High-level immediate languages get quite close to the spirit of natural languages in more than one respect, and we will have to single out the similarities and the differences not only to batch programming but also to natural languages.

Naur pointed out already in 1975 (with little success) that there is a need to shift attention away from programming languages toward the corresponding literature, i.e. toward the programs written in the languages.

Where does 'language' end and 'literature' begin? In an extensible language, where the extensions look exactly like original language constructs, the boundary is blurred and moving and depends on an individual user's current usage. A possible criterion is: if a particular construct is frequently used as a building block within others, then it belongs to the language. If it is never used in this way, then it belongs to the 'literature' written in the language. If we accept this criterion, then the core of a language for data analysis is completely general and contains negligibly few items *specific* to data analysis.

In ISP for example, REGRESS, the command for multiple linear regression, according to this criterion clearly belongs to the language. But if we abstract certain embellishments, regression is not specific to data analysis. Remember that APL has a special, built-in operator doing linear least squares fits. An ISP macro doing non-linear least squares is on the border line. The possibly misleading name 'macro' is historical, it denotes a full-fledged procedure written in ISP. Another large macro doing generalized linear models, all of GLIM and a little more, belongs to the 'literature'.

The issues are tightly interwoven, and the tale to be told is rather tangled. The discussion is arranged in several circles, beginning with a general discussion of languages. The following circles are concerned with interface issues, and finally with the requirements a language suitable for statistics ought to satisfy.

Natural languages and computing languages

> *What can we learn about programming languages from the social aspects of mathematics and natural languages? - Now why is this a relevant topic to consider? Well I definitely feel that there is something rotten in the realm of programming. There is a lot of discussion, but somehow I think that most of it misses the point. There are too many fads, too many quick solutions, a too wide gap between theory and practice.* (P.Naur (1975))

Sadly, almost 20 years later, Naur's comments still hold true.

Very little attention has been paid by systems developers to what we actually know about everyday natural languages.

It is less a question of borrowing directly from natural languages, but rather of borrowing methodology for looking at languages. For example, one ought to pay attention to the syntactic typology of computer languages in the same way one has investigated that of natural languages (compare Lehmann (1978)).

Natural languages

The recorded history of natural languages now spans approximately 5000 years. It is difficult to reconstruct precise reasons behind particular developments. But the recorded time span is long enough for natural selection to operate effectively. Here are a few random observations of potential relevance for the design and use of computer languages.

Language skills require extensive training. People stay with the language(s) they have learned, whenever this is possible, and they usually feel most comfortable with their mother's tongue.

Natural languages are mutually translatable into each other. Despite this (or because of this?) difference of language sometimes is the source of deep hate.

Spoken and written forms of a language are consistently being regarded as one and the same language. There are definite differences of style, but a literate person always was supposed to be able to read a text aloud and to take dictation.

Indentation, spaces, capitalization and punctuation serve as visual cues to indicate sentence structure in written texts (they are substitutes for absent phonetic cues). In natural languages, these items carry little, if any, information. The early alphabetic scripts did not use them at all.

Unpronounceable words are hard to memorize. The ancient Egyptian language is difficult to learn not because it is written in hieroglyphs, but because the vowels are not written and therefore unknown.

Human communication is two-way and depends on feedback. We use guesswork and common sense, and if necessary, we ask for clarification or confirmation.

Most alphabetically written languages have a redundancy factor of about 2, that is, a text can be reconstructed if up to about half of the letters have been erased (Shannon (1949)). If the redundancy deviates too much in either direction, comprehension suffers.

Natural languages avoid deep nesting, except if it is of a trivial chain-type.

The first known writing system was invented 5000 years ago in the Near East. It used pictorial signs ('icons'). Within a few decades, it developed into a full-scale, abstract writing system with hundreds of conventional signs for words (no longer recognizable as pictures of objects) and with phonetic complements to express grammatical relations. About 2000 years later, alphabetic writing systems were invented, allegedly because they were easier for the occasional users who had difficulties remembering several hundred signs.

Icons are successful means of communication in a multilingual society only if they are small in number, easily recognizable and completely standardized; international traffic signs may be the best example.

Some historical remarks on computing languages

Computer science is young and has little sense of history – even of its own history. The time horizon of CS graduates typically does not extend beyond their first programming course.

Serious computing began with numerical applications, programmed in raw machine language. I still remember that time (the mid-1950's); in essence, we then wrote programs by informally putting two levels of better readable languages on top of raw machine language: flowchart language and assembly language. We structured the program through flowcharts and then wrote it down by using symbolic names and addresses, instead of actual storage locations. Finally, this program was assembled into machine code by hand.

The really painful part was not the writing of a program, but making modifications to it. The problems with raw machine language were that it was more difficult to read than to write, and that adding a few lines of code usually implied making tricky non-local modifications.

Few people are aware that the first paper ever written on compiling and compilers was Rutishauser's Habilitationsschrift (1952). The primary purpose of ALGOL 57 (later to be redesigned into ALGOL 60) was the unambiguous description of numerical algorithms (whence the name – ALGOrithmic Language). Compilation by machine was only a secondary consideration, even though Rutishauser was the driving force behind the ALGOL effort, and Backus (the designer of the original FORmula TRANslation, 1953) was a member of the working group. The design of ALGOL 60 and its formal description ('Backus-Naur notation', Naur 1960) is most impressive; I believe it was Hoare who once said that ALGOL 60 was a notable improvement not only over its predecessors, but also its successors.

Though, ALGOL 60 had a gaping hole: the design committee had not been able to agree on string handling and hence on I/O. This was irrelevant with regard to the primary goal, but not for the secondary. Commercial ALGOL compilers for a long time were lacking, or poor and inefficient. An overly ambitious and overly complex redesign (ALGOL 68) effectively killed the ALGOL effort.

PASCAL (Wirth (1971)) was not a better language, but it was more successful because of the availability of portable compilers. Its problem was that it had been conceived too narrowly, the primary design goal was to teach programming through writing small, well structured programs. Now it is choking under its too narrow conception and its overly strong typing. The problems were evident from the beginning; after looking at its specifications in the early 70's I certainly would not have believed that it would survive into the 90's.

C was developed for the PDP-11 on the UNIX system in 1972. Its strengths are neatly summarized by Ritchie et al. (UNIX (1978, p. 1991)): "C is not a 'very high-level' language nor a big one and is not specialized to any particular area of application. Its generality and an absence of restrictions make it more convenient and effective for many tasks than supposedly more powerful languages. ... The language is sufficiently expressive and efficient to have completely displaced assembly language programming on UNIX."

The weakness of C lies in its crummy syntax, in its error-prone type declarations, and in its unnecessarily risky handling of pointers. As a consequence, it is hard to read and quite unsuited for the communication of algorithms between humans. If I am to venture a prophecy, these problems are serious enough that they will ultimately lead to its downfall and replacement. But this will not happen too soon, compare Naur's (1975) comment that 'even a language like Esperanto, that was recognized already in 1907 when it was but 20 years old to be poorly suited for its declared purpose, has retained its support to this day, 67 years later'.

It is very difficult for new programming languages to gain acceptance, and designing new languages has been out of fashion for a while now. Reluctance to learn yet another language is only part of the reason, it may also have to do with the fact that language designers never seem to learn from the mistakes corrected by their predecessors. The principal weakness of FORTRAN was and is its lack of dynamic arrays. ALGOL corrected this in a safe way, but PASCAL fell behind FORTRAN: in standard PASCAL one cannot even write a routine for transposing arbitrary arrays. PASCAL was a very 'safe' language, but C fell even behind FORTRAN in this respect.

The use of computing languages for the unambiguous description of algorithms, to be read also by humans, has been neglected for years by almost everybody. A notable exception is Donald Knuth (see his monumental works on the Art of Computer Programming (1968-) and on Computers and Typesetting (1986)).

I therefore think a good, readable, general purpose immediate language would fill more than one market gap.

From the random remarks made in this and in the preceding section I would like to distill a few maxims or rules for language design that should not be broken without good reason:

1. Aim for a general purpose language.
2. Keep close to the tested natural language style-sheet.
3. The language should not only be writable, but also readable.
4. Use pronounceable keywords.
5. Do not encode information into case and punctuation.
6. Avoid icons that are neither standardized nor self-explanatory.
7. Do not leave holes in a language design.
8. Make the language portable by providing good and portable interpreters or compilers, and by avoiding dependence on hardware and operating systems.
9. Beware of ambitious redesigns.

Object orientation and related structural issues

> *'Ergativity' is currently an 'in' term in linguistics. It is used by a wide variety of linguists, with a whole range of different meanings. As a result, much confusion exists at present about what an 'ergative' language is, and about the morphological, syntactic, and semantic consequences of such a characterization.* (R. M. W. Dixon (1979))

The current shibboleth in programming is OOPLA – Object Oriented Programming LAnguages. It is customarily described by invoking a number of magic words: objects, classes, inheritance, abstraction, encapsulation, polymorphism, and dynamic binding. Apart from that, there is much confusion: the above quote remains accurate if we replace 'ergative' by 'object oriented'.

The real killer hidden in the quote is that ergative natural languages are object oriented. They put the sentence focus on the direct object rather than on the subject of the sentence, as all the usual 'accusative' Western languages do. The tell-tale common pattern is that they treat the object of a transitive verb and the subject of an intransitive verb exactly alike, and different from the subject of a transitive sentence. Ergative languages may be in a minority, but completely unrelated ones are known from every corner of the globe; among the better known examples are Eskimo, Georgian, Basque, and among ancient near eastern languages, Sumerian and Hurrian. To us native speakers of 'accusative' languages, their syntax is very confusing at first. Some languages have switched in historical times and have become ergative, or have lost ergativity. Curiously, no language is 100% ergative, while there are fully accusative languages. Ergative languages seem to have more than their fair share of grammatical exceptions and irregularities.

The analogy between ergativity and object orientedness is not total. In computing languages, the shift of emphasis is not from subject to object, but from verb to object. Nevertheless, there are suggestive similarities. In particular, I wonder whether 100% object orientedness also leads to impossible languages.

Just like ergativity, object orientedness is a matter of degree. By acquiring some more object orientedness, C has turned into C++. SMALLTALK may be the most object oriented language ever created.

Object oriented languages may make things simpler for the programmer/ user by hiding complexity in a hierarchy of composite structures. The problem is that the complexity is still there, and it occasionally acts up.

Thus, a traditional SMALLTALK example is to print an object by sending it the message 'print yourself'. If the object happens to be composite, what does one expect to be printed:

- a list of the sub-objects?
- all sub-objects?
- the data sub-object, using the format sub-object for formatting?

All three possibilities make perfect sense, and there are some others. In essence, one needs to have at least three distinct variants of the print message, and at any time, one might have to add new ones. For the user, it is conceptually simpler to spell out things and tell the system exactly what he or she wants to be printed and how.

If one is improvising a data analysis in immediate mode, transient printing to the screen is one of the most-used facilities: one needs it to check whether the analysis is still on track, whether a particular format is appropriate, etc. Most of this activity is concerned with non-composite end objects, and object orientedness does not help much. On the contrary, it adds at least one more level that must be debugged: the correctness of an interactively created hierarchy of structures. In immediate languages, too much object orientedness may be just as harmful as too many goto's in batch programming.

ISP has the minimum amount of object orientedness an array language must have: the data objects carry their own types and dimensions with them. Clearly, it ought to have more. The question is how much. Here is an example.

With data graphics, data structures get too complex to be handled with ease on a command line: in addition to the data points, there may optionally be lines, areas, colors, symbols, subsets, point labels, variable labels, subset labels, etc. It must be possible to handle the whole as a single entity, but also to operate with ease on the parts, or on groups of the parts. In ISP, we prototyped such structures as macros, and this approach turned out to be extremely convenient and easy to use. We therefore felt that a streamlined, more sophisticated and more powerful approach in compiled code would be even better. Interestingly, this was not so, the more ambitious approach clashed with the KISS principle ('Keep It Simple, Stupid'): it was not simple and stupid enough to be used in immediate mode! In the macro version, the component objects had been linked simply through having a common family name (e.g. 'quake') followed by an underscore and a two-letter extension (quake_da, quake_li, ... for data, lines, etc.). There was no actual object 'quake'. In the ambitious version, there was one, sub-objects could be composite too, and the user could interactively create arbitrarily complex structures. However, even I myself seemed to have problems talking my own language fluently in immediate mode (batch programming was no problem), so I stopped the effort during alpha-testing. I still do not understand the exact cause of the problem. The probable reason is that one has to keep an overly large fraction of one's attention focused on remembering or looking up structures and on making distinctions between objects and sub-objects in a way one is psychologically not prepared to do, at least not in everyday life.

Somewhat more abstractly and generally, the complexity of an object oriented system increases with the product of the number of command verbs (more accurately: the available types of action) times the number of object classes. If both the actions and the object classes are user extensible, we have a serious complexity problem, overtaxing the user's memory. It should be emphasized that this is a problem peculiarly acute with immediate languages. Batch environments can tolerate structures that are an order of magnitude more complicated, but I suspect that all the current window systems are exceeding the limit.

Extremism and compromises, slogans and reality

> *Contradictory things [...] are characterised by identity, and consequently can coexist in an entity and transform themselves each into its opposite.* (Mao Tse-tung (1937))

In the introduction I have stressed that a language suitable for data analysis must make compromises. Unfortunately, reasonable, but working, compromises never are 'in'. An error in one direction is usually 'corrected' by erring even more into the opposite direction. Here are just a few examples, showing into what the early maxims of UNIX have been transformed.

Some of those maxims were (UNIX (1978, p. 1902f.)) to 'build afresh rather than complicate', and 'small is beautiful'. I would agree that these maxims are too extreme to be obeyed rigorously: once the installed base reaches a certain size, building afresh becomes difficult. Unfortunately, the systems that are currently being built are based not on relaxed versions of those maxims, but on the opposite extremes. For example, the most recent 'Definitive Guide to the X Window System' (X Window (1993)) comprises 11 volumes of close to 1000 pages each. In our terminology all this is 'language', not yet 'literature'. Homer, the Bible and Shakespeare together comprise about half as many pages...

The absurdity of such developments (and the difficulty to prevent them) was castigated early on by Hoare, even before X Windows were born. In his 1980 Turing Lecture he tells the story of the Emperor who, to avoid nakedness, ordered that each of his many new suits of clothes should simply be draped on top of the old, and of the consulting tailor, who was never able to convince his clients of his dawning realization that their clothes had no Emperor (who had slipped away to become swineherd in another story).

UNIX uses cryptic two-letter command names. The systems built on top indulge in equally unreadable long names such as XmuCopyISOLatin1Upper-ed.

One of the biggest advances of UNIX (listed as number one in Ritchie's retrospective, UNIX (1978, p. 1947)) was to get rid of a multiplicity of file structures. In UNIX, every file is regarded as a featureless, randomly addressable sequence of bytes. The systems now being built on top of UNIX (e.g. the already mentioned X-Windows, plus X-Tools, plus OSF Motif) grow veritable plantations of trees of classes of structures. When there are too many structures, the structure (if there still is any) is obscured.

The original gospel of Structured Programming was to adhere to a strict top-down design: plan the big structure of the application, then fill in the details. This did not work (it is humanly impossible to specify a complex system correctly in advance). The current gospel seems to be the bottom-up approach: design the menu interface first, then fill in the application. Also this approach does not work for the same reasons, and it has led to Potemkin villages: nice looking surfaces whose hidden flaws caused the applications to be still-born.

A rampant type of contradiction has to do with self(?)-delusion: use a name that is opposite to the facts. For example, after AT&T had introduced a so-called 'Portable UNIX', in an actual porting experiment 40% out of a total of 50000 lines had to be changed, namely 16000 lines of C and 4000 lines of assembly code (Jalics & Heines (1983)).

'User-friendliness' is a particularly insidious slogan. It is being used to split the computing community into a caste of priests (the programmers) who have access to the inner sanctum, and a caste of supposedly illiterate users clicking on icons. After the liberating period of the 1980's, when the people gained hands-on control of their own hardware, this is a reversion to the 1960s and 1970's, when the computer centers dictated what was good for the users.

Moral:
Watch out. Extremism breeds ignorance and fanaticism. Slogans usually clash with reality.

Computing languages and interface issues

But even if the effort of building unnecessarily large systems and the cost of memory to contain their code could be ignored, the real cost is hidden in the unseen efforts of the innumerable programmers trying desperately to understand them and use them effectively. (N. Wirth (1985))

Languages are for communication, and the primary purpose of a computing language is to provide efficient and effective communication between humans and machines, not to build a monument of one's ingenuity.

There are considerable differences between natural languages and languages used for man-machine communication. First of all, there is a basic asymmetry: a machine may possess some artificial intelligence, but in distinction to the human partner it is not supposed to have a free will. Moreover there are fundamental differences in the way information is processed by humans and by machines.

When we are talking about a language used to transmit human wishes to the machine, we mean exactly that, and not a machine-internal way of storing those wishes. This point often is misunderstood. Some system designers seem to be proud that in their system a complex statement such as

```
MyPanel:Panel -> InsertObject(
     CopyObject([(3,4) -> GetValue]))
```

can be generated with a few mouseclicks (Johnson et al. (1993)). Actually, it is a reason for worry, it only means that the statement being generated is in a language unsuitable for the human side of man-machine communication. In other words, one just is adding another suit of clothes to cover up an ugly earlier suit.

Computing languages are used not only for human-to-machine communication. All three other possibilities do occur too: machine-to-machine, machine-to-human and human-to-human. In view of the asymmetry, all need to be discussed separately.

Machine-to-machine communication occurs for example in connection with hardcopy: Postscript files are only rarely written or read by humans.

Machine-to-human communication is neglected by many systems. Some screen designs actually appear to minimize the data-to-ink ratio, compare Tufte (1983), who argued that in data graphics one should aim to maximize it, within reason. Machine response to human omissions or errors often is pathetic.

Human-to-human communication for example was behind the primary purpose of ALGOL 60: unambiguous description of numerical algorithms. A good computer language ought to be structured such that it can serve this purpose.

Human-to-machine clearly is the most important pathway. But one should not forget that communication is a two-way street. From time to time the human user needs and expects some feedback. In practice, this means that information is interchanged in chunks.

Computing languages can be classified according to the size of the chunks used for transmitting information to the computer. Usually, this is considered an interface issue, but really it is one of language, implying a distinct type of interface:

Small chunks (1-4 bits): menu interface,
Medium chunks (100 bits or so): command line interface,
Large chunks (1000 bits or more): program (batch) interface.

In any reasonably sophisticated system all three are needed, just as their metaphors are in natural language: phrasebook, dialogue, letter writing. All have their specific purposes, strengths and problems. To take the command line interface as a basis is a compromise: it makes possible to branch out to either side and provide a menu interface through a menu command, and batch programs through command procedures.

Some personal reminiscences may be in place here. In 1974, we only had access to a central batch computing facility, so we had to live with the batch paradigm. Our approach was to institute a rigorous programming discipline, with strict conventions about data interchange between procedures, so that we would gradually build up a software library of re-usable routines for statistics and data analysis.

This approach never quite worked. In retrospect, the reasons are clearer than they were then. We never achieved a critical mass of re-usable programs, and this had to do both with the size of the building blocks (we tended to make them too big), and with the time needed for programming and re-programming.

In 1975/76 we tried the menu approach. While working on his Ph.D. thesis, Werner Stuetzle wrote a system for the analysis of children's growth data, based on menus operated via the keyboard. The experiences with this otherwise quite successful system were: annoyingly much tree-climbing and too much rigidity. I decided that next time around I would go for a command language approach.

This time came in 1979, when together with two graduate students (David Donoho and Mathis Thoma) I started a research project on interactive graphical analysis of high-dimensional data. As an underlying data handling language, we revived and redesigned ISP, on whose original development Donoho had worked before with Peter Bloomfield at Princeton University.

Stuetzle on his part gave the batch paradigm another try: fast batch on Symbolics LISP machines with zero turnaround-time, first at Stanford University, and then at the University of Washington in Seattle.

The command line interface

(Admiral) *'If there should be war tomorrow, will the US or the USSR win?'*
(Computer, after 5 minutes) *'Yes'*
(Admiral) *'Yes what?'*
(Computer, after another 5 minutes) *'Yes, Sir'*

Command lines and sentences

The amount of information transmitted in a command line is comparable to that contained in a sentence of a natural language, so there is a ready-made metaphor: the sentence. A short discussion of analogies, similarities and dissimilarities is in order.

A traditional distinction is that computing languages must be more precise than natural languages. Natural languages rely on common sense and guesswork on part of the recipient of the message, and on occasional requests for clarification. In the case of immediate languages, this distinction loses its stringency. While common sense is difficult to achieve in a machine, intelligible error messages and requests for clarification or additional information are certainly feasible.

In natural languages, the typical sentence has three parts: Subject, Verb, Object, often in this order. Although Subject-last is unusual, all six possible orderings occur as the preferred order in some language. Word order is correlated in a subtle way with other aspects of syntax, see Lehmann (1978). Analogous typology studies for computing languages do not exist, but similar forces seem to be at work there too.

In computer languages, the subject of the sentence usually is implied (the user or the computer), but the object often is split into two, so that a typical command sentence again has three components: Verb, Input, Output. Again, each of the six possible orderings seems to occur as the preferred order of some language. Possibly, word order plays an even more significant role than in natural languages. For example, automated help facilities, such as prompting for omitted items, work best in traditional languages if the verb comes first, and best in object oriented languages if the input object comes first.

At present, two main syntax types dominate. One is the traditional command language, whose preferred word order is

Verb-Input-Output

as in 'copy x y', with the option of having multiple inputs and outputs. The other type is function style syntax, which seems to favor the word order

Output-Verb-Input

as in '$y = f(x)$', with the option of having multiple inputs, but only a single (possibly composite) output. The opposite order V-I-O, as in '$f(x) =: y$', is less common.

Some early utilities for the PDP-10 used the order V-O-I; I remember it because I found it awful.

As the name 'command' implies, the verb usually is an imperative, and occasional questions are disguised as orders to tell the answer.

In natural languages, a sentence corresponds to some unit of thought. Keeping in touch with the metaphor, also each command line should correspond to a (more or less) self-contained unit of thought. This suggests that the building-blocks of a command language ought to be chosen very carefully, neither too small, nor too big, so that they do not interfere with human thinking patterns. One should provide powerful, intuitive commands, and one should try to keep their number as small as possible, within reason. A large number of commands sometimes is advertised as a virtue of a system, while in reality it is a vice. It puts an unnecessary load on the user's memory, and hampers fluent use of a language.

By the way, in accordance with the 'one thought – one sentence – one command – one line' philosophy, the language should not make it possible to pack several commands onto the same line. While continuation lines occasionally may be needed, one should try to avoid them, their mere length makes them overly error-prone.

Consequences of immediate use

Command languages are designed for immediate use. This causes them to have characteristics very different from the customary batch programming languages.

First, an immediate language must be very robust with regard to errors. A moderately experienced user enters about a line per minute, and maybe a third of the lines are erroneous, ranging from mere typos to errors of thought, often to be discovered only several lines later. Errors are more frequent than in batch programming. It is necessary to design the language so that it does not invite errors. There are a number of quantitative empirical studies on programming errors in batch languages, see for example Litecky & Davis (1976), and note in particular their suggestion to deal with three of the four most error-prone features of COBOL by a relaxation of punctuation rules.

Then it must provide reasonable error messages, and easy recovery from errors. Division by zero, exponent overflow, and the like, should produce some form of NaN ('Not a Number') rather than a halt of execution. User errors must never produce a crash or hang-up. These are much tougher standards than those prevalent in current operating systems (see for example Miller et al. (1990), who were able to produce crashes or hang-ups with about a quarter of the utilities they tested).

Incidentally, error-resistance is a strong argument in favor of the classical command line syntax, and against function-style syntax. The problem with the latter lies in its capability for nesting. Nesting is nice because then the user does not have to think up names for temporaries. On the other hand, nesting usually leads to error-producing piles of parentheses. Nesting of functions with side effects is a recipe for disaster: users do not pay attention to the order of evaluation. Since traditional mathematical notation uses function style syntax, one should use it there, but avoid it otherwise.

Second, the language must provide carefully designed user-assistance. Whenever something can have a default, it must have one. The language must insert defaults for omitted arguments, or prompt for them, if none is available. More generally, if an operation looks sensible by analogy, it must be permissible. This eases the load on the user's memory.

Third, declarations do not work well. Because of memory limitations and lack of foresight, humans are unable to declare things in advance, before actually needing them. In compiled languages, the principal purpose of declarations is to furnish the compiler with information that otherwise would become available only at link or run time, so that it can produce more efficient code. Typically, when writing a program, the programmer will repeatedly jump back to the program header and add declarations. In immediate languages, jumping back is impossible, and what may look like an array declaration really is an executable statement, dynamically creating a new array filled with default zeros. Type declarations are replaced by inheritance rules: the left hand side of an assignment inherits its type and its dimensions from the right hand side. Integers are a subset of the real numbers, and should be treated as such (no distinction between short and long, signed and unsigned, please!). Side-effects of the lack of declarations are that immediate languages need dynamic memory management and a carefully designed garbage collection, and that highly optimized compilers are not feasible.

Forth, multi-line compound statements, looping and branching all belong to the batch realm. In immediate use, the incidence of typos is too high to make them practicable, they would need debugging. Whenever possible, the language therefore must eliminate loops by making them implicit. In other words, command languages ought to be array oriented, with implicit looping over all elements of the array.

Fifth, immediate languages need some degree of object-orientedness, both in data and in procedure objects. The question is how much. Clearly, all objects must carry hidden but accessible information about their own structure – data type and dimensions at the very least in the case of data objects, sophisticated built-in help in case of procedure objects. Though, excessive object-orientedness does not assist the user. Hiding structure increases complexity, if the user must access the hidden parts.

Sixth, the ability to keep legible records is crucial. In word processing, the final document is all that is needed. Similarly, in batch programming, each program version, when fully debugged, supersedes its antecedents. Data analysis is different: just as with financial transactions, the correctness of the end product cannot be checked without inspecting the path leading to it. Moreover, 'what if' types of analysis are very common: one repeats an analysis with minor changes. Thus, it is necessary to keep not only a legible trail suitable for auditing and compiling a report on the analysis, but also a trail that can be edited and re-executed.

The menu interface

The menu or WIMP interface (Windows-Icons-Menus-Pointing) is indispensable in graphics applications: a single mouse-click transfers some 20 bits of coordinate information very effectively. Admittedly, this is not the most typical use of this interface: the so-called graphical user interfaces (GUI's) are not graphical at all. Data analysis on the other hand uses graphics as an essential and effective method to exchange true quantitative graphical information in *both* directions between humans and computers.

The tiny amount of information (1-4 bits) transmitted in most elementary menu interactions is both an advantage and a liability. The main advantage is that a well-designed menu system offers tight guidance and can prevent user errors by not even offering the opportunity to commit them. This way, it can greatly assist the inexperienced or occasional user in doing an infrequent, standardized task.

Navigating a tree of menus can be very frustrating to a newcomer who easily gets lost in a labyrinth of branches, but also to an experienced user who knows where to go but cannot get there quickly. Few menu systems offer any means to retrace one's steps. This can be most aggravating for example when one is installing a major piece of software and overlooks the necessity to set some option. Typically, one then has to start from scratch, loses concentration, blunders again, and so on.

The greatest liability of menu systems is that one cannot easily automate repetitive steps. The maxim of early UNIX "Don't insist on interactive input" (UNIX (1978, p.1902)) should not have been forgotten!

Menu systems quite generally are not user-extensible. Attempts to provide at least some small extensions through macros are pathetic. The recent paper by Rawal (1992) is typical and may serve as an example. It describes a facility to capture a series of mouse and keyboard events into a macro and then to bind that macro to a key press event. Such extensions negate the advantages and strengths of the menu and of the mouse. The mere fact that somebody feels a need for such a wretched facility shows that there is something rotten with WIMP-based systems.

In applications like data analysis an almost equal liability is the lack of legible, editable and executable action trails. The problem lies with the smallness of the chunks of information, so by necessity there is extreme context-sensitivity. This makes it difficult to combine the small chunks into humanly intelligible records. The examples of macros contained in Rawal's paper illustrate the problem: each mouse-click is expanded into a line of text, but still, the meaning of that click is far from evident.

Saving an executable trail is easy, the problem is with editing it. In the mid-1980's we (Russell Almond, David Donoho, Peter Kempthorne, Ernesto Ramos, Len Roseman, Mathis Thoma and myself) were making a few movies to illustrate the use of high-interaction graphics in data analysis. After some trial and error we did this by capturing a complete trail of mouse and keyboard events. Then we edited the trail, and finally shot a movie, frame-by-frame, by running the edited trail overnight. Mathis Thoma wrote a sophisticated program to facilitate the editing part. The tricky part with editing was that one had to extract the intentions of the user and then reconstruct a new trail. Informally, a typical editing instruction might have been: 'Rotate the scene from position A to position B as in the original trail, but do it more smoothly, and in 7 instead of 5 seconds, so that the spoken commentary can be fitted in'.

The information transmitted in an elementary menu interaction typically is smaller than what would correspond to a natural unit of thought. An interesting consequence of this is that keyboard-based menu languages tend to develop into a weird kind of command language: short key sequences are used as units, without glancing at the menu between hitting the keys.

Thus, the key sequence 'F10 q q Ins' may take over the role of the word 'quit' in a traditional command language, and 'F10 p f Ins' that of the word 'print'!

The batch interface: programming environments

In traditional batch style programming, the user writes (or edits) a program, then compiles, links and executes it. Usually, many cycles through these activities are needed for testing and tuning before the program can be used in production. Batch programming per se clearly is needed and used in the final stages of prototyping and for preparing a producting system. This type of operation is traditional and well-understood. It usually is done under control of a make-like facility: after a change, the entire collection of modules is re-linked and re-started from scratch.

Here we are primarily concerned with the much less familiar batch-style paradigm in immediate computing, where computation is continued after a change without necessarily re-starting from scratch. It was introduced for program development by the AI community and was implemented primarily on SMALLTALK and LISP machines. In sophisticated applications like data analysis, this paradigm takes on a special flavor going beyond that of mere program development, since it is tightly coupled with actual, applied production work. The main difference is that the role of the programmer is taken over by that of the data analyst, and the focus is not on the program but on the data. At least in the research phase, what is programmed is dictated by ad hoc needs derived from the data at hand, rather than from program specifications: an exploratory session rarely follows the planned course more than 15-30 minutes, then one begins to improvise. Typically, one will work in immediate mode, dispatching pre-defined or previously prepared modules for immediate execution. But often, it becomes necessary to extend the system on the fly by writing short program modules, or by modifying existing modules. All those modules must be able to work together. Some of them are there to stay, for example there must be a large and stable kernel for doing standard tasks. Some ephemeral ones will be improvised for the task at hand. There must be a rigorous protocol for exchanging data and for documenting new modules, and there must be safeguards against a buggy module tearing down the rest.

Such an approach clearly can be effective only if there is a decent programming environment. By this, we mean an environment in which the user can create, edit and debug a program module in the midst of a session, without losing intermediate results generated earlier in the same session. The traditional compile-link-execute cycle must either take negligible time, or be completely eliminated. None of the classical programming languages

(FORTRAN, PASCAL, C, ...) can provide such an environment; apart from the AI languages LISP and SMALLTALK, only APL and a few versions of BASIC do.

The traditional distinction between operating system and programming language vanishes in a good programming environment.

Some data analysis systems have been built directly on top of immediate batch environments. For example, data analyses can be performed by modifying program modules and executing them as commands straight out of the editor built into the Symbolics LISP environment, see Stuetzle (1987). Though, because of its error-prone penchant for deep nesting, LISP is not overly well suited for immediate use of this kind. I suspect that these efforts will fade away together with their special purpose hardware.

Technically, this approach uses the immediate batch interface: a program module of an arbitrary size is dispatched out of the editor for execution. In practice however, it tends to coalesce with the command language approach: the complexity of frequently used larger modules is hidden behind small modules ('commands') invoking them. Conversely, command languages must be extensible on the fly, hence it is necessary to build a programming environment into the command language.

In any case, in an extensible language, new commands are batch constructs. They can be added either as interpreted command procedures ('soft commands') or as compiled code to be linked with the overall system ('hard commands'), or as separate main programs, to be invoked as a child process by a small command procedure responsible for mediating the data interchange. If command procedures are interpreted, looping is slow, but in data analysis this efficiency problem arises surprisingly rarely: with proper programming, most jobs can be done through fast, implicit looping over all array elements. Recursive calculations with scalars are among the few cases where this will not work.

From the linguistic point of view it does not matter whether the modules are interpreted, compiled, or subjected to run-time compilation, this is a matter of implementation.

Miscellaneous issues

On building blocks

> *Simple, elegant solutions are more effective, but they are harder to find than complex ones, and they require more time.* (N. Wirth (1985))

The choice of the right building blocks is absolutely essential in any computing language, both in the core of the language and in its extensions, where they take the form of library procedures. If the blocks are too small, that is, smaller than natural units of thought, we end up with an annoying kind of assembly language programming. If they are too large, they become straitjackets, impeding flexibility. For example, the X11 library (see X Window 1993) contains about 20 routines for dealing with the event queue. Most of them combine at least two unrelated actions and therefore tend to have unwanted side-effects. Five or six well-chosen simpler functions would do a better job with less aggravation. For example, there is no function that simply checks whether a particular event has happened, without waiting, and without modifying the contents of the queue.

A 'unit of thought' is a Protean concept; it depends on the current work of a particular user, who from time to time will combine an often-used combination into a new conceptual unit.

Once a building block has been included in a publicly distributed library, it is almost impossible to get rid of it without creating compatibility problems. The conclusion is: in case of doubt, leave it out.

At this point, I must respond to some issues raised by the referees: What is a powerful command? What about data base operations? What about a few examples to illustrate the ISP syntax?

One referee notes that the "keep things simple" strategy and its interpretation as "few, but powerful commands" in its last consequence is dangerous, since it may lead to commands overloaded with unwanted features und obscure options. As examples he mentions SAS procedures such as PROC UNIVARIATE. I agree; "powerful" should be interpreted as in early Unix: seemingly simple, but sophisticated. An example shall be given presently.

A statistical data base is nothing but a collection of a few arrays, occasionally very large ones, with rows corresponding to cases and columns to variables, interlinked in a non-trivial fashion. Any array oriented language deserving that name is well equipped for dealing with data bases. We found, just like the originators of C quoted earlier, that the generality of ISP made it more convenient and effective for many data base tasks than supposedly more powerful, specialized systems. Contrary to a referee's suggestion, I therefore would advocate against bulding a data analysis language around a programmable data base system, but would use the exactly opposite approach. Of course, we had to write a couple of data base macros. After several years of experience we recently distilled from these macros a powerful building block, the command *matchup*. Its syntax is

```
matchup a b > pointers [counts]
```

The input arguments a, b are (n,p) and (m,p) arrays, and the outputs are n- and m-vectors, respectively. For each row of a, *pointers* contains the row number of the first exactly matching row of b, or 0 if there is none, and the optional output *counts* contains the number of occurrences of the rows of b in a.

For example, if x contains the result of a survey and y the results of a follow-up, and if the first columns of x, y provide unique case identifications, then the following two lines update x by appending the results of the follow-up as additional columns:

```
matchup (x(*,1)) (y(*,1)) > i
glue/axis=2 x (y(i,2:0)) > x
```

Missing values are filled in automatically for those cases where there are no follow-up data.

Incidentally, the compiled version of this command is not faster than its macro prototypes (its speed is dominated by a $O((n+m)log(n+m))$ sort operation), but it uses substantially less temporary work space.

On names and notation

A computing language should be easy to write and to read; ease of reading is the more important of the two. A program, or a session transcript, must be easy to scan, even by somebody who is only moderately proficient in the language. Notational details can make a big difference, and it is important to keep a wise balance between terseness and redundance.

The *scope of names* is a notorious open question in programming languages. A safe and clean solution is to treat all names occurring in a procedure as local to that procedure, except if they are imported or exported in an explicit fashion. However, this turns out to be very bothersome in a language intended for immediate use, because one then may have to spell out too many names. An alternative solution (going back to ALGOL) is to treat all names as global, unless they have explicitly been made local. This permits selective hiding of information. The problem is, of course, that a forgotten local declaration can lead to nasty errors through an inadvertent modification of a global variable.

A basic decision is on the kind of syntax the language should follow. Operating system shells suggest a *classical command language.* Traditional mathematical notation suggests a *function style* approach. In the case of ISP, we went for a compromise: use both, but restrict the latter to built-in simple functions. A dogmatic insistence on purity only hampers the average user who is not a computer science professional.

The reason against using function style throughout has to do with its ability for nesting: nesting of complicated functions is error-prone in immediate mode, and it is difficult to provide graceful user assistance (e.g. prompts for omitted or erroneous arguments).

On the other hand, one should follow established models for expressing mathematical functions. After all, traditional mathematical notation has undergone several centuries of natural selection. For example,

$$x * (y + z)$$

is easier to read than the reverse Polish version

$$x\ y\ z+\ *$$

which, however, is easier to use on a calculator.

In case of doubt, one should use self-explanatory notation. For example, the meaning of the expression

$$z = \text{if } a > b \text{ then } a \text{ else } b$$

is reasonably easy to figure out, namely $z = max(a, b)$, while a C-style

$$z = (a > b)\ ?\ a : b;$$

is baffling to the uninitiated.

Book-keeping problems

With large multi-session data analyses, book-keeping problems get severe, comparable to those of large multi-person software projects. It is not sufficient to keep readable, editable and executable records, it must also be possible to retrieve specific information from those records. For example, if one learns about a serious error in a data set or a program module, it is necessary to find out which parts of the analysis are affected and must be repeated. This is considerably trickier than the mere use of a make-facility and involves AI-type programming. A system for script analysis must for example know about the syntax and semantics of the underlying data analysis language. For a brief discussion of the issues see Huber (1986b); a prototype 'Lab Assistant' was subsequently programmed in LISP by W. Vach (1987). His system was able to hunt down forward and backward logical dependencies, and to put up a warning signal, where a dependency might have been introduced by an invisible outside intervention (e.g. by the user entering a literal number read off the screen). To talk about languages: since we were aware that this was an AI project, we felt that we had to use an AI language. We first thought of PROLOG, but then we settled on LISP, because it was stabler. However, in retrospect I wonder whether it might not have been preferable to expand the data analysis language ISP itself, to improve its text handling capabilities to the point that it could handle its own session analysis, and to use the information already built into it rather than to duplicate its parser, syntax analyzer and semantics in LISP. Otherwise, it simply is not possible in the longer run to keep a session analysis system in step with a developing data analysis system.

Requirements for a general purpose immediate language, and issues specific to statistics and data analysis

A system suitable for statistics and interactive data analysis must satisfy all the requirements outlined so far for a general purpose immediate language. We can summarize them as follows:

1. It must have a simple command line syntax. The language must be easily writable and readable, and it should not be error-prone. Hence it should put little reliance on special symbols and punctuation (except for visual cues).

2. It must be extensible and moderately object oriented, with a completely general language kernel, and the same syntax for hard and soft commands (i.e. built-in commands and command procedures). It must offer a full interactive programming environment.

3. It must have full record keeping capabilities.

4. There must be decent, automatic on-line help, with prompts for missing arguments, rather than mere error messages.

5. There must be various primitives not only for numeric, but also for text operations.

6. It must offer interactive and programmed access to the outside world without losing the current environment.

7. In particular, it must offer flexible and efficient I/O facilities that can be adapted quickly and easily to handle arbitrary data structures, binary and otherwise.

8. It must offer all standard functions, linear algebra, some carefully selected basic statistical tools, random numbers, and the main probability distributions.

9. There must be a sensible default treatment of missing values.

10. It must have tightly integrated general purpose tools for high-level, high-interaction graphics and menu construction.

The only items specific to statistics occur in items (7) to (10).

Advanced statistical tools do not belong into the language itself, but into the 'literature' written in that language, and into utilities to be accessed through (6). This way, the linguistic demands made on the user are kept to a minimum: command procedures are written in the general purpose data analysis language, while programs accessed through (6) can be written in the user's favorite language, whatever that is. The gray zone between language core and literatur demands constant attention, if it is not to grow out of hand.

Some of our experiences with ISP relative to large-scale, non-standard data analysis deserve to be discussed in detail.

Graphics. Like all other data analysis systems, ISP had started out with no graphics (that is, with character graphics, 80x25 resolution). It was the first system to add and integrate high-interaction graphics. The graphics part was soon copied by other systems (e.g. MacSpin), but the need for integration usually was neglected or misunderstood. Without such integration, in particular without the ability to create and interchange complex subset structures forth and back between graphics and the general purpose, non-graphical data analysis language, interactive data graphics is little more than a video game.

Exploration graphics must be simple and fast to use, that is, it must rely mostly on defaults.

Presentation graphics in data analysis tends to have rather unpredictable and high demands. The standard graphics packages (Harvard Graphics, etc.) are not up to the task, especially not, if dozens or hundreds of similar pictures must be produced in the same uniform, externally specified format. In such cases, our users resorted to writing Postscript files via ISP macros.

Fast prototyping and production systems. With a flexible, integrated system one can prototype complicated, non-standard, special purpose applications within days or weeks. The prototype is developed through interaction between the prototype designer and the end-user, while the latter applies it to his or her actual data. This way, the user can begin to work with the prototype very early, which permits to correct mis-specifications and design errors before they become too costly. Since the prototype is patched together from generic building blocks, the seams between the blocks may show, but it is not a mere mock-up, it flies. At the end of the prototyping phase it can be used to do the actual production work.

However, production work with large real data means that the demands on the system increase: the more performance one gets, the more one can use. For example, we had to increase the workspace of ISP from 40'000 floating point numbers in 1982 on a VAX to 3'000'000 in 1992 on a PC with 32-bit DOS extenders. A fortunate aspect of array languages is that efficiency of interpretation relative to compilation improves with array size.

An experience derived from this type of work is: if the final production system is to be operated through menus, then the menu tree cannot be specified in advance, it must be constructed on the basis of the application, and together with it. In order that this could be done, a generic menu command became indispensable.

Another repeated experience derived from fast prototyping is that the language must be able to handle sophisticated data base management tasks. Large real-life problems always require a combination of data base management and data analysis, and there is a great diversity of such problems. The larger the data sets, the more complicated their structure tends to be. Neither data base management systems nor traditional statistics packages (SPSS, SAS, ...) are up to the task. Clients realize this and traditionally hire a programmer or a software firm to write a special purpose system adapted to their needs. Statisticians therefore rarely ever get to see the full scope of those problems. The client and the programmers on the other hand only see their application, on which they may work for years. Very often those software efforts go astray and get stuck, because without a working prototype the crucial early interaction between the client and the programmers is lacking. When the situation gets really desperate and one needs results in a hurry, a specialist for fast prototyping is called in. The hardest problems typically have to do with inadequate and changing specifications of the data structures. A flexible data analysis language is needed for the sole purpose of chasing that moving target!

Concluding remark

A language for data analysis must not only serve for *doing* data analyses, but also for *describing* them in an unambiguous, humanly legible form.

The core of a language for data analysis is completely general and contains negligibly few items *specific* to data analysis.

The combined significance of these two remarks is that any such language has the potential of developing into a *koine*, a common standard computer language. If well implemented, it could easily take over both the role of an operating system shell and of a very high level programming language. In view of Naur's comments, some of which have been quoted in Section 2, my optimism is muted. But it might happen if and when the towers of Babel erected on top of UNIX should crumble under their own weight.

References

Dixon, R. M.W. (1979). Ergativity. Language, Vol. 55, 59-138.

Hoare, C. A. R. (1981). The Emperor's Old Clothes. Comm. of the ACM, Vol. 24, 75-83.

Huber, P. J. (1986a). Data Analysis Implications for Command Language Design. In: Foundation for Human-Computer Communication. K. Hopper & I. A. Newman (Eds.). Elsevier, North-Holland.

Huber, P. J. (1986b). Environments for supporting statistical strategy. In: Artificial Intelligence and Statistics. W. A. Gale (Ed.). Addison-Wesley, Reading, MA.

Huber, P. J. & Huber-Buser, E. H. (1988). ISP: Why a Command Language? In: Fortschritte der Statistik-Software I, F. Faulbaum & H.-M. Uehlinger (Hrsg.). Gustav Fischer, Stuttgart.

Iverson, K. E. (1962). A Programming Language. Wiley, New York.

Jalics, P. J. & Heines, T. S. (1983). Transporting a portable operating system: UNIX to an IBM minicomputer. Comm. of the ACM, Vol. 26, 1066-1072.

Johnson, J. A., Nardi, B. A., Zarmer, C. L. & Miller, J. R. (1993). ACE: Building interactive graphical applications. Comm. of the ACM, Vol. 36, No.4, 41-55.

Knuth, D. (1968-). The Art of Computer Programming. Addison-Wesley, Reading MA.

Knuth, D. (1986). Computers and Typesetting. Addison-Wesley, Reading MA.

Lehmann, W. P. (1978). Syntactic Typology. Studies in the Phenomenology of Language. Univ. of Texas Press, Austin.

Litecky, C. R. & Davis, G. B. (1976). A study of errors, error-proneness, and error diagnosis in Cobol. Comm. of the ACM, Vol. 19, 33-37.

Mao Tse-tung (1937). On Contradiction Quoted from Mao Tse-tung: An Anthology of his Writings, edited by Anne Freemantle, The New American Library (1962), 240.

Miller, B. P., Fredriksen L. & So, B. (1990). An empirical study of the reliability of UNIX utilities. Comm. of the ACM, Vol. 33, No.12, 32-42.

Naur, P. (1960). Report on the algorithmic language ALGOL 60. Comm. of the ACM, Vol. 3, 299-314.

Naur, P. (1975). Programming languages, natural languages, and mathematics. Comm. of the ACM, Vol. 18, 676-682.

Press, W. H., Flannery, B. P., Teukolsky, S. A. & Vetterling, W. T. (1986). Numerical Recipes. Cambridge University Press, Cambridge

Shannon, C. E. (1949). The Mathematical Theory of Communication. U. of Illinois Press.

Rawal, K. (1992). A macro facility for X. The X Resource 1, Winter 1992, 6th Annual X Technical Conference. p. 133-142.

Rutishauser, H. (1952). Automatische Rechenplanfertigung bei programmgesteuerten Rechenmaschinen. Mitt. Inst. Angewandte Math. ETH Zürich. Nr. 3. Birkhäuser, Basel/Stuttgart.

Stuetzle, W. (1987). Plot Windows. J. Amer. Statist. Assoc., Vol. 82, 466-475.

Tufte, E. R. (1983). The Visual Display of Quantitative Information. Graphics Press, Cheshire CT.

Tukey, J. W. (1968). Is Statistics a Computing Science? In: The Future of Statistics, ed. D. G. Watts. Academic Press, New York and London.

UNIX Time Sharing System (1978). Bell Systems Technical Journal, Vol. 57 (1978), No. 6, Part 2.

Vach, W. (1987). The design of a 'Lab Assistant' for data analysis. Tech. Report.

Wirth, N. (1971). The programming language Pascal. Acta Informatica, Vol. 1, 35-63.

Wirth, N. (1985). From programming language design to computer construction. Comm. of the ACM, Vol. 28, 160-164.

X Window (1993). The Definitive Guide to the X Window System (various authors). O'Reilly & Associates, Inc., Sebastopol, CA.

A Brief History of S

Richard A. Becker
AT&T Bell Laboratories, Murray Hill, New Jersey 07974, USA

Key Words:
data analysis, data display, graphics, programming language, very-high level language, interactive computing, functional language, vector data.

Abstract
S is a programming language designed for interactive data analysis. It provides a computing environment with persistent data objects, an expressive and powerful language that operates on collections of data, and a wide range of graphics capabilities. This paper describes the evolution of the S language, from its origins at Bell Laboratories in 1976 to the present.

Introduction

The S language has been in use for more than 15 years now, and this appears to be a good time to recollect some of the events and influences that marked its development. The present paper covers material on the design of S that has also been addressed by Becker & Chambers (1984b), but the emphasis here is on historical development and less on technical computing details. Also, many important new ideas have come about since that work. Similarly, parts of Chambers (1992) discuss the history of S, but there the emphasis is on very recent developments and future directions.

Why should anyone care about the history of S? This sounds like the question people ask of history in general, and the answer is much the same — the study of the flow of ideas in S, in particular the introduction and dropping of various concepts and their origins, can help us to understand how software in general evolves. Some of the ideas that are so clear today were murky yesterday; how did the insights arise? What things were done well? What conditions helped S to grow into the strong system that it is now? Could we have anticipated the changes that took place and have implemented years ago something more like the current version of S?

This history of S should also be of interest to people who use the commercially available S-PLUS language, because S and S-PLUS are very closely related.

The Early Days (1976-80 Version of 'S')

S was not the first statistical computing language designed at Bell Laboratories, but it was the first one to be implemented. The pre-S language work dates from 1965 and is described in Becker & Chambers (1989). However, because it did not directly influence S, we will not elaborate here.

S grew up in 1975-1976 in the statistics research departments at Bell Laboratories. Up to that time, most of our statistical computing, both for research and routine data analysis work, was carried out using a a large, well-documented Fortran library known as SCS (Statistical Computing Subroutines). The SCS library had been developed in-house over the years to provide a flexible base for our research. Because the statistics research departments were charged with developing new methodology, often to deal with non-standard data situations, the ability to carry out just the right computations was paramount. Routine analysis was rare.

Commercial software did not fit well into our research environment. It often used a "shotgun" approach — print out everything that might be relevant to the problem at hand because it could be several hours before the statistician could get another set of output. This was reasonable when computing was done in a batch mode. However, we wanted to be able to interact with our data, using Exploratory Data Analysis (Tukey (1971)) techniques. In addition, commercial statistical software usually didn't compute what we wanted and was not set up to be modified. The SCS library provided excellent support for our simulations, large problems, monte-carlo, and non-standard analyses.

On the other hand, we did occasionally do simple computations. For example, suppose we wanted to carry out a linear regression given 20 x, y data points. The idea of writing a Fortran program that called library routines for something like this was unappealing. While the actual regression was done in a single subroutine call, the program had to do its own input and output, and time spent in I/O often dominated the actual computations. Even more importantly, the effort expended on programming was out of proportion to the size of the problem. An interactive facility could make such work much easier.

It was the realization that routine data analysis should not require writing Fortran programs that really got S going. Our initial goals were not lofty; we simply wanted an interactive interface to the algorithms in the SCS library.

In the Spring of 1976, John Chambers, Rick Becker, Doug Dunn and Paul Tukey held a series of meetings to discuss the idea. Graham Wilkinson, who was a visitor at the time, also participated in these early discussions. Gradually, we came to the conclusion that a full language was needed. The initial implementation was the work of Becker, Chambers, Dunn, along with Jean McRae and Judy Schilling.

The environment at Bell Laboratories in 1976 was unusual and certainly influenced the way S came together. Our computer center provided a Honeywell 645 computer running GCOS. Almost all statistical computing was done in Fortran, but there was a wide range of experience in using Fortran to implement portable algorithms, as well as non-numeric software tools. For example, the PFORT Verifier (Ryder & Hall (1974)) was a program written in Fortran to verify the portability of Fortran programs. Portability was a big concern then, since the Honeywell configuration was unlikely to be available to others and since the investment we made in software was considerable.

Bell Laboratories in 1976 also provided a unique combination of people, software tools, and ideas. John Chambers had been at Bell Laboratories for 10 years, had experimented with many approaches to statistical computing, and had codified his thoughts in a book (Chambers (1977)). Rick Becker came to Bell Labs in 1974 with experience in interactive computing from work with the Troll (NBER (1974)) system. The statistics research departments were heavily influenced by John Tukey and his approach to Exploratory Data Analysis (Tukey (1971)). Computer science research at Bell Laboratories was still actively working on the UNIX[1] Operating System (Ritchie & Thompson (1978)) and many of the tools from that work were available on GCOS. In particular, we had access to the C language, based on the Portable C Compiler (Ritchie et al. (1978)), the YACC and LEX tools for parsing and lexical analysis (Johnson & Lesk (1978)), the Ratfor language for writing structured Fortran (Kernighan (1975)), Struct, for turning Fortran into Ratfor (Baker (1977)), and the M4 macro processor (Kernighan & Ritchie (1977)).

An important precondition for an interactive computing system was a good graphics facility. Becker & Chambers (1976, 1977) (and Chambers (1975)) wrote a subroutine library called GR-Z to provide flexible, portable, and device-independent graphics. It used a computer-center supplied library, Grafpac, to produce output on batch devices such as microfilm recorders and printer/plotters. Interactive devices, such as Tektronix graphics terminals, line printers, and others were supported by device driver subroutines written for GR-Z. Its graphical parameters controlled graph layout and rendering,

[1]UNIX is a registered trademark of UNIX System Laboratories, Inc.

and were an extension of those provided by Grafpac. The system provided for graphic input as well as output, and became the cornerstone for graphics in S.

Another enabling experiment was one in separate compilation of portions of a system. Development of a statistical computing system under GCOS required the use of overlays, where portions of the instructions were brought into memory when they were needed. Unlike modern virtual memory systems, space on the Honeywell machine was severely limited, and compilation and linking was slow. Development of the system was likely to require lots of work on the various S functions, and it would be unacceptably slow and expensive to recompile or relink the entire system for every change in one of the functions. To get around this, we developed a pair of assembler language routines that implemented transfer vectors. The main program would ensure that jump instructions to various subroutines were placed at known fixed memory locations. Overlays could then be compiled and linked separately, knowing exactly where they could find various needed subroutines. Similarly, the overlays had a vector of transfer instructions to the overlay's functions stored in another set of fixed addresses. Thus, the main program could always be sure of the address to use when calling the subroutines. This separate compilation/linkage process fit right in with the notion that S would be based on a collection of functions; each function or related group of functions could reside in a single overlay.

A feature of the computing environment on GCOS was the QED text editor (Ritchie (1970)). It was used as a command interpreter, doing textual computations to build an operating system command and finally executing it. Thus, QED provided a primitive analog to the Unix system shell command interpreter. QED was used extensively behind the scenes, to control compilations and overlay loading; it provided the scaffolding that allowed us to build S.

Initial S Concepts

What were the basic ideas involved in S? Our primary goal was to bring interactive computing to bear on statistics and data analysis problems. S was designed as an interactive language based entirely on functions. A few of these functions would be implemented as prefix or infix operators (arithmetic, subscripting, creating sequences), but most would be written with a function name followed by a parenthesized argument list. Functions would be allowed an arbitrary number of positional and keyword arguments and would return a data structure that would allow a collection of named results. Arguments that were left out would have default values. The basic data structures would be vectors and a hierarchical structure, a combination of vectors and/or other such structures (Chambers (1978)). Entire collections of data would be referred to by a single name.

The power of S and its functions would never be used, however, unless the system were extremely flexible and provided the operations that our colleagues wanted. In order to access the computations that were presently available as subroutines in the SCS library (and those that would later be written) we planned to import subroutines written in a lower-level programming language (Fortran) and we planned to allow ordinary *users* to do the same thing. Of course, to import an algorithm would require some sort of interface, to translate between the internal S data structures and those of the language, as well as to regularize the calling sequence. Our model for this was a diagram that represented an algorithm as a circle, with a square interface routine wrapped around it, allowing it to fit into the square slot provided for S functions.

Another important notion that came early was that the language should be defined generally, based on a formal grammar, and have as few restrictions as possible. (We were all tired of Fortran's seemingly capricious restrictions, on the form of subscripts, etc.) English-like syntax had been tried in many other contexts and had generally failed to resemble a natural language in the end, so we were happy to have a simple, regular language. Other applications languages at the time seemed too complicated — although a formal grammar might describe the basic expression syntax, there were generally many other parts. We wanted the expression syntax to cover everything. General functions and data structures were the key to this simplicity. Functions that could take arbitrary numbers of arguments as input, and produce an arbitrarily complex data structure as output should be all the language would need. Infix operators for arithmetic and subscripting provided very natural expressions, but were simply syntactic sugar for underlying arithmetic and subscripting functions.

From the beginning, one of the most powerful operations in S was the subscripting operator. It comes in several flavors. First, a vector of numeric subscripts selects corresponding elements of an object. Even here there is a twist, because the vector of subscripts can contain repeats, making the result longer than the original. Negative subscripts may be an idea that originated with S: they describe which elements should *not* be selected. Logical subscripts select elements corresponding to `TRUE` values, and empty subscripts select everything. All of these subscripting operations generalize to multi-way arrays. In addition, any data structure can be subscripted as if it were a vector.[2]

[2]Later, for arrays, we added one further twist: a matrix subscript with k columns can be used to select individual elements of a k-way array.

S began with several notions about data. The most common side effect in S is accomplished by the assignment function; it gives a name to an S object and causes it to be stored. This storage is persistent — it lasts across S sessions.

The basic data structure in S is a vector of like-elements: numbers, character strings, or logical values. Although the notion of an *attribute* for an S object wasn't clearly implemented until the 1988 release, from the beginning S recognized that the primary vector of data was often accompanied by other values that described special properties of the data. For example, a matrix is just a vector of data along with an auxiliary vector named `Dim` that tells the dimensionality (number of rows and columns). Similarly, a time series has a `Tsp` attribute to tell the start time, end time, and number of observations per cycle. These vectors with attributes are known as *vector structures*, and this distinguishes S from most other systems. For example, in APL, everything is a multi-way array, while in Troll, everything is a time series. The LISP notion of a property list is probably the closest analogy to S attributes. The general treatment of vectors with other attributes in S makes data generalizations much easier and has naturally extended to an object-oriented implementation in recent releases (but that's for later).

Vector structures in S were treated specially in many computations. Most notably, subscripting was handled specially for arrays, with multiple subscripts allowed, one for each dimension. Similarly, time series were time aligned (influenced by Troll) for arithmetic operations.

Another data notion in the first implementation of S was the idea of a hierarchical structure, which contained a collection of other S objects.[3] Early S structures were made up of named *components*; they were treated specially, with the "$" operator selecting components from structures and functions for creating and modifying structures. (There was even a nascent understanding of the need to iterate over all elements of a structure, accomplished by allowing a number instead of a name when using the "$" operator.) By the 1988 release we recognized these structures as yet another form of vector, of mode *list*, with an attribute vector of names. This was a very powerful notion, and it integrated the treatment of hierarchical structures with that of ordinary vectors.

[3]In this discussion we will refer to S data structures as *objects*, even though this nomenclature came later. The connotations of the word are appropriate: an S object can contain arbitrary data, is self-describing, and stands on its own without requiring a name.

Implementation

Persistent data in S was stored in a system known as *scatter storage*, implemented in routines on the SCS library by Jack Warner. Earlier projects in statistics research had needed what amounted to a large flat file system, held in one operating-system file. Scatter storage provided that: give it a hunk of data, its length, and a name and it would store the data under the name. Given the name, it could retrieve the data.

In memory, S objects were stored in dynamically allocated space; S used SCS library Fortran-callable utilities that enabled a program to request space from the operating system and use that space for holding Fortran arrays.

The implementation was done in Fortran for many reasons. Primarily, there was little other choice. C was only beginning to get a start on the GCOS system and at the time, very little numerical work had been done in C. Assembler language was obviously difficult and non-portable. Also, by using Fortran, we were able to make good use of the routines already available on the SCS library. S was already a formidable implementation job and in the beginning we wanted to make the best use of the already-written components. (As S has been re-implemented, all of the reliance on SCS and most of the Fortran has gone away.) On the other hand, even though we were using Fortran, we were loath to give up modern control structures (IF/ELSE) and data structures. As a compromise, we used Ratfor along with the M4 macro processor to give us an implementation language that was quite reasonable, especially before the advent of Fortran 77; we called it the *algorithm language*.

As S was being built, there was a long-held belief at Bell Labs in the importance of software portability and we made a number of attempts to provide it. The Fortran code we used was intended to be portable, not only for numerical software, but also for character handling, where M4 was used to parameterize machine-specific constants such as the number of characters that could be stored in an integer. The GCOS environment certainly encouraged thoughts of portability, because it provided, on one machine, two different character sets (BCD and ASCII) and two different sets of system calls (for batch and time-sharing). To provide even more for portability, we tried to write code that would encapsulate operating-system dependencies: file opening, memory management, printing.

Because S is based so heavily on functions that carry out particular computational tasks, we realized as soon as we had a parser and executive program that it would be necessary to implement lots of functions. Although many of the functions had at their core an SCS subroutine, it soon became apparent that implementation of the necessary interface routines was repetitive, error-prone, and boring. As a result, a set of M4 macros, the *interface macros*, was developed to help speed the implementation of interface routines. We also developed some new algorithms for S; many of them, involving for example, linear algebra and random numbers, reflected ideas described in Chambers (1977).

From the beginning, S was designed to provide a complete environment for data analysis. We believed that the raw data would be read into S, operated on by various S functions, and results would be produced. Users would spend much more time computing on S objects than they did in getting them into or out of S. This is reflected in the relatively primitive input/output facilities for S. Basically, one function was provided to read a vector of data values into S from a file, and no provisions were made for accommodating Fortran-style formatted input/output.

From the beginning, S included interactive graphics. This was implemented by the GR-Z software. A separate overlay area was provided for the graphics device functions, so that the S user could perform appropriate graphics on a graphics terminal or at least line-printer style graphics on a character terminal.

Another early decision was that S should have documentation available online. A *detailed documentation* file was written for each function and described arguments, default values, usage, details of the algorithm, an example, and cross references to other documentation.

The S executive program was a very basic loop: read user input, parse it according to the S grammar, walk the parse tree evaluating the arguments and functions, and then print the result. The S executive used 20K 36-bit words, and the functions were executed in a 6K-word overlay area. This gave a reasonable time-sharing response time on GCOS for datasets of less than 10,000 data items.

Internal data structures were based on a 4-word header for S vectors, that gave a pointer to a character string name, a numeric code for the mode of the data, its length, and a pointer to a vector of appropriate values (Chambers, 1974). These headers are still around in recent implementations, although they no longer reflect the true data structure, since they make no explicit provision for attributes.

A Working Version of 'S'

By July, 1976, we decided to name the system. Acronyms were in abundance at Bell Laboratories, so it seemed sure that we would come up with one for our system, but no one seemed to be able to agree with any one else's suggestion: Interactive SCS (ISCS), Statistical Computing System (too confusing), Statistical Analysis System (already taken), etc. In the end, we decided that all of these names contained an 'S' and, with the C programming language as a precedent, decided to name the system 'S'. (We dropped the quotes by 1979).

By January, 1977, there was a manual describing Version 1.0 of S that included 30 pages of detailed documentation on the available functions. People who computed on the GCOS machine at Murray Hill were invited to use S, and a user community gradually began to form.

Work on S progressed quickly and by May, 1977, we had a revised implementation that used a new set of database management routines. Version 2.0 was introduced about a year later and included new features, such as missing values, looping, broader support for character data, and many more functions.

The earliest versions of S worked reasonably well and were used for statistical computing at Bell Labs. However, there were a number of problems that we hoped to address in the future. We saw the need for an increasing number of functions and, even with the interface macros, it was too hard to interface new subroutines to S. A goal was to enable users to interface their own subroutines. Our reliance on the QED editor to provide basic maintenance scripts for S was painful, because QED programs were difficult to write and almost impossible to read.

As S developed, the transfer vectors that pointed into the S executive would occasionally need updating. This necessitated a long process of relinking all of the S functions.

The basic parse-eval-print loop was successful and insulated users from most problems involving operating system crashes; when the computer went down, all computations carried out to that time were already saved. However, the master scatter storage file was sometimes corrupted by a crash that happened in the middle of updates, making the system somewhat fragile.

The UNIX Generation (1981 Brown Book)

Although the GCOS implementation of S proved its feasibility, it soon became apparent that growth of S required a different environment. Many of our colleagues in statistics research were using S for routine data analysis, but were still writing stand-alone Fortran programs for their research work. At the same time, word of S was getting to other researchers, both within the Bell System and at universities around the country. The time had come for S to expand its horizons.

Hardware Changes

In 1978, an opportunity came to perform a portability experiment with S. Portability was considered an important notion for S, so when the computer science research departments got an Interdata 8/32 computer to use in portability experiments for the UNIX operating system (Johnson & Ritchie (1978)), S naturally went along. The Interdata machine was a byte-oriented computer with 32-bit words. Prior to this, S work on UNIX was hampered by the 16-bit address space of machines that ran the UNIX system and by the lack of a good Fortran compiler. The latter problem was solved by a new Fortran 77 compiler (Feldman and Weinberger, 1978). As the UNIX research progressed on the Interdata machine, all sorts of useful tools and languages (C in particular) became available for use with S.

Gradually, we realized the advantages of using the UNIX system as the base for S. By the time that we had parallel implementations of S on GCOS and UNIX, it became apparent to us that porting S to a new operating system was going to be painful, no matter how much work was done to isolate system dependencies. That is because the utilities, the scaffolding needed to build S, would have to be rewritten for each new operating system. That was a big task. By writing those utilities once for the UNIX system and making use of the shell, the loader, *make*, and other tools, S could travel to any hardware by hitching a ride with UNIX. It was obvious even then that there would be far more interest in porting the UNIX operating system to new hardware than in porting S to a new system.

The link with the UNIX system provided other benefits as well. S could use the C language for implementation, since C was likely to be available and well debugged on any UNIX machine. Since the *f77* Fortran-77 compiler used the same code generator as C, inter-language calls were easy. S could rely on a consistent environment, with the shell command language (a real programming language to replace QED scripts), the *make* command for describing the system build process, known compilers and compiler options, a reasonably stable linker and libraries, and a hierarchical file system with very general file names.

The benefits of linking S and UNIX were obvious, but the Interdata environment was a one-time experiment and not hardware that would be widely used. The DEC PDP-11 machines were the natural home of the UNIX system at the time and we wanted to try S there. Unfortunately, these machines had a 16-bit architecture and the squeeze proved to be too much until the PDP-11/45 machine came out with 16-bit addressing for both instructions and data. S functions naturally ran as separate processes, but to fit on the PDP-11 we even made the parser into a separate process. Similarly, graphics was done by running the device function as a separate process linked by two-way pipes to the main S process. S put a rather severe strain on the resources of such a small machine, so it was obvious that the PDP-11 was not the ideal S vehicle, either.

Soon we had another opportunity to test our new model of piggyback portability. The UNIX system was ported to the DEC VAX computers, a new generation of 32-bit machines, where the size constraints of the PDP-11 wouldn't chafe so much. Freed from process size constraints, we tried a new implementation, where all of the functions were glued together into one large S process. That allowed the operating system to use demand paging to bring in from disk just the parts of S that were needed for execution. This worked well with S, because it is unlikely that a single S session will use more than a fraction of the S functions. The 32-bit architecture also allowed S to take on much bigger problems, with size limited only by the amount of memory the operating system would allow S to dynamically allocate. By eliminating all of the separately executable functions, disk usage was also much less, and the single S process could easily be relinked when basic S or system library routines changed.

By October, 1979, we had made a concerted effort to move all of our S functions to the UNIX version of S, making it our primary platform.

Changes to S

Perhaps the biggest change made possible by running S on the VAX machines was the ability to distribute it outside of Bell Labs research. Our first outside distribution was in 1980, but by 1981, source versions of S were made widely available. At the time, the Bell System was a regulated monopoly, and as such would often distribute software for a nominal fee for educational use. Source distribution not only allowed others access to S, but provided the statistics research departments a new vehicle for disseminating the results of our research. As we began wider distribution of S, we dropped version numbering in favor of dates; for example an S release might be identified as "June, 1980". Major releases were accompanied by completely new books; updates had revised versions of the online documentation.

As S became more widely used, it also gained some new features. One major addition was that of a portable macro processor, implemented by Ed Zayas, based on the macro processor from *Software Tools* (Kernighan & Plaugher (1976)).[4] Macros allowed users to extend the S language by combining existing functions and macros into entities designed specifically to carry out their own tasks. Once a macro was completed, it would carry out the computation and hide the implementation details.

[4] A macro facility was described in documentation for the earlier version of S, but it was seldom used.

As users began to use S for larger problems, it became apparent that the implicit looping that S used (iterating over all elements of a vector), was unable to handle every problem. It was necessary to add explicit looping constructs to the language. We did so with a general `for` expression:

```
for(index in values) expression
```

where the `index` variable would take on the series of values in turn, each time executing the `expression`.

Since one very common use of for loops was to iterate over the rows or columns of a matrix, the `apply` function was introduced. This function takes a matrix and the name of a function and applies that function in turn to each of the rows or columns of the matrix (or to general slices of an array) and does not require an explicit loop. We were able to implement `apply` so that the implicit looping was done with much greater efficiency than could be achieved by a loop in the S language.

Graphics were enhanced considerably at this time. New devices were supported, including pen plotters from Hewlett-Packard and Tektronix. Graphic input was available, to allow interaction with a plot. The user would manipulate the cursor or pen of the plotting device to signal a position. Underlying *graphical parameters* allowed control over both graphical layout and rendering. These parameters originated with the GR-Z graphics library, but extended naturally to S use. Layout parameters allowed specification of the number of plots per page (or screen), the size and shape of the plot region; these parameters were all set by the `par` function. Rendering parameters, controlling such things as line width, character size, and color, were allowed as arguments to any graphics functions, and their values were reset following the call to the graphics function.

Another innovation was the `diary` function, that allowed the user to activate (or deactivate) a recording of the expressions that were typed from the terminal. This recognized the importance of keeping track of what analysis had been performed and of the unique position of the system to provide a machine-readable record. This facility also made S comments more important. Syntactically, the remainder of a line following a "#" was ignored, thus allowing comments, but now, the comments could also be recorded in the diary file, giving a way of making annotations. In the interests of conserving disk space, the diary file was off by default.

At this time, S provided 3 directories[5] for holding datasets: the *working* directory, the *save* directory, and the *shared* directory. Any datasets created by assignments were stored in the working directory. Although it provided

[5]The name "directory" replaced the earlier usage "database" to emphasize that the S datasets were stored in a UNIX system directory.

storage that persisted from one S session to another, the thought was that the user would clean up the working directory from time to time, or perhaps even clear it out altogether. In order to preserve critical data (the raw data for an analysis and important computed results), the save directory was available. The function `save` was used to record datasets on the save directory.[6] Finally, the shared directory was the repository for some datasets that S provided — generally example datasets that were used in the manual.

A new data structure was introduced at this time: the category. It was designed to hold discrete (categorical) data and consisted of a vector of labels and an integer vector that indexed those labels. Computationally, the category acted like a vector of numbers, but it still retained the names of the original levels.

As S became more popular, demand for users to add their own algorithms began to grow, and we also felt the need to expand the basic pool of algorithms available within S. This lead to the creation of the *interface language*. This language was designed to describe new S functions, specifying the S data structures for arguments, their default values, simple computations, and the output data structure. It was built from several components. First, there was a "compiler" for the language, that used a *lex* program to turn the input syntax into a set of *m4* macro calls. This compiler was written by Graham McRae, a summer employee of Bell Labs. The macro calls emitted by the compiler were run through *m4* with a large collection of macro definitions, the interface macros, whose job was to produce the code that dealt with argument and data structure processing for S functions, producing a file in the algorithm language (described above). The algorithm language was then processed by *m4*, *ratfor*, and the Fortran compiler, in sequence, in order to produce executable code. As might be expected, the interface language had a number of quirks, each inherited from one of the many tools required to process it.

The macro processor was very popular, yet it was also the source of confusion for many users. While simple macros could be used by rote, a real understanding of macro processing was required for more complex tasks. Because macro processing was carried out as a preliminary to the S parser, users often needed to understand the difference between operations that were carried out at macro time by the macro processor and those that were carried out during execution by the S functions. The macro processor built the expressions that were given to the S parser and evaluator, but in so doing, occasionally needed its own special operations. Thus, the macro processor had separate assignments (the `?DEFINE` macro), conditionals (the

[6] We should have realized that people have a very hard time at cleaning up; the working data tended to be used for everything and the save directory was seldom used.

?IFELSE macro), and looping (the ?FOR macro). Although the initial "?" and the all-capital-letter names were meant as a cue that these operations were different from ordinary S "functions", the notion confused many users. While in many ways macros were able to provide a convenient way for users to store sequences of S expressions, there were annoying problems: macros could not be used by the apply function, for example.

Another example of the problems with macros was the use of storage. Often, a computation encapsulated in a macro required intermediate datasets, but it was important that those datasets not conflict with ones that the user had created explicitly. A naming convention, that named these temporaries with an initial "T", was adopted, but even this was insufficient. Suppose one macro called another macro. If the first macro had a temporary named Tx, it was important that the second macro not use the same name. Eventually, we developed a macro ?T that would include a level number in the generated name, to avoid these problems. Thus, ?T(x) would produce the name T1x in the first-level macro and T2x in the next level. Finally, some way was needed to clean up all of these temporary datasets at the end, so the function that removed objects was made to return a value so that the last step of a macro could remove temporaries and yet still return a value.

Overall, this first UNIX-based version of S worked well and its availability to a wider audience helped S grow. The next step was to polish it up, provide better documentation, and make it more available.

S to the World (1984 Brown Book)

In 1984, the book *S: An Interactive Environment for Data Analysis and Graphics* was published by Wadsworth (and later reviewed by Larntz (1986)). This book, along with an article in the *Communications of the ACM*, (Becker & Chambers (1984b)) described S as of 1984. The following year, Wadsworth published *Extending the S System*, describing the interface language and the process of building new S functions in detail.

Prior to 1984, S was distributed directly from the Computer Information Service at Bell Labs in New Jersey. Beginning in 1984, S source code was licensed, both educationally and commercially, by the AT&T Software Sales organization in Greensboro, North Carolina. S was always distributed in source form, in keeping with the small group of people working directly with it, i.e. Becker & Chambers (and, in 1984, Wilks). Especially after our experiences with the divergent GCOS and UNIX versions of S, we realized the importance of keeping only one copy of the source code. It was the only way that our small group could manage the software. We also knew that a source code distribution would allow people to make their own adjustments as they installed S on a variety of machines and we also knew that there was no way that we could provide binary distributions for even a handful of the hardware/software combinations that S users had. Our focus was on portability within the UNIX system (Becker (1984)). Also, remember that the S work was being done by researchers; S was not viewed as a commercial venture with support by Bell Laboratories. Thus, sites that used S needed source code to give them a way to administer and and maintain the system.

While a number of S features were expanded and cleaned up in the 1984 release, perhaps the more important influences were happening outside of that version. John Chambers moved to a different organization at Bell Laboratories and began work on a system known as QPE, the Quantitative Programming Environment (Chambers (1987)). The QPE work was based on S, but was intended to make it into a more general programming language for applications involving data. Up to now, S was known as a "statistical" language, primarily because of its origins in the statistics research departments, but in reality, most of its strengths were in general data manipulation, graphics, and exploratory data analysis. While millions of people with personal computers began using spreadsheets for understanding their own data, few of them thought they were doing anything like "statistics". The actual statistical functions in S were primarily for regression. There were, of course, functions to compute probabilities and quantiles for various distributions, but these were left as building blocks, not integrated with statistical applications. QPE was a natural notion to move S away from a perception that it was best used by a statistician.

For some time, Chambers worked on QPE by himself, rewriting the executive in C and generalizing the language. In 1986 QPE was ready to be used by others, and the question was how should QPE and S relate to one another? At the time, S had thousands of users, while QPE was still a research project; S had hundreds of functions, while QPE had few. A marriage was proposed: we would integrate the functions of S with the executive of QPE, gaining the best of both worlds.

New S (1988 Blue Book)

The combination of QPE and S produced a language that was called "New S". Because it was different from the 1984 version of S in many ways, there was a need for a new name.[7] In the Fall of 1984, just after publication of the Brown book, Allan Wilks joined Bell Laboratories and became a full partner in the S effort. The blue book, *The New S Language*, was authored

[7]eventually, the new S language was called simply S; the 1984 version became known as Old S and now it is hardly mentioned.

by Becker, Chambers, and Wilks, as was the New S software. Major changes to S occurred during the 1984-1988 period.

Functions and Syntax

The transition to New S was perhaps the most radical that S users experienced. The macro processor was gone, replaced by a new and general way of defining functions in the S language itself. This made functions into truly first-class S objects. Functions were named by assignment statements[8]

```
square <- function(x) x^ 2
```

S functions could now be passed to other functions (like `apply`), and could be the output of other computations. Because S data structures could contain functions, it no longer seemed appropriate to use the name "dataset"; from this point on, we referred to S "objects" (a suggestion of David Lubinsky).

One nice characteristic of S functions is that they are able to handle arbitrary numbers of arguments by a special argument named .... Any arguments not otherwise matched are matched by ... and can be passed down to other functions.

Internally, S functions are stored as parse trees, and explicit functions `parse` and `deparse` are provided to produce parse trees from text and to turn those trees back into character form. Parse trees are S objects, and can be manipulated by S functions. For example, the blue book contains a function that takes symbolic derivatives of S expressions.

Many functions were rewritten in the S language when New S was introduced. Functions like `apply` became easy to express in the S language itself, although these implementations were somewhat less efficient than the versions they replaced. In general, as functions are implemented in S, the users have an easier time of modifying and understanding them, because debugging is done in the interpreted S language (with the help of interactive browsing).

Unlike the transition from the 1981 to 1984 versions of S, the transition to New S was difficult for users. Their macros would no longer work. To help in this change, we provided a conversion utility, `MAC.to.FUN` that would attempt to convert straightforward macros to S functions.

New S provided a very uniform environment for functions. All S functions were represented by S objects, even those whose implementation was done

[8]The initial idea of functions in QPE did not include this form of function as an S object. Instead, the syntax was more akin to the way we defined macros: `FUNCTION square(x) x^ 2`.

internally in C code. An explicit notion of interfaces provided the key: there were functions to interface to internal code (`.Internal`), to algorithms in Fortran (`.Fortran`), to algorithms in C (`.C`), and even to old-S routines (`.S`). Along these same lines, the function `unix` provided an interface to programs that run in the UNIX system—the standard output of the process was returned as the value of the function. This was an outgrowth of successful early experiments in interfacing S and GLIM (Chambers & Schilling (1981)); we found it was quite feasible to have a stand-alone program carry out computations for S.

From the user's point of view, the most noticeable syntax change was in the way very simple expressions were interpreted. For example, with old-S, the expression

```
foo
```

would print the dataset named "foo" if one existed or else execute the function "foo" with no arguments. With new S, this expression would always print the object named "foo", even if it happened to be a function. To execute the function required parentheses:

```
foo()
```

In old-S, it had been possible to omit parentheses from even complicated expressions:

```
foo a,b
```

was once equivalent to

```
foo(a,b)
```

but it is now a syntax error.

Some functions familiar to old-S users had their names changed or were subsumed by new functions. These changes were described in the detailed documentation of the blue book under the name `Deprecated`. In some instances, the operations that the functions carried out were no longer needed; in others the operation changed enough that it was deemed reasonable to change the name to force the user to realize that something had changed. In other cases, fat was trimmed. For example, there was no need for the Fortran-style "**" operator for superscripts when "^ " was available and more suggestive.

Various functions were introduced in order to help in producing and debugging S functions. Most important was the `browser`, that allowed the user interactively to examine S data structures. It became an invaluable tool for understanding the state of a computation and was often sprinkled liberally throughout code under development. The function `trace` allowed selected functions to be monitored, with user-defined actions when the functions were executed. There was also an option that specified the functions to be called when S encountered an error or interrupt. Implementation of `trace` is particularly elegant; a traced function is modified to do the user actions and is placed in the session frame (retained for the duration of an S session) where it will be earlier on the search path when the S executive looks for it. Both the `browser` and `trace` functions are written entirely in S.

Data and Data Management

New S regularized the treatment of directories, by having a list of them, rather than the fixed sequence of working, save, and shared directories of Old S. The functions `get` and `assign` were extended to allow retrieval and storage of S objects on any of the attached directories. An S object, `.Search.list` was used to record the names of the directories on the search list.

A major change in data structures was in the separation of the notions of hierarchical structures and attributes of vectors. A vector of mode `list` represented a vector of other S objects. Such vectors could be subscripted and manipulated like other, atomic mode, vectors. Now, any vectors could have attributes: `names` to name the elements of the vector, `dim` to give dimensioning information for treating a vector as an array, etc. The notion of a vector structure was recognized as simply a vector with other attributes. Vectors were also allowed to have zero length, a convenience that regularizes some computations.[9]

A first hint at the object-oriented features of S came about with the 1988 release. The function `print` was made to recognize an attribute named `class`; if this attribute was present with a value, say, `xxx` and if there was a function named `print.xxx`, that function would be called to handle the printing.

Another feature of New S was the notion that S objects could be used to control the operation of S itself. In particular, there were four such objects: `.Program` expressed the basic parse/eval/print loop in S itself and could be replaced for special purposes; `.Search.list` gave a character vector describing the directories to be searched for S objects, `.Options` was a list controlling optional features such as the S prompt string, default sizes and limits for execution parameters, etc. `.Random.seed` recorded the current state of the random number generator. In addition, various functions named beginning with `sys.` contained information about the state of the S executive and its current evaluation. This allowed the `browser` to be applied to evaluation frames of running functions.

[9]The APL language is especially good at treating boundary cases such as zero-length vectors and we found that proper treatment of boundary cases often makes writing S functions easier.

The diary file of old-S was replaced by an audit file. The audit file furnishes both an audit trail by which the entire set of expressions given to S can be recreated, and it also is the basis for a history function that allows re-execution of expressions and for an externally implemented auditing facility (Becker & Chambers (1988)). There have been occasional complaints about the presence of the audit file, about its size, and all of the information therein describing which S objects were read and written by each expression, but generally it is recognized that the file contains important information about the path of the analysis. The external audit file processor can be used to recreate analyses, to find the genesis of objects, or to decide what should be executed following changes to data objects.

S data directories inherited a feature from commercial databases: commitment and backout upon error. As in the past, each S expression typed by the user is evaluated as a whole and the system starts afresh with each new expression. With the database backout facilities, S objects that are created during an expression are not written to the database until the expression successfully completes. This provides several advantages. First, because objects are kept in memory during expression execution, multiple use or assignments to one object do not require multiple disk access. Second, if an error occurs, the old data that was present before the error is preserved.

Another feature is that of in-memory databases, called *frames*, which hold name/value pairs in memory. A new frame is opened for each function that executes, providing a place to store temporary data until that function exits. When a function exits, its frame and all that was stored there is deallocated. There is also a frame associated with each top-level expression as well as one that lasts as long as the entire S session.

Implementation

New S was implemented primarily in C, as was the QPE executive. Internal routines, too, such as arithmetic and subscripting, were generally rewritten in C. The rewrite was necessary to support double-precision arithmetic. Code to support complex numbers was added. Many routines that in old-S had been implemented in the interface language were re-implemented in the S language, although graphics functions were largely left in the interface language and invoked through the `.S` function. Most of the Fortran that was present in S was rewritten, although basic numerical algorithms, such as those from LINPACK (Dongerra et al. (1979)) or EISPACK (Smith et al. (1976)) were left in Fortran.

The advent of New S also brought about a basic change in underlying algorithms and data. Now, everything is done in double precision, although the user is presented with a unified view of a data mode named `numeric`, which includes integer and floating point values. For special purposes (most notably for passing values down to C or Fortran algorithms), it is possible to specify the representation of numbers (integer, single, or double precision), but this is normally far from the user's mind.

S uses lazy evaluation. This means that an expression is not evaluated until S actually needs the value. For example, when a function is called, the values of its actual arguments are not computed until, somewhere in the function body, an expression uses the value.

Along with the ability to call C or Fortran algorithms, on some machines S is capable of reading object files (such as those produced by the C or Fortran compilers) and loading them directly into memory while S is executing. This means that new algorithms can be incorporated into S on the fly, with recompilation of only the changed program, and without relinking the S system itself. Dynamic loading like this is machine dependent, requiring a knowledge of object file formats and relocation procedures, so it only works on a select group of computers. It is always possible to link external subroutines with S by using the UNIX system loader.

Graphics functions in New S are very similar to those in earlier versions (for the most part, they are still implemented by old-S interface routines), however they have been regularized in many ways. A single function `plot.xy` is used to determine the x- and y-coordinates for a scatterplot. Utilizing the parse-tree for the plotting expression, S can provide meaningful labels for the plot axes, based on the expressions used to produce the x- and y-coordinates. Graphics devices are implemented by C routines that fill a data structure with pointers to primitive routines: draw a line, plot a string, clear the screen, etc.

As pen plotters and graphics terminals became less prevalent, new graphics device functions were brought into S. In particular, the most commonly used devices are now one for the X window system (`x11`), and for Adobe's PostScript™ language (`postscript`).

A novel portion of the blue book describes the semantics of S execution in the S language itself. We found that it was often easier to implement new ideas using the S language in order to try them out, and only after deciding that they were worthwhile would we recode them more efficiently in C.

The advent of New S and the blue book gave S a solid foundation as a computational language. The time was ripe for a major thrust into statistics.

Statistical Software in S (1991 White Book)

Because of its relatively recent publication, it is harder to reflect on the book, *Statistical Models in S* (Chambers & Hastie (1992)). Certainly, a number of important enhancements to the core of S were made in that time, although, for the most part, the Statistical Models work was a very large implementation of software primarily in the S language itself. Notice that statistics is not properly part of S itself; S is a computational language and environment for data analysis and graphics. The Statistical Models work is implemented in S and is not S itself.

The models work was done by ten authors, coordinated by John Chambers and Trevor Hastie. In some ways, this was a departure from the small-team model of S development. On the other hand, the work was based upon a common core and was carried out by researchers with intense personal interests in their particular portions. In addition, formal code reviews were held for all of the models software, helping to ensure that it was well-implemented and fit in with the rest of S.

The most fundamental contribution of the Models work is probably the modeling language, and it was achieved with a minimal change to the S language — the addition of the $\sim$ operator to create `formula` objects (Chambers (1993a)). The modeling language provided a way of unifying a wide range of statistical techniques, including linear models, analysis of variance, generalized linear models, generalized additive models, local regression models, tree-based models, and non-linear models.

The major change to S itself was the strong reliance on classes and methods. Generic functions were introduced; they call the function `UseMethod`, which tells the S executive to look for the `class` attribute of the generic function's argument and to dispatch the appropriate *method* function, the one whose name is made by concatenating the generic function name with the class name. By installing the object-oriented methods in this form much of S could remain as it was, with generics and methods being installed where they were needed. A form of inheritance was supplied by allowing the `class` attribute to be a character vector rather than just a single character string. If no method can be found to match the first class, S searches for a method that matches one of the later classes. Chambers (1993a) describes the ideas in much greater detail.

Another major addition to the language concerns data: the *data frame* class of objects provides a matrix-like data structure for variables that are not necessarily of the same mode (matrices must have a single mode). This enables a variety of observations to be recorded on the subjects: for example, data on medical patients could contain, for each patient, the name (a character string), sex (a factor), age group (an ordered factor), height and weight (numeric).

Another interesting inversion happened with S data directories; we once again refer to them as databases. This is because the notion of a UNIX directory became too confining. Now, there are several sorts of S databases: 1) UNIX system directories; 2) S objects that are attached as part of the search list; 3) a compiled portion of S memory; 4) a user-defined database. Because of this, we now think of methods that operate on S databases. There are methods to attach databases to the search list, to read and write objects, to remove them, to generate a list of objects in a database, and to detach them from the search list. The search list itself outgrew its early character-string implementation; it is now a true list.

The Future

This paper has described S up to the present, but will soon be out of date, as S continues to evolve. What is likely to happen?

We intend to address efficiency issues. New ideas in internal data structures may allow significant savings at run time by allowing earlier binding of operations to data. A better memory management model may use reference counts or other techniques to avoid extraneous copies of data, improving memory usage.

We hope to eliminate another of our side effects: graphics. Instead of calling graphics functions to produce output, graphics functions would instead produce a graphics object. The (automatically called) print method for a graphics object would plot it. This concept would allow plots to be saved, passed to other functions, modified, composed with one another, etc.

One general area that we would like to integrate into S is the area of high-interaction graphics with linked windows. Difficult research issues remain with this, including the problem of determining when a set of points on one plot corresponds to a set on another plot. There seems to be a conflict with this application and the general independence of S expressions.

In the long run, for portability reasons, we would like to eliminate Fortran from S, relying on C or perhaps C++ as the implementation language. One less language would make it easier (and less expensive) for people to compile S. But without Fortran, how do we make use of broadly available numerical algorithms written in Fortran? The *f2c* tool (Feldman et al. (1990)) holds some promise here, since it has been used successfully to install S on machines without Fortran.

When we built S, we intentionally did not provide a user interface with added features such as command editing or graphical input. We preferred instead to build a simple system that could support a variety of specialized user interfaces. The X window system (Scheifler & Gettys (1986)) may now provide a portable way to implement graphical user interfaces. Perhaps the time has come for serious experiments along these lines.

With the Statistical Models work, we did not press the object-oriented methodology to its logical conclusion. For example, we did not convert current time series, matrices, and categories into objects with `class` attributes, although we plan to do so in the future. This seems obvious in retrospect: many of the special cases in arithmetic computations and subscripting are caused by the need to deal with time-series, matrices, and categories. By using classes, we could handle these special cases in separate functions. Factors are one step in this direction; a factor is essentially a category with a class attribute and a variety of methods to provide appropriate operations on factors.

The future of S is increasingly influenced by people in the academic and business communities. The commercial success of the S-PLUS system from StatSci (1993) has brought the S language to thousands of people, including users of personal computers (reviewed by Therneau (1990)). Software related to S is distributed electronically via the Mike Meyer's Statlib archive, `statlib@lib.stat.cmu.edu`, and an electronic user group, `S-news@utstat.toronto.edu`, actively discusses S issues. Härdle (1990) has written a book describing smoothing with an implementation in S, and Venables & Ripley (1994) have written a statistics text based on S. Recently, Spector (1994) and Süsselbeck (1994) have written books that describe S, the latter in German.

Conclusions

Early on, we hinted that a study of the history of S could be helpful in understanding the software development process. Now, the time has come to try to draw some of these conclusions.

Much of the work in getting S to its current state followed an evolutionary process. Couldn't we have gotten there more quickly, more easily, and with less user pain by designing the language better in the first place? I think the answer is "No.", we could not have avoided some of the excursions we made.

S in 1976 was remarkably like S of 1993 — many of the basic ideas were in place. However, even if we had a blueprint for the full version of S in 1976, it would have been overwhelming, both to us as programmers and to the computing hardware of the day. Aiming too high may have caused us to give up in advance. It was useful to keep the project at a level that could be sustained by two or three people, because in that way, S evolved in a way that was aesthetically pleasing to all of us. S was not designed by committee. There is nothing like individual motivation — wanting to produce something for your personal use — to get it done right.

S was also the result of an evolution with plenty of feedback from our own and others' real use. The iteration of design, use, and feedback was important. It seemed that each time we completed a new revision to S and had a little while to rest, we then felt new energy that enabled us to think of forging ahead again.

Each version of S had limited goals. When we started, our initial idea was to produce an interactive version of the SCS library. Because it was limited, the goal also seemed attainable. As Brian Kernighan has stated "I never start on a project that I think will require more than a month." Of course, the projects do grow to require more time than that, but especially for research, it is good to feel like the time horizon is short and that there will be time for feedback and changes.

Another important concept is that we always emphasized the way the user would see the system. Our initial efforts gave only a vague thought toward machine efficiency and concentrated instead on human efficiency. It is a mistake to let hardware dominate the design of software; yes, it does have to be implemented and actually work, but it need not be totally controlled by current hardware. In the years that S has been around, machine speed has increased by several orders of magnitude, disk space is far more available, and costs have come down dramatically. It would have been unfortunate if we had allowed the performance of now-ancient hardware to affect the overall design of S. A good idea is to first make it work and then make it fast.

Efficiency in S has always been an elusive goal and has often been the part of the code that changes most from one version of S to another. Work on efficiency must be continuous because, as the system gets more efficient and runs on more powerful computers, our tendency and that of users is to implement more and more in the S language itself, causing efficiency to be a constant battle. We have also found more than once that it is not obvious which changes will improve efficiency. Our best advice is to implement and then measure. Chambers (1992) discusses some recent attempts to measure performance and Becker et al. (1993) describes a tool we devised to monitor memory usage.

Along these same lines, we tried from the beginning to make S as broad as possible, by eliminating constraints. S was built to use whatever cpu or memory resources the operating system would allow; nothing in S itself restricts the problem size. Similarly, S has no restrictions in its language — everything is an expression, built up from combinations of other expressions. Data structures are also general combinations of vectors, with no restrictions on how they can be put together.

A general principle is that the language should have minimal side effects. In S, the major side effects are assignments (that create persistent objects), printing, and graphics. Minimal side effects provide an important property: each expression is independent of the expressions that preceded it, aside from the data objects that it uses.

S also succeeded by its basic building-block approach. Within the broad context of the S language, we furnished a number of functions to carry out primitive computations, relying on the user to put them together in a way most appropriate to the problem at hand. Although this does not make for a system that can be used without thought, it does provide a powerful tool for those who know how to get what they want.

We have always *used* S ourselves. Although that may seem like a trite observation, it is very important to the way S has evolved. Often, the impetus for change in S is the result of using it to address a new problem. Without new problems to solve, S wouldn't grow. If we hadn't considered S our primary research tool, we would probably not have kept it up-to-date with the latest algorithms and statistical methods. S has always had new methodology available quickly, at least in part because it was designed to be extended by *users* as well as its originators.

S has done well, too, because of the synergy that comes from integrating tools. When developing an algorithm, the S user can use the data management and graphical routines to generate test data, to plot results, and generally to support the activity.

One of the characteristics that made S into what it has become is that each part of S is "owned" by someone. There has been no "market driven" component to S — each function is there because one of us wanted it there for our own use or because someone else convinced us that a particular function was necessary. Decisions about S have generally been by consensus (a necessity with two authors and highly desirable with three). Thus, S has the imprint of a few individuals who cared deeply and is not the bland work of a committee.

Because S came from a research environment, we have felt free to admit our mistakes and to fix them rather than propagate them forever. There have been a few painful conversions, most notably the transition from old-S to new S, but the alternative is to carry the baggage of old implementations along forever. It is better to feel the momentary pain of a conversion than to be burdened forever by older, less informed decisions. Perhaps, had S been a commercial venture, we would not have had the luxury of doing it again and again until we got it right.

A tools approach was certainly beneficial to the development of S. The software available within the UNIX system allowed us to build S quickly and portably. For example, the interface language compiler was a fairly easy project because of *lex*, *m4*, and *ratfor*. On the other hand, a tool built from other tools tends to inherit the idiosyncrasies of each of those tools. It was difficult to describe to users of the interface language just why there were so many reserved words in the language (see Appendix A of Becker & Chambers (1985)).

One way of assessing the evolution of a language is to watch for concepts that have been dropped over time. These notions are important clues to what was done wrong and later improved. In S, the notion of a dataset was dropped in favor of the idea of an object. This recognized the much more general role that S objects had in the language, holding not only data but functions and expressions. Similarly, our idea of a vector structure was approximately right, but was clarified by two more precise concepts: the (vector) list of S objects and the notion that any object could have attributes.

Steve Johnson once said that programming languages should never incorporate an untried feature. While we didn't slavishly follow that advice, most of the features present in S were influenced by some other languages. We often found a good concept in an existing system and adapted it to our purposes.

As an example, the basic interactive operation of S, the parse/eval/print loop, was a well-explored concept, occurring in APL, Troll, LISP, and PPL, among others. From APL we borrowed the concept of the multi-way array (although we did not make it our basic data structure), and the overall consistency of operations. The notion of a typeless language (with no declarations) was also present in APL and PPL. As the PPL manual (Taft (1972)) stated, "*type* is an attribute, not of *variables*, but rather of *values*."

We wanted the basic syntax of the language to be familiar, so we used standard arithmetic expressions such as those found in Fortran or C. We wanted to emphasize that the assignment operation was different from an equality test so, like Algol, we used a special operator. The specific "< –" operator came from PPL (and was a single character on some old terminals).

The basic notion that everything in S is an expression fits in very nicely with the formal grammar that describes S — a language statement is just an expression. Certainly C and Ratfor contributed the notion that a braced expression was an expression, thus allowing S to get by without need for special syntax like BEGIN/END or IF/FI or CASE/ESAC.

Some of the more innovative ideas in data structuring came from LISP: the lambda calculus form of function declarations, the storage of functions as objects in the language, the notion of functions as first-class objects, property lists attached to data.

Some pieces of syntax come from languages that are pretty far removed from the applications we foresaw with S. In particular, the IBM 360 Assembler and Job Control (JCL) languages helped to contribute the notions of arguments given by either positional or keyword notation, along with defaults for the arguments. The common thread between S and these languages was the need for functions that had many arguments, although a user would only specify some of them for any particular application.

More recently, we borrowed ideas about classes and inheritance from languages like C++, CLOS, and Smalltalk.

The Troll system contributed in many ways: the need for a time-series data type and special operations on time-series; interactive graphics for data analysis; persistent storage of datasets; the power of functions as a basic way of expressing computations.

The UNIX operating system also contributed a few ideas (as well as an excellent development platform): the use of a leading ! to indicate that a line should be executed by the operating system and the notion of having simple, well defined tools as the basis for computing. The UNIX shell also contributed to the notion of continuation lines. Like S, it knows when an expression is incomplete and responds with a special prompt to indicate this.

The origins of a few concepts are more mysterious. Perhaps they are unique to S in their precise implementation, but they were probably the result of some external influence. In particular, the syntax for the `for` loop: `for(i in x)` that allows looping over the elements of various data structures and doesn't force the creation of an integer subscript. Other programming languages have functions that can cope with arbitrary numbers of arguments, but the use of `...` in S has its own unique qualities.

One important non-technical concept was very important to the evolution of S. Almost from the beginning, S was made readily available in source form to AT&T organizations, to educational institutions, and to outside companies. This distribution of source allowed an S user community to grow and also gave the research community a vehicle for distributing software. At present there is an active S mailing list, consisting of hundreds of users around the world, asking questions, providing software, and contributing to the evolution of S. We hope that evolution continues for a long time.

Acknowledgements

I would like to thank John Chambers and Allan Wilks for comments on this manuscript. Over the years many people have been contributed to S in one way or another. Some of them have been mentioned here, but there are many others whose often considerable efforts could not be mentioned for lack of space. Among them are Doug Bates, Ron Baxter, Ray Brownrigg, Linda Clark, Bill Cleveland, Dick De Veaux, Bill Dunlap, Guy Fayet, Nick Fisher, Anne Freeny, Colin Goodall, Rich Heiberger, Ross Ihaka, Jean Louis Charpenteau, John Macdonald, Doug Martin, Bob McGill, Allen Mcintosh, Mike Meyer, Wayne Oldford, Daryl Pregibon, Matt Schiltz, Del Scott, Ritei Shibata, Bill Shugard, Ming Shyu, Masaai Sibuya, Kishore Singhal, Terry Therneau, Rob Tibshirani, Luke Tierney, and Alan Zaslavsky. Thank you. Unfortunately, I'm sure that other names that should appear here have been omitted. My apologies.

References

Baker, B.S. (1977), "An Algorithm for Structuring Flowgraphs", J. Assoc. for Comp. Machinery, Vol 24, No. 1, pp. 98-120.

Becker, R.A. (1978), "Portable Graphical Software for Data Analysis", Proc. Computer Science and Statistics: 11th Annual Symposium on the Interface, pp 92-95.

Becker, R.A. (1984), "Experiences with a Large Mixed-Language System Running Under the UNIX Operating System", Proc. USENIX Conference.

Becker, R.A. & Chambers, J.M. (1976), "On Structure and Portability in Graphics for Data Analysis", Proc. 9th Interface Symp. Computer Science and Statistics.

Becker, R.A. & Chambers, J.M. (1977), "GR-Z: A System of Graphical Subroutines for Data Analysis", Proc. 10th Interface Symp. on Statistics and Computing, 409-415.

Becker, R.A. & Chambers, J.M. (1978), 'S': A Language and System for Data Analysis, Bell Laboratories, September, 1978. (described "Version 2" of S).

Becker, R.A. & Chambers, J.M. (1981), S: A Language and System for Data Analysis, Bell Laboratories, January, 1981.

Becker, R.A. & Chambers, J.M. (1984), S: An Interactive Environment for Data Analysis and Graphics, Wadsworth Advanced Books Program, Belmont CA.

Becker, R.A. & Chambers, J.M. (1984b), "Design of the S System for Data Analysis", Communications of the ACM, Vol. 27, No. 5, pp. 486-495, May 1984.

Becker, R.A. & Chambers, J.M. (1985), Extending the S System, Wadsworth Advanced Books Program, Belmont CA.

Becker, R.A. & Chambers, J.M. (1988), "Auditing of Data Analyses", SIAM J. Sci. and Stat. Comp., pp. 747-760, Vol 9, No 4.

Becker, R.A. & Chambers, J.M. (1989), "Statistical Computing Environments: Past, Present and Future," Proc. Am. Stat. Assn. Sesquicentennial Invited Paper Sessions, pp. 245-258.

Becker, R.A., Chambers, J.M. & Wilks, A.R. (1988), The New S Language, Chapman and Hall, New York.

Becker, R.A., Chambers, J.M. & Wilks, A.R. (1993), "Dynamic Graphics as a Diagnostic Monitor for S", AT&T Bell Laboratories Statistics Research Report.

Chambers, J.M. (1974), "Exploratory Data Base Management Programs", Bell Laboratories Internal Memorandum, July 15, 1974.

Chambers, J.M. (1975), "Structured Computational Graphics for Data Analysis", Intl. Stat. Inst. Invited Paper, 1-9 IX 1975, Warszawa.

Chambers, J.M. (1977), Computational Methods for Data Analysis, Wiley, New York.

Chambers, J.M. (1978), "The Impact of General Data Structures on Statistical Computing", Bell Laboratories Internal Memorandum, May 20, 1978.

Chambers, J.M. (1980), "Statistical Computing: History and Trends", The American Statistician, Vol 34, pp 238-243.

Chambers, J.M. (1987), "Interfaces for a Quantitative Programming Environment", Comp. Sci. and Stat, 19th Symp. on the Interface, pp. 280-286.

Chambers, J.M. (1992), "Testing Software for (and with) Data Analysis", AT&T Bell Laboratories Statistics Research Report.

Chambers, J.M. (1993a), "Classes and Methods in S. I: Recent Developments", Computational Statistics, 8, 167-184

Chambers, J.M. (1993b), "Classes and Methods in S. II: Future Directions", Computational Statistics, 8, 185-196

Chambers, J.M. & Hastie, T. (eds.) (1992), Statistical Models in S, Chapman and Hall, New York.

Chambers, J.M. & Schilling, J.M. (1981), "S and GLIM: An Experiment in Interfacing Statistical Systems", Bell Laboratories Internal Memorandum, January 5, 1981.

Dongerra, J.J., Bunch, J.R., Moler, C.B. & Stewart, G.W. (1979), LINPACK Users Guide, Society for Industrial and Applied Mathematics, Philadelphia.

Feldman, S.I., Gay, D.M., Maimone, M.W. & Schryer, N.L. (1990), "A Fortran-to-C Converter, Computing Science Technical Report No. 149, AT&T Bell Laboratories.

Feldman, S. I. & Weinberger, P.J. (1978), "A Portable Fortran 77 Compiler", Bell Laboratories Technical Memorandum.

Härdle, W. (1990) Smoothing Techniques With Implementation in S, Springer Verlag, New York.

Iverson, K. E. (1991), "A Personal View of APL", IBM Systems Journal, Vol 30, No. 4, pp 582-593. (The entire issue of this journal is devoted to APL.)

Johnson, S.C. & Lesk, M.E. (1978), "Language Development Tools", The Bell System Technical Journal, Vol. 57, No. 6, Part 2, pp. 2155-2175, July-August 1978.

Johnson, S.C. & Ritchie, D.M. (1978), "Portability of C Programs and the UNIX System", The Bell System Technical Journal, Vol. 57, No. 6, Part 2, pp. 2021-2048, July-August 1978.

Kernighan, B.W. (1975), "RATFOR—A Preprocessor for a Rational Fortran", Software—Practice and Experience, Vol. 5, pp 395-406.

Kernighan, B.W. & Plauger, P.J. (1976), Software Tools, Addison Wesley.

Kernighan, B.W. & Ritchie, D.M. (1977), "The M4 Macro Processor", Bell Laboratories Technical Memorandum, April, 1977.

Larntz, K. (1986), "Review of S: An Interactive Environment for Data Analysis and Graphics", J. of the American Statistical Association, Vol. 81, pp. 251-252.

National Bureau of Economic Research (NBER, 1974), Troll Reference Manual, Installment 5, August 1974.

Ritchie, D.M. & Thompson, K. (1970), "QED Text Editor", Bell Laboratories Technical Memorandum.

Ritchie, D.M. & Thompson, K. (1978), "The UNIX Time-Sharing System", The Bell System Technical Journal, Vol. 57, No. 6, Part 2, pp. 1905-1929, July-August 1978.

Ritchie, D.M., Johnson, S.C., Lesk, M.E. & Kernighan, B.W. (1978), "The C Programming Language", The Bell System Technical Journal, Vol. 57, No. 6, Part 2, pp. 1991-2019, July-August 1978.

Ryder, B.G. (1974), "The PFORT Verifier", Software — Practice and Experience, Vol. 4, pp. 359-377.

Scheifler, R.W. & Gettys, J. (1986), "The X Window System", ACM Transactions on Graphics, Vol. 5, No. 2, pp. 79-109.

Smith, B.T., Boyle, J.M., Dongerra, J.J., Garbow, B.S., Ikebe, Y., Klema, V.C. & Moler, C.B. (1976), Matrix Eigensystem Routines—EISPACK Guide, Second Edition, Lecture Notes in Computer Science 6, Springer-Verlag, Berlin.

Spector, P. (1994), An Introduction to S and S-plus, Duxbury, Belmont, CA

Standish, T.A. (1969), "Some Features of PPL — A Polymorphic Programming Language", Proc. of Extensible Language Symposium, Christensen and Shaw, eds., SIGPLAN Notices, Vol 4, Aug. 1969.

StatSci (1993), S-PLUS Programmer's Manual, Version 3.1, Statistical Sciences, Seattle.

Süselbeck, B. (1993), S und S-PLUS, Eine Einführung in Programmierung und Anwendung, Gustav Fisher Verlag, Stuttgart.

Taft, E.A. (1972), PPL User's Manual Center for Research in Computing Technology, Harvard University, September, 1972.

Therneau, T.M. (1990), "Review of S-PLUS", The American Statistician, Vol. 44, No. 3, pp. 239-241.

Tukey, J.W. (1971), Exploratory Data Analysis, Limited Preliminary Edition. Published by Addison-Wesley in 1977.

Venables, W.N. & Ripley, B.D. (1994), Statistics with S, Springer-Verlag.

Practical Guidelines for Testing Statistical Software

Leland Wilkinson SYSTAT, Inc. 1800 Sherman Ave., Evanston, IL 60201 and
Department of Statistics, Northwestern University

Key Words:
accuracy, quality assurance, software testing, Longley data

Abstract
Traditional tests of the accuracy of statistical software have been based on a few limited paradigms for ordinary least squares regression. Test suites based on these criteria served the statistical computing community well when software was limited to a few simple procedures. Recent developments in statistical computing require both more and less sophisticated measures, however. We need tests for a broader variety of procedures and ones which are more likely to reveal incompetent programming. This paper summarizes these issues.

Introduction

Over twenty-five years ago, J.W. Longley (1967) submitted a simple regression problem to several widely used computer routines. The data were real economic series taken from government labor statistics, albeit blended in an imponderable linear equation. Longley's 16 case dataset looked innocuous but managed to clobber the most popular routines found in statistical computer libraries of that time. They can still foil some commercial statistics packages today.

Figure 1 shows a scatterplot matrix (SPLOM) of Longley's data. The ill-conditioning is apparent in the high correlations between predictors, particularly DEFLATOR, GNP, POPULATN, and TIME. And the bizarre distributions of ARMFORCE with the other predictors certainly merit further examination before linear modeling. A properly constructed prediction model obviously should not include all these predictors in their raw

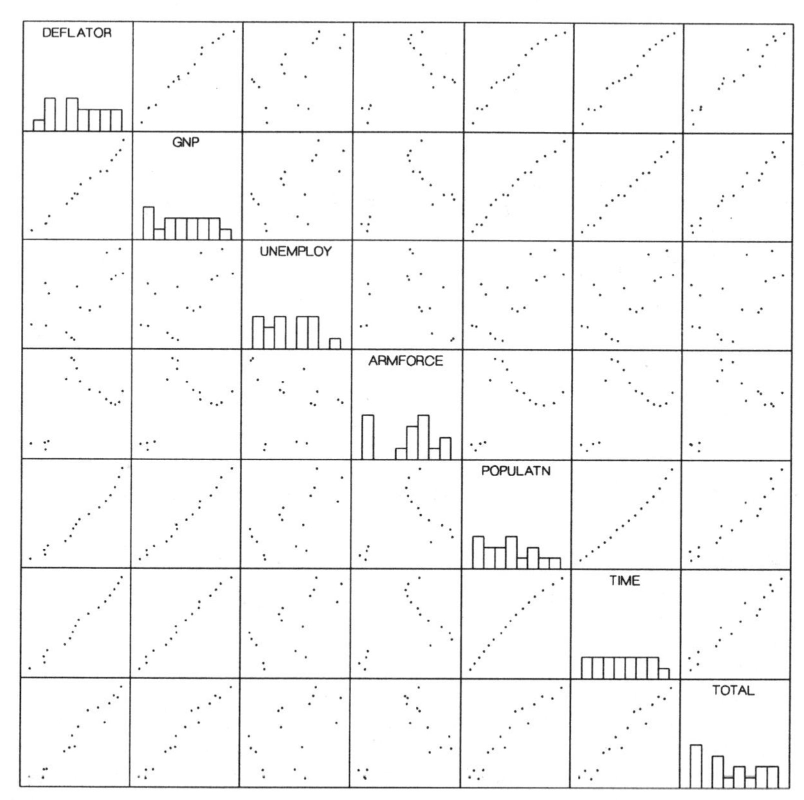

Figure 1: Scatterplot Matrix of Longley Data

form. Some have pointed to this ill-conditioning as evidence that the Longley data are artificial and unlikely to be encountered in practice. One need only talk to the technical support managers in any statistical software company to know that this claim is false, however. For better or worse, social scientists frequently attempt to fit data as ill-conditioned as Longley's and physical scientists often force ill-conditioned data through nonlinear models. Incredibly, it is not uncommon to see intercorrelations of .999 or greater among parameter estimates in published linear and nonlinear models. Software should be able to handle these situations and warn users when ill-conditioning can become a numerical or analytical problem.

Several years after Longley, Wampler (1970) formulated a conditioning index for pushing regression routines to their limits. Wampler's paradigm has been used many times since to compare the performance of computer packages, although it measures only one narrow aspect of algorithm performance.

In the light of these developments, the Statistical Computing Section of the American Statistical Association recommended a comprehensive study of the performance of statistical packages (Francis et al. (1975)). This project, subsequently modified and published in monograph form (Francis (1979, 1981)) was the first systematic attempt to evaluate the performance of the software used in academics and industry for critical statistical applications.

Soon after the release of microcomputer statistical packages in the early 1980's, many of the mainframe tests were applied at various times to micro packages (e.g. Simon & Lesage (1988)). The results were in general more alarming than for the mainframe packages. Many developers of microcomputer packages had not participated in the mainframe era of software development or were not trained in numerical analysis and statistics. The microcomputer developers quickly reacted to these reports by modifying their regression algorithms to deal with ill-conditioning. The Longley data soon ran correctly on all but the most amateurish microcomputer packages.

Because most of the accuracy tests focused on regression, the majority of statistical package algorithms continued to be untested by third parties. There were a few exceptions. Wilkinson & Dallal (1977) used a paradigm from Anscombe (1967) to reveal accuracy problems in widely used mainframe statistical package sample moments and discriminant analysis calculations.

Bernhard et al. (1988) assessed the performance of nonparametric procedures in SPSS, SAS, BMDP, and SYSTAT. The results were disappointing. None of the packages handled the tests well, either in more sophisticated areas such as exact probability computations or on simple tasks such as dealing with ties and missing data.

Another exception was a booklet, released by Wilkinson (1985), called *Statistics Quiz.* This booklet contained tests for graphics, data management, correlation, tabulation, as well as regression. The central feature of Wilkinson's booklet was a dataset called NASTY, shown in Table 1. The numbers in this dataset were extreme but not unreasonable. Each column was designed to reveal a different type of weakness in computational algorithms. LABEL contains character data. Well designed packages should be able to process character data and label cases intelligently. X contains simple integers. For summary statistics which compute exactly as integers, good packages should return integers (e.g. the mean should print as 5, not 4.999). Other operations which depend on integer values should not be confused by incorrect conversions from integer to floating point values. ZERO contains all zeros, a condition likely to cause zero divide and other singularity errors in badly designed software. MISS contains all missing values. Even packages which handle missing values may fail to deal correctly with a variable all of whose values are missing. BIG is a coefficient of variation problem. The significant variation is in the eighth digit. This means that programs in single precision will be incapable of analyzing this variable. Even some badly designed double precision programs may fail. It is possible, of course, to construct a similar variable to foil double precision programs, but the practicality of such a test would be questionable. Coefficients of variation in real data seldom are less than 10^{-10} (see Hampel et al. (1986), for a discussion of the prevalence of gross errors and other characteristics of real data). LITTLE presents similar problems to BIG. It is in the dataset to reveal formatting difficulties in output routines. HUGE and TINY are included to reveal formatting problems as well. Finally, ROUND is included to reveal rounding algorithms. The three popular options for displaying numbers are cutting (truncation), rounding up, and rounding even. Cutting is the same as rounding down. Or, if you wish, rounding up is the same as adding .5 and cutting. Cutting ROUND into one digit produces the numbers 0,1,2,3,4,5,6,7,8. Rounding up produces the integers 1,2,3,4,5,6,7,8,9. Rounding even produces 0,2,2,4,4,6,6,8,8.

Rounding even is more appropriate for binary register arithmetic or any accumulation process with finite accumulator because it is unbiased over many calculations. For formatting of output, however, the cutting or round up strategy is less confusing to users. It makes more sense to cut or round up when the numbers displayed will not be further aggregated because this does not introduce granularity where it does not exist. Cutting is used, for example, in the stem-and-leaf diagram (Tukey (1977)). Tukey recommends both cutting and rounding even, depending on the display purpose. In any case, programs should perform these operations consistently. It is not that difficult, when converting binary to decimal, to cut or round incorrectly (e.g. 0,2,3,4,5,6,7,8,9). Negative rounding can also cause problems.

Sawitzki (1993) ran a number of packages through Wilkinson's tests and found serious deficiencies in several widely used statistical programs as well as in the statistical functions of Microsoft EXCEL. The results for the spreadsheet are significant because it is quite possible that more basic statistical calculations are done worldwide in EXCEL than in all statistical packages combined.

Wilkinson also included the simple datasets shown in Table 2. He described them as follows: "The following problem was encountered by the National Park Service. There are two files. One (BOATS) has names of boats (NAME), the day they left port (DEPART), and the day they returned home (RETURN). The other file (WEATHER) has the daily temperature (TEMP) for each day (DAY) of the year, numbered from 1 to 365 during the year the boats were out. Now, neither file has the same number of records, of course, but the BOATS file may have multiple records for each boat, since each went on one or more cruises during the year. Your task is to create a file (CRUISE) with a separate record for each boat and the

LABEL	X	ZERO	MISS	BIG
ONE	1	0	*	99999991
TWO	2	0	*	99999992
THREE	3	0	*	99999993
FOUR	4	0	*	99999994
FIVE	5	0	*	99999995
SIX	6	0	*	99999996
SEVEN	7	0	*	99999997
EIGHT	8	0	*	99999998
NINE	9	0	*	99999999
	LITTLE	**HUGE**	**TINY**	**ROUND**
	.99999991	1.0E+12	1.0E-12	0.5
	.99999992	2.0E+12	2.0E-12	1.5
	.99999993	3.0E+12	3.0E-12	2.5
	.99999994	4.0E+12	4.0E-12	3.5
	.99999995	5.0E+12	5.0E-12	4.5
	.99999996	6.0E+12	6.0E-12	5.5
	.99999997	7.0E+12	7.0E-12	6.5
	.99999998	8.0E+12	8.0E-12	7.5
	.99999999	9.0E+12	9.0E-12	8.5

Table 1: NASTY Dataset from Wilkinson (1985)

average temperature (AVGTEMP) during its cruise. Warning: the weather for day 5 is missing. Ignore it in computing the average temperature."

Surprisingly, few statistical packages have been able to compute this simple problem. This is because their architecture has been based on a spreadsheet model: rows for cases and columns for variables. Those packages which do not include programming languages or sophisticated graphical relational operators cannot handle simple data management problems like this.

Several other investigations of performance outside the regression area have been published. Searle (1979) and Wittkowski (1992) examined analysis of variance output for several packages. They uncovered numerous anomalies, particularly in the choice of error terms for unbalanced and incomplete designs.

The contemporary milieu

Two recent dynamics of the rapidly changing computer market have impeded the quality testing of statistical software . The first is the fragmentation of the user community. The second is the lack of interest in accuracy

BOATS		
NAME	DEPART	RETURN
Nellie	1	10
Ajax	4	6
Queen	3	3
Ajax	2	3

WEATHER	
DAY	TEMP
1	48
2	40
3	45
4	52
6	44
7	48
8	49
9	50
10	52
11	50
12	49

Table 2: Data Management Problem from Wilkinson (1985)

among users.

In the 1970's and early 1980's, the market was governed by a few companies distributing packages on mainframe computers. Several smaller companies existed (see Francis (1983) for a brief survey), but the main user community was served by a select group of companies. During that time, the principal sources for high quality statistical code were few: Among them were UCLA's Health Sciences Computing Facility (developers of BMDP), whose code for discriminant analysis, factor analysis, and other procedures found its way into SPSS, SAS, and other packages. Another source was government laboratories, such as Argonne Labs in the U.S. and AERE Harwell and Numerical Algorithms Group in the UK, which produced certified numerical code. A few commercial organizations, such as IBM and IMSL, also produced certified code. The relative concentration of algorithm sources allowed more rapid monitoring and fixing of bugs and code defects.

In the last decade, published statistical algorithms have proliferated. While well designed algorithms have made their way into quality books and software libraries, there are still many sources which contain flawed code. The more reputable sources have contained warnings about their limitations, although many users have ignored these warnings. Press et al. (1988) for example, have written a widely used text on numerical procedures. They state in their introduction, "We call this book Numerical Recipes for several reasons. In one sense, this book is indeed a 'cookbook' on numerical computation. However there is an important distinction between a cookbook and a restaurant menu. The latter presents choices among complete dishes in each of which the individual flavors are blended and disguised. The former – and this book – reveals the individual ingredients and explains how they are prepared and combined."

Despite warnings to the contrary, pedagogical code has been adapted for commercial applications with little attention to production issues (underflows, zero-divides, etc.).

The second problem, lack of user interest in accuracy, is more serious than the first. Several major computer trends have contributed to this situation. The proliferation of graphics, user-friendliness, and GUI interfaces have led users to abandon their loyalties to single packages. They expect statistical packages to conform to the metaphors inherent in their word processing, graphics, and spreadsheet software. Microsoft and Apple have come to write the rules for statistical software as much as for the rest of the market. This perspective, coupled with the traditional fear most people have for computers ("Garbage in garbage out"), has led to a laissez-faire attitude about numerical methods. "If the graphs look professional, the numerics probably are too."

In the past, companies identified the statisticians behind them or on their boards, (e.g. Brown, Dixon, Engelman, Hartigan, Hill, Jennrich, and Sampson at BMDP; Gentle and Kennedy at IMSL; Nelder and others at NAG; Goodnight and Sall at SAS; Ryan at Minitab; Velleman at DataDesk; Polhemus at STATGRAPHICS; Becker, Chambers, Cleveland, Gale, Pregibon, Tukey, and Wilks for S; Hill, Engelman, and Wilkinson at SYSTAT). Some newer companies have refused to reveal the names of their developers. Unfortunately, reviewers of software have showed little interest in investigating the backgrounds of the companies and people producing commercial software. They would be surprised to discover that some companies selling widely advertised statistical software have no statisticians on their staff, despite their use of words like "lab" and "institute."

Two changes are needed in the testing of statistical software in order to improve this situation. First of all, we need to understand that selling statistical software is not like selling soft drinks. The reputation of companies is important in all business, but especially so in the area of technical software. We need to find a way to ferret out amateurish or sloppy developers. Second, we need to understand that statistical software quality is not simply a matter of computing regression examples to 15 digits of accuracy. Computing graphs correctly and accurately is just as critical as computing regression equations. We must expand our test suites.

How do we do this? I propose that we take a psychological approach. We need to understand the psychology of the opportunistic programmer. Opportunists motivated by money and fame take short cuts to achieve their goal. They use without modification "public domain" code from books, networks, and old computer tapes and libraries. Opportunists fail to recognize or hear about the well-known lore of the field, such as the spectral failings of the original RANDU uniform random number generator. And opportunists freely imitate other commercial packages by reverse engineering displays and interfaces without understanding their significance.

Practical tests

Instead of miscellaneous examples, I will present several maxims for helping one to devise effective tests. These maxims are generalizations and therefore often untrue. Some are, on the surface, mutually contradictory. I am presenting them to stimulate thinking about accuracy, however. The central principle of these maxims is that tests of statistics packages should focus on areas likely to affect users of real datasets. Artificial benchmarks are needed, but it is now time to move away from tiny datasets constructed to reveal specific numerical problems and toward larger and more realistic datasets which represent everyday applications more closely.

The maxims underscore another problem. Frozen standards will not improve performance of living software. Users and reviewers who make up tests unanticipated by developers will be more successful. For a forum on many of these issues, see Eddy et al. (1991).

1.Borrowed code is a ticking bomb. The most likely place to trip up a poorly written package is in the places where code has been taken from textbooks or public sources. The reasons are obvious. Those who borrow computer code from others usually don't know what it was intended to do and where its limitations are.

For example, the FORTRAN code for a clustering tree display routine in one PC statistical package was taken from a textbook. The authors did not realize that the code applied only to ultrametric trees. Their program produced incorrect displays for several types of hierarchical clustering.

How do we reveal borrowed code? One way is to find known bugs. Popular sources of carelessly borrowed code – despite their copyrights – are *ACM Algorithms* and *Applied Statistics* algorithms (published by the Royal Statistical Society). These sources usually publish updates and corrections that are not always seen by those who hastily appropriate actual computer code.

Another place to look is in asymptotic performance of algorithms. Many published algorithms, even reviewed ones, fail to implement comprehensive error trapping. For example, probability distribution routines often have difficulty with extreme parameter values. Try computing beta values for

B(.001,.001,.001) and B(.999,.001,.001). Figure 2 shows this distribution in the lower left corner. The cumulative distribution is almost a step function with these parameter values. It is tricky at these boundaries to asymptote properly (to 0 and 1 respectively). For packages lacking cumulative distribution functions, datasets can be constructed to test these distributions in the analytic sections (e.g. ANOVAs or t tests). Poor asymptotic performance

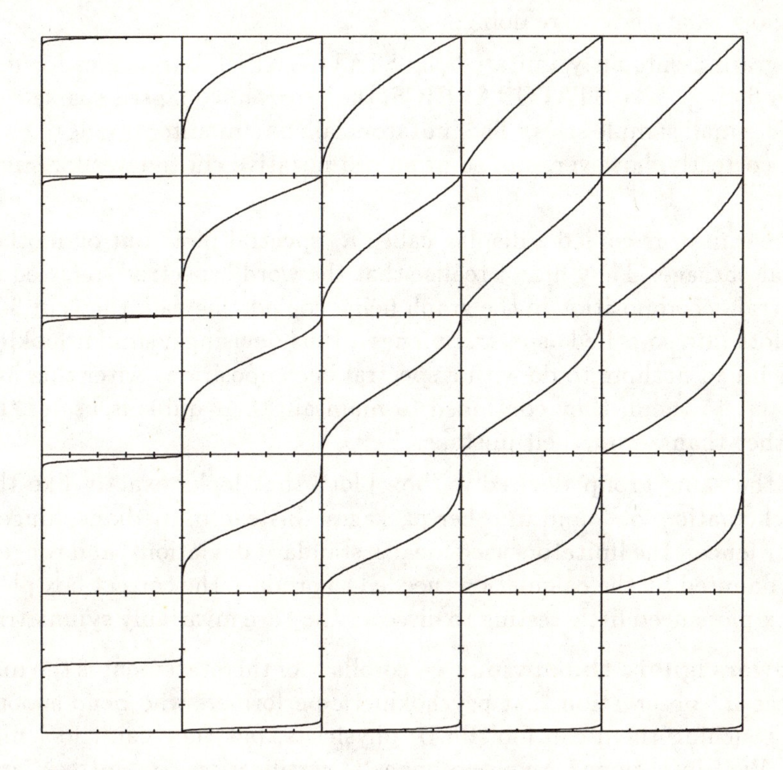

Figure 2: Beta Cumulative Distribution Function
Rows and Columns Show Beta Parameter from .01 to .99

can also be a symptom of borrowing approximate textbook formulas instead of using more accurate series. For example, one package prints probabilities for the null distribution of the correlation coefficient in a handy "statistics calculator." It refuses to calculate a probability for a value larger than .98.

2.Simple things are dangerous. This is a corollary of the old mechanic's saying that the most dangerous tool is the one least feared. For example graphs seem to be the easiest part of a statistical package. They are the last place reviewers look for accuracy. Wilkinson's NASTY dataset has been used on numerical and some graphics routines. It should be used to test other applications which seem obvious. Any routines requiring sample moments should be subjected to the same tests used to evaluate moments calculations themselves.

3.Imitation is neither sincere nor flattering. Everyone's work depends on the prior contributions of others, but true imitators don't know what they are copying. We should beware of packages which do ordinary things exactly the same way other packages do. This often means the imitators didn't know what they were doing.

One program assiduously imitated the STATGRAPHICS routines for distribution fitting. Like STATGRAPHICS, the program collapsed sparse categories for small sample sizes. The imitators failed to monitor the degrees of freedom correctly, however, and so produced negative chi-squares for small samples.

The same imitators copied a display called a "spectral plot" out of another statistical package. They didn't realize that the word "spectral" referred to the spectral decomposition in the graph being copied. Instead, they cut 3-D scatterplots into so-called "spectral planes", thus devising a similar looking plot which had nothing to do with a spectral decomposition. After this was pointed out to them, they continued to maintain their dubious use of the word rather than admit their mistake.

Finally, the same group devised a "box plot" that looks exactly like the Tukey schematic plot found in other packages. Instead of medians, hinges, and outer fences, the imitators used means, standard deviations, and ranges, perhaps daunted by the calculations needed to produce the correct box plot. Their box plots need little testing to discover they are invariably symmetric.

4.Reviewers ignore the obvious. A corollary of this statement is the magician Randi's observation that psychokinetic performers who bend spoons without touching them can fool Ph.D. physicists, but they can't fool magicians. Well intentioned reviewers, panels, certification committees, and others evaluating statistical software often focus intensively on a tiny part of a package while ignoring obvious tests of performance elsewhere. For example, nonlinear regression programs have been tested extensively using datasets from the numerical function minimization literature, but no reviewer has run the Longley data through nonlinear regression packages. Few nonlinear packages can compute the linear coefficients and their standard errors correctly for this dataset.

Other obvious pitfalls are easy to test. Submit a constant vector to a time series routine which requires first order differencing and observe subsequent calculations. Submit perfectly collinear data to a linear regression routine. Submit fully saturated single degree of freedom models to ANOVA routines. Feed an identity matrix to a principal components or factor analysis program.

5.Easy now, hard later. Incredibly, most purchasers of expensive statistical software decide to buy on an impulse. Either they are not spending their own money, or they are desperate. Reviewers, similarly, are most impressed by features discovered in the first few minutes of using a package. Researchers at Apple Computer have found that the first five minutes of exposure to a software package determines whether it will be bought and/or used.

The problem with this instant gratification is that software which shines in the first five minutes can tarnish when needed for more complex problems later. Teitel (1986) organized a symposium on how statistical packages handle data management. He extended an open invitation to statistical software developers to solve his test examples. Only BMDP, PRODAS, P-STAT, SAS, SPSS, and SYSTAT successfully completed the problems. Some highly regarded packages did not even enter his contest because they could not handle the data structures. Wilkinson's simple data problem in Table 2 still cannot be handled by most commercial packages. An easy test would be to ask the developer to show how this problem can be solved. Presumably, developers should have the simplest solution, if it exists at all in their package.

Serious data management capabilities are not the exclusive province of command based packages. There is nothing in graphic-user-interface systems which prevents their doing powerful data manipulation. Nevertheless, most menued packages have avoided this area because of the programming difficulties.

6.The results matter more than the algorithm. Statistical package developers cannot always be trusted to document their algorithms correctly. There are many reasons for this. Often algorithms are more complex than a simple set of formulas will show. And sometimes developers deliberately conceal important details in order to keep them from competitors. Sometimes documentation departments are staffed by non-programmers or non-statisticians. Finally, companies can lie to impress customers. One company stated that a particular algorithm had been used in its software. This algorithm appeared in another package which it had imitated, but did not exist in the imitation.

Numerical analysts often favor a particular algorithm because of its theoretical performance on certain problems, while they ignore other areas. While error analysis, pioneered by Wilkinson (1965), can be extremely valuable for standardized comparisons, it does not solve all the performance assessment issues. Some reviewers have compounded this mistake by favoring one package over another after comparing documentation only. Even superior algorithms can be coded improperly.

The only sure comparison is on real data examples which are designed to exploit known weaknesses of algorithms. This way, incorrect choices of machine constants and iterative parameters will be exposed.

7. No one knows everything. A corollary of this maxim is that no one is interested in everything. A package which passes one test with flying colors may be substandard elsewhere. Companies and individuals tend to specialize. Eminent time-series statisticians, for example, may know little about experimental design. This is why one cannot generalize about the performance of a package from simple tests of regression accuracy.

One developer, wishing to implement probabilities for the Kolmogorov-Smirnov test, apparently stored a set of tabulated values from a textbook. Whenever values outside the table – *in any direction* – are input, the program reports the result as not significant (n.s.).

8. Accuracy is not important. What matters is how an algorithm handles inaccuracy, not accuracy itself. The principal point of Velleman & Francis (1975) and Beaton et al. (1976) is that computing results to many digits on artificial test datasets is not useful in itself. The measurement errors in real datasets usually far exceed the size of the errors produced by computational problems. More important is the performance of a routine across a wide set of real datasets, both large and small. For any algorithm, it is possible to construct a short artificial dataset to exploit its weaknesses. Tests like those in Wilkinson and Dallal's and Longley's papers, on the other hand, are ones that any well-designed package should be able to handle easily.

An extreme case of numerical silliness has appeared recently in the marketing literature of one statistics company. The correlation routines of this company's statistics package were found to be seriously inaccurate on the Wilkinson & Dallal (1977) tests. In response, the company produced a revised version which uses extended precision arithmetic. It now includes in its advertising an example containing variables whose coefficient of variation is less than 10^{-14}. The major statistics packages yield the same Pearson correlation on this dataset to within four digits. The company apparently thinks the remaining difference is of practical importance. Those who make up such examples should be challenged to provide real data that have a similar coefficient of variation and whose measurement error is known to be less than the variation recorded.

Computing accurate estimates on a trick dataset is no guarantee that a package is accurate. Like performance benchmarks - Whetstones, Dhrystones, etc. - accuracy benchmarks can vary widely and depend on complex assumptions. Furthermore, using extended precision with a bad algorithm will not always correct accuracy problems.

Computing accurate estimates for the Longley data is simple. Packages which cannot do this should not be tolerated. Even worse, however, is a package which computes artificial examples correctly but fails to detect mathematical collinearity. Such a package can produce completely false results.

Conclusion

It would be nice to have a test suite of problems for certifying statistical packages. Unfortunately, certification cannot keep up with new developments in computer science and statistics. A more promising approach is for users to construct specialized examples in various applications areas. This was the approach of Wilkinson & Dallal (1977) and, more recently, Wittkowski (1992). News of these tests spreads quickly and causes developers to pay attention.

More importantly, let's develop process models for statistical analysis which include real world datasets and typical applications. To assess regression performance, for example, we need to develop tests which reveal how much effort it takes to perform transformations, refit models, recognize outliers, stratify by subgroups, produce professional quality reports and graphs, and manage many large datasets.

The time of statisticians in white lab coats writing certified code has passed, if indeed it ever existed. Users need to discard myths of this sort and be especially wary of companies which use a cloud of mathematical and statistical jargon to disguise incompetence. If users don't investigate or care about the quality of statistical software, they will get they packages they deserve.

Note

MaryAnn Hill and Laszlo Engelman contributed valuable suggestions.

References

Anscombe, F.J. (1967). Topics in the investigation of linear relations fitted by the method of least squares. Journal of the Royal Statistical Society, Series B, 29, 1-52.

Beaton, A. E., Rubin, D.B. & Barone, J.L. (1976). The acceptability of regression solutions: Another look at computational accuracy. Journal of the American Statistical Association, 71, 158-168.

Bernhard, G., Alle, M., Herbold, M. & Meyers, W. (1988). Investigation on the reliability of some elementary nonparametric methods in statistical analysis systems. Statistical Software Newsletter, 14, 19-26.

Eddy, W.F., Howe, S.E., Ryan, B.F., Teitel, R.F. & Young, F. (1991). The future of statistical software: Proceedings of a forum. Washington, DC: National Academy Press.

Francis, I., Heiberger, R.M., and Velleman, P.F. (1975). Criteria and considerations in the evaluation of statistical program packages. The American Statistician, 29, 52-56.

Francis, I. (1979). A comparative review of statistical software. International Association for Statistical Computing, Voorburg, Netherlands.

Francis, I. (1981). Statistical software: A comparative review. North-Holland, New York.

Francis, I. (1983). A survey of statistical software. Computational Statistics and Data Analysis, 1, 17-27.

Hampel, F.R., Ronchetti, E.M., Rousseeuw, P.J. & Stahel, W.A. (1986). Robust statistics: The approach based on influence functions. New York: John Wiley & Sons, Inc.

Longley, J.W. (1967). An appraisal of least-squares for the electronic computer from the point of view of the user. Journal of the American Statistical Association, 62, 819-841.

Press, W.H., Flannery, B.P., Teukolsky, S.A. & Vetterling, W.T. (1988). Numerical Recipes. Cambridge: Cambridge University Press.

Sawitzki, G. (1993). Numerical reliability of data analysis systems. Paper presented at the 25th meeting, SIG Computational Statistics of the International Biometrical Society, Reisensburg, Germany.

Searle, S.R. (1979). Annotated computer output for analysis of variance of unequal- subclass-numbers data. The American Statistician, 33, 222.

Simon, D. & Lesage, J.P. (1988). Benchmarking numerical accuracy of statistical algorithms. Computational Statistics & Data Analysis, 7, 197-209.

Teitel, R.F. (1986). Benchmarking vendor packages. In T.J. Boardman (ed.), Computer Science and Statistics: Proceedings of the 18th Symposium on the Interface. American Statistical Association, 193-229.

Velleman, P.F. & Francis, I. (1975). Measuring statistical accuracy of regression programs. In J.W. Frane (ed.), Computer Science and Statistics: Proceedings of the 8th Symposium on the Interface. UCLA Health Sciences Computing Facility, 122-127.

Wampler, R.H. (1970). A report on the accuracy of some widely used least squares computer programs. Journal of the Americal Statistical Association, 65, 549-565.

Wilkinson, J.H. (1965). The Algebraic Eigenvalue Problem. Oxford: Clarendon Press.

Wilkinson, L. (1985). Statistics Quiz. Evanston, IL: SYSTAT, Inc.

Wilkinson, L. & Dallal, G.E. (1977). Accuracy of sample moments calculations among widely used statistical programs. The American Statistician, 31, 128-131.

Wittkowski, K.M. (1992). Statistical analysis of unbalanced and incomplete designs - experiences with BMDP and SAS. Computational Statistics & Data Analysis, 14, 119-124.

On the Choice and Implementation of Pseudorandom Number Generators

Jürgen Lehn, Stefan Rettig
Technical University of Darmstadt, Department of Mathematics, Schloßgartenstraße 7, D — 64289 Darmstadt, Germany

Key Words:
pseudorandom number generators, linear congruential generators, lattice structure, Minkowski reduced bases, matrix generators, inversive congruential generators

Abstract
Stochastic simulation has become very popular since it is one of the easiest things one can do with a stochastic model and the computers needed for simulation methods are available everywhere. This paper is intended for all those who use these methods in their work and need some knowledge about pseudorandom numbers which are fundamental in stochastic simulation. Therefore, the most frequently used pseudorandom number generators are discussed and we describe how good generators can be selected and implemented correctly. The paper contains also several remarks on recent trends in pseudorandom number generation.

Introduction

In most applications of stochastic simulation the randomness comes from a sequence of so–called standard pseudorandom numbers $u_0, u_1, u_2, \ldots$, i. e. a sequence of reals in the unit interval generated by a deterministic algorithm and looking at first glance like a sequence which stems from independent identically distributed random variables $U_0, U_1, U_2, \ldots$ having a uniform distribution on the unit interval. In a second step these standard pseudorandom numbers are often transformed in order to fit a prescribed non–uniform distribution. There is a great variety of algorithms for solving this problem when special distributions as binomial distributions, Poisson distributions, normal distributions, beta distributions, gamma distributions, and so on, are given as target distributions. The monograph Devroye (1986) is devoted to these methods. This paper deals only with algorithms producing sequences of standard pseudorandom numbers. Such algorithms play the role of the source of randomness in most applications of stochastic simulation methods. They are called pseudorandom number generators.

Many users of simulation do not think about how their standard pseudorandom numbers are produced. They merely call the standard functions of the computer software available. This attitude is dangerous since the strength of simulation results depends heavily on the "stochastic quality" of the underlying sequence of pseudorandom numbers and it is known that many pseudorandom number generators in use have serious defects (see e.g. Dieter (1986, p.10), Ripley (1987, p.17), Park & Miller (1988)). Using a poor or defective pseudorandom number generator can invalidate all simulation results.

Pseudorandom numbers should be distinguished from the so–called quasi–random numbers. While a sequence of pseudorandom numbers should reflect both features of a sequence of independent uniformly distributed random variables, namely uniform distribution and stochastic independence, a sequence of quasi–random numbers satisfies only the requirement of uniformity and may exhibit dependencies which are in contrast to the idea of stochastic independence. Thc main ficld of application for quasi–random numbers are Monte Carlo integration and optimization methods.

More about this difference can be found in the excellent monograph Niederreiter (1992a) which deals with pseudorandom numbers as well as quasi–random numbers and describes the state of the art in both fields. The reader is also referred to the book Ripley (1987) which contains a lot of information about pseudorandom number generators including Fortran 77 subroutines for implementing and testing. An elementary introduction to pseudorandom number generation is given in Schmitz & Lehmann (1976). More about generators and about applications of stochastic simulation may be found in Afflerbach & Lehn (1986), Dieter & Pflug (1992) and in the survey papers Eichenauer–Herrmann (1992), Lehn (1992a), Lehn (1992b), Niederreiter (1991), Niederreiter (1992b).

Historical Examples

In the early days of stochastic simulation people used physical processes (e.g. tossing coins, rolling dice, and spinning roulette wheels) for generating sequences of randomly varying numbers. Later on tables of millions of digits were produced from electronic noise and recorded on tapes. However, in the 1940s when the first simulations were performed by electronic computers it turned out that physical methods did not fit well to computer programming. Even later on, when tables of random numbers were available on punched cards or tapes the simulations were too slow and awkward. Therefore, recursive generation methods have been developed in order to produce sequences of randomly varying numbers to be used as input for simulations.

Example 1: The Middle Square Method

One of the earliest methods was the middle square method suggested by John von Neumann in the early 1940s. Starting from a real u_0 between 0 and 1 with an even number k of digits behind the decimal point this number has to be squared. The result u_0^2 is a real with $2k$ digits behind the decimal point. Now the middle k digits have to be extracted and are used as the digits behind the decimal point of the real u_1 between 0 and 1. Then this procedure is repeated and a sequence $u_0, u_1, u_2, \ldots$ is obtained. Let $k = 4$ and $u_0 = 0.3572$. Then $u_0^2 = 0.12\underline{7591}84$ and $u_1 = 0.\underline{7591}$. By continuing this process the following sequence is generated

$$u_0 = 0.3572,\ u_1 = 0.7591,\ u_2 = 0.6232,\ u_3 = 0.8378,\ u_4 = 0.1908,\ \ldots$$

An investigation of this sequence shows that up to $u_{16} = 0.6100$ a more or less randomly varying sequence is obtained but then the numbers

$$u_{16} = 0.6100,\ u_{17} = 0.2100,\ u_{18} = 0.4100,\ u_{19} = 0.8100,\ u_{20} = 0.6100,\ \ldots$$

appear and since $u_{16} = u_{20}$ the same four reals are produced again and again (see Figure 1).

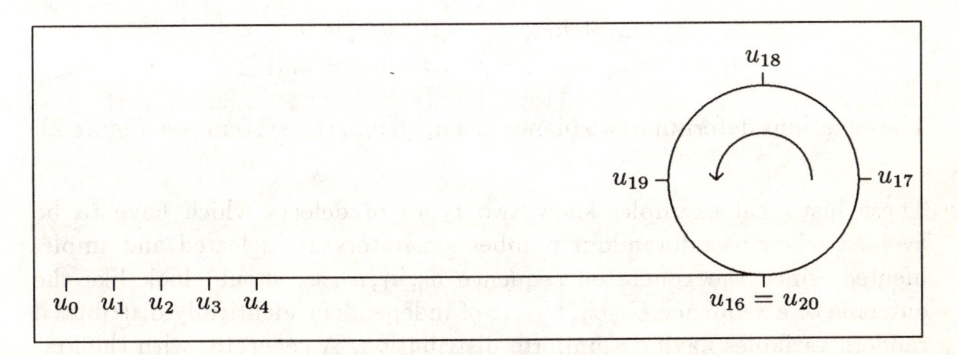

Figure 1

This example exhibits that the middle square method may reach a very short sequence that repeats itself. This is, of course, a defect which makes the method inacceptable for generating the input of a simulation.

Example 2: The Fibonacci Method

Another method of producing sequences to be used as random input for simulations is motivated by Fibonacci's recurrence $a_i = a_{i-1} + a_{i-2}, a_0 = 0, a_1 = 1$, which defines the famous Fibonacci sequence. It is given by the formula

$$u_i \equiv u_{i-1} + u_{i-2} \quad (mod\, 1)$$

and has been applied since the 1950s. Starting e. g. with $u_0 = 0.32759831$ and $u_1 = 0.13794822$ one obtains the sequence

$$\begin{array}{lll} u_0 = 0.32759831, & u_4 = 0.06904128, & u_8 = 0.15569065, \\ u_1 = 0.13794822, & u_5 = 0.67253603, & u_9 = 0.56980399, \\ u_2 = 0.46554653, & u_6 = 0.74157731, & u_{10} = 0.72549464, \\ u_3 = 0.60349475, & u_7 = 0.41411334, & \ldots \quad \ldots \quad \ldots \end{array}$$

The sequences generated by this method look randomly varying, but a more thorough analysis exhibits a serious defect which makes them unsuitable for simulations. If one considers the sequence of triplets

$$(u_0, u_1, u_2),\ (u_1, u_2, u_3),\ (u_2, u_3, u_4),\ (u_3, u_4, u_5),\ \ldots$$

it turns out that the corresponding points in the unit cube are carried by only two planes given in the following figure. This contradicts the idea of randomness, since a truly random sequence determines points which are distributed irregularly in the unit cube and do not concentrate upon only two planes. The effect is due to the underlying additive formula $u_i \equiv u_{i-1} + u_{i-2} \quad (mod\, 1)$ which means

$$\begin{array}{lll} u_i = u_{i-1} + u_{i-2} & ,\text{if} & u_{i-1} + u_{i-2} < 1 \\ u_i = u_{i-1} + u_{i-2} - 1 & ,\text{if} & u_{i-1} + u_{i-2} \geq 1 \end{array}$$

The equations determine two planes in a $u_{i-2}/u_{i-1}/u_i$–system (see Figure 2).

These historical examples show two types of defects which have to be avoided when pseudorandom number generators are selected and implemented, since the generated sequence $u_0, u_1, u_2, \ldots$ should look like the outcome of a sequence $U_0, U_1, U_2, \ldots$ of independent identically distributed random variables having a uniform distribution. A generator with the first defect does not even fulfil the requirement of the uniform distribution because within a very short sequence repeating itself the same few numbers occur again and again. A generator with the second defect is not in accordance with stochastic independence since the concentration of the triplets upon only a few planes exhibits strong dependencies.

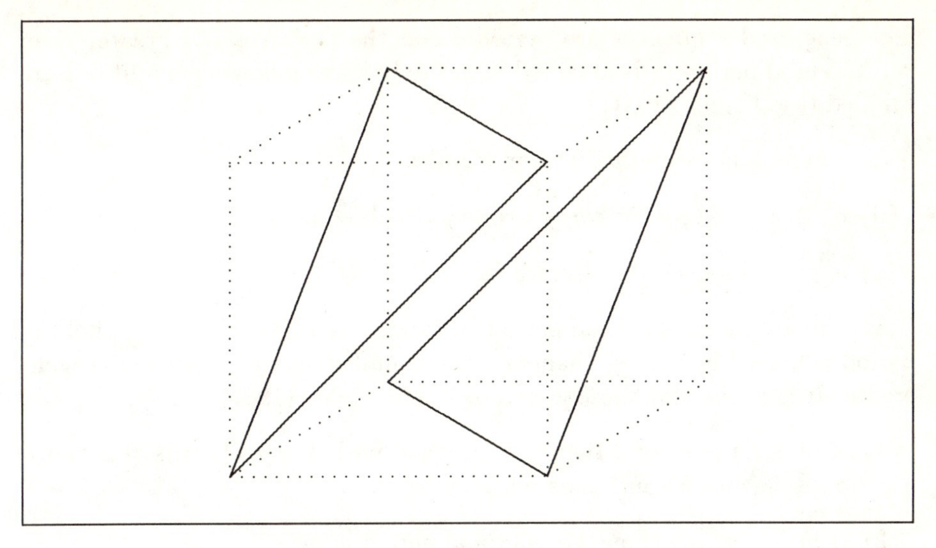

Figure 2

Therefore, good generators should produce sequences without short subsequences repeating themselves. A second requirement which has to be fulfilled by a good generator is a good uniform distribution in higher dimensions, e. g. the k-tuples $(u_0, \ldots, u_{k-1}), (u_1, \ldots, u_k), \ldots$ do not concentrate upon a too small number of hyperplanes in $[0,1]^k$. In what follows, these are the basic principles for the choice and the implementation of good pseudorandom number generators.

Linear Congruential Generators with Maximal Period

The most frequently used method of producing standard pseudorandom numbers is the linear congruential method which goes back to Lehmer (1951) and Rotenberg (1960). A linear congruential generator is defined by

$$x_i \equiv a \cdot x_{i-1} + b \pmod{m}$$

for a multiplier a, a shift b, a seed x_0, and a modulus m, which is a large integer. The other parameters x_0, a, and b are also integers which are taken between 0 and $m-1$. If $b = 0$, then the generator is called multiplicative congruential. The standard pseudorandom numbers $u_0, u_1, u_2, \ldots$ are obtained by

$$u_i = x_i / m$$

The generated sequences are periodic and the period is not greater than m. Maximal period m is attained if and only if the following conditions are fulfilled (see Knuth (1981, Chapter 3)):

(i) There is no prime p dividing b and m.

(ii) $a \equiv 1 \ (mod\ p)$ for every prime p dividing m.

(iii) $a \equiv 1 \ (mod\ 4)$ if 4 divides m.

Now it follows that maximal period m cannot be attained if $b = 0$, i.e. the period achieved is smaller than m for any multiplicative congruential generator. It is known for these generators (see Ripley (1987)):

1) If m is a power of 2 then the maximal period is $m/4$. It is attained if $a \equiv 5 \ (mod\ 8)$ and x_0 is odd.

2) If m is a prime then the maximal period is $m - 1$.

Conditions on a which guarantee that in case 2) period $m - 1$ is attained are also given in Ripley (1987) (see Theorem 2.3, Theorem 2.4). From these conditions it follows e. g. that the multiplicative congruential generator

$$x_i \equiv 16807 \cdot x_{i-1} \quad (mod\ 2^{31} - 1)$$

with the Mersenne prime $2^{31} - 1$ as modulus has maximal period $2^{31} - 2$.

If generators with maximal period are chosen the first defect described above (short subsequences repeating themselves) cannot occur. Therefore, maximal period guarantees the basic requirement for good approximation to uniformity of the standard pseudorandom numbers, because e. g. in the case $b \neq 0$ each integer between 0 and $m - 1$ is produced exactly once in a period.

Example 3: TI–59 Pocket Calculator

The software manual of the standard software module of the TI–59 pocket calculator says that the pseudorandom numbers are generated by

$$x_i \equiv 24298 \cdot x_{i-1} + 99991 \quad (mod\ 199017)$$

One may easily check that the conditions (i) — (iii) stated above are fulfilled ($m = 199017 = 3^7 \cdot 7 \cdot 13$), i. e. for this generator maximal period $m = 199017$ is attained. Therefore, users were extremely surprised when they observed short subsequences consisting of only 406 numbers repeating themselves (see Afflerbach (1986, p.26)). The reason for this strange behaviour of the pseudorandom number sequence was a mistake in the source code. Instead of integer arithmetic which has to be applied for an exact implementation floating–point arithmetic has been used. Therefore, sequences were produced which behaved quite differently from the sequence theoretically determined by the generator.

This example shows us how important it is that by the implementation of linear congruential generators exact integer calculations are guaranteed. Above all a sufficient word length has to be provided. The following example shows again that floating–point arithmetic may cause terrible defects of type "short subsequences repeating themselves" as observed in Example 1.

Example 4: HP–25 Pocket Calculator

The pseudorandom numbers of the HP–25 pocket calculator are generated by the algorithm

$$u_i \equiv (u_{i-1} + 3)^5 \quad (mod\, 1)$$

With a seed u_0, $0 < u_0 < 1$, the floating–point arithmetic creates sequences which behave as demonstrated in Example 1. After a more or less random behaviour they enter a short subsequence repeating itself. The HP–25 calculator may enter such a cycle of only 29 numbers.

This effect can be attributed to the rounding errors of the floating–point arithmetic. The generators of the BASIC interpreters on the Commodore and Apple microcomputers (CBM PET 2001 Series and Apple II europlus) are further examples for generators having this defect. They are described and analysed in Afflerbach (1985) and Ripley (1987,p.18). In Afflerbach (1985) it is shown that they may reach short cycles of period 202 and 703, respectively. Examples 3 and 4 elucidate why congruential generators defined by integer calculations (and division by the modulus) should be preferred to congruential generators *mod* 1. More about the implementation of pseudorandom number generators can be found in Gentle (1990) and Ripley (1990).

Lattice Structure and Beyer Ratios

Since maximal period is the first requirement a linear congruential generator has to fulfil, only generators with this property are considered in the sequel. Due to the linearity of the recursion formula the linear congruential generator

$$x_i \equiv a \cdot x_{i-1} + b \quad (mod\, m)$$

creates a strong linear pattern in $\mathbb{R}^k$ for any dimension k. The points $(x_0, \ldots, x_{k-1}), (x_1, \ldots, x_k), \ldots$ determine a certain sublattice of $\mathbb{Z}^k$ in most cases shifted away from the origin. Two typical situations are illustrated by the following example referring to dimension $k = 2$. Generators are chosen with same modulus m and shift b but different multipliers.

Example 5: Two–dimensional Lattice Structure

The points $(x_0, x_1), (x_1, x_2), \ldots$ produced by the generator with multiplier $a = 133$ are better distributed within the square while in case of multiplier $a = 109$ the points concentrate upon only 4 straight lines (see Figure 3). Therefore, with $a = 133$ one is in a much better situation than with $a = 109$.

$x_i \equiv 133 \cdot x_{i-1} + 5\ (mod\ 216)$ $\qquad$ $x_i \equiv 109 \cdot x_{i-1} + 5\ (mod\ 216)$

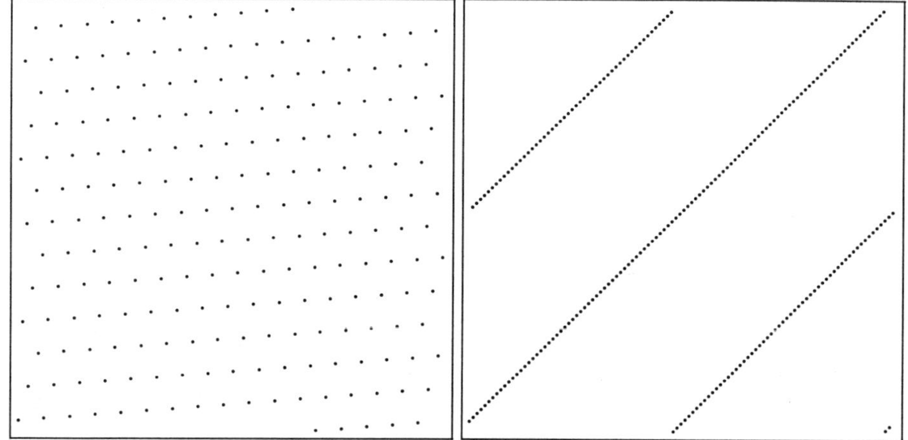

Figure 3

This example illustrates the basic idea for selecting good multipliers. It should be mentioned that the shift b does not influence the shape of the lattice. It is only its location with respect to the origin which depends on b.

The strong lattice patterns shown in Figure 3 appear in any dimension k. The shifted lattice determined by the points $(x_0, \ldots, x_{k-1}), (x_1, \ldots, x_k), \ldots$ is given by the following basis:

$$
\begin{aligned}
\vec{c}_1 &= (1, a, a^2, \ldots, a^{k-1})^T \\
\vec{c}_2 &= (0, m, 0, 0, \ldots, 0)^T \\
\vec{c}_3 &= (0, 0, m, 0, \ldots, 0)^T \\
&\vdots \\
\vec{c}_k &= (0, 0, 0, 0, \ldots, m)^T
\end{aligned}
$$

This vector basis provides all information needed about the lattice. It enables one to decide whether the points $(x_0, \ldots, x_{k-1})$, $(x_1, \ldots, x_k), \ldots$ concentrate upon a small number of hyperplanes in $\mathbb{R}^k$ or are well spread in the hypercube. But the following example illustrates that this information cannot be drawn directly from the basis $\vec{c}_1, \ldots, \vec{c}_k$. Another basis describing the unit cell of the lattice has to be calculated from the given one. This Minkowski reduced basis $\vec{e}_1, \ldots, \vec{e}_k$ is named after H. Minkowski who introduced it in the theory of quadratic forms.

Example 6: Lattice Basis and Minkowski Reduced Basis

$$x_i \equiv 5 \cdot x_{i-1} + 1 \; (mod\; 16)$$

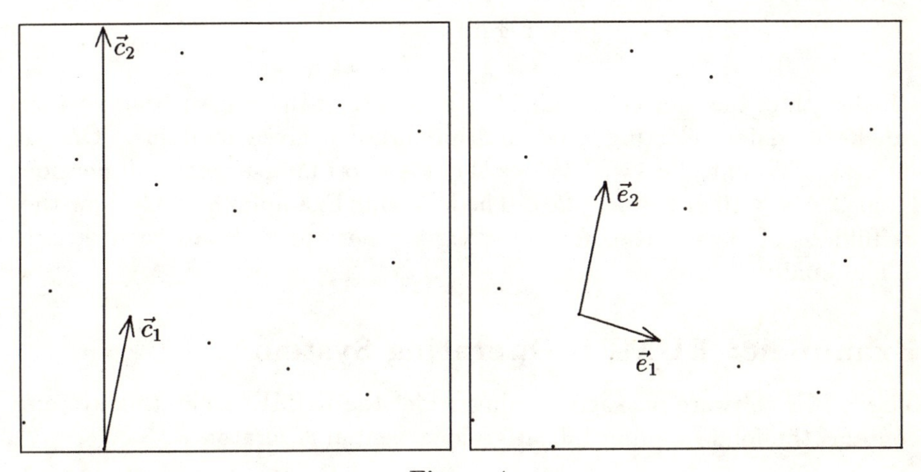

Figure 4

In Afflerbach & Grothe (1985) an algorithm is given for calculating the $\vec{e}_1, \ldots, \vec{e}_k$ from the $\vec{c}_1, \ldots, \vec{c}_k$.

Example 7: The Unit Cell of a Lattice

The pictures in Figure 5 make clear that unit cells consisting of vectors of approximately the same length indicate a good lattice structure, i.e. the points do not concentrate upon a too small number of hyperplanes. Since $\vec{e}_1, \ldots, \vec{e}_k$ is in some sense a basis of shortest vectors describing the unit cell of the lattice and $|\vec{e}_1| \leq |\vec{e}_2| \leq \ldots \leq |\vec{e}_k|$ holds, the Beyer ratio in dimension k is

$$q_k = |\vec{e}_1| / |\vec{e}_k|$$

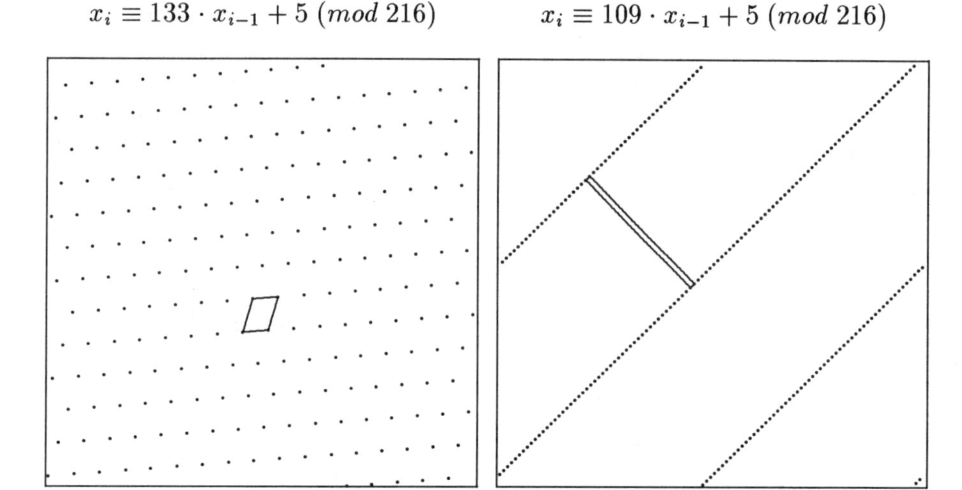

Figure 5

(the length of the shortest divided by the length of the longest basis vector) can be used for selecting good multipliers for a given modulus. One is interested in multipliers with values of q_k close to 1 up to a certain dimension (e. g. $2 \le k \le 10$ or $2 \le k \le 20$). The following Examples 8 and 9 show the usefullness of Beyer ratios for detecting a generator with lattice structure of bad quality.

Example 8: EUMEL Operating System

The ELAN software package running under the EUMEL operating system contains the following multiplicative congruential generator

$$x_i \equiv 4095 \cdot x_{i-1} \quad (mod\ 16777213)$$

This generator has maximal period. However, the values of the Beyer ratios $q_k, k = 2, \ldots, 20$, are

$q_2 = 0.999512$	$q_6 = 0.004549$	$q_{11} = 0.051270$	$q_{16} = 0.660350$
$q_3 = 0.001036$	$q_7 = 0.006507$	$q_{12} = 0.074549$	$q_{17} = 0.702483$
$q_4 = 0.001749$	$q_8 = 0.013988$	$q_{13} = 0.162552$	$q_{18} = 0.777791$
$q_5 = 0.003209$	$q_9 = 0.016620$	$q_{14} = 0.205310$	$q_{19} = 0.783237$
	$q_{10} = 0.032324$	$q_{15} = 0.348934$	$q_{20} = 0.782089$

These values indicate that the generator has a bad lattice structure in dimensions $k > 2$. Indeed, for dimension $k = 3$ the points (x_0, x_1, x_2), $(x_1, x_2, x_3), \ldots$ concentrate upon only 4 planes. The choice $a = 4095$ originates from a recommendation given in Marsaglia (1972) that the multiplier

a should be chosen nearby the square root of the modulus m. This choice always leads to a high value of the Beyer ratio in dimension $k = 2$ but as Example 8 shows to bad lattice structures in higher dimensions.

Example 9: Turbo Pascal 3.0

In Turbo Pascal 3.0 the linear congruential generator

$$x_i \equiv 129 \cdot x_{i-1} + 907\,633\,385 \quad (mod\ 2^{32})$$

is implemented. In many dimensions the Beyer ratios are smaller than 0.1:

$q_2 = 0.000004$	$q_6 = 0.272765$	$q_{11} = 0.099120$	$q_{16} = 0.066985$
$q_3 = 0.000500$	$q_7 = 0.163709$	$q_{12} = 0.086858$	$q_{17} = 0.065265$
$q_4 = 0.064443$	$q_8 = 0.152722$	$q_{13} = 0.083144$	$q_{18} = 0.062685$
$q_5 = 0.727009$	$q_9 = 0.116312$	$q_{14} = 0.076092$	$q_{19} = 0.061043$
	$q_{10} = 0.101079$	$q_{15} = 0.071381$	$q_{20} = 0.060136$

It can be observed that the generated points $(x_0, x_1), (x_1, x_2), \ldots$ concentrate upon only 130 straight lines which is a too small number compared with the generator's period being $2^{32} = 4\,294\,967\,296$.

In 1968 Marsaglia published a paper (Marsaglia (1968)) entitled "Random numbers fall mainly in the planes" in which he described the strong pattern of random number sequences produced by linear congruential generators. Since this time the lattice structure of linear congruential generators has been studied extensively (see Niederreiter (1992a) or Knuth (1981)). As an alternative to the Beyer ratios mentioned above the maximal distance between adjacent parallel hyperplanes carrying the generated lattice in $\mathbb{R}^k$ (spectral test) can be taken as criterion for selecting good multipliers a (see Dieter (1986) and Dieter (1975)). Other methods of analysing the lattice structure are to determine the minimal number of parallel hyperplanes or to calculate the minimal distance between the generated points $(x_0, \ldots, x_{k-1}), (x_1, \ldots, x_k), \ldots$ or to estimate the so-called discrepancy (see Niederreiter (1992a) or Fishman & Moore (1986)). An exhaustive search for good multipliers a in the cases $m = 2^{31} - 1$ and $m = 2^{32}$ as well as a partial search with respect to $m = 2^{48}$ has been carried out by Fishman and Moore. The results in Fishman (1990) and Fishman & Moore (1986) show that, among others, the following generators can be recommended for stochastic simulation methods:

$$x_i \equiv 1\,226\,874\,159 \cdot x_{i-1} \quad (mod\ 2^{31} - 1)$$
$$x_i \equiv 1\,732\,073\,221 \cdot x_{i-1} \quad (mod\ 2^{32})$$
$$x_i \equiv 55\,151\,000\,561\,141 \cdot x_{i-1} \quad (mod\ 2^{48})$$

The main reason why most software packages contain linear congruential generators is the simplicity of the recursion formula which allows a theoretical analysis of the produced sequences. It should be mentioned here, that a more complicated recursion does not necessarily improve the stochastic quality of the resulting sequences (see e. g. the middle square method given as Example 1 or Algorithm K in Knuth (1981, p.4)). Recursion formulae should be applied for which theoretical results are known. Users of simulation should not trust a random number generator for which theoretical investigations are impossible.

Applications

In this section the generators implemented in some frequently used software packages are analysed by calculating the Beyer ratios q_k of their lattice structures. Following a recommendation of Dieter given in Dieter (1986, p.13) the Beyer ratios of generators being acceptable should not be smaller than 0.1. It turns out that the generators described in the following examples fulfil this condition for all considered dimensions k.

Example 10: Statistical Software Systems SAS and GAUSS

In the statistical software system SAS, the software package GAUSS, and the subroutine library IMSL pseudorandom numbers are produced by the multiplicative congruential generator

$$x_i \equiv 397\,204\,094 \cdot x_{i-1} \quad (mod\; 2^{31} - 1)$$

The Beyer ratios q_k are in this case:

$q_2 = 0.354859$	$q_6 = 0.580860$	$q_{11} = 0.686250$	$q_{16} = 0.806387$
$q_3 = 0.380569$	$q_7 = 0.583453$	$q_{12} = 0.705595$	$q_{17} = 0.870312$
$q_4 = 0.667559$	$q_8 = 0.847557$	$q_{13} = 0.676128$	$q_{18} = 0.902248$
$q_5 = 0.758132$	$q_9 = 0.542997$	$q_{14} = 0.729357$	$q_{19} = 0.889961$
	$q_{10} = 0.628898$	$q_{15} = 0.767592$	$q_{20} = 0.866047$

Example 11: Subroutine Library IMSL and ESSL

A second generator in the subroutine library IMSL which is also implemented in the subroutine library ESSL is a multiplicative one with modulus $m = 2^{31} - 1$ and multiplier $a = 16\,807$:

$$x_i \equiv 16\,807 \cdot x_{i-1} \quad (mod\; 2^{31} - 1)$$

The Beyer ratios of this generator are:

$q_2 = 0.131506$	$q_6 = 0.599588$	$q_{11} = 0.705486$	$q_{16} = 0.812962$
$q_3 = 0.295328$	$q_7 = 0.544748$	$q_{12} = 0.631099$	$q_{17} = 0.851923$
$q_4 = 0.483303$	$q_8 = 0.430915$	$q_{13} = 0.744402$	$q_{18} = 0.869532$
$q_5 = 0.598703$	$q_9 = 0.621043$	$q_{14} = 0.720045$	$q_{19} = 0.828231$
	$q_{10} = 0.542142$	$q_{15} = 0.796718$	$q_{20} = 0.856324$

The third multiplicative congruential generator in the subroutine library IMSL is

$$x_i \equiv 950\,706\,376 \cdot x_{i-1} \quad (mod\ 2^{31} - 1)$$

The Beyer ratios of this generator are given by

$q_2 = 0.821209$	$q_6 = 0.824334$	$q_{11} = 0.872052$	$q_{16} = 0.711620$
$q_3 = 0.928709$	$q_7 = 0.611807$	$q_{12} = 0.855170$	$q_{17} = 0.824052$
$q_4 = 0.906989$	$q_8 = 0.554430$	$q_{13} = 0.814036$	$q_{18} = 0.884454$
$q_5 = 0.767440$	$q_9 = 0.570492$	$q_{14} = 0.738331$	$q_{19} = 0.818921$
	$q_{10} = 0.782042$	$q_{15} = 0.782285$	$q_{20} = 0.790583$

Example 12: Turbo Pascal 4.0

The pseudorandom number generator implemented in Turbo Pascal version 3.0 has bad lattice structures in most dimensions (see Example 9). Therefore, in version 4.0 of the software this generator has been replaced by the linear congruential generator

$$x_i \equiv 134\,775\,813 \cdot x_{i-1} + 1 \quad (mod\ 2^{32})$$

The Beyer ratios show that this generator is better suited for stochastic simulation methods:

$q_2 = 0.623354$	$q_6 = 0.812116$	$q_{11} = 0.790194$	$q_{16} = 0.796969$
$q_3 = 0.262422$	$q_7 = 0.328201$	$q_{12} = 0.743478$	$q_{17} = 0.882844$
$q_4 = 0.410870$	$q_8 = 0.512759$	$q_{13} = 0.858313$	$q_{18} = 0.846230$
$q_5 = 0.435128$	$q_9 = 0.668839$	$q_{14} = 0.803445$	$q_{19} = 0.878513$
	$q_{10} = 0.700798$	$q_{15} = 0.864230$	$q_{20} = 0.873087$

Example 13: The Wichmann–Hill Generator

In Wichmann & Hill (1982) a pseudorandom number generator is proposed which combines the three multiplicative congruential generators

$$\begin{aligned} x_i &\equiv 171 \cdot x_{i-1} \quad (mod\ 30269) \\ y_i &\equiv 172 \cdot y_{i-1} \quad (mod\ 30307) \\ z_i &\equiv 170 \cdot z_{i-1} \quad (mod\ 30323) \end{aligned}$$

by addition modulo 1, i. e. the standard pseudorandom numbers are obtained by the operation

$$u_i \equiv x_i/30269 + y_i/30307 + z_i/30323 \; (mod\ 1)$$

This generator has a period of approximately $6.96 \cdot 10^{12}$ and is equivalent to a multiplicative congruential generator with modulus $m = 30269 \cdot 30307 \cdot 30323$ and multiplier $a = 1\,655\,425\,264\,690$ (see Zeisel (1986)). In any dimension k the points $(x_0, \ldots, x_{k-1}), (x_1, \ldots, x_k), \ldots$ generated by this method are carried by a lattice. The Beyer ratios of these lattices have values greater than 0.45 up to dimension 20. But since the lattices are not fully occupied the generator cannot be classified as good only by calculating the values of the Beyer ratios. An advantage of the generator is that a short word length is sufficient for an exact integer arithmetic implementation (see Wichmann & Hill (1982) for the code).

Matrix Generators and Nonlinear Methods

As a generalization of the linear congruential method the multiple-recursive generator

$$x_i \equiv a_1 \cdot x_{i-1} + \ldots + a_r \cdot x_{i-r} \; (mod\ p)$$

for recursion depth r and prime modulus p has been introduced by Knuth (1981). The multipliers $a_1, \ldots, a_r$ and the starting values $x_0, \ldots, x_{r-1}$ are integers between 0 and $p - 1$, not all equal to zero. This generator attains maximal period $p^r - 1$, if a certain algebraic condition is fulfilled by the polynomial $P(\lambda) = \lambda^r - a_1\lambda^{r-1} - \ldots - a_{r-1}\lambda - a_r$ (see Knuth (1981)).

The matrix generator (see Grothe (1986), Niederreiter (1986)) generalizes this method. It is defined by

$$\vec{x}_i \equiv A \cdot \vec{x}_{i-1} \quad (mod\ p)$$

where $\vec{x}_i$ is a r–vector with integer components, A a $r \times r$ matrix with integer entries, and $\vec{x}_0 \neq \vec{0}$ a starting vector with integer components, all between 0 and $p - 1$. Maximal period $p^r - 1$ is attained if the so–called characteristic polynomial of the matrix A has a certain property (see Grothe (1986), Niederreiter (1986)). The advantage of these generators is the long period $p^r - 1$. Since these generators create also lattices they can be analysed by calculating Beyer ratios. In his thesis Grothe carried out a search for good $r \times r$ matrices up to $r = 6$ for different prime moduli p (see Grothe (1988)).

Example 14: Matrix Generator

For $r = 4$ and $p = 2^{31} - 1$ the matrix generator

$$\vec{x}_i \equiv A \cdot \vec{x}_{i-1} \quad (mod\ 2^{31} - 1)$$

with

$$A = \begin{pmatrix} 881\,864\,634 & 572\,866\,339 & 1\,548\,588\,836 & 899\,073\,391 \\ 1\,845\,529\,955 & 1\,677\,206\,044 & 546\,800\,766 & 1\,424\,614\,466 \\ 1\,957\,793\,824 & 1\,981\,544\,750 & 283\,441\,529 & 1\,447\,458\,052 \\ 972\,879\,486 & 1\,856\,122\,070 & 1\,822\,799\,467 & 1\,116\,871\,587 \end{pmatrix}$$

has a period greater than 10^{37} and was recommended by Grothe (1988).

All generators considered so far are defined by linear recursions. Therefore, they create strong linear patterns which can be described by lattice bases. This enables one to analyse the stochastic quality of the produced random numbers very efficiently. On the other hand the strong linear patterns have been regarded as disadvantage (see Eichenauer & Lehn (1986)) and criticized by Marsaglia in his famous 1968 paper (Marsaglia (1968)). In his opinion the lattice structure is "a defect caused by the linearity that cannot be removed by adjusting the starting value, multiplier, or modulus". Therefore, in an attempt to get rid of these linear patterns the inversive congruential method was introduced in Eichenauer-Herrmann & Lehn (1986). The inversive congruential generator is given by

$$x_i \equiv a \cdot \bar{x}_{i-1} + b \quad (mod\ p)$$

where p is a prime modulus and the starting value x_0 as well as a and b are integers between 0 and $p-1$. Observe that for any $x \in \{1, 2, \ldots, p-1\}$ there is exactly one $\bar{x} \in \{1, 2, \ldots, p-1\}$ with $x \cdot \bar{x} \equiv 1\ (mod\ p)$, the multiplicative inverse of x in the finite field of order p. If additionally $\bar{0} = 0$, then the generator is well defined. There are fast algorithms for calculating the multiplicative inverse $\bar{x}$ from $x \neq 0$, e. g. the Euclidean algorithm (cf. Knuth (1981, Section 4.5)). These generators have been studied extensively during the last years. A lot of deep mathematical results indicating the good stochastic quality of pseudorandom numbers generated by these generators can be found in the survey paper of Eichenauer–Herrmann (Eichenauer–Herrmann (1992)) and in Chapter 8 of Niederreiter's book (Niederreiter (1992a)). The patterns created by these "inversively generated" pseudorandom numbers are extremely nonlinear. It holds that any hyperplane in $\mathbb{R}^k$ contains at most k points $(x_i, x_{i+1}, \ldots, x_{i+k-1})$ with $x_i \cdot \ldots \cdot x_{i+k-2} \neq 0$. This property makes the inversive congruential generators very attractive for simulations, e. g. in Stochastic Geometry. From the paper "Inversive congruential pseudorandom numbers avoid the planes" (Eichenauer–Herrmann (1991)) where this nonlinearity result has been proved it can be derived that for these generators it does not hold what Marsaglia found for pseudorandom numbers produced by linear congruential generators, namely that they "fall mainly in the planes".

Acknowledgement

The authors would like to thank H. Grothe and R. Weilbächer for calculating the Beyer ratios and a referee for valuable comments improving the paper.

References

Afflerbach, L. (1985). The pseudo-random number generators in Commodore and Apple microcomputers. Statist. Papers 26, 321–333.

Afflerbach, L. (1986). Die Gitterstruktur gleichverteilter Zufallsvektoren. In: Afflerbach & Lehn (1986), 21–28.

Afflerbach, L. & Grothe, H. (1985). Calculation of Minkowski reduced lattice bases. Computing 35, 269–276.

Afflerbach, L. & Lehn, J., eds. (1986). Zufallszahlen und Simulationen. Teubner, Stuttgart.

Devroye, L.P. (1986). Non-Uniform Random Variate Generation. Springer, New York.

Dieter, U. (1975). How to calculate shortest vectors in a lattice. Math. Comp. 29, 827–833.

Dieter, U. (1986). Probleme bei der Erzeugung gleichverteilter Zufallszahlen. In: Afflerbach & Lehn (1986), 7–20.

Dieter, U. & Pflug, G., eds. (1992). Simulation and Optimization: Proceedings of the International Workshop on Computationally Intensive Methods in Simulation and Optimization. Springer Lecture Notes in Economics and Mathematical Systems 374.

Eichenauer-Herrmann, J. (1991). Inversive congruential pseudorandom numbers avoid the planes. Math. Comp. 56, 297–301.

Eichenauer-Herrmann, J. (1992). Inversive congruential pseudorandom numbers: a tutorial. Int. Statist. Rev. 60, 167–176.

Eichenauer-Herrmann, J. & Lehn, J. (1986). A non-linear congruential pseudo random number generator. Statist. Papers 27, 315–326.

Fishman, G.S. (1990) Multiplicative congruential random number generators with modulus 2^{β}: An exhaustive analysis for $\beta = 32$ and a partial analysis for $\beta = 48$. Math. Comp. 54, 331–344.

Fishman, G.S. & Moore, L.R. (1986). An exhaustive analysis of multiplicative congruential random number generators with modulus $2^{31} - 1$. SIAM J. Sci. Statist. Comput. 7, 24–45. Erratum, ibid. 7, 1058.

Gentle, J.E. (1990). Computer implementation of random number generators. In: Lehn & Neunzert (1990), 119–125.

Grothe, H. (1986). Matrix–Generatoren. In: Afflerbach & Lehn (1986), 29–34.

Grothe, H. (1988). Matrix–Generatoren zur Erzeugung gleichverteilter Pseudozufallsvektoren. Thesis, Darmstadt.

Knuth, D.E. (1981). The Art of Computer Programming, Vol. 2: Seminumerical Algorithms (2nd ed.). Addison–Wesley, Reading.

Lehmer, D.H. (1951). Mathematical methods in large–scale computing units. In: Proceedings of 2nd Symposium on Large–Scale Digital Calculating Machinery, Harvard University Press, Cambridge, 141–146.

Lehn, J. (1992a). Special methods for pseudorandom number generation. In: Jöckel, K.–H., Rothe, G., Sendler, W., eds. (1992). Bootstrapping and Related Techniques. Springer Lecture Notes in Economics and Mathematical Systems 376, 13–19.

Lehn, J. (1992b). Pseudorandom number generators. In: Gritzmann, P., Hettich, R., Horst, R., Sachs, E., eds. (1992). Operations Research '91. Physica–Verlag, Heidelberg, 9–13.

Lehn, J. & Neunzert, H., eds. (1990). Random Numbers and Simulation. J. Comp. Appl. Math 31 (Special Issue).

Marsaglia, G. (1968). Random numbers fall mainly in the planes. Proc. Nat. Acad. Sci. U.S.A. 61, 25–28.

Marsaglia, G. (1972). The structure of linear congruential sequences. In: Zaremba, S.K., ed. (1972). Applications of Number Theory to Numerical Analysis. Academic Press, New York, 249–285.

Niederreiter, H. (1986). A pseudorandom vector generator based on finite field arithmetic. Math. Japonica 31, 759–774.

Niederreiter, H. (1991). Recent trends in random number and random vector generation. Ann. Operations Research 31, 323–346.

Niederreiter, H. (1992a). Random Number Generation and Quasi–Monte Carlo Methods. SIAM, Philadelphia.

Niederreiter, H. (1992b). Nonlinear methods for pseudorandom number and vector generation. In: Dieter & Pflug (1992), 145–153.

Park, S.K. & Miller, K.W. (1988). Random number generators: Good ones are hard to find. Communications on the ACM, 31, 1192-1201

Ripley, B.D. (1987). Stochastic Simulation. John Wiley & Sons, New York.

Ripley, B.D. (1990) Thoughts on pseudorandom number generators. In: Lehn & Neunzert (1990), 153–163.

Rotenberg, A. (1960). A new pseudo–random number generator. J. Assoc. Comput. Mach. 7, 75–77.

Schmitz, N. & Lehmann, F. (1976). Monte–Carlo–Methoden I: Erzeugen und Testen von Zufallszahlen. A. Hain, Meisenheim.

Wichmann, B.A. & Hill, I.D. (1982). An efficient and portable pseudo–random number generator. Appl. Statist. 31, 188–190. Corrections, ibid. 33 (1984), 123.

Zeisel, H. (1986). A remark on Algorithm AS 183. An efficient and portable pseudo–random number generator. Appl. Statist. 35, 89.

Seven Stages of Bootstrap

Rudolf Beran
University of California, Berkeley, Department of Statistics, Berkeley, CA 94720, U.S.A.

Key Words:
coverage probability, iterated bootstrap, two-step bootstrap, Monte Carlo, high-dimensional parameter

Abstract
This essay is organized around the theoretical and computational problem of constructing bootstrap confidence sets, with forays into related topics. The seven section headings are: Introduction; The Bootstrap World; Bootstrap Confidence Sets; Computing Bootstrap Confidence Sets; Quality of Bootstrap Confidence Sets; Iterated and Two-step Bootstrap; Further Resources.

Introduction

Bradley Efron's 1979 paper on the bootstrap in Statistics gained the immediate interest of his peers for several historical reasons. First, the bootstrap promised to extend formal statistical inference to situations too complex for existing methodology. By the late 1970's, theoretical statisticians had recognized that classical formulations of statistics, whether frequentist or Bayesian or otherwise, did not provide a reasonable way to analyze the large data sets arising in the computer age. This awkward defensive position made theoreticians receptive to the bootstrap, as well as to other data analytic ideas that seemed less model-dependent than classical statistical theory.

Second, by the 1970's, developments in theoretical statistics had provided tools that soon proved powerful in analysing the behavior of bootstrap procedures. For instance, the theory of robust statistics accustomed researchers to working with continuous or differentiable statistical functionals. This prepared the way for the later interpretation of bootstrap distributions as statistical functionals. The need to quantify contamination neighborhoods in robustness studies drew attention to metrics for probability measures. Huber (1981) presents both developments in robust statistics. Several decades of work on asymptotic optimality theory, culminating in the early 1970's with the local asymptotic minimax bound and with Hájek's convolution theorem, encouraged statisticians to think about weak convergence of triangular arrays. Ibragimov and Has'minskii (1981) give a comprehensive account. Edgeworth expansions and saddlepoint approximations saw a revival in the 1970's that is summarized by Hall (1992). These various theoretical ideas were well-suited to studying the convergence in probability of bootstrap distributions and the asymptotic properties of bootstrap procedures.

Third, bootstrap-like methods were natural as computers proliferated. From the 1960's onwards, some data analysts, not all statisticians, began experimenting with Monte Carlo simulations from fitted distributions. These resampling experiments were based more on intuition rather than on logical analysis and were published outside the main-stream statistical journals. Since a careful historical study has not yet been done, it is possible that the origins of the resampling idea are substantially older. (After all, the paired comparisons design in Statistics can be traced back to the philosopher Carneades, head of the Academy in Athens around 150 B. C., who argued that the different fortunes of twins disproves the efficacy of astrology). An essential contribution of Efron's (1979) paper was to formulate the bootstrap idea, as an intellectual object that could be studied theoretically, and give it a name.

The purpose of this essay is to introduce the bootstrap, to indicate when and in what sense it works, to discuss basic questions of implementation, and to illustrate the main points by example. The exposition is organized around the construction of bootstrap confidence sets—an application where bootstrap methods already enjoy considerable success—with forays into related topics. The last section contains suggestions for further reading.

The Bootstrap World

We recall that a statistical model for a sample $X = (X_1, \dots, X_n)$ consists of a family of distributions, written $\{P_{n,\theta}: \theta \in \Theta\}$. One member of this model, the true distribution, is considered to generate probability samples similar to the observed data. However, the value of the parameter θ that identifies the true distribution is not known to the statistician. We suppose that the parameter space Θ is metric, but do not require it to be finite dimensional.

Bootstrap methods are a particular application of simulation ideas to the problem of statistical inference. From the sample X, we construct an estimator $\hat{\theta}_n = \hat{\theta}_n(X)$ that converges in $P_{n,\theta}$-probability to θ, for some convergence concept to be chosen. The bootstrap idea is then to:

- Create an artificial world in which the true parameter value is $\hat{\theta}_n$ and the sample X^* is generated from the fitted model $P_{n,\hat{\theta}_n}$. That is, the conditional distribution of X^*, given the data X, is $P_{n,\hat{\theta}_n}$.

- Act as if sampling distributions computed in the artificial world are accurate approximations to the corresponding true (but unknown) sampling distributions.

The *original world* of the statistician's model consists of the observable X whose distribution is $P_{n,\theta}$. The *bootstrap world* consists of the observable X^* whose conditional distribution, given X, is $P_{n,\hat{\theta}_n}$. In the original world, the distribution of X is unknown. However, in the bootstrap world, the distribution of X^* is fully known. Thus, any sampling distribution in the bootstrap world can be computed, at least in principle.

This brief description omits several important issues. First, for each statistical model, there may be many possible bootstrap worlds, each corresponding to a different choice of the estimator $\hat{\theta}_n$. Only some choices may be successful. Second, the plug-in method for constructing the model distribution in the bootstrap world can be generalized, and sometimes must be. When a high or infinite dimensional θ lacks a consistent estimator in a natural metric, it may still be possible to construct a useful bootstrap world that mimics only relevant aspects of the model in the original world. Time series analysis and curve estimation provide leading examples; see Mammen (1992) and Janas (1993) as well as Example 1 in the next section. Third, computation of sampling distributions in the bootstrap world often involves Monte Carlo approximations, whose design raises further issues. Fourth, bootstrap methods are rarely exact; their theoretical justification typically rests on asymptotics under which the bootstrap world converges to the original world. These points will be developed further as the essay proceeds.

Bootstrap Confidence Sets

Suppose we wish to construct a confidence set for the parametric function $\tau = \tau(\theta)$. Classical theory advises us to find a pivot—a function of the sample X and of τ whose distribution under the model $P_{n,\theta}$ is continuous and completely known. Archetypal are confidence intervals for the mean of a $N(\mu, \sigma^2)$ distribution when location μ and scale σ are unknown. Here $\theta = (\mu, \sigma)$, the parametric function $\tau(\theta) = \mu$, and the pivot is the t-statistic, whose sampling distribution does not depend on the unknown θ. Though important as an ideal case, the exact pivotal technique is rarely available. It already fails to generate confidence intervals for the difference of two normal means in the Behrens-Fisher problem, for lack of a pivot.

Bootstrap ideas permit generalizing the pivotal method. Let $R_n(X, \tau)$ be a function of the sample and of τ, whose distribution under the model $P_{n,\theta}$ is denoted by $H_n(\theta)$. Because it need not be a pivot, but plays an analogous role, we call R_n a *root.* A plausible estimator of the root's sampling distribution is then the ***bootstrap distribution*** $\hat{H}_{n,B} = H_n(\hat{\theta}_n)$. This bootstrap distribution has two complementary mathematical interpretations:

- As defined, $\hat{H}_{n,B}$ is a random probability measure, the natural plug-in estimator of the sampling distribution of $R_n(X, \tau)$. From this viewpoint, $\hat{H}_{n,B}$ is a statistical functional that depends on the sample only through $\hat{\theta}_n$.

- Alternatively, $\hat{H}_{n,B}$ is the conditional distribution of $R_n(X^*, \tau(\hat{\theta}_n))$ given the sample X. In other words, $\hat{H}_{n,B}$ is the distribution of the root R_n in the bootstrap world described at the end of Section 2.

The interpretation as conditional distribution leads readily to Monte Carlo approximations for a bootstrap distribution (see Section 4). The interpretation as statistical functional is the starting point in developing asymptotic theory for bootstrap procedures, as we shall see next.

Suppose that, for some convergence concept in the parameter space Θ, both of the following conditions hold, for every $\theta \in \Theta$:

A. The estimator $\hat{\theta}_n$ converges in probability to θ as n increases.

B. For any sequence $\{\theta_n\}$ that converges to θ, the sampling distribution $H_n(\theta_n)$ converges weakly to the limit $H(\theta)$.

Then, the bootstrap distribution $\hat{H}_{n,B}$ also converges weakly, in probability, to the limit distribution $H(\theta)$. Though apparently very simple, this reasoning provides a template for checking the consistency of bootstrap estimators. The skill lies in choosing the convergence concept so as to achieve both conditions A and B.

We can now construct bootstrap confidence sets by analogy with the classical pivotal method. Let $\hat{H}_{n,B}^{-1}(\alpha)$ denote the α-th quantile of the bootstrap distribution and let T denote the space of possible values for the parametric function $\tau = \tau(\theta)$. Define the ***bootstrap confidence set*** for τ to be

$$C_{n,B} = \{t \in T \colon R_n(X, t) \leq \hat{H}_{n,B}^{-1}(\alpha)\}. \tag{1}$$

If conditions A and B above hold and if the limiting distribution $H(\theta)$ is continuous at its α-th quantile, then the coverage probability $P_{n,\theta}(C_{n,B} \ni \theta)$

converges to α as n tends to infinity. The following application to Stein confidence sets illustrates two key aspects of the bootstrap method: its remarkable power and the care often needed to harness this power when the dimension of θ is not small relative to sample size.

EXAMPLE 1. We observe the time-series $X = (X_1, \ldots, X_n)$, which is related to the signal $\theta = (\theta_1, \ldots, \theta_n)$ by the following model: the distribution of X is normal with mean vector θ and with covariance matrix identity. The parametric function τ of interest is the signal θ itself. The classical confidence set of level α for θ is a sphere centered at X, with radius determined by the chi-squared distribution having n degrees of freedom. Let $|\cdot|$ denote Euclidean norm. A Stein confidence set is a sphere centered at the Stein estimator

$$\hat{\theta}_{n,S} = [1 - (n-2)/|X|^2]X. \tag{2}$$

The root that is used to determine the radius of a Stein confidence set is

$$R_n(X, \theta) = n^{-1/2}\{|\hat{\theta}_{n,S} - \theta|^2 - [n - (n-2)^2/|X|^2]\}, \tag{3}$$

which compares the loss of the Stein estimator with an unbiased estimator of its risk. This approach to confidence sets for θ was proposed at the end of Stein (1981). By invariance under the orthogonal group, the sampling distribution of the root (3) depends on θ only through $|\theta|$, and so may be written in the form $H_n(|\theta|^2/n)$.

Let $\{\theta_n \in R^n, n \geq 1\}$ denote any sequence such that $|\theta_n|^2/n \to a$, a finite non-negative constant. Then $H_n(|\theta_n|^2/n)$ converges weakly to a normal distribution with mean 0 and variance

$$\sigma^2(a) = 2 - 4a/(1+a)^2. \tag{4}$$

This is condition B for this example. To meet condition A requires a careful choice of the estimator of $\hat{\theta}_n$, such as

$$\hat{\theta}_{n,CL} = [1 - (n-2)/|X|^2]_+^{1/2} X. \tag{5}$$

Note the square root in (5), unlike in (2). The essential point is that, under the sequence $\{\theta_n\}$ described above, the estimators $\{|\hat{\theta}_{n,CL}|^2/n\}$ converge in probability to a, the limiting value of $\{|\theta_n|^2/n\}$. Consequently, the bootstrap distribution $\hat{H}_{n,B} = H_n(|\hat{\theta}_{n,CL}|^2/n)$ converges to the same $N(0, \sigma^2(a))$ limit as does the actual sampling distribution of the root.

On the other hand, the plausible alternative estimators $H_n(|\hat{\theta}_{n,S}|^2/n)$ and $H_n(|X|^2/n)$ both converge weakly, in probability, to the wrong limits. In the successful bootstrap world for this problem, the conditional distribution of X^* is $N(\hat{\theta}_{n,CL}, I)$, not $N(X, I)$ or $N(\hat{\theta}_{n,S}, I)$.

The bootstrap confidence set $C_{n,B}$ in this example is just the sphere centered at the Stein estimator $\hat{\theta}_{n,S}$ with radius

$$\hat{d}_{n,B} = [n - (n-2)^2/|X|^2 + n^{1/2}\hat{H}_{n,B}^{-1}(\alpha)]_+^{1/2}. \tag{6}$$

By the reasoning sketched above, the coverage probability of this bootstrap Stein confidence set is asymptotically α, in the uniform sense that

$$\lim_{n\to\infty} \sup_{|\theta|^2 \le nc} |P_{n,\theta}(C_{n,B} \ni \theta) - \alpha| = 0 \tag{7}$$

for every positive finite c. For more on bootstrap Stein confidence sets, see Beran (1993).

A very different approach to constructing bootstrap confidence sets is Efron's BC_a method. This is suited to one-dimensional parametric functions τ. The asymptotic relationship between the BC_a method and the root-based method described above is discussed in Hall (1992).

Computing Bootstrap Confidence Sets

Only rarely does a bootstrap distribution $\hat{H}_{n,B}$ have a closed form distribution. Strategies for computing the quantile $\hat{H}_{n,B}^{-1}$ fall into two broad catcgorics: Monte Carlo approximations on the one hand; Edgeworth expansions or saddlepoint approximations on the other hand. Computers are potentially useful in doing the algebra of the analytic approximations as well as in performing Monte Carlo simulations. However, the computational emphasis to date has been on Monte Carlo algorithms.

The simplest, and very general, Monte Carlo approach is to to construct, in the bootstrap world, M conditionally independent repetitions $X_1^*, \ldots X_M^*$ of the original experiment. The conditional distribution of each bootstrap sample X_j^*, given X, is $P_{n,\hat{\theta}_n}$. The empirical distribution of the values $\{R_n(X_j^*, \hat{\theta}_n) : 1 \le j \le M\}$ then converges to the theoretical bootstrap distribution $\hat{H}_{n,B}$ as M increases. This approximation technique, whose origins lie in Monte Carlo tests, is responsible for the name *resampling* method that is sometimes used imprecisely as a synonym for bootstrap method. In reality, resampling is only one of the ways to approximate a bootstrap distribution.

How many bootstrap samples should we use when resampling? The answer to this question is twofold, as was pointed out by Hall (1986). On the one hand, to achieve accurate coverage probability, we should choose the number of bootstrap samples M so that $k/(M+1) = \alpha$ for some integer k; and then use the k-th order statistic of the values $\{R_n(X_j^*, \hat{\theta}_n)\}$ as the critical value for the numerical implementation of $C_{n,B}$. Then, the coverage probability of this Monte Carlo version of $C_{n,B}$, evaluated under the joint distribution of the sample X and of the artificial samples $\{X_j^*: 1 \leq j \leq M\}$, is α plus a term that goes to zero as n increases. That coverage probability can be accurate for large values of n, when M is small but chosen as above, is useful in debugging a simulation study of bootstrap confidence sets.

On the other hand, the Monte Carlo approximation to the theoretical confidence set $C_{n,B}$ is a randomized procedure. Unless M is large, the computed critical value, and consequently the computed confidence set, will depend strongly upon the realization of the artificial samples $\{X_j^*: 1 \leq j \leq M\}$. To limit the amount of randomization, writers on the bootstrap have moved, with time, from the suggestion that M be of order $O(10^2)$ to the recommendation that M be as large as possible and preferably at least of order $O(10^3)$.

Several authors have investigated more efficient Monte Carlo schemes for approximating bootstrap distributions. Most successful in the bootstrap context have been importance sampling (Johns (1988)), balanced resampling (Davison et al. (1986)), and antithetic sampling (Snijders (1984)). Appendix II of Hall (1992) compares the relative efficiencies, when M is large, of these methods for approximating a bootstrap distribution function or quantile.

The discussion above pretends that random number generators produce realizations of independent, identically distributed random variables. This assumption is, at best, a rough approximation. A more satisfactory analysis of Monte Carlo approximations to bootstrap confidence sets is an open problem.

Edgeworth approximations to bootstrap distributions have proved valuable in studying the asymptotic properties of bootstrap confidence sets (Hall (1992)). As a practical means for determining bootstrap critical values, Edgeworth expansions suffer from relative inaccuracy in their tails as well as algebraic cumbersomeness. Saddlepoint approximations to bootstrap distributions, initiated by Davison and Hinkley (1988), appear to be more accurate, but currently lack convenient implementation outside the simplest cases.

Quality of Bootstrap Confidence Sets

A good confidence set is both reliable and selective. By reliability, we mean that the coverage probability is accurate; by selectivity we mean that the confidence set is not too large. Keeping a confidence set small, among all those of coverage probability α, is a fundamental design question, a matter of picking the root well. Achieving accurate coverage probability is then the simpler matter of constructing a good critical value for the chosen root. General criteria for picking a root include: minimizing $P_{n,\theta}(C_{n,B} \ni \theta')$ for $\theta' \neq \theta$, as Neyman proposed; or minimizing a geometrical risk such as $E_\theta \sup\{|t - \theta| : t \in C_{n,B}\}$. The bootstrap Stein confidence set in Example 1 has smaller geometrical risk, at every α and for sufficiently large n, than does the classical confidence sphere centered at X (Beran (1993)).

Bootstrap theory has made significant progress in understanding how to control coverage probability once the root is chosen. A number of important examples exhibit the following structure. The left continuous distribution function $H_n(\cdot, \theta)$ of the root admits an asymptotic expansion

$$H_n(x, \theta) = H_A(x, \theta) + n^{-k/2} h(x, \theta) + O(n^{-(k+1)/2}), \tag{8}$$

where the first two terms on the right hand side are smooth functions of θ, k is a positive integer, and the asymptotic distribution function $H_A(x, \theta)$ is continuous and strictly monotone in x. In this setting, a competitor to the bootstrap confidence set $C_{n,B}$ is the *asymptotic confidence set* for τ:

$$C_{n,A} = \{t \in T : R_n(X, t) \leq H_A^{-1}(\alpha, \hat{\theta}_n)\}. \tag{9}$$

Like $C_{n,B}$, the asymptotic coverage probability of $C_{n,A}$ is α.

To compare rates-of-convergence of the coverage probabilities to α, suppose that the estimators $\{\hat{\theta}_n\}$ are $n^{-1/2}$-consistent. By heuristic argument, as in Beran (1988b), we find:

- If the asymptotic distribution H_A of the root depends on θ, then the coverage probabilities of $C_{n,A}$ and $C_{n,B}$ converge to α at the same rates.

- If the asymptotic distribution H_A does *not* depend on θ, then the coverage probability of $C_{n,B}$ converges to α faster than does the coverage probability of $C_{n,A}$.

In the first case, both the asymptotic and bootstrap approaches estimate the leading term of the expansion (8). In the second case, the bootstrap approach successfully estimates the second term in the expansion (the leading term is now known); however the simple asymptotic approach continues to estimate only the first term, having no information about the second term. The asymptotic approach might be refined by using a two term Cornish-Fisher expansion to generate the critical value in (9). In practice, this refinement may not be easy. The bootstrap approach is attractively intelligent in its handling of both cases without technical intervention by the statistician. Hall (1992) has placed the heuristics above on a rigorous footing, in a certain more specialized setting.

EXAMPLE 2. As an instance of the case most favorable to bootstrapping, let us consider the Behrens-Fisher problem—devising a confidence interval for the difference between two means when the variances in two independent normal samples are unknown and possibly unequal. We take as root the t-statistic constructed from the difference of the two sample means. The limiting distribution of this root, under the normal model, is standard normal. Bootstrapping from the fitted normal model for the two samples yields a confidence set that is asymptotically equivalent and numerically close to Welch's solution (Beran (1988b)). Moreover, if n denotes the combined sample size, the error in coverage probability of both the Welch and the bootstrap confidence sets is of order $O(n^{-2})$. By contrast, the asymptotic confidence set based on the normal limiting distribution of the t-statistic incurs a coverage probability error of order $O(n^{-1})$.

EXAMPLE 1 (continued). In this Stein confidence set problem, the limiting normal distribution of the root depends upon the unknown parameter through the limiting value of $|\theta|^2/n$. The asymptotic variance of the root (3) is estimated consistently by

$$\hat{\sigma}_n^2 = \sigma^2(|\hat{\theta}_{n,CL}|^2/n) \tag{10}$$

for σ^2 defined in (4). The bootstrap Stein confidence set $C_{n,B}$ was described in earlier, around equation (6). The corresponding asymptotic Stein confidence set is the sphere centered at $\hat{\theta}_{n,S}$ with radius

$$\hat{d}_{n,A} = [n - (n-2)^2/|X|^2 + n^{1/2}\hat{\sigma}_n\Phi^{-1}(\alpha)]_+^{1/2}. \tag{11}$$

Here the coverage probability errors of $C_{n,A}$ and $C_{n,B}$ are both of order $O(n^{-1/2})$, as shown in Beran (1993). Figure 1 plots, for $n = 19$, the coverage probabilities of $C_{n,A}$ (diamonds) and $C_{n,B}$ (crosses) against the normalized noncentrality parameter $|\theta|^2/n$. The intended coverage probability is $\alpha = .90$; each bootstrap critical value is computed from 199 bootstrap samples by the method described in the previous section; and the coverage probabilities themselves are estimates based on 20,000 pseudo-random normal samples. The marked changes that occur in coverage probability as the normalized noncentrality parameter increases from 0 to 2 reflect variations in the asymptotic skewness and in the slope of the asymptotic variance of the root.

To improve coverage probability accuracy of the Stein confidence set $C_{n,B}$, we can pursue a more sophisticated strategy: First transform the root in a one-to-one way so that its asymptotic distribution does *not* depend on the unknown parameter; and then construct the bootstrap confidence set based

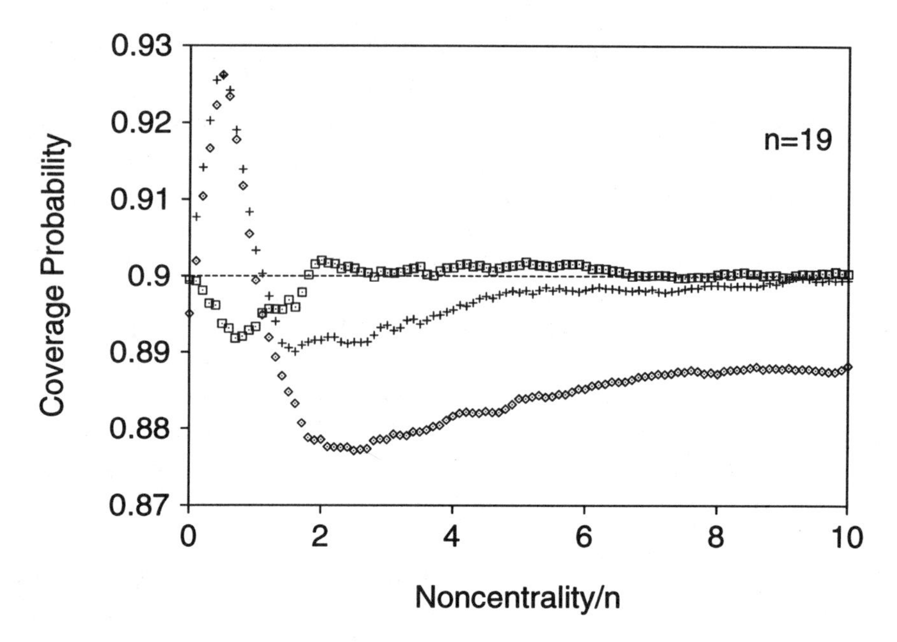

Figure 1: Coverage probabilities in Example 1 of $C_{n,A}$ (diamonds), of $C_{n,B}$ (crosses), and of $C_{n,TB}$ (squares) when α is .90 and n is 19.

on the transformed root. Studentizing, as was done implicitly in Example 2, is an instance of such transformation. However, studentizing does not work well for moderate values of n in Example 1 or in other cases where the distribution of the root is substantially non-normal. More successful in Example 1 is the use of a variance stabilizing transformation. Instead of (3), consider the root

$$R_{n,T}(X,\theta) = n^{1/2}\{g[|\hat{\theta}_{n,S} - \theta|^2/n] - g[1 - (n-2)^2/(n|X|^2)]\}, \qquad (12)$$

where

$$g(u) = 2^{-1}\log[-2 + 4u + 2^{3/2}(2u^2 - 2u + 1)^{1/2}]. \qquad (13)$$

The limiting distribution of root (12) is standard normal, in view of (4). Let $C_{n,TB}$ denote the transformed bootstrap Stein confidence set that is based on $R_{n,T}(X,\theta)$. The coverage probability error in $C_{n,TB}$ is of order $O(n^{-1})$, a significant improvement over $C_{n,A}$ and $C_{n,B}$ that is borne out by the coverage probabilities (squares) plotted in Figure 1.

Iterated and Two-step Bootstrap

We can use the bootstrap itself to transform a root $R_n(X,\tau)$ into a new root whose limiting distribution does not depend on the unknown parameter.

Let $\hat{H}_{n,B}(\cdot)$ denote the left continuous bootstrap distribution function of the root R_n and define

$$R_{n,B}(X,\tau) = \hat{H}_{n,B}(R_n(X,\tau)) = H_n(R_n(X,\tau),\hat{\theta}_n). \tag{14}$$

When the limiting distribution of R_n is continuous, the limiting distribution of the new root $R_{n,B}$ is typically Uniform (0,1). Let $C_{n,BB}$ denote the bootstrap confidence set based on $R_{n,B}$. If $\hat{H}_{n,BB}$ denotes the bootstrap distribution of $R_{n,B}(X,\tau)$, then

$$C_{n,BB} = \{t \in T: R_n(X,t) \leq \hat{H}_{n,B}^{-1}[\hat{H}_{n,BB}^{-1}(\alpha)]\}. \tag{15}$$

In the light of the preceding section, we expect that the coverage probability of $C_{n,BB}$ converges to α at a faster rate than the coverage probability of $C_{n,B}$. This often turns out to be the case, as argued in Beran (1988b) and elsewhere. The transformation (14) is called *prepivoting*, because it maps the original root into one that is more nearly pivotal when n is large.

Construction of $C_{n,BB}$ involves two bootstrap worlds. In the *first bootstrap world*, as described at the beginning of this essay, the true parameter is $\hat{\theta}_n$ and we observe an artificial sample X^* whose conditional distribution, given X, is $P_{n,\hat{\theta}_n}$. Write θ_n^* for $\hat{\theta}_n(X^*)$, the recalculation of the estimator in the first bootstrap world. In the *second bootstrap world*, the true parameter is θ_n^* and we observe an artificial sample X^{**} whose conditional distribution, given X and X^*, is P_{n,θ_n^*}. Then

- The conditional distribution of $R_n^* = R_n(X^*,\tau(\hat{\theta}_n))$, given X, is the bootstrap distribution $\hat{H}_{n,B}$.

- The conditional distribution of $R_{n,B}^* = R_{n,B}(X^*,\tau(\hat{\theta}_n))$, given X, is the bootstrap distribution $\hat{H}_{n,BB}$. Moreover, by (14),

$$R_{n,B}^* = H_n(R_n^*,\theta_n^*) = P(R_n^{**} < R_n^*|X,X^*), \tag{16}$$

  where $R_n^{**} = R_n(X^{**},\tau(\theta_n^*))$.

From this we see that practical computation of $C_{n,BB}$ generally requires a double nested Monte Carlo algorithm. The inner level of this algorithm approximates $H_{n,B}$, while both levels are needed to approximate $H_{n,BB}$. For further details, see Beran (1988b). Constructing the second bootstrap world is often called iterated or double bootstrapping. The underlying idea is that differences between the first bootstrap world and the original world (which are unknown) approximately equal corresponding differences between the second bootstrap world and the first bootstrap world (which are computable).

Prepivoting is not the only use for iterated bootstrapping. Other inferential problems, such as bias reduction, can benefit from repeated bootstrapping, as discussed by Hall and Martin (1988). Alternative constructions of iterated bootstrap confidence sets, asymptotically equivalent to those derived from prepivoting, are treated by Hall (1992).

Superficially similar to double bootstrapping, but different logically and much less intensive computationally, is two-step bootstrapping. Two-step bootstrapping provides a way to extend the classical Tukey and Scheffé simultaneous confidence sets from normal linear models to general models. Suppose that the parametric function τ has *components* labelled by an index set U; that is $\tau(\theta) = \{\tau_u(\theta): u \in U\}$. For each u, let $C_{n,u}$ denote a confidence set for the component τ_u. By simultaneously asserting the confidence sets $\{C_{n,u}\}$, we obtain a simultaneous confidence set C_n for the family of parametric functions $\{\tau_u\}$. The problem is to construct the component confidence sets $\{C_{n,u}\}$ in such a way that

$$P_{n,\theta}(C_{n,u} \ni \tau_u) \text{ is the same for every } u \in U \tag{17}$$

and

$$P_{n,\theta}(C_n \ni \tau) = \alpha. \tag{18}$$

Suppose that $R_{n,u} = R_{n,u}(X, \tau_u)$ is a root for the component parametric function τ_u. Let $H_{n,u}(\cdot, \theta)$ and $H_n(\cdot, \theta)$ denote the left-continuous distribution functions of $R_{n,u}$ and of $\sup_u H_{n,u}(R_{n,u}, \theta)$ respectively. The corresponding bootstrap estimators for these two distributions are then $\hat{H}_{n,u,B} = H_{n,u}(\cdot, \hat{\theta}_n)$ and $\hat{H}_{n,B} = H_n(\cdot, \hat{\theta}_n)$. Define the critical values

$$\hat{d}_{n,u} = \hat{H}_{n,u,B}^{-1}[\hat{H}_{n,B}^{-1}(\alpha)]. \tag{19}$$

Let T_u and T denote, respectively, the ranges of $\tau_u(\theta)$ and $\tau(\theta)$. Every point in the range set T can be written in component form $t = \{t_u\}$, where t_u lies in T_u. Define a bootstrap confidence set for τ_u by

$$C_{n,u,B} = \{t_u \in T_u: R_{n,u}(X, t_u) \leq \hat{d}_{n,u}\}. \tag{20}$$

Simultaneously asserting these component confidence sets generates the following bootstrap simultaneous confidence set for τ:

$$C_{n,B} = \{t \in T: R_{n,u}(X, t_u) \leq \hat{d}_{n,u} \text{ for every } u \in U\}. \tag{21}$$

Asymptotically in n, the confidence set $C_{n,B}$ satisfies the overall coverage probability condition (18); and the confidence sets $\{C_{n,u,B}\}$ satisfy the *balance* condition (17). Regularity conditions that ensure the validity of these conclusions are analogous to conditions A and B in Section 3. Beran (1988a) gives particulars. Interestingly, the Tukey and Scheffé simultaneous confidence intervals in the normal linear model are special cases of the bootstrap confidence set (21). These classical procedures satisfy (17) and (18) exactly.

Since the definition of simultaneous confidence set $C_{n,B}$ involves only the first bootstrap world, a Monte Carlo approximation to the critical values (20) requires only one round of resampling. Indeed, $\hat{H}_{n,u,B}$ and $\hat{H}_{n,B}$ are just the conditional distributions of $R_{n,u}(X^*, \tau_u(\hat{\theta}_n))$ and of $\sup_u H_{n,u}(R_{n,u}(X^*, \tau_u(\hat{\theta}_n), \hat{\theta}_n)$, given X. Computational difficulties can arise when the index set U is not finite. However, in practice we are usually interested in only a finite number of parametric functions. Iterated bootstrapping can be used to improve the rate at which the simultaneous confidence set approaches properties (17) and (18) as n increases. For details, see Beran (1990).

Further Resources

In this short account, we have sketched only how bootstrap methods may be used to construct reliable confidence sets. Significant progress has occurred in several additional directions, including: bootstrap tests; bootstrap prediction regions; bootstrap confidence sets for models where the dimension of the parameter space is high relative to sample size (Example 1 illustrates this situation); bootstrap inference based on nonparametric regression estimators or density estimators; bootstrap inference for spectral density estimators. Further information on these and other bootstrap developments may be found in the following sources:

Monographs. Efron & Tibshirani (1993) give a wide-ranging relatively nonmathematical introduction to the bootstrap and its application. Hall (1992) uses Edgeworth expansions to study higher-order asymptotic properties of bootstrap methods; the appendices treat other important aspects of bootstrap theory. Each chapter ends with brief bibliographical notes citing related work by other authors. Mammen (1992) develops higher-order bootstrap analyses without Edgeworth expansions; bootstrap worlds for models where the dimension of the parameter space is large relative to sample size (the *wild* bootstrap); and bootstrap methods for M-estimators in such circumstances. Beran & Ducharme (1991) records six introductory lectures on bootstrap inference. Efron (1982) raises several problems that remain incompletely solved.

Survey papers. Surveys of bootstrap theory, which reflect the state of knowledge at the time of writing, include: Hinkley (1988), DiCiccio & Romano (1988), and Beran (1984). The Trier proceedings volume (Jöckel et al. (1992)) contains papers on random number generation as well as on bootstrap theory and applications. A second bootstrap proceedings volume is Billard & LePage (1992). The dissertation of Janas (1992) discusses bootstrap methods for time series.

Acknowledgements

Grant support from the National Science Foundation, notably DMS 9224868, and the hospitality of Sonderforschungsbereich 123 at Universität Heidelberg gave me time to think about bootstrap methods over an extended period. This essay could not have been written otherwise.

References

Beran, R. (1984). Bootstrap methods in statistics. Über. d. Dt. Math. Verein. 86, 212–225.

Beran, R. (1988a). Balanced simultaneous confidence sets. J. Amer. Statist. Assoc. 83, 679–686.

Beran, R. (1988b). Prepivoting test statistics: A bootstrap view of asymptotic refinements. J. Amer. Statist. Assoc. 83, 686–697.

Beran, R. (1990). Refining bootstrap simultaneous confidence sets. J.Amer. Statist. Assoc. 85, 417–426.

Beran, R. (1993). Stein confidence sets and the bootstrap. Preprint.

Beran, R. & Ducharme, G. (1991). Asymptotic Theory for Bootstrap Methods in Statistics. Publications CRM, Université de Montréal.

Billard, L. & Lepage. R. (eds.) (1992). Exploring the Limits of Bootstrap. Wiley, New York.

Davison, A.C. & Hinkley, D.V. (1988). Saddlepoint approximations in resampling methods. Biometrika 75, 417–431.

Davison, A.C., Hinkley, D.V. & Schechtman, E. (1986). Efficient bootstrap simulation. Biometrika 73, 555–566.

Diciccio, T.J. & Romano, J.P. (1988). A review of bootstrap confidence intervals (with discussion). J. Roy. Statist. Soc. Ser. B 50, 338–354.

Efron, B. (1979). Bootstrap methods: Another look at the jackknife. Ann. Math. Statist. 7, 1–26.

Efron, B. (1982). The Jackknife, the Bootstrap, and Other Resampling Plans. SIAM, Philadelphia.

Efron, B. & Tibshirani, R.J. (1993). An Introduction to the Bootstrap. Chapman and Hall, New York

Hall, P. (1986). On the number of bootstrap simulations required to construct a confidence interval. Ann. Statist. 14, 1453–1462.

Hall, P. (1992). The Bootstrap and Edgeworth Expansion. Springer, New York.

Hall, P. & Martin, M.A. (1988). On bootstrap resampling and iteration. Biometrika 75, 661–671.

Hinkley, D.V. (1988). Bootstrap methods (with discussion). J. Roy. Statist. Soc. Ser. B 50, 321–337.

Huber, P.J. (1981). *Robust Statistics.* Wiley, New York.

Ibragimov, I.A. & Has'minskii, R.Z. (1981). *Statistical Estimation: Asymptotic Theory.* Springer, New York.

Janas, D. (1993). *Bootstrap Procedures for Time Series.* Shaker, Aachen.

Jöckel, K.-H., Rothe, G., and Sendler, W. (eds.) (1992). Bootstrapping and Related Techniques. Lecture Notes in Economics and Mathematical Systems 367. Springer, Berlin.

Johns, M.V. Jr (1988). Importance sampling for bootstrap confidence intervals. J. Amer. Statist. Assoc. 83, 709–714.

Mammen, E. (1992). When Does Bootstrap Work? Lecture Notes in Statistics 77. Springer, New York.

Snijders, T.A.B. (1984). Antithetic variates for Monte Carlo estimation of probabilities. Statist. Neerland. 38, 55–73.

Stein, C. (1981). Estimation of the mean of a normal distribution. *Ann. Statist.* **9**, 1135-1151.

Special Resampling Techniques in Categorical Data Analysis

Iris Pigeot
University of Dortmund, Department of Statistics
Vogelpothsweg 87, D–44221 Dortmund, Germany [1]

Key Words:
categorical data, jackknife, contingency tables, odds ratio

Abstract
In this paper, different applications of resampling techniques in categorical data analysis are reviewed with particular emphasis on the jackknife principle. As a special topic, jackknife estimators of the common odds ratio are considered in the situation of several 2×2 tables. Asymptotic and finite–sample properties of such estimators are summarized. Finally, an overview of current research in this field is given.

Introduction

In the last three decades, the advance in computer technology has strongly influenced among others the kind of statistical research.

Statistical methods have been developed taking advantage of high–speed computers. Methods such as the bootstrap as well as the jackknife and other resampling plans (Efron (1982)) would not have been feasible in a reasonable way without support of this special tool. But particularly these resampling techniques can be very effective when being used for estimating measures of error such as bias or variance. In this paper, the jackknife is applied to get a bias–corrected estimator of the common odds ratio and to estimate the asymptotic variance of a classical estimator of this parameter.

In addition, extensive Monte–Carlo experiments and large numerical studies can nowadays be carried out without taking an immense amount of time.

[1]This research was partially supported by the Benningsen–Foerder–Grant of Nordrhein–Westfalen.

Such studies yield further insight e.g. into the discussion of common statistical methods under deviations from standard assumptions. They are also informative when investigating estimators w.r.t. their finite–sample behaviour if otherwise only their asymptotic properties are known as it is the case for the special topic discussed in this paper.

The particular emphasis of this paper is on the application of resampling techniques in the analysis of categorical data (c.f. Cox (1970); Fleiss (1973); Plackett (1981)). In many practical problems, only categorical information on certain variables can be obtained. For instance, when examining the influence of a risk factor on a disease of interest very often the status of disease can only be recorded in certain categories. Thus, categorical data forms an important part of statistical information, especially in medicine.

In the following section, different applications in particular of the jackknife and the bootstrap are reviewed without going too far into detail. A more detailed overview of results concerning jackknife estimators of a common odds ratio in several 2×2 tables is given in the third section. Here, asymptotic properties as well as finite–sample results gained from simulation studies are presented. Some findings concerning jackknife variances and corresponding confidence intervals are added. The last section deals with a computer program for calculating the discussed jackknife estimators and with some special aspects under current research.

Applications of resampling techniques in categorical data analysis

The intention of this section is to give a short and rough review of research in categorical data analysis where resampling techniques are applied. That means, not all papers are taken into account and the results of those presented here are not given in detail.

Three resampling techniques are quite often used: the cross–validation, the jackknife and the bootstrap.

Using cross–validation, originally the sample has to be randomly divided into two halves, where the first half is used for model fitting and the second for model checking, i.e. the results obtained from the first half are used to explain the behaviour of the data in the second half. Typical applications are e.g. prediction and discrimination.

This original idea can also be changed by not dividing the sample into two halves, but by leaving out one observation at a time. Then, the model is fitted using the remaining observations and it is checked by predicting the point excluded before.

The jackknife which will be described in more detail in the next section is also based on the idea of 'leaving–out–one'. Therefore, as Efron (1982, p. 49) points out, the nowadays use of cross–validation is often mixed up with the jackknife. Typically, the jackknife is considered, however, in relation to some statistic $\hat{\vartheta}$ which is to be jackknifed. That means, the statistic is recomputed after one observation or a whole group of observations has been omitted from the data at a time, which yields the statistic $\hat{\vartheta}_i$ based on the reduced sample. The index i denotes the i–th observation out of N or the i–th group out of K, respectively, excluded from the sample. Then, in a second step $\hat{\vartheta}$ and $\hat{\vartheta}_i$ are combined to get the corresponding jackknife statistic (c.f. the third section).

Let us now assume that a certain measure of error, denoted by R, of a statistic $\hat{\vartheta}(X_1, \ldots, X_N)$ with X_i i.i.d. according to some unknown probability distribution F can be written as a function solely of F, provided that the sample size N and the statistic $\hat{\vartheta}$ are given, i.e. $R = T(F)$. Then, the bootstrap estimate of R is simply this function T evaluated at $F = \hat{F}$ with $\hat{F}$ denoting the empirical distribution function, i.e. $\hat{R}_{Boot} = T(\hat{F})$. Since the function $T(F)$ is often not available in a closed form, a Monte–Carlo algorithm has to be used for calculating $\hat{R}_{Boot}$. As one step of this algorithm, a bootstrap sample of $\hat{F}$ is drawn, denoted by $X_1^*, \ldots, X_N^*$, and $\hat{\vartheta}^* = \hat{\vartheta}(X_1^*, \ldots, X_N^*)$ is calculated. This step is replicated quite a large number of times and, based on such obtained bootstrap replications, the bootstrap estimate of $T(F)$ can be approximated. For a more detailed discussion of the bootstrap, we refer to Efron (1982, Chapter 5).

These methods as well as the well–known delta method are closely related from a theoretical point of view, but they usually differ in the results they yield in practical applications.

As already indicated, cross–validation plays an important role in discriminant analysis. It is for instance used in its original form by Mosteller & Wallace (1963), who are interested in solving 'the authorship question of *The Federalist* papers': Two authors, namely Hamilton and Madison, published anonymously quite a number of papers about political matters in *The Federalist.* Some of these papers could not be ascribed clearly to one of these authors. Therefore, Mosteller and Wallace try to discriminate between the writings of Hamilton and Madison using word counts as the variable for discrimination. The same subject is adressed by Mosteller & Tukey (1977, p. 148ff) as an example mainly for the application of the jackknife, but also for cross–validation and a combination of these two techniques. They jackknife among others the discriminant function for assessing its variability in terms of the variability of its individual coefficients. For this purpose, the authors omit 'one observation' at a time where one observation consists of a pair of papers by Hamilton and Madison.

Besides these concrete applications, Lachenbruch & Mickey (1968) compare different estimators of error rates in discriminant analysis in a more general setting. Thus, they consider the leave–out–one approach, which is described as an application of the jackknife by Cochran (1968) in a comment to this paper. The authors derive two different methods based on this approach, which both behave well in a Monte–Carlo study carried out by Lachenbruch and Mickey.

Efron (1979, Chap. 4) discusses the estimation of error rates in discriminant analysis, too, but also without a special application to categorical data. Nevertheless, especially because of the general assumptions underlying his investigations, his work should be mentioned here. Efron focuses on the derivation of a bootstrap estimate of error rates which is shown to behave better than the leave–out–one approach in two examples considered there.

Another approach in discriminant analysis uses a kernel method of density estimation as tool. Aitchison & Aitken (1976) are especially interested in a multivariate binary discrimination, but they also mention an extension of their method to categorical data in general. They use the leaving–out–one approach to determine a suitable value of the bandwidth, also known as smoothing parameter, as part of the kernel estimation. Here, we can notice again some confusion about this 'leaving–out–one' method, which is denoted as 'jackknife' by the above authors. It should be better referred to as 'cross–validation' as it is also pointed out by Titterington (1980) who compares different approaches for deriving a kernel estimator of probability density functions for categorical data including the proposal of Aitchison and Aitken. The consistency of the kernel estimator derived by Aitchison and Aitken is shown by Bowman (1980).

Moreover, Bowman et al. (1984) consider kernel–based estimators of probabilities for categorical data. They discuss asymptotic properties for cross–validation in terms of conditions on the underlying loss function. As it can be seen from their results, cross–validation seems to be sensitive to the shape of the loss function, but with a correct choice of the loss function the cross–validatory approach can be regarded as performing well w.r.t. its asymptotic properties. For a comparison of different methods of density estimation in case of ordered categorical data we refer to Titterington & Bowman (1985). They use a cross–validatory criterion to choose the smoothing parameter when the properties of the resulting estimator are to be judged in terms of its accuracy of prediction.

This aspect of prediction based on cross–validation for categorical data is discussed more extensively by Stone (1974). Given frequency data $n = (n_1, \ldots, n_k)$ belonging to k categories with a total number of N items associated with a multinomial indicator δ_i expressing the category occupied by the i–th item, Stone (1974) considers the problem of predicting the category for an additional item. Two different loss functions are used to evaluate a certain prediction. Consider e.g. a quadratic loss and a class of predictors given as

$$\mathcal{P} = \{\alpha\lambda + (1-\alpha)n/N;\ \alpha \in [0,1]\}, \tag{1}$$

where λ is a prescribed k vector with components all greater or equal to zero and adding up to one. Then, a cross-validatory approach is applied to get a criterion for choosing α. This idea is considered for different types of (1), i.e. for instance for another loss function than the quadratic. Based on six examples, Stone (1974) concludes that none of the different methods for choosing α yields uniformly the best results. For an application of the bootstrap in a prediction problem see Efron & Gong (1983).

Wahrendorf (1984, p. 18f, p. 42ff) suggests the use of the bootstrap in epidemiological studies for estimating e.g. the variability of a given measure of the relationship between a status of exposure and a certain disease. As an example he considers the Mantel–Haenszel estimator $\hat{\psi}_{MH}$ (Mantel & Haenszel (1959)) of a common odds ratio in case–control–studies (c.f. the third section). Using the data of the Ille–et–Vilaine study of oesophageal cancer (Tuyns et al. (1977)) with a binary risk factor given as the average daily alcohol consumption being dichotomized and the additional confounder age stratified in four groups, Wahrendorf derives the bootstrap distribution of $\hat{\psi}_{MH}$ using a Monte–Carlo algorithm.

Another application of the bootstrap in medical research is given by Wahrendorf & Brown (1980). This concerns nonparametric modelling of independent joint action of two drugs (c.f. Hewlett & Plackett (1959)). The authors use the bootstrap to estimate a measure of agreement between observed data and the derived model in this case of three independent binomial random variables. In addition, an example is given to illustrate this technique.

The bootstrap can also be used to distort the distribution of data for privacy protection. In the situation, where the data is ordered in several 2×2 tables, Pigeot et al. (1992) investigate the effect of replacing the original data set by a synthetic one. The latter is gained from bootstrap samples, where especially the idea of the parametric bootstrap is involved. This method is checked using an example. It can be seen that already ten bootstrap replications are sufficient to come close to the original data. In particular, small cell counts are exactly reproduced. Thus, there is nearly no effect of privacy protection, especially if the used technique is known. In addition, the obtained estimates based on the synthetic data set compared with those based on the original one are not convincing.

Bowden & Sim (1992) consider another possibility for using the bootstrap in this field. They perturb the data with a random noise generated by bootstrapping from the original empirical distribution. Bowden and Sim explore the trade–off between privacy protection based on the bootstrap and the efficiency of estimation in an example and report some simulation results which are regarded as 'encouraging' by the authors.

In a paper by Liou & Yu (1991), the bootstrap is applied in psychological research, especially for deriving confidence intervals and estimates of the standard deviation of ability estimators in item response theory. For this purpose, the authors assume the n items of a psychological test to be randomly selected from a set of items being at disposal. In addition, they assume the following logistic model for the binary response variable U_i, $i = 1, \ldots, n$:

$$P_i(U_i = 1|\theta) = c_i + \frac{1 - c_i}{1 + \exp\{-1.7a_i(\theta - b_i)\}} \tag{2}$$

with a_i, b_i, c_i, $i = 1, \ldots, n$, known constants. The bootstrap is now used to determine the statistical accuracy of estimators of θ.

Efron & Tibshirani (1986) discuss several applications of the bootstrap. Among others they consider a study of leukemia remission times in mice where a treatment and a control group is taken into account. Since some of the remission times are censored, Cox's well–known proportional hazards model is used and the parameter β of the hazard function is estimated through maximization of the partial likelihood. Here, the authors derive the bootstrap estimate of the standard error of $\hat{\beta}$ and a confidence interval for β based on the bootstrap distribution, which 'should be interpreted with some caution'.

Let us now present some special applications of the jackknife. Gladen (1979) considers animal experiments in toxicological work where the effects of treatments on the fetuses or the offspring of pregnant animals are investigated. Under the assumption of two groups, one treated and one control group, it is of special interest to estimate the proportion affected in these groups conditional on litter size. Gladen uses e.g. a beta–binomial model, i.e. the affections of the fetuses are supposed to be independently binomially distributed with parameter p where p follows a beta distribution. She first derives a jackknife estimator for the proportion in each group before developing a two–sample jackknife estimator. These proposals and corresponding variance estimators are studied w.r.t. their large–sample behaviour and in practical examples.

The two– and three–sample jackknife is also investigated by Farewell (1978) as illustration of his suggestion to reflect the structure of a given data set in the jackknife calculation. More precisely, given the situation that e.g. the two samples of interest show a $t:1$ ratio of sample sizes, this ratio should be taken into account by removing t observations from the larger sample and one from the other. Based on simulation results and some examples, Farewell points out that the idea of maintaining any pattern in the data when creating the subsamples yields an improved jackknife procedure.

Another application in medical research concerns the problem of testing dose–response effects. Salsburg (1971) derives a test statistic by jackknifing the covariance estimator of dose versus response in case that the responses are binomially distributed with a success probability close to zero or one. Simulation results indicate that the test is somewhat conservative and has low power. These results are criticized by Frawley (1974), who shows that the procedure proposed by Salsburg has better power than originally reported. He also presents an alternative test where another variance estimator is used. This approach seems to yield in general a more powerful procedure.

Woodward & Schucany (1977) use the jackknife principle in a study concerning consumers' preferences for two competing brands, where a triangle taste test is combined with a preference test. In particular, they apply the jackknife for assessing the variability of the maximum–likelihood estimator of the conditional probability that a person prefers a certain brand out of two given that he/she can distinguish between the two products.

Simonoff (1986) takes account of the problem to estimate the variance of goodness–of–fit statistics under a composite null hypothesis in a sparse multinomial data model. The bootstrap estimate based on the empirical distribution function, a parametric bootstrap, a scaled and centered version of the ordinary jackknife variance estimator, and a jackknife approach in which the cell entries rather than the single observations are omitted one at a time (c.f. the third section; Breslow & Liang (1982)) are compared by Monte–Carlo experiments. The bootstrap based on the empirical distribution function shows a poor behaviour. Good results are observed for the transformed ordinary jackknife, which should also be preferred over the parametric bootstrap and the jackknife based on the idea of omitting the cell entries.

Simonoff (1986) also considers a general approach for jackknifing functions of multinomial frequencies. First and second order jackknife estimators of this type are investigated w.r.t. their asymptotic properties by Parr & Tolley (1982). Using the fact, that the multinomial parameter vector is a vector–valued U–statistic, the asymptotic normality of the first order jackknife estimator and the consistency of the corresponding variance estimator follow from the theorems proven in Arvesen (1969). Parr and Tolley also show the asymptotic normality of the second order jackknife estimator and discuss the advantages and disadvantages of the jackknife approach compared with the delta–method. Furthermore, they give two applications of the jackknife: Shannon's index of diversity and the logarithm of the odds ratio. Jackknifing such indices of diversity is also pursued by Zahl (1977), Adams & McCune (1979) as well as by Heltshe & Forrester (1983).

The second example of Parr & Tolley (1982) is considered by quite a number of other authors. Thus, Fleiss & Davies (1982) use the odds ratio as illustration for the applicability of the jackknife in the same context as in the paper of Parr & Tolley (1982). In addition, Fleiss and Davies derive a jackknife estimator of a measure of concordance, the so–called Kappa coefficient, and present the jackknife estimate of its variance. A closely related jackknife estimator of an extended Kappa coefficient is introduced by Kraemer (1980) to obtain a confidence interval. This is constructed from a statistic based an the jackknifed Kappa coefficient and standardized with a suitable jackknifed variance estimator. Due to the results of Arvesen (1969) and Arvesen & Schmitz (1970), this statistic is assumed to be approximately distributed as a t random variable.

The odds ratio is again of interest in Camilli & Smith (1990). The authors derive a test for detecting biased items based on the jackknifed logarithm of the Mantel–Haenszel estimator of a common odds ratio, where again complete subsets are omitted one at a time. Using an example, this test is compared with the Mantel–Haenszel chi–square test and another proposal of Camilli and Smith denoted as randomization test. The authors conclude that 'the Mantel–Haenszel chi–square test reflects most of the information about bias' in their example and that the jackknife approach does not show an essentially different behaviour from that classical test.

Jewell (1984) investigates the bias of several point estimators of the odds ratio and its logarithm when arising from matched–pair data. Among others, he applies the jackknife principle for reducing the bias. A bootstrap version of an odds ratio estimator is also addressed to but not pursued in detail because of the additional computational effort.

The last paper to be mentioned here presents a jackknife variance estimator of the logarithm of the Mantel–Haenszel estimator in stratified 2×2 tables. To derive this jackknife variance, Breslow & Liang (1982) jackknife $\ln(\hat{\psi}_{MH})$ by omitting complete 2×2 tables one at a time. This approach and a corresponding one based on omitting the individual observations due to Pigeot (1991a) will be discussed in more detail in the following section.

Finally, it should again be mentioned that the above review does not claim to represent a complete overview of all articles in this field, but to give a rough idea of the variety of applications of resampling techniques in categorical data analysis.

Jackknife estimators of a common odds ratio

Before applying the jackknife principle in a concrete situation, i.e. here to estimators of a common odds ratio, let us first reproduce the basic idea of this method originally introduced by Quenouille (1949, 1956) and reviewed e.g. in Miller (1974).

A jackknife estimator J of a real–valued parameter ϑ based on a sample of size N is given as the arithmetic mean of so–called pseudo–values, denoted by J_i, $i = 1, \ldots, N$. Consider a particular estimator $\hat{\vartheta}$ of ϑ, omit the i–th observation and recalculate $\hat{\vartheta}$ now based on a sample of size $N-1$, then the pseudo–values are based on $\hat{\vartheta}$, itself, and on such recalculated estimates, in the following denoted by $\hat{\vartheta}_i$, i.e.

$$\begin{aligned} J_i &= J_i^N(\hat{\vartheta}) = N\hat{\vartheta} - (N-1)\hat{\vartheta}_i, \quad i = 1, \ldots, N \\ J &= \frac{1}{N}\sum_{i=1}^{N} J_i. \end{aligned} \tag{3}$$

The 'jackknife variance' $V(J)$ (Tukey, 1958) is given as

$$V(J) = \frac{1}{N(N-1)}\sum_{i=1}^{N}(J_i - J)^2. \tag{4}$$

In the literature, we find two general approaches (Breslow & Liang (1982); Pigeot (1991a)) for constructing jackknife estimators of a common odds ratio ψ in the situation of a series of K 2×2 tables, $1 < K < \infty$, where the cell entries are observations from K independent pairs of independent binomial random variables X_{0k}, X_{1k} with parameters N_{0k}, p_{0k} and N_{1k}, p_{1k}, $k = 1, \ldots, K$, respectively. Assume further that there exists a finite common odds ratio ψ, i.e. $\psi_k = \psi$ with ψ_k defined as $\{p_{1k}(1-p_{0k})\}/\{p_{0k}(1-p_{1k})\}$, $k = 1, \ldots, K$.

The two approaches to be mentioned here differ in what is perceived as 'one observation' when calculating the pseudo–values. Breslow & Liang (1982) consider each 2×2 table as one observation to be omitted. Jackknife estimators based on this concept are in the following referred to as type I. The other approach (Pigeot (1991a)) uses the fact that each binomial variate X_{jk}, $j = 0, 1$, $k = 1, \ldots, K$, is a sum of independent Bernoulli trials. Hence, the outcome of each trial can be regarded as one observation. This idea is already used e.g. by Parr & Tolley (1982) and by Fleiss & Davies (1982) in jackknifing functions of multinomial frequencies.

When jackknifing odds ratio estimators the above authors pursue different aims. Breslow & Liang (1982) use their approach primarily for deriving a suitable variance estimator of the Mantel-Haenszel estimator (see (7)) in the case where the number of tables is large, whereas Pigeot (1991a) is mainly interested in a bias reduction of the usually biased estimators of ψ. Since such a bias reduction should also be achieved if the number of tables is small, she focuses her interest on the second jackknife approach briefly described above. In the following both approaches are discussed in more detail.

Let $\hat{\psi}$ be any estimator of ψ, then the jackknife estimator of type I results in

$$J^I = \frac{1}{K}\sum_{k=1}^{K} J_k^I \quad \text{with } J_k^I = K\hat{\psi} - (K-1)\hat{\psi}_k^I \tag{5}$$

with $\hat{\psi}_k^I$ being of the same type as $\hat{\psi}$ but calculated after omitting the k-th table, $k = 1, \dots, K$. The jackknife variance $V(J^I)$ can be calculated as

$$V(J^I) = \frac{1}{K(K-1)}\sum_{k=1}^{K}(J_k^I - J^I)^2. \tag{6}$$

For illustrative purposes, consider the well–known and widely used Mantel–Haenszel estimator $\hat{\psi}_{MH}$ of ψ (Mantel & Haenszel (1959)), which is defined as

$$\hat{\psi}_{MH} = \frac{\sum_{k=1}^{K} \hat{p}_{1k}\hat{q}_{0k}N_{1k}N_{0k}/N_k}{\sum_{k=1}^{K} \hat{p}_{0k}\hat{q}_{1k}N_{1k}N_{0k}/N_k} \tag{7}$$

with $N_k = N_{0k} + N_{1k}$, $\hat{p}_{jk} = X_{jk}/N_{jk}$, $\hat{q}_{jk} = 1 - \hat{p}_{jk}$, $j = 0, 1$, $k = 1, \dots, K$. Then, J_{MH}^I can be derived as

$$J_{MH}^I = K\hat{\psi}_{MH} - \frac{K-1}{K}\sum_{k=1}^{K}\hat{\psi}_{MH,k}^I \tag{8}$$

$$\text{with } \hat{\psi}_{MH,k}^I = \frac{\sum_{i=1, i\neq k}^{K} \hat{p}_{1i}\hat{q}_{0i}N_{1i}N_{0i}/N_i}{\sum_{i=1, i\neq k}^{K} \hat{p}_{0i}\hat{q}_{1i}N_{1i}N_{0i}/N_i}. \tag{9}$$

It should be mentioned that Breslow and Liang jackknife the logarithm of the Mantel–Haenszel estimator. Then, an estimator of ψ can be obtained by using an exponential transformation. Let us denote this transformed jackknife estimator by $\hat{\psi}_{BL} = \exp(J_{BL})$ with

$$\hat{\psi}_{BL} = \exp\left(K \ln \hat{\psi}_{MH} - \frac{K-1}{K} \sum_{k=1}^{K} \ln \hat{\psi}_{MH,k}^{I}\right). \tag{10}$$

The application of the second jackknife approach to an estimator $\hat{\psi}$ of ψ can be simplified by noting that although there are N single observations and therefore N pseudo–values, there are only $4K$ distinct values. Marking the cells of each 2×2 table by a, b, c, d as usual, the $4K$ distinct pseudo–values can be defined as $J_k^a, \ldots, J_k^d$ with $J_k^a = N\hat{\psi} - (N-1)\hat{\psi}_k^a$, J_k^b, J_k^c, and J_k^d defined analogously. Again, $\hat{\psi}_k^a, \ldots, \hat{\psi}_k^d$ are of the same type as $\hat{\psi}$, but they are calculated after reducing the number of observations in the appropriate cells of the k–th table by one. Each of these distinct pseudo–values exists exactly X_{1k}, $N_{1k} - X_{1k}$, X_{0k}, and $N_{0k} - X_{0k}$–times, respectively, which yields the following jackknife estimator of type II based on $\hat{\psi}$ (Pigeot (1993))

$$\begin{aligned} J^{II} &= N\hat{\psi} - \frac{N-1}{N} \sum_{k=1}^{K} \Big[X_{1k}\hat{\psi}_k^a + (N_{1k} - X_{1k})\hat{\psi}_k^b + X_{0k}\hat{\psi}_k^c \\ &\qquad + (N_{0k} - X_{0k})\hat{\psi}_k^d\Big] \end{aligned} \tag{11}$$

with the jackknife variance $V(J^{II})$:

$$\begin{aligned} V(J^{II}) &= \frac{N-1}{N} \sum_{k=1}^{K} \Big[X_{1k}(\hat{\psi}_k^a)^2 + (N_{1k} - X_{1k})(\hat{\psi}_k^b)^2 + X_{0k}(\hat{\psi}_k^c)^2 \\ &\qquad + (N_{0k} - X_{0k})(\hat{\psi}_k^d)^2\Big] - \frac{1}{N-1}(J^{II} - N\hat{\psi})^2. \end{aligned} \tag{12}$$

Let us again illustrate this approach by applying it to the Mantel–Haenszel estimator. According to (11), we get J_{MH}^{II} based on $\hat{\psi}_{MH}$ by just replacing $\hat{\psi}$ with $\hat{\psi}_{MH}$, i.e.

$$\begin{aligned} J_{MH}^{II} &= N\hat{\psi}_{MH} - \frac{N-1}{N} \sum_{k=1}^{K} \Big[X_{1k}\hat{\psi}_{MH,k}^a + (N_{1k} - X_{1k})\hat{\psi}_{MH,k}^b \\ &\qquad + X_{0k}\hat{\psi}_{MH,k}^c + (N_{0k} - X_{0k})\hat{\psi}_{MH,k}^d\Big], \end{aligned} \tag{13}$$

where

$$\hat{\psi}_{MH,k}^a = \frac{(X_{1k} - 1)(N_{0k} - X_{0k})\dfrac{1}{N_k - 1} + \displaystyle\sum_{i=1, i \neq k}^{K} \hat{p}_{1i}\hat{q}_{0i}N_{1i}N_{0i}/N_i}{(N_{1k} - X_{1k})X_{0k}\dfrac{1}{N_k - 1} + \displaystyle\sum_{i=1, i \neq k}^{K} \hat{p}_{0i}\hat{q}_{1i}N_{1i}N_{0i}/N_i} \tag{14}$$

and $\hat{\psi}^b_{MH,k}, \dots, \hat{\psi}^d_{MH,k}$ defined analogously. From (14), it can be clearly seen that the number of observations in the a–cell of the k–th table, given by X_{1k}, has been reduced by one. Consequently, the total number of observations in that table results in $N_k - 1$.

Asymptotic properties

The asymptotic properties of the jackknife estimators from above depend on those of the original estimators of ψ. Thus, before further investigating their behaviour we have first to fix the particular estimators to be jackknifed.

These are on the one hand the Mantel–Haenszel estimator and transformed versions of $\hat{\psi}_{MH}$ given as $g^{-1}(\hat{\psi}_{MH})$ with $g \in G$ and G consisting of all nonnegative and real valued functions, which are strictly monotone and at least once differentiable. On the other hand, the jackknife principle is applied to the transformed elements $g^{-1}(\hat{\psi}_g)$ of a class of asymptotically efficient noniterative estimators of ψ (Pigeot (1990)), where $\hat{\psi}_g$ is defined as

$$\hat{\psi}_g = g\left(\sum_{k=1}^{K} \hat{w}_k g^{-1}(\hat{\psi}_k)/\hat{w}\right), \quad g \in G, \tag{15}$$

with

$$\hat{w}_k^{-1} = (N_{1k}\hat{p}_{1k}\hat{q}_{1k})^{-1} + (N_{0k}\hat{p}_{0k}\hat{q}_{0k})^{-1}, \qquad \hat{w} = \sum_{k=1}^{K} \hat{w}_k,$$

$$\hat{\psi}_k = \hat{p}_{1k}\hat{q}_{0k}/\hat{p}_{0k}\hat{q}_{1k}, \qquad k = 1, \dots, K.$$

This class contains e.g. the well–known estimator $\hat{\psi}_W$ suggested by Woolf (1955), which can be seen by choosing the function g as the exponential function.

Similarly to the Breslow–Liang estimator, the jackknife estimators based on $g^{-1}(\hat{\psi}_{MH})$ or $g^{-1}(\hat{\psi}_g)$, respectively, have to be transformed by the function g to obtain an estimator of ψ itself. It should be remarked, that the transformed estimators $g^{-1}(\hat{\psi}_{MH})$ and $g^{-1}(\hat{\psi}_g)$ are especially considered as a generalization of $\ln(\hat{\psi}_{MH})$ and $\ln(\hat{\psi}_W)$, respectively.

All asymptotic properties to be mentioned below are valid under the assumption of increasing sample sizes, but a fixed number of tables. More precisely, it is assumed

$$\begin{gathered} N_{jk} = \lambda_{jk} N, \quad N = \textstyle\sum_{k=1}^{K} N_k, \quad N \to \infty, \\ \textstyle\sum_{k=1}^{K} \lambda_k = 1, \quad \lambda_k = \lambda_{1k} + \lambda_{0k}, \quad j = 0, 1, \ k = 1, \dots, K. \end{gathered} \tag{16}$$

Consistency of the point estimators can be shown under the following slightly weaker condition:

$$N_{jk} = \lambda_{jk} N + o(N), \quad N \to \infty, \quad 0 < \lambda_{jk} < 1,$$
$$\textstyle\sum_{k=1}^{K} \lambda_k = 1, \quad \lambda_k = \lambda_{1k} + \lambda_{0k}, \quad j = 0, 1, \; k = 1, \ldots, K. \tag{17}$$

Let us first discuss the jackknife estimators of type I based on $\hat{\psi}_{MH}$ and on $\hat{\psi}_{id}$, where $id(x) = x$ denotes the identity. Pigeot (1993) proves that J^{I}_{MH} as well as J^{I}_{id} are consistent estimators of ψ and asymptotically normal with asymptotic expectation ψ and appropriate asymptotic variances, denoted in the following by $\text{var}^{A}(.)$. Both jackknife estimators are in general asymptotically inefficient. In the degenerate case, however, where there is no need to stratify, i.e. $p_{jk} = p_j$, $j = 0, 1$, $N_{1k}/N_k = \varrho$, $\varrho > 0$, $N_k = N/K$, $k = 1, \ldots, K$, asymptotic efficiency of both estimators is shown. These results can be generalized to the case where the jackknife is applied to $g^{-1}(\hat{\psi}_{MH})$ and $g^{-1}(\hat{\psi}_{g})$, $g \in G$ (Pigeot (1992, 1993)). That means especially, that $\hat{\psi}_{BL}$ is consistent and asymptotically normal, too.

The jackknife estimators of type II based on $\hat{\psi}_{MH}$ and $\hat{\psi}_{id}$ turn out to be consistent and asymptotically normal with asymptotic expectation ψ and asymptotic variances identical to those of the original estimators (Pigeot (1993)). Thus, J^{II}_{id} is asymptotically efficient and the conditions for J^{II}_{MH} to be asymptotically efficient are the same as those for $\hat{\psi}_{MH}$ derived by Tarone, Gart & Hauck (1983).

Originally, Breslow & Liang (1982) jackknife $\ln(\hat{\psi}_{MH})$ to get a suitable estimator of its asymptotic variance in the so–called sparse–data model with an increasing number of tables. Nevertheless, Pigeot & Strugholtz (1992) investigate $V(J_{BL})$ as well as $V(J^{II}_{MH})$ under the asymptotic condition (16). It turns out that $V(J^{II}_{MH})$ is a consistent estimator of $\text{var}^{A}\hat{\psi}_{MH}$ and of $\text{var}^{A}(J^{II}_{MH})$. The jackknife variance $V(J_{BL})$, however, is shown to be in general not consistent for $\text{var}^{A}(\ln \hat{\psi}_{MH})$ or for $\text{var}^{A}(J_{BL})$.

This is not surprising since in the asymptotic model (16) or (17) the number of tables remains fixed, which means a fixed number of pseudo-values for the jackknife variance of type I. However, this fact does not matter when looking at the jackknife estimators of type I of ψ, because here the pseudo-values themselves converge to ψ for increasing N and so does the arithmetic mean (5).

In addition, the above authors suggest different confidence intervals of ψ based on jackknife estimators and different variance estimators. These intervals are discussed there from a theoretical point of view as well as w.r.t. their finite–sample behaviour investigated in Monte–Carlo experiments.

Finite–sample results

Monte–Carlo experiments and a numerical study (Pigeot (1991a, b, c)) have been carried out to investigate the finite–sample properties of the above jackknife estimators of a common odds ratio.

For the experimental design (reported in detail in Pigeot (1991b)), a varying number of tables ($K = 2, 5, 10$) with different sample sizes and with a balanced as well as an unbalanced ratio of cases to controls has been considered. The common odds ratio has been chosen as 1, 1.7, 3.5, 5, and 10 in combination with five different sets of p_{0k}, $k = 1, \ldots, K$.

For each of the 240 constellations of the involved parameters, 1000 simulation runs have been carried out comparing $\hat{\psi}_{MH}$, $\hat{\psi}_W$, $\hat{\psi}_{BL}$, J_{MH}^{II}, and $\exp(J_W^I)$, the transformed jackknife estimator of type I based on $\ln(\hat{\psi}_W)$, w.r.t. their sample bias, standard error, and mean squared error (MSE). From the simulation results, a clear dominance of J_{MH}^{II} w.r.t. the bias and MSE can be noticed especially with small sample sizes. The jackknife estimators of type I, originally proposed for the sparse–data model, behave indeed better with an increasing number of tables, while the clear dominance of J_{MH}^{II} gets lost. Furthermore, it should be mentioned that J_{MH}^{II} tends to underestimate ψ. This tendency gets serious if a large number of tables contains only a few observations in the b or c–cells. Thus, it seems as if the bias correction obtained by the second jackknife approach is somewhat too strong.

Concerning the classical estimators $\hat{\psi}_{MH}$ and $\hat{\psi}_W$, this study confirms the results already gained in other Monte–Carlo studies. Roughly spoken, $\hat{\psi}_{MH}$ shows a smaller bias than $\hat{\psi}_W$ (especially with small values of ψ) and $\hat{\psi}_W$ turns out to be preferable w.r.t. the MSE (especially with large values of ψ). The same tendency can be observed for the corresponding jackknife estimators of type I.

A numerical comparison of the asymptotic relative efficiencies (ARE) of $\hat{\psi}_{MH}$, $\hat{\psi}_{BL}$, and $g(J_g^I)$ shows very high AREs over a wide range of situations likely to occur in practice. Since $\hat{\psi}_{MH}$ as well as J_{MH}^{II} are asymptotically efficient for $\psi = 1$, best results can be observed for these estimators with small values of ψ. Similarly to the results of the simulation study, the jackknife estimators of type I behave better for a large number of 2×2 tables. In general, it can be seen that the ARE of $\hat{\psi}_{BL}$ is lower than those of $\hat{\psi}_{MH}$ and of $g(J_g^I)$ in most cases, where in turn that of $\hat{\psi}_{MH}$ is often lower than that of $g(J_g^I)$ for large values of ψ and a large number of tables.

The experimental design briefly described above has also been used in a Monte–Carlo comparison of variance estimators of $\hat{\psi}_{MH}$ and of different confidence intervals (Pigeot & Strugholtz (1992)). Here, only a short sketch of the obtained results is given. It can be stated that $V(J_{MH}^{II})$ shows better results than the empirical version of the asymptotic variance of $\hat{\psi}_{MH}$ derived by Hauck (1979). The variance estimator suggested by Robins et al. (1986), however, outperforms $V(J_{MH}^{II})$ in nearly all situations considered. But the jackknife variance yields good results when used for the construction of confidence intervals.

Concluding remarks

From the review in the second section, it can be seen that in most cases resampling techniques have only been used as aid for deriving e.g. variance estimators. In contrast to this, the jackknife estimators presented in the preceding section are by themselves of interest. Especially the jackknife estimator of type II based on $\hat{\psi}_{MH}$ offers a real alternative to classical noniterative estimators of ψ as it can clearly be seen from the simulation results.

Although the numerical calculation of the jackknife estimators of ψ usually takes less than a second CPU time on a 486/50 PC, calculating for example J_{MH}^{II} is quite burdensome. In addition, resampling techniques are typically not included in standard software packages. Therefore, a computer program has been developed for calculating jackknife estimators of both types based on $\hat{\psi}_{MH}$ and $\ln(\hat{\psi}_W)$. Appropriate variance estimators can also be obtained as well as estimates of ψ based on classical approaches. According to the assumptions underlying the above estimators, the program only allows for a binary risk factor and a binary disease variable. A maximum number of 100 2×2 tables can be handled. In addition to the estimates of the common odds ratio, the program lists all tables involved including the margins combined with the individually estimated odds ratios. The program is written in Turbo Pascal. A compiled version to run on a personal computer can be obtained from the author on request.

Current work in this field concerns on the one hand a revision of the computer program. On the other hand, the asymptotic behaviour of the jackknife estimators and of the jackknife variances for an increasing number of tables is under research. Under the assumptions of such a sparse–data model, the finite–sample performance of these estimators will be investigated by Monte–Carlo experiments.

References

Adams, J. E. & McCune, E. D. (1979). Shannon's index of diversity and the generalized jackknife. In: J. F. Grassle, G. P. Patil, W. K. Smith, C. Taillie, eds. Satellite Program in Statistical Ecology 6, International Cooperative Publishing House, Burtonsville, Md, 117–131.

Aitchison, J.& Aitken, C. G. G. (1976). Multivariate binary discrimination by the kernel method. Biometrika 63, 413–420.

Arvesen, J. N. (1969). Jackknifing U–statistics. Ann. Math. Statist. 40, 2076–2100.

Arvesen, J. N. & Schmitz, T. H. (1970). Robust procedures for variance component problems using the jackknife. Biometrics 26, 677–686.

Bowden, R. J. & Sim, A. B. (1992). The privacy bootstrap. J. Bus. Econ. Statist. 10, 337–345.

Bowman, A. W. (1980). A note on consistency of the kernel method for the analysis of categorical data. Biometrika 67, 682–684.

Bowman, A. W., Hall, P. & Titterington, D.M. (1984). Cross–validation in nonparametric estimation of probabilities and probability densities. Biometrika 71, 341–351.

Breslow, N. E. & Liang, K. Y. (1982). The variance of the Mantel–Haenszel estimator. Biometrics 38, 943–952.

Camilli, G. & Smith, J. K. (1990). Comparison of the Mantel–Haenszel test with a randomized and a jackknife test for detecting biased items. J.Educat. Statist. 15, 53–67.

Cochran, W. G. (1968). Commentary on 'Estimation of error rates in discriminant analysis'. Technometrics 10, 204–205.

Cox, D. R. (1970). The analysis of binary data, Methuen, London.

Efron, B. (1979). Bootstrap methods: Another look at the jackknife. Ann.Statist. 7, 1–26.

Efron, B. (1982). The jackknife, the bootstrap and other resampling plans, CBMS Monogr. No 38 SIAM, Philadelphia.

Efron, B. & Gong, G. (1983). A leisurely look at the bootstrap, the jackknife, and cross–validation. American Statistician 37, 36–52.

Efron, B. & Tibshirani, R. (1986). Bootstrap methods for standard errors, confidence intervals, and other measures of statistical accuracy. Statist. Science 1, 54–77.

Farewell, V. T. (1978). Jackknife estimation with structured data. Biometrika 65, 444–447.

Fleiss, J. L. (1973). Statistical methods for rates and proportions, Wiley, New York.

Fleiss, J. L. & Davies, M. (1982). Jackknifing functions of multinomial frequencies, with an application to a measure of concordance. Amer. J. Epidemiol. 115, 841–845.

Frawley, W. H. (1974). Using the jackknife in testing dose responses in proportions near zero or one – revisited. Biometrics 30, 539–545.

Gladen, B. (1979). The use of the jackknife to estimate proportions from toxicological data in the presence of litter effects. J. Amer. Statist. Assoc. 74, 278–283.

Hauck, W. W. (1979). The large sample variance of the Mantel–Haenszel estimator of a common odds ratio. Biometrics 35, 817–819.

Heltshe, D. & Forrester, N. E. (1983). Estimating species richness using the jackknife procedure. Biometrics 39, 1–11.

Hewlett, P. S. & Plackett, R. L. (1959). A unified theory for quantal responses to mixtures of drugs: non–interactive action. Biometrics 15, 591–610.

Jewell, N. P. (1984). Small–sample bias of point estimators of the odds ratio from matched sets. Biometrics 40, 421–435.

Kraemer, H. C. (1980). Extension of the Kappa coefficient. Biometrics 36, 207–216.

Lachenbruch, P. A. & Mickey, M. R. (1968). Estimation of error rates in discriminant analysis. Technometrics 10, 1–11.

Liou, M. & Yu, L. (1991). Assessing statistical accuracy in ability estimation: a bootstrap approach. Psychometrika 56, 55–67.

Mantel, N. & Haenszel, W. (1959). Statistical aspects of the analysis of data from retrospective studies of disease. J. Nat. Can. Inst. 22, 719–748.

Miller, R. G. (1974). The jackknife - a review. Biometrika 61, 1–15.

Mosteller, F. & Tukey, J. W. (1977). Data analysis and regression, Addison–Wesley Publishing Company, Inc., Reading, Menlo Park.

Mosteller, F. & Wallace, D. L. (1963). Inference in an authorship problem – A comparative study of discrimination methods applied to the authorship of the disputed Federalist papers. J. Amer. Statist. Assoc. 58, 275–309.

Parr, W. C. & Tolley, H. D. (1982). Jackknifing in categorical data analysis. Austral. J. Statist. 24, 67–79.

Pigeot, I. (1990). A class of asymptotically efficient noniterative estimators of a common odds ratio. Biometrika 77, 420–423.

Pigeot, I. (1991a). A jackknife estimator of a combined odds ratio. Biometrics 47, 373–381.

Pigeot, I. (1991b). A simulation study of estimators of a common odds ratio in several 2×2 tables. J. Statist. Comp. Simulation 38, 65–82.

Pigeot, I. (1991c). Asymptotic relative efficiency of some jackknife estimators of a common odds ratio. Biom. J. 33, 305–316.

Pigeot, I. (1992). Jackknifing estimators of a common odds ratio from several 2×2 tables. In: K.–H. Jöckel, G. Rothe, W. Sendler, eds. Bootstrapping and related techniques, Proceedings, Trier, FRG, 1990. Lecture Notes in Economics and Mathematical Systems, Springer–Verlag, Berlin, Heidelberg, 203–212.

Pigeot, I. (1993). Asymptotic properties of several jackknife estimators of a common odds ratio. To appear in: Journal of Applied Statistical Science.

Pigeot, I. & Strugholtz, H. (1992). Jackknife variances and confidence intervals of the Mantel–Haenszel estimator. Forschungsbericht 92/6, Fachbereich Statistik der Universität Dortmund.

Pigeot, I., Schach, E. & Schach, S. (1992). Auswirkungen von Anonymisierungsverfahren auf Risikoschätzungen in epidemiologischen Studien. In: W. van Eimeren, K. Überla, K. Ulm, eds. Gesundheit und Umwelt, 36. Jahrestagung der GMDS, München, September 1991. Medizinische Informatik, Biometrie und Epidemiologie, Springer–Verlag, Berlin, Heidelberg, 52–56.

Plackett, R. L. (1981). The analysis of categorical data, 2nd ed., Charles Griffin & Co Ltd., London, High Wycombe.

Quenouille, M. H. (1949). Approximate tests of correlation in time–series. J. Roy. Statist. Soc. B 11, 68–84.

Quenouille, M. H. (1956). Notes on bias in estimation. Biometrika 43, 353–360.

Robins, J., Breslow, N. E. & Greenland, S. (1986). Estimators of the Mantel–Haenszel variance consistent in both sparse data and large–strata limiting models. Biometrics 42, 311–323.

Salsburg, D. (1971). Testing dose responses on proportions near zero or one with the jackknife. Biometrics 27, 1035–1041.

Simonoff, J. S. (1986). Jackknifing and bootstrapping goodness–of–fit statistics in sparse multinomials. J. Amer. Statist. Assoc. 81, 1005–1011.

Stone, M. (1974). Cross–validation and multinomial prediction. Biometrika 61, 509–515.

Tarone, R. E., Gart, J. J. & Hauck, W. W. (1983). On the asymptotic inefficiency of certain noniterative estimators of a common relative risk or odds ratio. Biometrika 70, 519–522.

Titterington, D. M. (1980). A comparative study of kernel–based density estimates for categorical data. Technometrics 22, 259–268.

Titterington, D. M. & Bowman, A. W. (1985). A comparative study of smoothing procedures for ordered categorical data. J. Statist. Comp. Simulation 21, 291–312.

Tukey, J. W. (1958). Bias and confidence in not–quite large samples. (Abstract). Ann. Math. Statist. 29, 614.

Tuyns, A. J., Péquignot, G. & Jensen, O. M. (1977). Le cancer de l'oesophage en Ille–et–Vilaine en fonction des niveaux de consommation d'alcool et de tabac. Bull. Cancer 64, 45–60.

Wahrendorf, J. (1984). Interaktive Effekte in epidemiologischen, klinischen und experimentellen Studien. Habilitationsschrift. Fakultät für Theoretische Medizin der Ruprecht-Karls-Universität, Heidelberg.

Wahrendorf, J. & Brown, C. C. (1980). Bootstrapping a basic inequality in the analysis of joint action of two drugs. Biometrics 36, 653–657.

Woodward, W. A. & Schucany, W. R. (1977). Combination of a preference pattern with the triangle taste test. Biometrics 33, 31–39.

Woolf, B. (1955). On estimating the relation between blood group and disease. Ann. Hum. Genetics 19, 251–253.

Zahl, S. (1977). Jackknifing an index of diversity. Ecology 58, 907–913.

Statistical Problems in Planning, Conduct and Analysis of Epidemiological Studies

Karl-Heinz Jöckel
Bremen Institute for Prevention Research and Social Medicine (BIPS), Grünenstr. 120, D-28199 Bremen, Germany

Key Words:
epidemiology, validity, sensitivity analysis, Monte Carlo tests, computer science, statistics

Abstract
The contribution of a statistician to an epidemiologic study is often (mis)understood as preparing the statistical analyses of the data. This paper tries to emphasize the important role and responsibility a statistician takes over when he or she gets involved into such a study: starting from the planning phase over data collection to the final statistical analysis, statistical expertise is needed in all phases of the study. This point of view is demonstrated by examples with specific reference to the observational nature of epidemiologic studies. Computerized support taking the interrelationship between the phases into account would be highly desirable. This support should also include the possibility of making sensivity analyses to check the impact of model assumptions, analysis procedures and data specialities on the obtained results.

Introduction

According to a popular definition epidemiology is concerned with the study of the distribution of diseases and their determinants in (human) populations. So the primary units of concern are groups of individuals, not separate individuals. If, however, a statistician is confronted with epidemiologic data, he or she has to realize that the major point of interest is to draw statistical inference from some sample of the population to the general population in which the study has been conducted. The second question that arises then is whether the observed exposure-disease relationship also holds to humans in general, see (Kleinbaum et.al. (1982)) for an extended discussion on internal and external validity. Since most of the epidemiologic studies are observational without any possibility of a random assignment of exposure, valid inference may only be drawn if both, the random component (sample from the population) and the deterministic component (i.e. bias) are adequately addressed. It should be stressed that in most practical situations the latter is of major concern. Bias, that may artificially decrease or increase a disease-exposure association, is mostly introduced during the planning and data collection stage of a study. So the statistician has to engage him- or herself also into this part of a study and not to rely on the possibility of posthoc corrections by statistical methods, since these are only valid if the assumptions of the underlying model are met. As a consequence of this observation the present paper tries to address some of the problems that arise during the three stages of an epidemiologic study: planning, conduct and analysis, although the primary interest lies in statistical analysis. Most epidemiologic studies are relatively large (with usually several hundreds or even thousands of individuals enrolled) and several hundred variables so that computational statistics plays an essential role which also is tried to be covered in this paper.

The paper is organized as follows: after setting up the general framework in the next section, the following sections are devoted to specific problems and possible solutions. The last section contains a discussion and an outlook for further research.

A general framework

Following usual terminology, for a set of individuals from a fixed population, an outcome variable, e.g. disease status D and a set of exposure variables (risk factors) $X_1, \ldots, X_p$ are observed. In the simplest case where there is only a binary outcome ($D = 1$ if diseased, 0 otherwise) and only a dichotomous exposure variable ($E = 1$ if exposed, 0 otherwise), the whole study may be expressed in a 2×2 table where a, b, c, d denote the entries of the table and $n = a+b+c+d$ is the total sample size, see table 1. It should be emphasized that the major study types may be understood in this simple setting:

Cross-sectional study: In this situation only n is fixed in advance. If all individuals are randomly drawn from a population then any reasonable statistical measure of association $r(a, b, c, d)$ correctly reflects the correlation in the population between **reported** exposure and disease status at the time of the conduct of the study. However, this does not necessarily carry over to the true status of exposure, nor does a positive association imply a causal relationship.

	Disease		
Exposure	$D = 0$	$D = 1$	Total
$E = 0$	a	b	$a + b$
$E = 1$	c	d	$c + d$
Total	$a + c$	$b + d$	$a + b + c + d = n$

Table 1: Schematic outcome of an epidemiologic study for dichotomous exposure and outcome variables

Case-control study: In this study type a preassigned number of cases ($D = 1$) and controls ($D = 0$) is sampled, so that $a + c$ and $b + d$ have to be regarded as fixed quantities, whereas the distribution of c given $a + c$ and d given $b + d$ are the random quantities of interest.

Cohort study: In this design a defined population of exposed ($E = 1$) and non-exposed ($E = 0$) individuals is compared with respect to later disease status, so that $a + b$ and $c + d$ are considered to be fixed.

Generalizations to more complex designs ($k \times l$ tables, stratification) seem to be obvious. If the exposure variables are continuous, or there are too many of them, regression approaches may be appropriate, relating the probablity of disease to exposure via

$$P(D = 1|x) = F_D\left(\alpha_0 + \sum_{i=1}^{p} \alpha_i \cdot x_i\right) .$$

A prominent example is the so-called logistic regression with

$$F_D(z) = \frac{1}{1 + e^{-z}} .$$

If the outcome is quantified via time to event (disease or death, loss to follow-up) survival models come into play.

Relevant measures of association in terms of table 1 are the relative risk

$$RR = \frac{d}{c + d} : \frac{b}{a + b}$$

and the odds ratio

$$OR = \frac{a \cdot d}{b \cdot c} ,$$

the former being estimable only from cross-sectional (more precisely as a prevalence ratio) and cohort studies, whereas the latter is a good approximation to the relative risk if we are concerned with a rare disease. It may be estimated from all study types. Except for intervention studies where E may be allocated randomly, all the study types and their generalizations are observational, so that the way the sample is constituted is of major concern.

The planning stage

Given an anticipated outcome disease association to be investigated in a study (we do not want to go into the details of the derivation of such a hypothesis) the appropriate design has to be decided. Whereas in principle computerized support (like knowledge-based systems) might be desirable, in practice the study design has to be chosen on the basis of practical considerations like available number of cases, pre-knowledge of exposure prevalence in the general population or specific subpopulations. For the moment we will assume that the principal decision on the design has been taken. Then the natural question on sample size (which may be critical for the design decision itself) arises. In standard text books easily applicable formulae may be found that relate prevalence of exposure in the general population or incidence of disease, statistical error probabilities and detectable risk differences (or ratios) to required sample size. These formulae usually are derived under simple assumptions like those given in the 2×2 table in the second section. Sometimes deviations from these simple assumptions are allowed like frequency or individual matching in case-control sampling, see e.g. (Gail (1973); Munoz & Rosner (1984); Parker & Bregmann (1986); Fleiss (1988)). If, however, the assumptions of these (extended) models do not fit into the side conditions of the designated design, ad hoc procedures have to be developed to meet the specific situation. Such a situation arose in the planning stage of a case-control study on the impact of air pollution on lung cancer, where matching for sex and age was planned. However, relevant confounding factors like smoking and occupational exposure could not be accounted for by the design but only by the method of analysis. The questions that arose were: (1) Is there a considerable bias by ignoring confounding of the unmatched variables? (2) What is the difference in sample size for different evaluation models depending on the extent of control of confounders, given that the anticipated effect of air pollution is small as compared to that of the potential confounder smoking and occupational exposures?

Details of the theoretical derivations may be found in (Drescher et al. (1990)). For the practical situation where a small risk has to be detected ($RR \cong 1.15$) it turned out that the strength of the association between risk factor and confounders may be neglected whereas the relative risk of the confounders had a high impact on the required sample sizes. Furthermore stratified sampling appeared to be preferable to individual matching.

Even if the question of necessary sample size is clarified, there remains the question of the accuracy of exposure and disease assessment. It has to be recalled that sample size calculations are based on the assumption that the assessments are error free. In practice, however, misclassification (differential and non-differential), has to be anticipated that will bias the study results. It belongs to the task of a statistician engaged in the planning of an epidemiological study to determine the impact of this misclassification. Sometimes it is possible to perform validation studies within the context of the major study. If so, the design has to be established in such a way that the results of the validation can be adequately incorporated into the analysis, see (Schill et al. (1993)) for an example.

Data collection

Having in mind the caveats of the introduction and the preceeding section, data collection is obviously a decisive step in the conduct of an epidemiologic study, since the way the data are sampled and the information is obtained are strongly influential on the outcome of the study. Any unbalanced assessment of exposure and/or disease will definitely result in a bias that in practice is not easily removed by statistical analysis or even detected. This is why quality control and quality assurance of the data collection step requires special attention by the responsible statistician. Areas that may be of special concern include

- digit preference with respect to numerical reading (like blood pressure)
- quality control of diagnostic criteria (like reference pathologists or panel decisions)
- standardization of laboratory measurements
- validity and reliability of interview information (use of tape recorders, re-interviews, next-of-kin interviews, biological markers like cotinine for assessment of smoking status)
- data quality (e.g. double data entry)
- assessment of interviewer drifts (like duration of interviews, number of items addressed by the interviewer)

It has to be stressed that the data collection step is a process involving a lot of activities of different individuals. Consequently, data control is a continuous task over the whole time of data collection, see (Jöckel & Zachcial (1992)) for an example from the German National Health Examination Survey. This also means that quality control has to be understood as a measurement taken in defined time intervals on the data collected so far, where appropriate stepped reaction has to be given to the field staff.

If for example an observer shows a marked digit preference an immediate training program and/or the decision is needed whether he or she has to be removed from the team. The other decision that has to be based on statistical considerations is whether or not the observations obtained so far by this observer will probably be removed from the final analysis so that additional recruitment of individuals is necessary. From the remarks made so far it should have been made clear that in this phase of a study statistical expertise is strongly required for an adequate management of the study, where every decision that is taken has to be regarded with respect to the possible impact on the outcome of the study.

At the end of the data collection step an overall quality assessment is required that decides which part of the data collected so far will enter the final analysis. This decision has to be taken without any reference to the main study hypothesis. As an example consider an observer whose blood pressure measurements are considerably higher as compared to the observers within a region although during the conduct of the study no relevant preference nor any outlying position could be detected.

Statistical analysis

After data collection and all quality procedures have been finished the final data set for analysis is supplied. Of course during the statistical analysis errors will be detected that usually lead to corrections of the data set, e.g. on the basis of the questionnaire. However, at some point in time the data set has to be regarded essentially final and will be the basis of all further analyses. In theory a predesigned statistical procedure will be used to do the final analysis with relation to the main hypotheses. In practice, however, the problem of unforeseen specialities of the data arises:

- a confounder that has not been anticipated at the planning stage has been more or less established by other researchers. This will necessarily lead to an extension of the statistical model
- bias sources are detected during the course of analysis, see as an example (Becher & Jöckel (1992)) for dealing with the problem of different biases for different control groups within a case-control-setting
- missing values play a more important role than previously anticipated
- validation was found necessary during the collection of the data, see (Schill et al. (1993)) for a solution to the two latter problems.

Although at least partially sufficient solutions exist to overcome the above mentioned difficulties, it has again to be stressed that these procedures rely on more or less restrictive assumptions that hardly can be tested in practice.

Consequently avoiding these difficulties or at least most of them in advance would be more desirable.

However, there will remain problems and the statistician will end up with a data set that meets some of the requirements of the theoretical assumptions needed for the designated statistical analysis procedures whereas others are not met.

In this case he/she would like to know which impact the violations of the assumptions might have on the statistical results presented (estimates, p-values, confidence limits etc.). One possible solution to this problem is offered by the use of so-called Monte Carlo or simulation techniques, see (Jöckel (1980)), in order to try to quantify the impact of certain deviations from the underlying model. These techniques of sensitivity analysis seemingly are widely neglected in the statistical literature whereas a specific procedure, namely the so-called Bootstrap (Efron (1979)), attracted some practical and a lot of theoretical statistical considerations. Although developed as a theoretical tool, the practical Bootstrap requires Monte Carlo simulation. It offers the possibility of hypothesis testing and confidence interval estimation and advocates the possibility of model discrimination, see (Hall & Wilson (1991); Becher (1993)) for some controversities on this subject. In essence, however, it turns out that the application of the Bootstrap again requires theoretical knowledge, that presents it as one method among others and not as a general key to all statistical problems presented by a real-life epidemiologic data set.

The Bootstrap is closely related to another method, namely that of Monte Carlo tests, see (Jöckel (1991)), that is based on theoretically much more sound grounds. If we consider a situation where the null hypothesis H_0 has been formulated in statistical terms and a test statistic T (small values of T pointing into the direction of the alternative) is given such that

$$Prob(T \leq t|H_0) \leq P(T \leq t) \tag{1}$$

for one specific P out of the null hypothesis, then an intuitively appealing approach would be to simulate $t_1, \ldots, t_m$ distributed according to $P(T)$ and compare the observed value of the test statistic t with $t_{k:m}$, the kth order statistic of the simulated values, where the significance level is $\alpha = \frac{k}{m+1}$. It may be shown that the test that rejects the null hypothesis if $t \leq t_{k:m}$ is an exact level α-test for H_0 versus the alternative and under weak assumptions has desirable statistical power properties, see (Jöckel (1986)). However, this approach requires the checking of (1) for a valid application.

A simple example may serve as an illustration of the method: in a lung cancer case-control study for all never smokers and all occasional smokers (smokers having smoked less than half a year or less than one cigarette per day) the exposure to passive smoking (environmental tobacco smoke) has been assessed, see (Jöckel (1991a)). The distribution of exposure by sex and smoking status (never/occasional) is given in table 2.

Smoking Status	Exposure[1]	Males Controls	Males Cases	Females Controls	Females Cases
never	NO	24	1	19	8
	YES	14	5	13	11
occasional	NO	20	1	6	2
	YES	12	3	7	2

$OR_{MH} = 2.62 \quad P = 2.1\% \quad P(\text{MONTE CARLO}) = 1.3\%$

Table 2: Lung Cancer Risk[2] of Passive Smoking[3]

[1] Exposure = YES if cumulated lifelong exposures is in upper quartile of the distribution for any source (childhood, spouse, other people at home, occupational, in vehicles, other public places)

[2] Preliminary results from an ongoing case-control study

[3] Persons having never smoked regularly for at least 1/2 year at least one cigarette/day

The Mantel-Haenszel estimate of the common odds ratio is given by $OR_{MH} = 2.62$, the standard cutpoint of a statistical analysis system (SAS) claims a p-value of 2.1%. Conditional on the number of cases and controls within the sex-smoking status strata a Monte Carlo test is easily applied yielding a p-value estimate of 1.3% based on 9999 simulations. Since for simulation sizes of this order the power and efficiency of the Monte Carlo test is generally good, see (Jöckel (1986)), this result may be interpreted as a confirmation of the standard analysis that the effect of passive smoking is significant at the 5% level.

Discussion

If statisticians try to contribute to the interdisciplinary research process in epidemiology, they have to face the challenge that their skills are needed in each step of the process. Very often the non-technical aspects will be of utmost importance and the statistician should try to stand the temptation to overemphazise analytical aspects. Although the question of correct p-values in the example of the preceeding section is certainly important, the overall quality of the case-control study is much more influenced by the degree the wellknown sources of bias (see e.g. Lee (1988); Schlesselman (1982)) could be avoided. On the other hand this observation should not be misinterpreted as an underestimation of the value of correct analyses: What we need is a stringent and flexible concept to plan studies, to collect data, to clean and to analyze them. At the first glance this last sentence seems to be a contradiction in itself. How can we combine stringency and flexibility?

As has already been said several times, real-life research, especially in observational studies, will pose unforseen problems, so flexibility is necessary. Since, however, the management of these problems may be highly influential on the results, the management has to be stringent in the sence that

- similar problems are treated similar
- problems are treated in concordance with other decisions and the design
- the consequence of the management for the next stages of the study are considered, necessary changes to the study design are made.

It seems to be obvious that computerized support for such a flexible and stringent management is highly desirable. In order to improve epidemiologic studies, statisticians, computer scientists and epidemiologists should cooperate to design software systems that could support planning, conduct and analysis. Special emphasis should be given to the interrelation of these steps. Of course the best available statistical procedures should be incorporated, also offering the possibility of sensitivity analyses. This also implies that more research is needed on Monte Carlo statistical inference, see (Jöckel et al. (1992a)) for additional references.

References

Becher, H. & Jöckel, K.-H. (1990) Bias Adjustment with Polychotomous Logistic Regression in matched Case-Control-Studies with two Control Groups. Biometrical Journal 32, 801-816

Becher, H. (1993) Bootstrap Hypothesis Testing Procedures. Biometrics 49, 1201-1204

Drescher, K., Timm, J. & Jöckel, K.-H. (1990) The Design of Case-Control-Studies: The Effect of Confounding on Sample Size Requirements. Statistics in Medicine 9, 765-776

Efron, B. (1979) Bootstrap Methods: Another Look at the Jackknife. The Annals of Statistics 7, 1-26

Fleiss, J. L. (1988) The Theoretical and Practical Effects of Pairwise Matching on the Required Sample Size of a Case-Control Study. Proceedings of the 14th International Biometric Conference. Namur, 18-23 July 1988

Gail, M.H. (1973) The Determination of Sample Size for Trials Involving Several independent 2×2 tables. Journal of Chronic Diseases 26, 669-673

Hall, P. & Wilson, S.R. (1991) Two Guidelines for Bootstrap Hypothesis Testing. Biometrics 47, 757-762

Jöckel, K.-H. (1980) Einige Aspekte zur Beurteilung statistischer Verfahren mittels Simulation. EDV in Medizin und Biologie 11, 83-89

Jöckel, K.-H. (1986) Finite Sample Properties and Asymptotic Effiency of Monte Carlo Tests. The Annals of Statistics 14, 336-347

Jöckel, K.-H. (1991) Monte Carlo Techniques and Hypothesis Testing. The Frontiers of Statistical Computation, Simulation. & Modeling (Volume I of the Proceedings of ICOSCO-1. (The First International Conference on Statistical Computing), 21-42, NELSON, P.R. (ed.) American Science Press

Jöckel, K.-H. (1991a) Passivrauchen - Bewertung der epidemiologischen Befunde. Krebserzeugende Stoffe in der Umwelt Herkunft, Messung, Risiko, Minimierung, Tagung Mannheim, 23.-25. April 1991. Kommission Reinhaltung der Luft im VDI und DIN (VDI Berichte 888). VDI-Verlag Düsseldorf

Jöckel, K.-H. & Zachcial, M.(1992) Studienübergreifende Qualitätssicherung in einer multizentrischen epidemiologischen Studie - Schwerpunkt Surveydaten. In: Schach, S., Trenckler, G. (eds.): Data Analysis and Statistical Inference. Festschrift in Honour of Prof. Dr. Friedhelm Eicker. Verlag Josef Eul, Bergisch Gladbach, Köln

Jöckel, K.-H., Rothe, G. & Sendler, W. (1992a, eds.) Bootstrapping and Related Techniques. Proceedings of an International Conference, Trier, 04.-08.06.1990. Lecture Notes in Economics and Mathematical Systems 326, Springer Verlag Berlin Heidelberg

Kleinbaum, D.G., Kupper, L.L. & Morgenstern, H. (1982) Epidemiologic Research. Wadsworth, London

Lee, P.N. (1988) Misclassification of Smoking Habits and Passive Smoking: A Review of the Evidence. Springer Verlag, Berlin

Munoz, A. & Rosner, B. (1984) Power and Sample Size for a Collection of 2x2 tables. Biometrics 40, 995-1004

Parker, R.A. & Bregman, D.J. (1986) Sample Size for Individually Matched Case-Control Studies. Biometrics 42, 919-926

Schill, W., Jöckel, K.-H., Drescher, K. & Timm, J. (1993) Logistic Analysis in Case-Control-Studies under Validation Sub-sampling. Biometrika 80, 339-352

Schlesselman, J.J. (1982) Case-Control Studies: Design, Conduct, Analysis. Oxford University Press, Oxford

Computer Aided Design of Experiments

Dieter Rasch[1], Paul Darius[2]
[1] Wageningen Agricultural University, Department of Mathematics, Dreijenlaan 4, 6703 HA Wageningen, The Netherlands
[2] Katholieke Universiteit Leuven, Laboratory of Statistics and Experimental Design, Kard. Mercierlaan 92, B-3001, Leuven-Heverlee, Belgium

Key Words:
experimental design, sample size determination, optimal design, factorial design, block design, mixture design, variety trials

Abstract
Software for statistical analysis is far more widespread than software supporting the statistical design of experiments. Yet many separate programs have been developed that deal specifically with design problems, and many of the well-known packages have added modules for experimental design. Here we try to give a comparative overview of about 30 software programs (or sets of programs, or parts of a broader software package) that deal specifically with one or more aspects of the design process.

Introduction

The meaning of "design of experiments", "planning of experiments" or "experimental design" differs from author to author and changed also during the development of statistics. Planning or designing an experiment contains further many non-statistical components. But also for these components the experimental worker needs help. Experimental design may be considered to be the second of two brothers. The "crown prince" (we may use this term even if the father, Sir Ronald, may not be a king but is at least a Lord) or first born, "statistical analysis", is in a very powerful position with many more followers and proud of its mother, probability theory (with grandparents measure theory and sampling theory). Older ancestors such as actuarial theory and governmental statistics are nowadays often forgotten.

Experimental design suffers from all the disadvantages of a second born son and step-child and a non-mathematical mother, variety testing (Fisher (1935)). Further relatives include combinatorial analysis, finite geometry and optimization techniques.

But experimental design is also much more disliked than data analysis even by "experimental workers" [in Sir Ronalds words, Fisher (1925)]. The reason is that an awful lot of questions must be answered about:

- the exact aim of the experiment,
- the precision required or demanded, such as the risks of tests and other decisions
- the universe or population to which the results should apply

before an experiment can be designed thoroughly.

To answer such questions, experimental workers must often make what they consider to be too many important decisions during the pre-experimental phase. It is much easier to define the experimental question after the experiment has been performed and all the data are available or even have been analyzed, and this means neglecting all the design problems. Yet, a careful consideration of the design options is a necessity. Bad design decisions can (and often do) lead to insurmountable problems in the analysis.

In spite of the importance of the design phase, currently software support for design cannot be compared to software support for analysis. This situation is changing. In this paper we try to give an overview of what kind of software is currently available. We hope that the growing availability of software tools for design will be an incentive to experimental workers to pay more attention to the design phase of their experiments.

Usually the process of gaining knowledge in empirical sciences consists of the following steps:

I Formulation of the problem,

II Definition of the precision required

III Model selection,

IV Designing an experiment

V Performing the experiment

VI Analyzing the experiment

VII Interpretation of the results

The order is not necessarily strictly sequential: there might be some iteration between steps I to IV, and furthermore there is a natural break between step IV and VI. The steps are clearly not independent: each of these steps should be performed based on the outcome of the preceding steps. Moreover, some of the steps may be given a rather similar structure: it has been argued that each kind of analysis has or could have its counterpart in the planning and design phase (Rasch (1992), O'Brien & Muller (1992)). There is one branch in statistics where the designing and data processing phases are interrelated as well in the theory as also in computing - sequential design (or analysis). Until now very few program packages assist sequential sampling and offer good stopping rules. We will not include this type of software in the review and only mention the PEST package (for information contact Whitehead), the program EaSt (for information see Cytel) and a program developed by Schneider (1992) called TRIQ (see BIORAT).

Although there is an interrelation between the steps, this is not adequately reflected in current software. Some steps (e.g. I and III) hardly have any software support, other steps are supported with (sometimes) very different software tools. Support for some of the steps in many of the well known general statistical packages is marginal or lacking. Outside the packages, the software tools that exist are very different, despite the fact that they are supposed to feed their results to each other. The tools that take a more integrated approach are generally based on the expert and dialogue system paradigm : they contain a certain amount of non-computational statistical knowledge and use it to assist the user in one of the following : selection of appropriate techniques and options, chaining of subtasks, interpretation of the results (see the section on expert and dialogue systems).

Branches of experimental design

Before an experimental research program can be started, the problem should be formulated in the language of that science in which the experiment has to be performed. If an optimization of the experimental work and a statistical analysis are desirable, a statistical model has to be selected first. This usually is done in several steps. General questions, like :

- what are the factors influencing a result, are there noisy (environmental) factors?
- what are the responses to be observed?
- are the factors qualitative or quantitative?
- what is the experimental region (range of factors)?

- how should noisy factors be taken into account (blocking, covariables)?
- what are the experimental units?

and others have to be answered.

In factorial experiments the number of factors should be as small as possible, on the other hand no factor with possible influence should be forgotten. Even if such questions are not of a statistical or mathematical kind, it would be good to have software which helps young researchers to use experience of others and not to forget important questions. This software can, in addition to its other roles, play the role of a check-list.

In the sequel a rough classification of methods and branches of experimental design is given. To save space we use **M**: for a short description of the mathematical methods used in a branch and **C**: to denote the column in table 1 under which the corresponding software is mentioned.

Construction of designs

Most programs deal with one or some of the following groups of designs:

- balanced incomplete block (BIB) designs
- partially balanced incomplete block designs (PBIB)
- factorial designs
- central composite designs
- row-column-designs
- mixture designs

M: Combinatorial analysis (finite geometries, Galois theory, difference sets, orthogonal arrays) but also search procedures.
C: CONSTRUCTION

Sample size determination

Determining minimal sample sizes for given precision requirements including investigating the robustness of the sample sizes against violations of the model assumptions. The corresponding software often offers the possibility to calculate the precision (a.o. power) for a maximum possible sample size.

M: Techniques used are based on knowledge about non-central distributions and asymptotic properties of special statistics.
C: SAMPLE SIZE

Optimum choice of the design structure in linear and nonlinear models

The branch includes the optimum choice of design matrices i.e. the optimum allocation of observations in linear but also in nonlinear regression or analysis of variance but also the optimum choice of factor levels and subclass numbers in model II and in mixed models of analysis of variance and covariance analysis.

M: Solutions by analytical methods for discrete and continuous (D-, A-, G-, E-) optimum designs; Integer nonlinear optimization, Search procedures (for instance Mitchell's (1974) Detmax) for exact designs.
C: ALLOCATION

Joint optimization of the size and the structure of an experiment.

If an overall risk function is defined the separate determination of the optimum structure (for instance of the D-optimum design for a given size n of the experiment like in Kiefer's (1959) theory) and the minimum size n of the experiment is inadequate.

M: Functional analysis, nonlinear discrete optimisation.
C: Not especially mentioned, only WACH and NREG are in this group.

All these branches should also be represented in design-software packages. While the statistical packages for data analysis cover a wide range of the statistical methodology and some like SAS or Statgraphics offer also programs for constructing factorial designs, the design-software is often restricted to special areas. Those design packages covering several areas are listed in the last column (SEVERAL) of table 1.

Kinds of software investigated

There are several kinds of statistical software as well in design as in analysis. The spectrum goes from software solving a very specific problem like "determination of the sample size for comparing two means" to software for both analysis and design of experiments in the form of very large packages. Design software is under rapid development so that we probably missed some programs. The authors will be thankful for information about further programs.

We define the following groups for the design area :

Special programs:
i.e. programs solving small special problems.
Abbreviation: SPEC

Programs solving special types of problems:
i.e. programs solving some problems of the same kind like constructing BIB
Abbreviation: TYPE

Programs for a design area:
i.e. programs solving globally the problems of one of the groups defined in the section on branches of experimental design
Abbreviation: AREA

Description of the software mentioned in table 1

What follows is a short description of some programs and packages classified in table 1. Each description contains one or more references and/or the address of persons or distributors are given who can help with more information. If not otherwise stated the software runs on IBM PC or compatibles and is written in English.

Software for modelling and formulating precision requirements

DATACHAIN is a pre-processing package. It provides a fullscreen, menu-driven facility for the entry of experimental information and creation of data collection forms. DATACHAIN also acts as a front-end to the statistical packages SAS and GENSTAT. [Williams].

ECHIP is a software tool for the design of experiments. A proven training program in this methodology is also available. Using ECHIP one can examine the effect of over 20 variables to determine those which most strongly control the desired outputs. [Koch & Haag (1992)]

ANOV is, as a part of CADEMO (see the description of CADEMO in the section on expert and dialogue systems), menu driven and allows the choice of the appropriate classification and model in the analysis of variance as well as sample size determination. [BIORAT]

Branch of Experimental Design			
Kind of prog.	MODELLING	PRECISION	CONSTRUCTION
SPEC	DATACHAIN ECHIP		BOX-B CAMOS FACTORIAL-DESIGN RANCODE
TYPE	ANOV EXPERIMEN- TAL DESIGN NREG WACH	LEDA PREC WIBI	ALPHA ALPHAGEN APO DESIGN-EASE DESIGN- & TAGUCHI-KIT FACTEX KEYFINDER MIXTURE-DESIGN OC-STAT
AREA	LPRO	CADEMO	ANLA (KBUB, FAK2,ZEZU) Design Expert DESIGN OF EXPERI- MENTS in MINITAB DESIGN OF EXPERI- MENTS in Statgraphics DSIGNX Experimental Design of Statistica FIEL (FEVE) GENEME GOSSET Industrial Experimental Design of SPlus NAM RS/DISCOVER SOLO Experimental Design STATITCF

Table 1: An overview of the packages discussed in this paper

Branch of Experimental Design			
Kind of prog.	ALLOCATION	SAMPLE SIZE	SEVERAL
SPEC	ECHIP LIRE	DESIGN POWER ECHIP N PLANUNG POWER	
TYPE	NREG OPTEX	AUWA EGRET SIZ EX-SAMPLE LEDA NSURV RANKSEL SOLO Power Analysis STAT-power	
AREA			CADEMO DAEDALUS DEXPERT DEXTER STAVEX

Table 1: An overview of the packages discussed in this paper (continuation)

EXPERIMENTAL DESIGN is an expert system program using decision analysis techniques to help researchers to determine which of 17 types of experimental designs are most appropriate for their research projects. Recommended designs include comparison of means, Latin Squares, factorial, fractional factorial, saturated fractional factorial, hierarchical, Plackett-Burman, central composite, Box-Behnken, simplex mixture and fixed size sequential simplex designs. [Koch & Haag (1992)]

WACH, a part of CADEMO, provides a model choice between 10 growth functions using one of four possible criteria. A more general analogue program for model choice in regression analysis is NREG. (Rasch et al. (1992))

LPRO is a module complex of CADEMO. It provides a menu-driven facility for finding the appropriate statistical model and procedure for a given experimental or observational problem. The dialogue can be arranged at different levels and help functions can be invoked to show explanations of unknown terms on the screen. For instance, the discrimination between "Regression Analysis" and "Contingency Table Analysis" can either be done directly by people knowing these branches or by others using several practical examples given in the menu and comparing them with their own problem. [Rasch, Guiard & Nürnberg (1992)].

The formulation of precision requirements is sustained by the module PREC of FIEL (see the separate description of FIEL), a part of CADEMO, by the module complex WIBI of CADEMO which offers precision requirements for designing experiments in bioassay and by LEDA (another part of CADEMO) in the field of statistical quality control and survival problems.[BIORAT]

Software for constructing designs

In this chapter different types of programs are described. There are programs constructing special designs, there are packages for the construction of all designs needed in an area and there are further parts in large analysis packages mainly for the construction of factorial designs.

Programs for constructing special designs

BOX-B is a pair of computer programs using Box-Behnken experimental designs to investigate three-factor and four-factor response surfaces. BOX-B is based on three-level designs that can detect and describe curvature effects that two-level designs cannot. It gives the user information on the linear and quadratic effects as well as two-factor interactions. It generates and prints out a specific design. [Koch & Haag (1992)]

CAMOS is a program to construct optimal factorial designs for optimality criteria defined by the user. [Osyczka (1992)]

FACTORIAL-DESIGN implements classical full factorial designs that are often used to (1) determine the most significant factors, (2) model the influence of the factors on the response, and finally, (3) choose the optimum combination of factor levels. FACTORIAL-DESIGN is useful for acquiring data to fit a first order polynomial model with all two-factor interaction terms. Factorial designs works with both qualitative factors and quantitative factors. [Koch & Haag (1992)]

RANCODE is a special purpose program for preparing random code lists, sealed envelopes and stick-on labels for the randomisation of patients. Randomisation can be done for up to 20 centers with up to 16,000 cases in one run; lists may be stored for editorial work. [IDV]

Programs for constructing special types of designs

ALPHA is a design generation package and a major upgrade to the package ALPHAGEN. The package constructs efficient experimental designs for applications such as field variety trials, laboratory and glasshouse experiments. There are options for incomplete block designs, row-column designs and latinized designs. Randomisation is carried out appropriate for the particular blocking structure. The package has facilities for restricted treatment randomisation (when the treatments have a nested structure) and the production of resolvable designs with unequal block sizes. Extra facilities include options to

- construct latinized alpha designs for use with contiguous replicates
- construct alphalpha designs to cater for two-dimensional within replicate variation
- search for near optimal row-column designs
- impose a nested treatment structure into the randomized field plan of alpha and alphalpha designs
- construct randomized field plans using treatment labels rather than a number sequence
- generate designs with unequal block sizes for situations where the number of treatments is not a multiple of the block size.

ALPHA also produces a log file containing details of the interactive session. [Williams]

ALPHAGEN is a package for the construction and randomisation of alpha designs and row-column designs. These designs are useful for situations where within replicate control of experimental variation is required, such as field variety trials.

The class of resolvable incomplete block designs called alpha designs was developed because of the lack of suitable designs for use in field variety trials. Square and rectangular lattice designs were available for v = sk treatments in r replicates and incomplete blocks of size k = s or s-1 respectively, but designs were required for other values of k. [Talbot]

APO is a procedure for the construction of special factorial designs with a reduced number of experimental units. [SYSTEGRA]

DESIGN-EASE provides the construction and analysis of 2 level factorial, fractional factorial and Plackett-Burman designs. [STATEASE]

Further we should mention DESIGN-KIT and TAGUCHI-KIT. [for both contact Hasselaar]

FACTEX is a SAS/QC procedure and provides the construction of orthogonal designs (especially factorial designs) possibly within blocks. The number of levels per factor is arbitrary (≥ 2) and may differ from factor to factor. In the case that the design is symmetric (equal numbers of factor levels for all factors) for a given resolution the smallest possible design is constructed. Randomisation is offered. [SAS]

KEYFINDER is a menu-driven Prolog program for generating, randomizing and tabulating factorial designs in general situations. It uses search procedures to generate fractional-replicate and blocked designs meeting the user's detailed a priori specifications vis-a-vis design dimensions and aliasing/confounding properties. [Zemroch (1992)],[Zemroch]

MIXTURE-DESIGN is a pair of computer programs that aids researchers in setting up and analyzing three- or four-component mixtures with a minimum of 7 or 11 factor combinations, respectively. MIXTURE-DESIGN leads the user through the process of setting up the design by quickly eliciting the user-specified overall constant (the 100% constraint), names of the factors, and their low and high levels. It then generates factor combinations which include vertex, edge, and centroid points. [Koch and Haag (1992)]

OC-STAT provides the construction of simple factorial designs for different alias-structures. [SYSTAT]

A program developed by Lorenz (1990) can be used to construct optimum mixture designs.

Programs for the construction of several types of designs

The programs described in this section are listed in row 3 of the third column in table 1.

ANLA is a dialogue system (a part of CADEMO) for the construction of block and factorial designs and randomisation. It contains five module complexes :

- KBUB for constructing BIBs
- TBUB for constructing PBIBs
- ZUZU for random assignment of treatments to experimental units and randomisation of designs constructed by other modules
- FAK2 for constructing factorial and fractional factorial designs for predetermined confounding structures
- ZEZU for constructing central composite designs.

ANLA is available in English and German. [BIORAT]

The KBUB module of ANLA provides the construction of balanced incomplete block designs (BIB) with v treatments and block size k with the minimum number r of replications for $2 \leq k \leq v-1$ and $v \leq 25$.

The FAK2 module of ANLA provides the construction of complete or fractional factorial designs with up to nine factors at two levels each. The user specifies which effects (main effects or interactions) could be confounded and obtains then the corresponding design. The user may want to estimate:

either all existing interactions individually (this leads to a complete 2^p-design)

or all threefold and lower interactions individually

or all twofold interactions individually

or not all twofold interactions individually, but all these separately from the main effects

or not all main effects separately from the twofold interactions

Plackett-Burman and Hartley designs are also offered. [Malig (1992)]

ZEZU is another module of ANLA and provides the construction of central composite designs (CCD). For a specified number of factors up to nine a choice between ordinary CCD, rotatable CCD, orthogonal CCD and orthogonal and rotatable CCD is possible. If for instance a CCD with 9 factors has to be constructed a minimum of x experimental units is needed (x = 83 for an ordinary, x = 147 for a rotatable, x = 200 for an orthogonal and rotatable CCD) [Malig (1992)].

DESIGN-EXPERT provides the construction of response surface designs as well as mixture designs. [STATEASE]

The MINITAB package includes facilities to construct factorial designs and response surface designs, and to perform analysis of means. [MINITAB]

DESIGN OF EXPERIMENTS is a module in release 5 of STATGRAPHICS. It allows the construction of 2^p (fractional) factorial designs for $2 \leq p \leq 16$ and central composite designs with p factors for $2 \leq p \leq 8$. Further central composite, Box-Behnken and Plackett-Burman designs can be constructed. [STATGRAPHICS]

In the agricultural research of the UK DSIGNX is used. It provides the construction of most of the designs needed in variety testing and other fields such as

- randomized blocks
- incomplete block designs
- Latin Squares and other row-column designs
- split plot designs

- factorial design (with and without confounding and with and without blocking).

[A. Mann].

The Experimental Designs module in STATISTICA/w contains facilities to design: (1) two-level fractional factorials, (2) central composite designs, (3) variations of Latin Squares, and (4) Taguchi robust designs. All designs can be randomized, saved to a file, or printed (as data entry forms). [StatSoft]

FIEL (FEVE) is a part of CADEMO. FIEL (Field experiments) contains several modules :

- PREC is a module which helps to find the precision requirements for the problem (as the minimum difference of practical interest and the risks of a test) and derives the number of experimental units needed to fulfil these requirements.

- UNIT helps to determine the size of the plots and the replications needed to fulfil the precision requirements determined by PREC. The other modules (DSGN, FACT, IBLO, STAN) help to construct specific designs with the number of replications calculated with UNIT.

- STAN constructs a standard design, DSGN a completely (unrestricted) randomised design.

- FACT provides the construction of Latin Squares, Latin Rectangles and of split-plot designs with one, two or three treatment factors each. The number of treatments or factor level combinations is limited to 22, the number r of replications to 12.

- IBLO constructs incomplete block designs with a maximum of 200 treatments.

- FARM assists in planning a field trial under practical farming conditions.

FIEL is available in English and German. [Harpke (1989), Thomas und Schrumpff (1989a,1989b), Verdooren & Rasch (1992), Dörfel (1993)) [BIORAT]

GENEME is a program package for the construction of two- and three-level factorial designs, asymmetrical factorial designs, central composite designs, mixture designs and lattice designs. It is written in French. [Mathieu]

GOSSET is a general-purpose program for constructing experimental designs. Variables may be continuous or discrete, qualitative or quantitative, e.g. range over spheres and cubes or in regions defined by linear equalities or inequalities. General models can be handled. The program searches for I-, A-, D- or E- optimal designs, or the same given that one experiment will be lost; or for a packing if no model is available. The errors may be correlated or uncorrelated. Some design points can be specified in advance, so recursive designs can be obtained. A library of several thousand designs is included. The program runs under UNIX. (Hardin & Sloane (1992), (1993)], [Sloane]

The Industrial Experimental Design module of S-PLUS provides the construction of (fractional) factorial designs, central composite and Taguchi-type designs.[Statistical Sciences]

We also mention here a couple of programs elaborated in Australia by N. Nguyen in collaboration with E. Williams. Even if they are not united in a package we list them under the heading NAM in table 1 because they have a common structure and a common author (see Nguyen in the address list).

NAM consists of:

- RIB: A program for constructing resolvable incomplete block designs with up to 100 treatments.
- BIB: A program for constructing optimal resolvable/nonresolvable incomplete block designs with up to 100 treatments.
- CIB: A program for constructing cyclic block designs with up to 256 treatments. [Nguyen & Miller (1992)]
- RRC: A program for constructing optimal resolvable row-column designs. RRC is also a module of ALPHA (see the separate description of ALPHA).
- CAR: A program for constructing optimal nonresolvable row-column designs. [Nguyen & Williams (1993)]

RS/DISCOVER software helps the user to determine the objective of the experiment and define the important experimental conditions. The system supports a screening objective, for determining which factors significantly affect responses, and a Response Surface Modeling objective, for determining functional relationships between factors and responses. Experimental conditions include characteristics of factors and responses, type of experimental design, class of factor/response relationship to consider, and desired number of experimental runs. Throughout the definition process, RS/Discover recommends appropriate choices to assist the user. During the design and worksheet generation stage, RS/Discover creates an experimental design suitable for the user-specified experimental conditions. The design may be a classical screening or response surface design, or it may be a computer-generated optimal design. [BBN]

SOLO Experimental Design helps with the construction of the same types of design as in the S-PLUS module. [BMDP]

For STATITCF, a package which is available in French and Spanish, information is available from Tranchefort.

Software for optimum allocation (constructing optimal designs)

ECHIP provides the construction of D-optimal designs. [Koch & Haag (1992)]

LIRE is a program for designing experiments in linear regression models $y = \beta_0 + \beta_1 x + e$ with one cause variable. It gives the possibility to calculate:

- D-optimal designs for estimating the vector (β_0, β_1) taking costs into account
- sample sizes for estimating β_1 (point or interval estimation)
- G-optimal designs taking costs into account and minimal sample sizes
- sample sizes for testing hypotheses for β_1 and for comparing two regression lines. [BIORAT]

NREG (a part of CADEMO) provides a model choice within a subset of ten isotonic non-linear regression functions by one of four model choice criteria. For each of the functions a module for design (and analysis) is available. Beside determining the minimum sample size for an optimal design, (locally) optimal designs can be constructed for minimizing

- the determinant of the asymptotic covariance matrix
- the asymptotic variance of the estimator of any function parameter

For a given design the values of any optimality criterion can be calculated and compared with that of the optimal design. [BIORAT].

OPTEX is a SAS procedure and generates designs based on A-optimality or D-optimality. OPTEX is not part of SAS/STAT, but of the quality control product SAS/QC software. [SAS]

Software for sample size and power calculation

There are several programs calculating sample sizes or the power of a test for a given size. Most of them are limited to one- or two-sample problems for estimating or testing means (often for normal distributions) such as
DESIGN POWER [Sayn & Merkle (1989)],
ECHIP [Koch & Haag (1992)],
N [IDV, Statistical Cons. Inc.],
PLANUNG [DKFZ],
POWER [Sayn & Merkle (1989)].

Other programs calculate sample sizes in a broader set of situations. They are described in the remainder of this section. There are only a few programs for the sample size determination for selection procedures. We describe two commercially available programs: RANKSEL and the module SELE (AUWA) of CADEMO.

Some of the previously discussed software, e.g. UNIT, LIRE and NREG, include facilities for calculating power.

AUWA determines sample sizes for the indifference zone formulation of the selection problem for the following cases with independent populations:

- selecting the $t < a$ largest of the means from a normal populations with common and known variances,
- selecting the largest of the means of a normal populations with common, unknown variances,
- selecting the largest of the probabilities of a binomial distributions [BIORAT].

EGRET SIZ provides sample size and power for a special set of nonlinear regression models. The package emphasizes ease-of-use and sophisticated models. It is menu-driven and screen-oriented, with online context-sensitive help available at all times and free technical support. The statistical models are principally for survival analysis and discrete outcome variables, including logistic regression for unmatched case:control studies, logistic regression, Poisson regression with person-time data, and Cox proportional hazards regression, for which one can specify hazard, censoring, and accrual functions, test for effect (using dummy variables) or for trend across levels. It helps to fit a generic 'sample size versus power' curve to the Monte Carlo results to provide empirical estimates. [Koch & Haag (1992)]

EX-SAMPLE computes sample sizes for about 60 types of analyses. It also includes procedures to adjust for response rates, contamination, design effects, multiple comparisons and other methodological issues. [Idea Works]

LEDA provides sample size determination for survival problems based on the exponential and the Weibull distribution. [Schmidt (1993)]

NSURV is a program for planning and interpretation of studies concerning sample size and related topics. For time-to-occurrence data it offers: Different target parameters: sample size N, beta, alpha, smallest detectable difference. Two alternatives: one-sided and two-sided. Situations: test for difference and equivalence. Different specifications of difference: either difference or quotient, lambda (hazard), median, mean, proportion. Various group size ratios; balanced and unbalanced. OC-values available where reasonable. Different models : observation to the time-to-occurrence, observation to time T, common to all units, suspended recruitment and observation to time T, different for each unit, suspended recruitment to time T and observation to $R > T$, different for each unit. [IDV]

RANKSEL it its most recent version allows the determination of sample sizes for selecting the $t < a$ largest means from a set of $a < 999$ means of independent normal populations with common and known variances for the indifference zone selection problem (Edwards (1985)), [Edwards].

SOLO power analysis offers sample size and power calculations for testing hypotheses about means (one- two- and k-sample problems), probabilities, correlation and regression coefficients. [BMDP]

STAT-Power is a program for sample size determination and power calculation for testing hypotheses about means, parameters of regression models and probabilities and for constructing confidence intervals for these parameters. [iec ProGAMMA]

Expert and dialogue systems for several branches of experimental designs

We don't want to discuss here whether a special software package is an expert system or not. Two of the packages discussed here have been developed by collaboration of one of the authors of this article each. Related references are Grüger & Ostermann (1989), Nachtsheim et al. (1990), Govaerts et al. (1992), Prat et al. (1992).

CADEMO

CADEMO was and is developed by several agricultural and industrial research centres and universities. This results in several sub-systems for special fields of applications, e.g. statistical quality control, population genetics, field experiments, survival analysis, growth curve analysis or bioassay but also in systems covering special statistical branches like Analysis of Variance, Regression Analysis, Selection Procedures, Confidence Estimation and

Hypothesis Testing. CADEMO provides the model choice as well as design of experiments and gives hints for the analysis. It sustains especially the steps I - IV mentioned above.

The model selection step with the dialogue system CADEMO involves the interactive selection of a suitable statistical model and a method of analysis for a specific problem from a set of possible models or methods. CADEMO can be used to plan a statistical experiment. This includes in particular

- construction of design as:
  - balanced incomplete block design
  - factorial designs
  - central composite designs
- sample size determination (for more than 400 specific problems) and inferences concerning means, variances, probabilities, in Bioassay, in quality control, in population genetics, in survival testing a.o.
- optimum choice of design matrices in linear and nonlinear models (optimum allocation of observations in linear and nonlinear regression)

CADEMO is a modular system, consisting of a guiding program and several module complexes like WIBI (for Bioassay), POPG (for population genetics), FIEL (FEVE) for field experiments, LEDA [for statistical quality control and renewal (survival) problems], ANLA, MIWA, AUWA, NREG, WACH and VARZ. Some of these have already been described above. CADEMO includes explanation components. For instance, if at any stage of the dialogue the user comes across an unfamiliar term, he/she can call a dictionary on the display. Furthermore one can call a list of literature of the statistical background. The path of the dialogue is documented and one can track back the system to any point of the dialogue and make new decisions. After the CADEMO-Session one can get a protocol with the path and the results of planning the experiments. The module complexes are available in English, German and some of them in Spanish. [Rasch et al. (1992)] [BIORAT]

DAEDALUS

DAEDALUS is a system running within the SAS environment. It offers integrated and partly knowledge-based support for the steps I to VII mentioned in the introduction. It is built around the TAXSY inference engine [Darius (1990)]. Its scope is limited to relatively simple experiments, and it is targeted towards agricultural researchers. The current prototype has a user interface in Dutch, but a multilingual version will soon be available for SAS under Windows. (Darius et al. (1990))

DEXPERT

DEXPERT is an expert system, built using KEE and SAS, for the design and analysis of experiments. It can deal with quite complicated types of designs, and also assists in data collection, analysis and interpretation. (Lorenzen et al. (1992)).

DEXTER

Dexter is a guide to selecting the best design for an industrial screening experiment. It searches through a catalogue of designs and compares the user's requirements to design properties based on the number of experimental factors, the number of three-level versus two-level factors, the number of two-factor interactions, and the sample size. Each design is given a score based on its properties and the top four designs and their properties are reported to the user. Specific patterns of two factor interactions are checked against designs in the catalogue using a graph isomorphism algorithm. [Haaland et al. (1990)] [Haaland]

STAVEX

STAVEX is an expert system, running under MS Windows, for the design and evaluation of experiments. It contains the knowledge of experts in industrial statistics, and was designed to reduce experimental effort by means of strategically planned experiments. It integrates statistical experimental design, statistical evaluation of results, and guidance to follow-up designs. [Flühler]

Summary and conclusions

After defining the task and the branches of experimental design we described programs which provide some assistance to experimenters in designing their experiments. We only described the contents of the programs but did not evaluate them.

We find that there is now a substantial amount of new software tools available. The area of software support for experimental design has clearly been recognized as lacking the necessary tools. Unfortunately the programs available are relatively scattered, some of them offer help only for very limited situations.

Finally we propose a list of properties which could serve as a guide-line for developers of future tools.

Experimental design software should:

- closely be connected with the modelling phase
- use a unified and generally accepted terminology
- provide the results of the design phase to the analysis phase
- not only consider single aspects of designing an experiment (integrate power analysis in the construction phase)
- be written in a language close to the experimenter
- provide facilities to take into account typical resource constraints (cost-optimal designs).

Acknowledgement

The authors thank Mr. B. Schäfer from STATCON for helpful comments. We also thank the participants of the 'Concerted Action' "Harmonizing the aims and objectives of quantitative methods in agricultural research and production with emphasis on environmentally sensitive farming system for optimizing and justifying decision making processes" of the European Community research programme for information about programs.

References

Darius, P.L. (1990) A toolbox for adding knowledge-based modules to existing statistical software. Ann. Math. and Artific. Intell-2, 109-117.

Darius, P.L., Duchateau, L. & Nys, M. (1990). A knowledge-based environment for the statistical management of experimental data. Compstat, Short Communications, 63-65.

Dörfel, H. (1993). Biometrie im Feldversuchswesen von Deutschland in: Geidel, H. & Lorenz, R.J. 40 Jahre Biometrie in Deutschland, Landwirtschaftsverlag Münster-Hiltrup, 49-63.

Edwards, H.P. (1985). RANKSEL – An interactive Computer Package on Ranking and Selecting Procedures. in: Dudewicz, E.J. (ed.) The Frontiers of Modern Statistical Inference Procedures, Amer. Sciences Press, Columbus, Ohio.

Fisher, R.A. (1925). Statistical Methods for Research Workers, Oliver and Boyd, Edinburgh

Fisher, R.A. (1935). The Design of Experiments, Oliver and Boyd, Edinburgh

Govaerts, B., De Toffol, P. Dutron, Y. & Florins, P. (1992). SEPSOL: a knowledge-based system for experimental design. In: Koenig, S. (ed) Computational Statistics Vol 3, 31-36. Presses Academiques Neuchatel.

Grüger J. & Ostermann, R. (1989). Construction and Integration of a Statistical Expert System for Binomial Experiments, Statistical Software Newsletter, 12, 124-128

Guiard, V. & Rasch, D. (1990). Das Expertensystem CADEMO: Computer Aided Design of Experiments and Modelling. Zeitschrift für Klinische Medizin 45 15, 1349-1351.

Guiard, V (1992). Designing biological assays by means of the module complex WIBI in: Faulbaum, F. (ed.) Softstat '91 - Advances in Statistical Software 3, Fischer Verl. Stuttgart, 285-292.

Haaland P.D., Lusth J.C., Liddle R.F. & J.W. Curry (1990). DEXTER: a guide in selecting the best design for an industrial screening experiment. Ann. Math. and Artific. Intell-2, 179-197.

Hardin, R.H. & Sloane, N.J.A. (1992). GOSSET: a general-purpose program for constructing experimental designs. ATT Bell Labs, Statistics Report, 110.

Hardin, R.H. and Sloane, N.J.A. (1993). A new approach to the construction of optimal designs. J. Statist. Plann. Inference. In press.

Harpke, K. (1989). Installation und Benutzeroberfläche von CADEMO-FEVE. Feldversuchswesen, Berlin, 6, 24-36.

Kiefer, J. (1959). Optimum experimental designs. Jour. Roy. Statist. Soc. B. 21, 272-319.

Koch, A. & Haag, U. (1992). The Statistical Software Guide '92/'93. Statistical Software Newsletter, Comp. Statist. & Data Analysis 15, 241-262.

Lorenz, G. (1990) Ein Programm zur Konstruktion von Mischungsplänen. Vortrag auf der 27. Sitzung der Interessengemeinschaft Mathematische Statistik der Deutschen Statistischen Gesellschaft, Ahrenshoop, 5-9 -11-1990.

Lorenzen, T.J., Truss, L.T., Spangler, W.S., Corpus,W.T. & Parker, A.B. (1992). DEXPERT: An expert system for the design of experiments. Statistics and Computing 2, 47-75.

Malig, H.J. (1992). Interactive construction of central composite designs for response surface modelling. In: Faulbaum, F. (ed.) Softstat '91 - Advances in Statistical Software 3, Fischer Verl. Stuttgart, 293-300.

Mitchell, T.J. (1974). An algorithm for the construction of "D-optimal" experimental designs. Technometrics 16, 203-210.

Nachtsheim, C.J., Johnson, P.E., Kotnour, K.D., Meyer, R.K. & Zualkernan, I.A. (1990). Expert Systems for the design of experiments. In: Gosh, S. (ed), Statistical design and analysis of industrial experiments. New York: Marcel Dekker.

Nguyen, N.K. and Miller, A.J. (1992). A review of some exchange algorithms for constructing discrete D-optimal designs. CSDS 14, 489-498.

Nguyen, N.K. and Williams, E.R. (1993). Construction of row-column designs by computers. CSIRO-IAPP Biometrics Unit Report BU-9301N.

O'Brien, R.G. and Muller, K.E. (1992) . Unified Power Analysis for t Tests through Multivariate Hypotheses. In: Edwards, L.K (ed), Applied Analysis of Variance in the Behavioral Sciences. New York: Marcel Dekker.

Osyczka, A. (1992). Computer Aided Multicriterion Optimization System (CAMOS). Int. Software Publ. Krakow.

Prat, A., Catot, J.M., Lores, J., Galmes, J., Riba, A. & Sanjeevan, K. (1992). Construction of a Statistical KBFE for Experimental Design Using the Tools and Techniques Developed in the Focus Project. In: Dodge, Y. & Whittaker, J. (eds), Computational Statistics Vol 2, 245-250. Physica-Verlag.

Rasch, D., Guiard, V., Nürnberg, G., Rudolph, P.E. & Teuscher, F. (1987). The Expert System CADEMO. Computer Aided Design of Experiments and Modelling. Statistical Software Newsletter 13, 107-114.

Rasch, D., Nürnberg, G. & Busch, K. (1988). CADEMO - Ein Expertensystem zur Versuchsplanung. In: Faulbaum,F. und Uehlinger, M. (Hrsg.). Fortschritte der Statistik-Software 1, Gustav Fischer, Stuttgart, 193-201.

Rasch, D., Guiard, V. & Nürnberg, G. (1990). Present and Planned Future of the Expert System CADEMO. In: F. Faulbaum; R. Haux & K-H. Jöckel (Hrsg.): SOFTSTAT '89 Fortschritte der Statistik-Software 2. Gustav Fischer, Stuttgart, 332-339.

Rasch, D. (1990a). Constructing BIBDs by an expert system. Journal of Combinatorial Mathematics and Combinatorial Computing 8, 27-29.

Rasch, D. (1990b). Experimental Design using a Biometric Workstation. Proc. of the XVth International Biometric Conference A, Budapest 2.7.-6.7.1990. vol. Invited papers 153-156.

Rasch, D. (1990c). The expert system CADEMO as a tool of the teaching of statistics. International Statistical Education Newsletter 14 2, 13-14.

Rasch, D. (1991). Thc CADEMO Program System for Model Choice and Design of Experiments. Use of Computers in Agriculture. Proceedings of a Symposium of the UN Economic Commission for Europe (ECE), held in Leipzig from 15.10.-19.10.1990. Bonn, 176-191.

Rasch, D., Guiard, V. & Nürnberg, G. (1992). Statistische Versuchsplanung - Einführung in die Methoden und Anwendung des Dialogsystems CADEMO. G. Fischer Verlag, Stuttgart, Jena, New York.

Rasch, D. (1992). Software for designing experiments. In: Faulbaum (ed.) SoftStat'91, Advances in Statistical Software 3 Fischer Verlag Stuttgart, 301-308.

Sayn, H. & Merkle, W. (1989). Statistical Software for Sample Size Estimation: Power, Design-Power, and IFNS. Statist. Software Newsletter, 15, 56-59.

Schmidt, K. (1993). Sample size determination in survival analysis. Softstat, March 1993, Heidelberg.

Schneider, B. (1992). An interactive computer program for design and monitoring of sequential clinical trials. Proc. Int. Biom. Conf. Hamilton, New Zealand, 7-11 dec., Invited Paper Vol. 237-246.

Thomas, E. & Schrumpff, M. (1989a) Die Planung und Konstruktion von Standardanlagen im landwirtschaftlichen Feldversuchswesen. Feldversuchswesen, Berlin, 6, 74-82.

Thomas, E. & Schrumpff, M. (1989b). Die Planung von Grobversuchen (Produktionsexperimenten) mit Hilfe des Modulkomplexes FEVE. Feldversuchswesen, Berlin, 6, 83-89.

Verdooren, L.R. & D. Rasch (1992). A Dialogue system for PC's for designing field trials Proc. Intern. Biometric Conference, Hamilton New-Zealand, 7-11 dec. vol. 2.

Zemroch, P.J. (1992). KEYFINDER - A complete toolkit for generating fractional-replicate and blocked factorial designs. In: Dodge, Y. & Whittaker, J. (eds.). Computational Statistics Vol 2, 263-268. Physica-Verlag.

Addresses for detailed information and of distributors

BBN Software Products, 10 Fawcett Street, Cambridge, MA 02238, USA

BMDP Statistical Software Inc., 1440 Sepulveda Blvd Suite 316, Los Angeles CA 90025, USA

BIORAT GmbH, im. Rost. Innov. u. Gründerzentr., Joachim Jungius Str. 9, D-18059 ROSTOCK, GERMANY

Cytel Software Corporation, 675 Massachusetts Ave., Cambridge, MA 02139, USA

DKFZ, Inst. für Dokumentation, Information und Statistik, Im Neuenheimer Feld 280, D-W-6900 Heidelberg 1, GERMANY

H.Edwards, Department of Statistics, Massey University, Palmerston North, New Zealand

H. Flühler, Mathematical Applications, CIBA-GEIGY AG., R-1008 Z 236, CH-4002 Basel, SWITZERLAND

P. Haaland, Becton Dickinson Research Center, P.O.Box 12016, Research Triangle Park, North Carolina, 27709, USA

M. Hasselaar, Centre for Quantitative Methods, P.O.Box 513, 5600 MB Eindhoven, THE NETHERLANDS

IDEA Works, Inc., 607 Jackson Street, Columbia, MO 65203, USA

IDV-Datenanalyse und Versuchsplanung, Wessobrunner Str. 6, D-8035 Gauting, GERMANY

iec - ProGAMMA, P.O. Box 841, 9700 AV Groningen, THE NETHERLANDS

Alec Mann, Scottish Agricultural Statistics Service, The University of Edinburgh, The King's Buildings, Edinburgh EH9 3JZ, SCOTLAND, UNITED KINGDOM

Mathieu, D., L.P.R.A.I., I.U.T. Avenue Gaston Berger, F-13625 Aix-en-Provence Cedex, FRANCE

MINITAB Inc., 3081 Enterprise Drive, State College, PA 16801-3008, USA

Nam-Ky Nguyen, Biometrics Unit, Inst. of Animal Production and Processing, CSIRO, Private Bag 10, Clayton Vic. 3168, AUSTRALIA

SAS Institute Inc., SAS Campus Drive, Cary, N.C. 27513, USA

Sloane, N.J.A., ATT Bell Labs, Room 2C-376, 600 Mountain Ave, Murray Hill N.J. 07974, USA

STATCON, Mündener Str. 1, D-W 3430 Witzenhausen, GERMANY

STATEASE Inc., For Europe: QD Consulting, 68 Station Road, Steeple Morden, Royston, Herts SG8 ONS, UNITED KINGDOM

STATGRAPHICS, Manugistics Inc., 2115 East Jefferson Street, Rockville, Maryland 20852, USA

Statistical Consultants, Inc., Park Plaza Office Building, 462, East High Street, Lexington, KY 40508, USA

Statistical Sciences, Inc., 1700 Westlake Ave N, Suite 500, Seattle WA 98109, USA

StatSoft, 2325 E. 13th St., Tulsa, OK 74104, USA

SYSTEGRA, Vertriebsgesellschaft für Automation und Sicherheitstechnik mbH, Frankfurter Str. 63-69, 6236 Eschborn, GERMANY

SYSTAT, Inc., 1800 Sherman Ave, Evanston, Illinois 60201, USA

Mike Talbot, Scottish Agricultural Statistics Service, The University of Edinburgh, The King's Buildings, Edinburgh EH9 3JZ, SCOTLAND, UNITED KINGDOM

Jean Tranchefort, Service des Etudes, Statistiques, I.T.C.F., 91720 - Boigneville, FRANCE

Prof. Dr. John Whitehead, University of Reading, Department of Applied Statistics, Whiteknights, P.O. Box 217, Reading, RG6 2AN, ENGLAND, UNITED KINGDOM

Emlyn R. Williams, CSIRO Division of Forestry, P.O. Box 4008, Queen Victoria Terrace, Canberra ACT 2600, AUSTRALIA

P.J. Zemroch, Shell Research Ltd, Thornton Research Centre, P.O. Box 1, Chester CH1 3SH, ENGLAND, UNITED KINGDOM

Knowledge-Based Systems in Statistics: A Tutorial Overview with Examples

Uwe Haag
University of Heidelberg, Institute for Medical Biometry and Informatics, Im Neuenheimer Feld 305, D-69120 Heidelberg, F.R.G.

Key Words:
heuristic strategies, inference mechanism, knowledge-based system, knowledge representation, statistical expert system

Abstract
This paper presents a tutorial introduction to knowledge-based systems in statistics. The objective is not to provide a complete overview of the relevant research, but to impart a better understanding of aims, roots, and problems of this promising research field.
Since the terms knowledge-based system or expert system are open to various interpretations, the paper starts with an introduction to the fundamental ideas behind them. The subsequent sections summarize the development of methods, tools, and systems from the early roots to more recent work. Two systems are reviewed in more detail in order to demonstrate the state of the art. This article concludes with a discussion of lessons that have been learnt in this field, and of future challenges.

The idea

Specific expert activity

The original research idea of knowledge-based systems aimed at the development of sophisticated software that is able to *perform activities that otherwise require* a single, or team of, *human experts.* In the statistical area, the expert can be characterized as a person with specialized knowledge in data analysis, in general a trained statistician. This intention coined the term expert system.

The characterization of what is an expert and his/her specific activity is an important problem if this idea is used for a definition. Assuming this problem is solved, how should the definition be interpreted? One could read that the expert should be fully replaced in all expert tasks, but obviously, this would claim too much. It seems sufficient to state that the expert is relieved of only parts of specific activities.

Nowadays however, many software systems undertake expert tasks, even 'conventional' systems that never claimed to be an expert system. Consider statistical analysis systems as an example. Strictly regarded, this is an activity that was performed by specialized humans originally. This may explain why many developers state their software contains an expert system component, although computer scientists contradict.

Not only knowledge-based, but also conventional software is replacing human experts. Moreover, software is able to solve problems that cannot be tackled without computers, for example, complex calculation or control processes, sometimes in real time.

Since the paradigm of performing expert activities alone is insufficient for a definition, a closer look is necessary to differentiate conventional software systems from knowledge-based.

Separating knowledge and control structure

Conventional programs use a specific method for the treatment of a certain class of problems. For instance, the class of problems might be the calculation of a particular statistical test. Any actual data to be tested represents one problem of this class. Such an *algorithm* in conventional programming is typically organized as a sequence of logical comparisons and operations to be executed according to preceding comparison results. A problem-specific control structure with branches, loops and subroutine calls implicitly contains the knowledge how to solve the problem. Any other problem class needs another specific algorithm.

The paradigm of knowledge-based systems, on the contrary, uses one universal method for solving various classes of problems. The knowledge for solving problems of a particular class is *explicitly declared* external from the algorithm using a predefined formal *knowledge representation*. A specific problem is described by a set of additional statements. The task of the universal method, usually called *inference method*, consists then in the solution of the problem with the help of the problem-specific knowledge and of the general knowledge about the problem class. Thus, one characteristic of knowledge-based systems is the *separation of knowledge* about the solution on the one hand, *from the control structure* on the other. The term knowledge-based system has its roots in this paradigm.

Note that knowledge-based systems make extensive use of symbol manipulations. That is why you would not apply this technology to classical calculations such as the example of statistical tests above. However, it is well suited for abstract tasks like the selection of appropriate analysis methods, where a method should be matched with a statistical model.

The problem of a definition

Unfortunately, a definition based on the separation of knowledge and control structure is also problematic. Some systems – although correctly assigned to this research field – do not fulfil this condition. It is a widespread experience that the consequent realization of this paradigm is very important, but also very difficult.

This definition problem can be recognized in many text books on knowledge-based systems, and also in the literature on knowledge-based systems in statistics (Darius (1988), Haux (1989) with discussion, Naeve (1990), Wittkowski (1990)). Although observing a broad agreement with respect to the domain's boundaries, an adequate definition seems difficult. Two basic characteristics of knowledge-based systems have been identified however: first, modelling of expertise and second, separating knowledge from control structure. In the subsequent sections, some further characteristics are mentioned, mostly concerning methods and tools that have been developed in this field and are often used. These supplementary characteristics are optional in actual systems, but help to define the area.

The roots

From Turing to knowledge-based systems

This excursion to the roots shows that knowledge-based systems are no new development. Conversely, this research field has rather a long tradition in computer science. So far, the history of knowledge-based systems *in statistics* does not run much longer than 10 years.

One of the first stimuli already dates from 1950. The mathematician Alan Turing, who was more interested in theoretical aspects of computing machines than in the engineering part, raised the question whether machines can think (Turing (1950)). He proposed a game in which a person communicates via type-writer, with another man on the one hand, and with a computer on the other. The test person should decide which of the two communication partners is the computer. This game is today well known as the Turing test. Remark that the crucial point is the *behaviour* of the computer without any regard to the methods it uses. It is not important whether the computer works in the same way like human cognitive processes.

Following Turing's proposals, researchers began to construct games programs (e.g. chess) (Newell et al. (1958)) and automatic translators (Oettinger (1963)). Newell et al. (1957) also considered *heuristic strategies* in a program called 'Logic Theorist'. They were the first who accepted that a program need not always find a solution. This idea led to a program with the optimistic name 'general problem solver – GPS' (Newell & Simon (1961)). The researchers observed some analogies between the reasoning processes of the GPS and of man. They even considered whether human properties such as intuition could be understood as GPS-like heuristic strategies. This was the starting point of extended research leading to worthwhile experiments. In the same way, this optimism initiated many exaggerated and misleading ideas of 'intelligent' computers, not only in the past, and not only by outsiders to the field.

The first criticism of this efforts was soon raised. Weizenbaum (1966) developed a program for conversation between man and computer called ELIZA, and used it to simulate (or rather to make a parody of) a psychotherapeutic session. He did not at all aim to replace a psychologist since the program idea was very simple. ELIZA recognized patterns in the sentences that had been entered and used them to construct a response. To his great astonishment, some people developed an emotional relationship and confided personal problems to the program. In fact, some psychologists believed that a future version of the system could be used to replace human experts. Through this experience, Weizenbaum became an active campaigner against the 'artificial intelligencia' and his vision of omnipotent computers (Weizenbaum (1976)).

Not much later, the first systems appeared that would today be called expert systems. DENDRAL, for example, was a program to identify chemical structures based on mass spectroscopy data (Buchanan et al. (1969), Feigenbaum et al. (1971)). This program was discussed within the researchers' community since it (together with other systems) represented a turning-point to systems that, indeed, have capabilities that are characteristic of human experts, but without necessarily simulating human behaviour. In the subsequent years, two schools of thought can be identified, one aiming at a better understanding of human cognitive capabilities by simulating human behaviour, and the other aiming at the construction of useful tools. The research field of knowledge-based systems is to a large extent assigned to the second school. Today, both streams yet exist, but additional diversification superseded this classification.

Refer to Dreyfus (1979) for a more detailed overview of these roots. Particularly, this paper represents a well-founded criticism of this research field and sets out its borders.

The ideas find their way to the statistical domain

Already in the seventies, proposals for more sophisticated statistical programs were made. Finney, in his presidential address at the Royal Statistical Society, said (Finney (1974), p. 17/18): "A more ambitious idea [...] is the writing of a range of statistical programs with built-in statistical advice to guide the user in avoiding the more serious mistakes of methodology."

In the late seventies, statisticians more and more criticized statistical analysis systems (Nelder (1977)). Some observed the parallel developments in the field of knowledge-based systems and promoted their use for the improvement of data analysis (Campbell & Woodings (1981), Chambers (1981)). Scientists (mainly statisticians) started the development of knowledge-based systems in statistics. The first systems were soon available, for instance, STATPATH (Portier & Lai (1983)), REX (Gale & Pregibon (1983), Gale (1986b)), PANOS (Wittkowski (1985, 1992)), GLIMPSE (Wolstenholme & Nelder (1986)). RX (Blum (1980, 1982)) should also be mentioned in this list of early systems, although it was more of a tool for inferring knowledge from clinical databases than a system aiming at ameliorating statistical analysis.

Conclusion

When looking back to the roots of today's knowledge-based systems in statistics, one can observe that the vision was far from reality. For this reason, professionals mostly were moderating their aims; the dream of omnipotent robots replacing mankind seems far, and research concentrates on the development of dedicated systems for practical use.

The term expert system has grown very popular. At no time were there as many people working in this domain as today. Therefore, the term is very often misused, sometimes thoughtlessly or unknowingly, but sometimes even deliberately, pursuing well-directed marketing or financial interests.

The last forty years of research have led to useful software engineering methods and tools, especially for systems with a considerable proportion of symbol manipulation. From an inverse point of view, these developments characterize knowledge-based systems and therefore, ease the definition problem.

Methods and tools

Knowledge representation

Benefits ...

Before having a closer look at available systems, the ideas of knowledge representation and inference are explained in more detail. A comprehensive overview of common methods and tools would go beyond the scope of this paper. Therefore, this section concentrates on some basic topics and examples. The fundamental idea of separating knowledge that can solve a problem from the program's control structure has already been mentioned. Several benefits should follow from this general design idea.

First of all, programs should be easier to develop. The software engineer need not care about the control structure. Assuming that there is already a universal inference method, the programming process ideally stops with the problem's specification, in other words, with the declaration of relevant knowledge.

Neglecting for reasons of simplicity that any such declaration has to follow certain formal regulations, it could for example be stated: Given n observations $x_1, ..., x_n$ of a random variable X that are identical and independently normal distributed with unknown mean μ and with variance σ^2, the hypotheses H_0: $\mu = 0$ vs. H_1: $\mu \neq 0$ can be examined with a one sample t-test. This knowledge element can be formulated independently from any others. Knowledge about the practical application of the test in real situations (examination of the assumptions, robustness, calculation of the test, and so on) can automatically be linked by the inference method. This represents a very high level programming style and programs should be very compact.

Consequently, testing of programs should become much easier. Whereas in conventional programs the complete algorithms have to be examined (that include much control information), the declarative programmer only reviews the knowledge base. As long as the represented knowledge is correct, the inference method guarantees the correctness of its conclusions. Programs additionally are easier to understand, maintain and adapt since knowledge browsing replaces detailed code inspection.

Also, knowledge-based systems are said to be highly interactive. Since knowledge is explicitly declared, such systems are able to ask the user for missing knowledge elements and to give explanations of their reasoning processes. This latter point is very important since the proposals of the system become transparent.

... and problems

The features listed above are very impressive. The reader may ask why such systems are not developed and used more often. There are various reasons:

First, knowledge-based systems cannot be built for all problems. Their application is limited to fields in which symbolic computation can be adequately used. This is an important restriction, although there are a lot of application fields in statistics: trial design, selection of appropriate statistical methods, guided data analysis, teaching, and so on. But one would not seriously use it for numeric algorithms like the calculation of an empiric variance or even of a general linear model.

A further problem is computational complexity, especially of the inference method. Existing methods are often NP-complete, i.e. no algorithm is known where the relationship between knowledge base size and inference time is better than exponential. This does not necessarily represent a crucial point, but has to be considered. Possible consequences include restricting the search depth or using heuristics. In both cases, the risk of unsolvable problems cannot be avoided. Computational complexity is the main argument of Streitberg (1988) for the non-viability of statistical expert systems.

Although implemented on a deterministic machine, indeterminacy is to some extent inherent in knowledge-based systems since the inference method's links, and thus the flow of control, depends on the internal organization (and order) of the knowledge base. Two extreme cases of indeterminate behaviour have to be considered in the inference method: first, the stopping problem (whether the system finds a solution within limited time and storage resources, especially without coming in an endless loop), and secondly, the problem of dealing with conflicts (when the knowledge base contains contradictive facts).

Currently, much research is invested in solving the problems of knowledge acquisition (e.g. Breuker & Wielinga (1989); Musen (1989); Schreiber et al. (1993)). Experts in knowledge acquisition agree that human specifications are never complete, sufficiently structured or consistent. However, these specifications have to be transformed into a formal model.

There is evidence that the problem outlines can be adequately dealt with, but this still remains a challenge.

Example: Rule-based knowledge representation

Various formal representations of knowledge have been developed, some of them logically equivalent. Probably the oldest representation is based on *rules*, apparently a very natural way to express knowledge. Rules therefore are the best way of demonstrating the idea of formal knowledge representation, although current systems usually apply more advanced representations.

Rules are composed of a condition part and a conclusion part. If the condition part of a rule is evaluated to true, the conclusion part becomes equally true. Both condition and conclusion part follow the same formal representation.

A sample rule from a statistical application could state: if a variable is of ordinal scale then it follows a discrete distribution. More complex rules can be built with the help of conjunction, disjunction, and negation. This sample rule is still written informally. For implementation in a formal model, it has to be reformulated. For this purpose *first-order predicate calculus* is appropriate:

A *term* is either a constant, a variable or a compound term. A *constant* is either a number or a word starting with a lower case letter. A *variable* is written as a word starting with an upper case letter. A *functor* is notated as a cursively printed word. *Compound terms* are defined recursively: if $t_1,...,t_n$ are terms and if f is a n-ary functor, then $f(t_1,...,t_n)$ is also a term (a compound term).

Predicate symbols are notated as words starting with a lower case letter. If p is a n-ary predicate symbol and if $t_1,...,t_n$ are terms, then $p(t_1,...,t_n)$ is an *atomic formula.* Atomic formulas are the axioms of predicate calculus.

Well-formed formulas are atomic formulas or negations, conjunctions, disjunctions, implications and equivalence relations of well-formed formulas. If X is a free variable and α is a well-formed formula, then $\exists X : \alpha$ and $\forall X : \alpha$ are well-formed with quantifiers $\exists$ and $\forall$.

Rules correspond to implications in predicate calculus. However, the notation presented here is still too powerful for implementation. Fortunately, there are normalized forms of well-formed formulas, so called *clauses.* If $\alpha_1, ..., \alpha_n$ and $\beta_1, ..., \beta_m$ are atomic formulas then a clause is defined as $\alpha_1 \wedge ... \wedge \alpha_n \Rightarrow \beta_1 \vee ... \vee \beta_m$. There is an algorithm to transform general first-order predicate formulas into clauses.

The above sample rule can be directly expressed in clauses. For example, a two-ary predicate 'scale' and a two-ary predicate 'distribution' might be appropriate. The rule can then be notated as 'scale(X,ordinal) $\Rightarrow$ distribution(X,discrete)'.

To aid knowledge structuring, existing systems often use much more complex representations than presented here, see e.g. Jackson (1990) for an introduction. In addition, special representations are used to model uncertainty. Note also, no system action can be adequately described in logical terms, i.e. in terms of 'true' and 'false'. For instance, printing a result is not a logical action.

Inference

Inference methods for rule-based knowledge representation

The inference method to be used depends on the formal knowledge representation method it works upon. In the case of rule-based systems, there are two basic strategies.

On the one hand, the inference method could start with the set of known facts and from them infer additional facts with the help of the rules in the knowledge base. This strategy is called *data-driven inference* or *forward chaining.*

On the other hand, the inference method could begin with a goal or a set of goals that should be proved. If this focus is not directly given in the form of facts, the inference method matches its goals with the conclusion part of the rules in the knowledge base. The condition part of relevant rules becomes a subgoal that is evaluated with the help of the knowledge base. This procedure is repeated until the final goal is proven or until no new rules can be used. This is called *goal-driven inference* or *backward chaining.*

Example: Goal-driven inference based on the resolution principle

One special backward chaining method for rules in the form of clauses is based on the *resolution principle* (Robinson (1965)). Resolution is an inference rule, prescribing how a step of logical inference follows from others. Applying this inference rule allows the purely automatic proving of theorems.

Another notation for clauses is introduced before taking a closer look at the resolution rule. Since $(\alpha \Rightarrow \beta) \Leftrightarrow (\neg\alpha \vee \beta)$ the implication symbol can be replaced by negation and disjunction clauses. Clauses can therefore be equally expressed in their disjunctive normal form $\neg\alpha_1 \vee \ldots \vee \neg\alpha_n \vee \beta_1 \vee \ldots \vee \beta_m$.

Next, we need a definition of *unification.* Unification is a method for matching well-formed formulas. More precisely, given two formulas, a *substitution* is sought that will make them identical. A substitution is a function of a set of variables in terms. Two formulas s and t are unifiable if there exists a substitution Ψ such that $\Psi(s) = \Psi(t)$. In that case, Ψ is called a unifier of s and t and $\Psi(s)$ is called a unification of s and t.

Given the two rules 'if a variable is of ordinal scale then it is discretely distributed' and 'if a variable is discretely distributed, then so are its derived variables' the following new rule should be inferred: 'If a variable is of ordinal scale, then all its derived variables are discretely distributed.'

The first job is to put these three statements into a logical form and next, to transform them into clauses. The clauses in disjunctive normal form are for example:

(1) $\neg$scale(X,ordinal) $\vee$ distribution(X,discrete)

(2) $\neg$distribution(Y,discrete) $\vee$ $\neg$is_derived(Y,*transformation*(Y))
$\vee$ distribution(*transformation*(Y),discrete)

(3) $\neg$scale(Z,ordinal) $\vee$ $\neg$is_derived(Z,*transformation*(Z))
$\vee$ distribution(*transformation*(Z),discrete)

The resolution rule is: If clause α contains some atomic formulas s and clause β contains the negation of some atomic formulas t, and if s and t are unifiable by a substitution Ψ, then a *resolvent* of α and β is generated by combining the clauses from α and β, removing formulas s and t and applying the substitution Ψ to the remaining terms. If clauses α and β have a resolvent γ, then γ may be inferred from α and β.

In the example, let s be 'distribution (X, discrete)', and let t be 'distribution (Y, discrete)'. Clause (1) contains s and (2) contains the negation of t. Formulas s and t are unifiable under the substitution $\Psi = \{X \rightarrow Z, Y \rightarrow Z\}$. Removing s from (1) and the negation of t from (2) and applying the substitution Ψ to the rest of the terms in (1) and (2), the resulting resolvent is equivalent to (3) and, therefore, resolution has made the intended inference.

In general, the proof of a goal requires more than one resolution step. Therefore, the resolution rule is combined with a search strategy to find the expressions to be unified next. It often appears that a unification succeeds on a first step, but fails on a later one. In these cases, the inference method *backtracks* to the last unification it has done and tries to find an alternate unification. This procedure is repeated until the goal is proven or until there are no more alternatives.

In the last case, the goal is unprovable. Usually, it is assumed that the knowledge base is complete and consequently, the goal is considered to be false. This is called the *closed world assumption.*

Tools

Programming languages and shell systems

Various programming languages have been used for building knowledge-based systems. Since the requirements of actual implementations differ from each other, any programming language may be considered, even classical procedural languages. This point is emphasized, because many people believe that only specialized languages for symbolic manipulation are adequate. Actually, these languages resulted from the research sketched above and therefore represent potentially good candidates for the implementation of knowledge-based systems. Examples are the functional language LISP, developed as early as 1958 (Winston & Horn (1989)), or the object-oriented language Smalltalk (Goldberg & Robson (1989)).

The same can be said concerning so-called *shell systems*. Shell systems provide one or more methods for knowledge representation with appropriate inference methods, but no knowledge. Such 'empty' systems can be filled with knowledge and thus, can be used to develop a knowledge-based system relatively easily. The question is whether the representations provided by shell systems are always adequate to the application domain, which distributors would like to make us believe. In passing it should be mentioned that there also exists specialized hardware to support dedicated symbolic languages.

Example: The programming language Prolog

Instead of listing a large number of tools, the widespread language Prolog is looked at in more detail. Prolog stands for *pro*gramming in *log*ic since it is based on first-order predicate calculus. It uses a special form of clauses, so-called Horn clauses, for knowledge representation. Horn clauses are clauses with at most one atomic formula β on the right hand side, i.e. at most one formula that is not negated in disjunctive normal form.

A clause with both negated and not negated atomic formulas is called a *rule*. If there is only the conclusion part of a clause it is called a *fact*. Finally, Prolog *queries* are a disjunction of negated atomic formulas only. A Prolog predicate having several clauses with the same predicate name and arity corresponds to a disjunction of Horn clauses.

The Prolog interpreter is based on the inference method that was discussed in the preceding section. If the interpreter fails to prove a goal it responds with 'no' (false). In contrast to the answer 'yes' (true) indicating that the goal is proved, the answer 'no' only indicates that the fact is false under the closed world assumption. In actual applications this assumption is seldom justified.

Prolog provides a set of predefined functors and predicates. For example, in order to represent lists (i.e. ordered sets) syntactically correct as terms, the functor '.' is used. Its two arguments stand for head (i.e. first element) and tail of the list. The list with the elements 'a' and 'b' is then internally represented as .(a, .(b, [ ])) where [ ] denotes the empty list. There are predefined predicates for input and output, for the administration of the knowledge base, for the classification and manipulation of terms, for arithmetic operations, and many more.

Since Prolog provides its own inference mechanism, the user need not add much control information. With respect to the time problems mentioned, and in order to extend the capabilities of the inference method, a set of predefined predicates influencing the control flow has been defined. For example, such predicates can be used to reduce the search space and thus, prevent the inference method from generating undesired solutions.

The reader who is interested in Prolog could refer to the introductory textbook of Clocksin & Mellish (1987), or to one of more advanced books of Sterling & Shapiro (1987); O'Keefe (1990). Lloyd (1987) deals with logic programming theory.

Sample systems

Overview

Haux (1989) gives a critical overview of the first decade's applications in the field of knowledge-based systems in statistics. More recent reports can be found in EUROSTAT (1989, 1992), Momirovíc, Mildner (1990), Hand (1991), Hatabian & Augendre (1991), and Faulbaum (1992). The most current and extensive overview is the one by Gale et al. (1993).

A commercial group around Brent developed a set of seven systems based on an expert system shell that are distributed under the name Methodologist's Toolchest Professional^{TM}. Brent (1989c) is an article describing one of these systems; user guides of the five systems developed first are Brent et al. (1988), Brent (1989a,b,d), Brent, Mirielli (1989). Rödel & Wilke (1990) focused on the selection and execution of tests for independence in bivariate distributions. They tested several shell systems for implementation, but in the end used a rule-based approach realized in Prolog. The system was announced to be commercially available.

Chowdhury et al. (1990) demonstrate the application of the MAXITAB system (Chowdhury et al. (1989)) for deriving knowledge from large medical data bases. Graber (1989) describes a knowledge-based system for clinical trials, but only specifies detailly and implements one prototype module for randomization with the help of a shell system. Jirků & Havránek (1990) describe a frame and rule-based system for exploratory data analysis. Locarek (1988) developed a language for the construction of statistical analysis systems including elements of structured programming, of functional programming and of knowledge representation, and then used it to develop a prototype system.

Govaerts et al. (1992) report about a system for the design of experiments mainly in chemistry. Lehmann & Shortliffe (1991) propose a Bayesian framework for the construction of knowledge-based systems in statistics and demonstrate this approach with a system for the interpretation of randomized clinical trials. Marino et al. (1992) implemented a system for multiexponential model discrimination. EVER is a knowledge-based system developed at Bern University supporting all phases of statistical analysis from initial planning to final interpretation for problems up to two samples (Matthäus (1992)). Prat et al. (1992) built a knowledge-based front-end for the design and analysis of fractional factorial experimental designs with factors at two levels. Wyatt et al. (1993, 1994) built a knowledge-based system for the design of clinical trials that also includes medical knowledge.

Adèr (1992) proposes a rule-based shell system for the development of knowledge-based systems in statistics. Fröschl (1992b) focuses on theoretical aspects of formalizing statistical expertise.

The work of two research groups, the CADEMO and the WAMASTEX group, is reviewed in more detail. One important selection criterion was that the systems are developed for several years. A second was that the two approaches are very different and thus, could be used to demonstrate different aspects of related research. An implicit criterion was of course that they were known to the author.

The CADEMO system

Introduction

CADEMO (Computer-Aided Design for Experiments and Modelling) was developed under the leadership of Dieter Rasch at the Rostock Academy of Agriculture, at this time GDR, in a cooperation with experts from many other institutions (Rasch et al. (1987, 1992)). The system, realized with Turbo Pascal^{TM} on Personal Computers, is still being extended and updated. It is commercially available.

The CADEMO system aims at the planning phase of statistical experiments. It provides help for many general aspects of experimental design, such as the selection of appropriate methods for analysis or sample size determination, but also for very specific topics, for example planning of bioassay experiments.

User interaction

The developers state that very little statistical knowledge is necessary for using CADEMO. The user need not be an expert. However, I would recommend the system for users with advanced experience in statistical analysis only.

The system presents screens with text or graphical information. On most screens, the user has to decide upon a question by selecting one of several predefined answers. On some screens he should enter one or several parameter values, for example a variance estimate. In the system's terminology, this is called a dialogue.

The user can also ask for explanations of statistical terms or for relevant literature, display a general usage help, display the path that was taken according to the user's decisions, track back to any point of this path, and some more functions.

Knowledge representation and inference

There is no explicit knowledge representation. A module consists of a set of screens with text information, associated dictionary of statistical terms, literature references, and algorithms for specific calculations, e.g. sample size determination. The screens are organized as a directed graph, in most cases as a tree.

The inference method works as follows: Depending on the user's input, the next screen is displayed, but the user could also decide to go back to an earlier decision and restart from there. The degree of knowledge about a specific problem depends on the depth in the screen graph. At the terminal nodes of the graph, the system presents its proposals, for example an analysis method together with the required sample size.

One sample knowledge element, for estimating the variance of a normal distributed parameter for the purpose of sample size determination, can be expressed as follows: if the smallest and largest values of the parameter that can be realized in the population are known, then estimate the variance from the formula LARGEST – SMALLEST = 6 STANDARD DEVIATIONS.

Knowledge acquisition

The developers use a tool for generating screens and their connections. The main task of the knowledge engineer is structuring questions and possible answers into a dialogue. A screen within this dialogue can include questions or text information. Eventually, the expert specifies calculations to be implemented in algorithms and in which situation to activate them.

Discussion

The first point to be discussed is whether CADEMO provides the two basic characteristics of a knowledge-based system. CADEMO is modeling human expertise and undertakes expert tasks, but there is no explicit declaration of knowledge independent from the control structure. Readers might ask why this system is shown here as an example although it is not knowledge-based. There are two answers. First, few of the available systems really fulfil the definition above. Second and more important, the aims of CADEMO are a typical example where a knowledge-based approach would have been appropriate. It is additionally one of the few systems that is commercially available, and has more than a handful of installations.

We should next discuss whether the intuitive approach used for CADEMO has disadvantages in comparison to the knowledge-based approach. The main criticism is that the proposals of the system are not completely transparent. Note for example, that the sample knowledge element above only exists as a formula in an algorithm and is not explicitly declared (I inferred it with the help of the final sample size calculated from this variance estimate). The user has no opportunity to display the knowledge that was applied, and, thus, to verify the system's proposals. He cannot receive any automatic explanations about strategies used, immediate consequences of a decision, or how the system's proposals were inferred. However, such information would have been useful for individual decisions on a specific screen. Only experts already familiar with the problems of experimental design can (sometimes) guess the strategies used from the results obtained.

My proposals to ameliorate the CADEMO system consist in (1) giving more help text explaining the consequences of a decision, possibly in a separate window displayed only on user request, and (2) fully describing the strategies used in the system's documentation. However, much of this work could have been saved if a knowledge-based approach had been used originally.

The quality of the knowledge implemented in CADEMO is generally good. One could discuss whether the sample rule above should be applied when there is no other information available, or whether one should rather perform a pilot trial to estimate the variance. (This is the proposal of CADEMO if one states that there is no information at all.) The decision of an expert would probably in addition consider the costs and expected benefits of such a pilot trial.

To summarize, CADEMO is one of the leading systems for the planning of experiments, it is comprehensive, and as far as I have observed them, most of its proposals are reliable. The basic research aim is developing an actual tool for practical use. It is astonishing how well this has been attained with very restricted hardware and software resources originally.

The only criticism is that an intuitive approach has been used instead of a knowledge-based one.

The WAMASTEX system

Introduction

WAMASTEX is the name of a software project of Vienna University aiming to formalize statistical knowledge primarily in the fields of univariate descriptive statistics (Dorda et al. (1988), Dorda et al. (1990)) and of treatment effects in paired sample data (Fröschl (1992a)). The earlier work at least, is available for physicians and medical statisticians at Vienna University Hospital serving as the test site, but is not commercially distributed. The system is programmed using tools of the statistical analysis system SAS^{TM}. It is developed and tested on a mainframe but also runs on Personal Computers.

User interaction

The WAMASTEX project aims to support inexperienced users primarily. The current user interface will be further developed. The basic design idea is that within a so-called 'flow graph' of analysis, the user answers questions asked by the system. Succeeding actions are determined through the user's responses. At some points in the flow graph, the complete set of system actions is made available.

Knowledge representation and inference

Knowledge in WAMASTEX is organized in the form of a decision tree. Nodes are user decision points and data decision points. At the latter, decisions are automatically inferred from the data. This is similar to the concept of the CADEMO system consisting of dialogues and algorithms. One important difference is that in WAMASTEX, the information collected about a problem is stored external from the flow graph. Another difference is that in CADEMO – with respect to the specific tasks such as sample size calculation – algorithms are mostly rather short. In WAMASTEX, automatic decisions can be rather complex and are sometimes based largely on heuristic strategies. They use their own inference and representation mechanisms.

For example, in the case of analysing treatment effects in paired sample data, based on a suggestion of Cox and Snell (1981), not only a standard model is considered, but also all linear transformation models, i.e. models with shifts in expectation as well as possible effects on variances (scaling), and one non-linear transformation model. After specifying the problem and before running the analysis, the complete model evaluation is done automatically without any user interaction. All possible models to be selected from are evaluated with the help of heuristic rules.

Two sample rules are based on the following considerations (Fröschl (1992a), p.16): Independence of the two measured variables is rather unlikely in experiments for such treatment effects. Whatsoever, samples without any detectable correlation should be singled out. Therefore, WAMASTEX tests the hypothesis that Pearson's correlation coefficient equals zero. If this hypothesis is not rejected with $\alpha = 5\%$, a similar test based on Spearman's correlation coefficient is done additionally. The sample rules in this context are:

- Abort model evaluation if none of the two hypotheses is rejected.
- If the first hypothesis is accepted, but the second rejected, then there must be a non-linear relationship (if there is any at all).

All rules are automatically evaluated and applied by the system. Depending on the result of a rule evaluation and of its meaning, the fit of the different models in accordance to any rule is calculated by adding scores 0, ± 1, ± 10 or ± 100. In the case of the second sample rule, all models assuming at most a simple shift in location receive a score of -100, and the (only) model assuming a non-linear relationship receives a score of +100. After checking all rules, the models with the highest scores are presented for user selection, who can choose any other model, however. There is no definite declaration on whether one certain model is 'the correct one'.

Knowledge acquisition

There are no tools for knowledge acquisition within WAMASTEX. A knowledge editor is used for reviewing and updating the main decision graph, especially to look for cycles in this graph.

It should be mentioned additionally that the WAMASTEX group, in order to validate and improve their knowledge base of descriptive statistics, asked experts to classify 16 sample data sets. The responses were analysed for the heuristics which were applied, and the experts' proposals were compared with the system's results. Another evaluation checked the system's capabilities with the help of simulated data.

Discussion

Again, the first question to be discussed is whether WAMASTEX has the two basic characteristics of a knowledge-based system. The system performs typical activities of statistical experts. The condition of explicit knowledge representation independent from the control structure is at least partly fulfilled. Knowledge that is specific for a problem is stored externally from the flow graph. The flow graph itself is represented as a SAS data set, i.e. in a standardized format. Recall also the example of model evaluation above: Knowledge is made explicit in the form of rules following a certain format (although conditions are of course evaluated with the help of external algorithms). There is an inference algorithm that assigns scores to the different models and eventually comes to an conclusion.

Unfortunately, there is no explanation component implemented at the moment. For example, if users would like to know, why their preferred model is not proposed, or why the proposed model is considered best by the system, they must refer to the technical report and then manually recheck the conditions in order to make the system's decision transparent. This is a quite an elaborate procedure. Adding such an explanation component should not be too difficult if the implementation separates inference from knowledge.

Secondly, I would like to discuss the quality of the knowledge represented. Consider the sample rules about correlation above. Since pretests are used for the evaluation, the usual problems arise: Is the power sufficient for performing the pretests? This is important since conclusions are also drawn in case the hypotheses are not rejected. Is an α-level of 5% always appropriate? Should p-values be adjusted, as multiple tests are performed? Does the decision not largely depend on the sample size? Thus, can these tests really be interpreted automatically?

Another approach could do without full automation and rely on more user interaction instead. The system could print the parametric and rank correlation coefficients together with a scatter plot of the data. This allows for a judgement of the researcher about the association without a need for testing. With respect to innocent users, the decision could be supported by displaying a set of typical reference plots to choose from. This strategy evitates the above mentioned theoretical problems. Of course, a consultation is more time-consuming than with the original decision procedure. A further advantage of the current approach is found in the possibility of evaluating the system with the help of simulations (as it was initiated by the WAMASTEX developers). Evaluation of a decision strategy that allows for user interaction is much more extensive.

The basic assumption behind the sample rules is that there should at least be a weak correlation between the measured variables. This, like any heuristics, remains subject to discussion. Particularly, the correlation coefficient is an indicator for association. The expert statistician would probably more carefully look at the data in the absence of any correlation, but one cannot conclude independence. The fact that WAMASTEX aborts model evaluation in this case cannot be criticized, since the system is unable to make a qualified model evaluation. The assumption of a non-linear relationship, if only Spearman's rank correlation coefficient significantly differs from zero, is based on the following property: Spearman's correlation coefficient is better suited for curved associations than Pearson's correlation coefficient which is a measure of linear association.

Fröschl et al. (1994) write: "However, since model assessment cannot be forgone practically, it must rely on heuristic criteria to a considerable degree (though this approach, of course, is prone to various theoretical objections). It remains to be seen, if the WAMASTEX approach strikes an acceptable balance between decision-theoretic feasibility and practical problem coverage." The use of heuristic-based decision strategies will always result in a controversial discussion. It is a contribution of the WAMASTEX project to make this discussion possible.

In conclusion, although its development is not concluded, WAMASTEX is an interesting research project with promising ideas concerning the formalization of statistical knowledge that should be examined more closely. The research aim is more formalizing, recompiling and automating statistical knowledge than the provision of a tool for routine use. The implementation has some limitations due to the tools chosen and the aim of extensive automation without user interaction.

The challenge

State of the art

Currently, a handful of knowledge-based systems in statistics are commercially available, together with some ten well developed prototypes. Not all of these systems are still being developed and maintained.

There are various reasons for the limited spread of such systems. Common named deficiencies of existing systems are (Haux (1989), Molenaar (1988), Wittkowski (1990)):

- insufficient formal knowledge representation,
- insufficient quality of the knowledge,
- earlier experiences are ignored,
- restricted portability,
- prototype character.

A critical reviewer could observe that the complexity of the problems is often underestimated, and the hopes of researchers are too optimistic in the short run. There is increasing evidence that this judgement is more and more accepted in the research community. What then is the future for knowledge-based systems in statistics?

Future challenges

Concerning the first deficiency, there has been much experience with knowledge representation in general. This experience should be combined with specific experience in statistics in order to identify appropriate representations for statistical knowledge. Certainly, there is no universal representation form suitable for any statistical problem (for example, experimental design and data analysis are quite different tasks with quite different knowledge), but I am convinced that a specific solution could be found for most of the problems attacked in the past or future.

The quality assurance of the knowledge represented is probably the main problem of today's knowledge-based systems, since individual knowledge is heuristic and contradictory. Tools for structured knowledge acquisition and editing could support this task, but in the end, experienced statisticians must be asked to formalize their knowledge, and to critically evaluate the results inferred from this knowledge.

Concerning the issue of earlier experiences it can only be stated that any experience is worthful. This earlier work bore fruit, so why should it be ignored?

Portability is an important aim of software development, but there are serious problems since tools are often very different among computational environments. Developers should check whether the environment they use is not too specialized, i.e. can easily be made available at other sites.

The fact that most of the available systems are only prototypes can be changed only if research groups work continuously in this field, are provided with both statistical and computational skills, and with appropriate resources in hardware, software and staff.

In conclusion, the methodology of knowledge-based systems is promising for the field of statistics. Not only could users benefit from more sophisticated systems, but also the subject of statistics might benefit if its knowledge is further structured and made explicit.

Acknowledgements

The author thanks Karl Fröschl, Armin Koch, Jeremy Wyatt, and the reviewers for very useful comments and criticism.

Trademarks

Data Collection Selection, Designer Research, EX-SAMPLE, Measurement & Scaling Strategist, Methodologist's Toolchest Professional, and Statistical Navigator are trademarks of The Idea Works, Inc.; SAS is a registered trademark of SAS Institute, Inc.; Smalltalk-80 is a trademark of ParcPlace Systems; Turbo Pascal is a registered trademark of Borland International, Inc.

References

Adèr, H.J. (1992): Protoshell: an empty shell to develop statistical knowledge based systems. In: Koenig (1992), 4-15.

ASA – American Statistical Association (1983): ASA Proceedings of the Statistical Computing Section. Washington: American Statistical Association.

Barra, J.R., Brodeau, F., Romier, G. & Cutsem, B. van (eds) (1977): Recent developments in statistics. Amsterdam: North Holland.

Blum, R.L. (1980): Automating the study of clinical hypotheses on a time-oriented database: The RX project. In: Lindberg, Kaihara (1980), 456-460.

Blum, R.L. (1982): Discovery and representation of causal relationships from a large time-oriented clinical data base: The RX project. Berlin: Springer.

Brent, E.E. (1989a): Data Collection Selection^{TM}: An expert system to assist in the selection of appropriate data collection procedures, version 1.0. Columbia, MO: The Idea Works, Inc.

Brent, E.E. (1989b): Measurement & Scaling Strategist^{TM}: An expert system to assist in the design of questionnaires. Columbia, MO: The Idea Works, Inc.

Brent, E.E. (1989c): Statistical expert systems: an example. Journal of Statistical Computation and Simulation 31, 103-110.

Brent, E.E. (1989d): Statistical Navigator^{TM}: An expert system to assist in selecting appropriate statistical analyses, Version 1.1. Columbia, MO: The Idea Works, Inc.

Brent, E.E. & Mirielli, E. (1989): Designer Research^{TM}: An expert system to assist in designing research projects. Columbia, MO: The Idea Works, Inc.

Brent, E.E., Scott, J.K. & Spencer, J.C. (1988): EX-SAMPLE^{TM}: An expert system to assist in designing sampling plans. Columbia, MO: The Idea Works, Inc.

Breuker, J. & Wielinga, B. (1989): Models of expertise in knowledge acquisition. In: Guida, Tasso (1989), 265-295.

Buchanan, B.G., Sutherland, G.L. & Feigenbaum, E.A. (1969): Heuristic DENDRAL: a program for generating explanatory hypotheses in organic chemistry. In: Meltzer, Michie (1969), 209-254.

Campbell, N.A. & Woodings, T.L. (1981): Improved diagnostic output from statistical packages. In: ISI (1981), invited paper, I.P. 18.1.

Chambers, J.M. (1981): Some thoughts on expert software. In: Eddy (1981), 36-40.

Chowdhury, S., Linnarsson, R., Wallgren, A., Wallgren, B. & Wigertz, O. (1990): Extracting knowledge from a large primary health care database using a knowledge-based statistical approach. Journal of Medical Systems 14, 213-225.

Chowdhury, S., Wigertz, O. & Sundgren, B. (1989): A knowledge-based system for data analysis and interpretation. Methods of Information in Medicine 28, 6-13.

Clocksin, W.F. & Mellish, C.S. (1987): Programming in Prolog. 3rd ed., Berlin: Springer.

Cox, D.R. & Snell, E.J. (1981): Applied statistics: Principles and examples. London: Chapman and Hall.

Darius, P.L. (1988): Statistical expert systems: Some implementation and experimentation aspects. Österreichische Zeitschrift für Statistik und Informatik 18, 181-185.

Dodge, Y. & Whittacker, J. (eds) (1992): Computational Statistics. Volume 2. Proceedings of the 10th Symposium on Computational Statistics. COMPSTAT'92. Heidelberg: Physica.

Dorda, W., Fröschl, K.A. & Grossmann, W. (1990): WAMASTEX – Heuristic guidance for statistical analysis. In: Momirovíc K, Mildner V (eds): COMPSTAT 1990. Heidelberg: Physica, 93-98.

Dorda, W., Schneider, B., Fröschl, K.A. & Grossmann, W. (1988): Ein statistisches Expertensystem-Modell am Beispiel deskriptiver Statistik. Österreichische Zeitschrift für Statistik und Informatik 18, 172-180.

Dreyfus, H.L. (1979): What computers can't do: the limits of artificial intelligence. Revised ed., New York: Harper & Row.

Eddy, W.F. (ed.) (1981): Computer science and statistics: Proceedings of the 13th Symposium on the Interface. New York: Springer.

EUROSTAT (ed.) (1989): Development of statistical expert systems. Proceedings of the seminar held in Luxembourg in December 1987. Luxembourg: Office for Official Publications of the European Communities.

EUROSTAT (ed.) (1992): New technologies and techniques for statistics. Proceedings of the conference, Bonn, 24 to 26 February 1992. Luxembourg: Office for Official Publications of the European Communities.

Faulbaum, F. (ed.) (1992): SoftStat'91. Advances in statistical software 3. Stuttgart: Fischer.

Faulbaum, F., Haux, R. & Jöckel, K.-H. (eds) (1990): SoftStat'89. Fortschritte der Statistik- Software 2. Stuttgart: Fischer.

Feigenbaum, E.A., Buchanan, B.G. & Lederberg, J. (1971): On generality and problem solving: A case study using the DENDRAL program. In: Meltzer, Michie (1971), 165-190.

Feigenbaum, E.A. & Feldman, J. (eds) (1963): Computers and thought. New York: McGraw- Hill.

Finney, D.J. (1974): Problems, Data, and Inference (with comments). Journal of the Royal Statistical Society A137, 1-23.

Fröschl, K.A. (1992a): A formal model evaluation approach to the analysis of treatment effects in paired sample data. Technical report. University of Vienna, Institute of Statistics and Computer Science.

Fröschl, K.A. (1992b): Formalizing Statistical Analysis: Approaches and Prospects. In: Pichler, Moreno-Diaz (1992), 225-238.

Fröschl, K.A., Grossmann, W., Dorda, W. & Sachs, P. (1994): The WAMASTEX user-interaction design. The case of metric paired sample data. Presented at SoftStat'93 and submitted to: Faulbaum, F. (ed.) (1994): SoftStat'93. Advances in statistical software 4. Stuttgart: Fischer (to appear).

Gale, W.A. (ed.) (1986a): Artificial intelligence and statistics. Reading: Addison-Wesley.

Gale, W.A. (1986b): REX review. In: Gale (1986a), 173-227.

Gale, W.A., Hand, D.J. & Kelly, A.E. (1993): statistical applications of Artificial Intelligence. In: Rao, C.R. (ed.): Handbook of Statistics, Vol.9: Computational Statistics, 535-576.

Gale, W.A. & Pregibon, D. (1983): An expert system for regression analysis. In: Heiner, Sacher & Wilkinson (1983), 110-117.

Goldberg, A. & Robson, D. (1989): Smalltalk-80^{TM}: the language. Reading: Addison- Wesley.

Govaerts, B., De Toffol, B., Dutron, Y. & Florins, P. (1992): SEPSOL: a knowledge-based system for experimental design. In: Koenig (1992), 31-36.

Graber, A. (1989): Ein Expertensystem (ES) zur Planung und Durchführung kontrollierter klinischer Therapiestudien (KKT). Dissertation, Universität Innsbruck.

Guida, G. & Tasso, C. (eds) (1989): Topics in expert system design. Amsterdam: Elsevier.

Hand, D. (ed.) (1991): AI and computer power: The impact on statistics. Middlesex, UK: Unicom Seminars Ltd.

Hatabian, H. & Augendre, H. (1991): Expert Systems as a tool for statistics: a review. Applied Stochastic Models and Data Analysis 7, 183-194.

Haux, R. (ed.) (1986): Expert systems in statistics. Stuttgart: Fischer.

Haux, R. (1989): Statistische Expertensysteme (with discussion). Biometrie und Informatik in Medizin und Biologie 20, 3-65.

Heiner, K.W., Sacher,. R.S. & Wilkinson JW (eds) (1983): Computer science and statistics: Proceedings of the 14th symposium on the interface. New York: Springer.

ISI - International Statistical Institute (1981): Bulletin of the International Statistical Institute, Proceedings of the 43rd session. Buenos Aires.

Jackson, P. (1990): Introduction to expert systems. 2nd ed., Wokingham: Addison-Wesley.

Jirků, P. & Havránek, T. (1990): Constructing an experimental expert system for exploratory data analysis. Computational Statistics Quarterly 4, 283-297.

Koenig, S. (ed.) (1992): Computational Statistics. Volume 3. Proceedings of the 10th Symposium on Computational Statistics. COMPSTAT'92. Neuchâtel: Presses Académiques Neuchâtel.

Lehmann, H.P. & Shortliffe, E.H. (1991): THOMAS: building Bayesian statistical expert systems to aid in clinical decision making. Computer Methods and Programs in Biomedicine 35, 251-260.

Lindberg, D.A.B. & Kaihara, S. (eds) (1980): MEDINFO 80. Amsterdam: North Holland.

Lloyd, J.W. (1987): Foundations of logic programming. 2nd ed., Berlin: Springer.

Locarek, H. (1988): Wissensbasierte Systeme zur Durchführung statistischer Analysen. Konzeption und Implementation eines Prototypen. Frankfurt: Peter Lang.

Marino, A.T., Distefano, J.J. & Landaw, E.M. (1992): DIMSUM: an expert system for multiexponential model discrimination. American Journal of Physiology 262, E546-E556.

Matthäus, D. (1992): EVER: Expertensystem für den statistischen Vergleich von Erhebungen von der Versuchsplanung bis zur Resultatinterpretation. Lizentiatsarbeit, Universität Bern.

Meltzer, B. & Michie, D. (eds) (1969): Machine Intelligence 4. Edinburgh: Edinburgh University Press.

Meltzer, B. & Michie, D. (eds) (1971): Machine Intelligence 6. Edinburgh: Edinburgh University Press.

Molenaar, I.V. (1988): To exist or not to exist. A Comment on Statistical Expert Systems. Letter to the editor. Statistical Software Newsletter 14(3), 127-128.

Momirovíc, K. & Mildner, V. (eds) (1990): COMPSTAT 1990. Heidelberg: Physica.

Musen, M.A. (1989): Automated generation of model-based knowledge-acquisition tools. London: Pitman.

Naeve, P. (1990): Is there something new that can be learned or solved by building statistical expert systems? In: Faulbaum, Haux, Jöckel (1990), 291-299.

Nelder, J.A. (1977): Intelligent programs, the next stage in statistical computing. In: Barra et al. (1977), 79-86.

Newell, A., Shaw, J.C. & Simon, H.A. (1957): Empirical explorations with the Logic Theory Machine: a case study in heuristics. Proceedings of the Western Joint Computer Conference (WJCC) 15, 218-239. Reprinted in: Feigenbaum & Feldman (1963), 109-133.

Newell, A., Shaw, J.C. & Simon, H.A. (1958): Chess-playing programs and the problem of complexity. IBM Journal of Research and Development 2, 320-335. Reprinted in: Feigenbaum & Feldman (1963), 39-70.

Newell, A. & Simon, H.A. (1961): Computer simulation of human thinking. Science 134, 2011-2017.

Oettinger, A.G. (1963): The state of the art of automatic language translation: An appraisal. Beiträge zur Sprachkunde und Informationsverarbeitung 1(2), 17-29.

O'Keefe, R.A. (1990): The craft of Prolog. Cambridge, MA: MIT Press.

Pichler, F. & Moreno-Diaz, R. (eds) (1992): Computer aided systems theory – EUROCAST'91. A selection of papers from the second international workshop. Berlin: Springer.

Portier, K.M. & Lai, P.-Y. (1983): A statistical expert system for analysis determination. In: ASA (1983), 309-311.

Prat, A., Catot, J.M., Lores, J., Galmes, J., Riba, A. & Sanjeevan, K. (1992): Construction of a statistical KBFE for eperimental design using the tools and techniques developed in the Focus project (Esprit II num. 2620). In: Dodge, Whittacker (1992), 245-250.

Rasch, D., Guiard, V. & Nürnberg, G. (1992): Statistische Versuchsplanung. Einführung in die Methoden und Anwendung des Dialogsystems CADEMO. Stuttgart: Fischer.

Rasch, D., Guiard, V., Nürnberg, G., Rudolph, E. & Teuscher, F. (1987): The expert system CADEMO: Computer-aided design of experiments and modelling. Statistical Software Newsletter 13(3), 107-114.

Robinson, J.A. (1965): A machine-oriented logic based on the resolution principle. Journal of the ACM 12(1), 23-41.

Rödel, E. & Wilke, R. (1990): A knowledge-based system for testing bivariate dependence. Statistical Software Newsletter 16, 7-12.

Schreiber, G., Wielinga, B. & Breuker, J. (eds) (1993): KADS. A principled approach to knowledge-based system development. London: Academic Press.

Sterling, L. & Shapiro, E. (1987): The art of Prolog: advanced programming techniques. 4th ed., Cambridge, MA: MIT Press.

Streitberg, B. (1988): On the nonexistence of expert systems – critical remarks on artificial intelligence in statistics (with discussion). Statistical Software Newsletter 14, 55-74.

Turing, A.M. (1950): Computing Machinery and Intelligence. Mind 59, 433-460. Reprinted in: Feigenbaum, Feldman (1963), 11-35.

Weizenbaum, J. (1966): ELIZA – A computer program for the study of natural language communication between man and machine. Communications of the ACM 9, 36-45.

Weizenbaum, J. (1976): Computer power and human reason: from judgement to calculation. San Francisco: WH Freeman.

Winston, P.H. & Horn, B.K.P. (1989): LISP. 3rd ed., Reading: Addison-Wesley.

Wittkowski, K.M. (1985): Ein Expertensystem zur Datenhaltung und Methodenauswahl für statistische Anwendungen. Dissertation, Universität Stuttgart.

Wittkowski, K.M. (1990): Statistical knowledge-based systems – critical remarks and requirements for approval. In: Momirovíc, Mildner (1990), 49-56.

Wittkowski, K.M. (1992): A knowledge-based interface to statistical analysis systems – report on the current status of the MS-DOS implementation. In: Faulbaum (1992), 57-64.

Wolstenholme, D.E. & Nelder, J.A. (1986): A front end for GLIM. In: Haux (1986), 155-177.

Wyatt, J., Altman, D., Heathfield, H. & Pantin, C. (1994): Development of Design-a-Trial, a knowledge-based critiquing system for authors of clinical trial protocols. Computer Methods and Programs in Biomedicine (to appear).

Wyatt, J., Altman, D., Pantin, C., Heathfield, H., Cuff, R., Foster, N. & Peters, M. & Singleton, V. (1993): Design-a-Trial: a knowledge-based aid for writers of clinical trial protocols. In: Reichert, A., Sadan, B.A., Bengtsson, S., Bryant, J., Piccolo, U. (eds) (1993): Proceedings of MIE'93. London: Freund, 68-73.

Diagnostic Plots for One-Dimensional Data

Günther Sawitzki
StatLab Heidelberg, Im Neuenheimer Feld 294,
D 69120 Heidelberg, Germany

Key Words:
diagnostic plots, data analysis, excess mass, shorth plot, density estimation

Abstract
How do we draw a distribution on the line? We give a survey of some well known and some recent proposals to present such a distribution, based on sample data. We claim: a diagnostic plot is only as good as the hard statistical theory that is supporting it. To make this precise, one has to ask for the underlying functionals, study their stochastic behaviour and ask for the natural metrics associated to a plot. We try to illustrate this point of view for some examples.

Introduction

Though the general contribution of diagnostic plots to statistics is accepted, sometimes diagnostic plots seem more of a fashion than a tool. There are uncountable possibilities to design diagnostic plots, not all being of equal use. Diagnostic plots can and should be judged the same way as any other statistical method. We have to ask: What is their power? What is their reliability? While we may have to stay with accidental notes or examples for some time now, in the end a diagnostic plot is only as good as the hard statistical theory that is supporting it.

For many diagnostic plots, we are still far from having this theory. For some plots, we have to ask: What precisely are they trying to diagnose? How do we judge their reliability or confidence? For other plots, we know at least the statistical methods they are related to. In this paper, we consider plots as views on probability measures: We relate plots to functionals, operating on probability measures. If we have a functional defining a plot, we can proceed in three steps. We can ask which features are exhibited by the functional, and which are collapsed. As a second step, we can analyze what is retained by the empirical version and which stochastic fluctuation is to be expected. Third, we can optimize the functional and its empirical version to gain maximal power.

Where possible, we try to indicate classical tests related to the plot. If these tests meet the core of the plot, the power of the plot may be identified with and judged by the power of these tests. The associated functional may even indicate a notion of distance, or a metric, associated with the plot. We can use this to find natural neighbourhoods of a given empirical plot, leading to confidence sets of compatible models.

We restrict ourselves to a very modest case: Assuming a continuous distribution on the real line, we look at diagnostic plots based on a sample from this (unknown) distribution. We exclude some of the more difficult questions: we assume independent sample points with identical distribution. So we do not look at plots for the diagnosis of dependency, trend, heteroscedasticity or other inhomogeneities. As is to be expected, the opportunity is taken to advocate some new plots: the silhouette, the densitogram, and the shorth plot.

Diagnostic plots, what do we need them for?

We use diagnostic plots to investigate a data set by itself (a descriptive problem), or in comparison to a model distribution or model family (the one-sample-problem), or to compare two data sets (the two-sample-problem).

In a classical framework, we may want to apply a specific method, like regression or analysis of variance. The validity of this method will depend on conditions, often on conditions which in principle cannot be verified. The rôle of a diagnostic plot is that of a detector: Since we are unable to verify the preconditions, we use diagnostic plots to have at least a warning instrument.

In other situations, we may still be exploring. We have not settled on a specific model or method, but are looking at what the data are telling us. In the next step we select a certain model or decide to apply a certain method. Here diagnostic plots are a means to navigate through the models or methods at our disposition, and should be considered a model selection tool.

In any case diagnostic plots could be considered in a decision framework, either as filtering out bad situations after applying a model/method, or as selecting a model/method to be applied afterwards. It would be most appropriate to judge diagnostic plots as one step in an analysis process. But still too little is known about the interplay between use of diagnostic plots and application of formal models/methods.

The data we feed into diagnostic plots are rarely raw data. Often we use diagnostic plots on residuals. Of course the conditions we have to check refer to the error terms. The residuals are only some (model dependent) estimators of the errors. So the true story will be more complex than the i.i.d. simplification told here.

Where no model dependence is included initially, we still have had some choice how to measure the data: what we consider to be the data is a result of our choice of a measurement process. This can be a practical choice, or this may be culture dependent. Even in simple examples it may be more than just a linear change of scale (for example, energy consumption in a car is measured in miles per gallon in the U.S.A., and as litre per 100 km in Europe). Sometimes detection of the "proper" scale is the major achievement. For example the Weber-Fechner law in psychology tells us that the amount of enery which must be added to a stimulus to produce a detectable difference is proportional to the energy level of the stimulus. Hence using a logarithmic scale may be more appropriate for perception experiments than a linear energy scale. Choice of (nonlinear) scale may be a major application of diagnostic plots. Identifying the shape of a distribution is equivalent to finding a way to transform it to some model distribution.

What do we look for in diagnostic plots?

We use diagnostic plots to check for special features revealed by or inherent in the data. Of course, these checks are useful only if we know how presence or absence of these features affects the statistical methods we are going to apply (but then, if we do not know this, it would be wise not to apply these methods at all). Usually, for ordinary statistical applications, there are only few features we have to check. Here is a short check list:

Missing or censored data. Contrary to what classical statistics would like to see, real data sets usually contain registrations meaning "below detection level", "not recorded", "too large". In survival analysis, respecting missing or censored data is a mark of the trade. Although missing or censored data are a pending challenge in practical statistics, we will not deal with this problem here.

Discretization. Usually, all data we record are discretized (truncated or rounded to some finite precision, for example). For methods based on ranks, this may lead to ties, with appropriate corrections being well known. For other methods, these effects are often grossly ignored, although it would be easy to take them into account in tests of t or F type.

Multimodality. Sometimes, multimodality is a hint to a factor which separates the modes and should be included in the analysis. In other cases, as for instance in psychological preferences and choices, multimodality may be an inherent feature. Classical methods have notorious pitfalls if multi-modal distributions are involved.

Symmetry and skewness. In best cases, skewness is an indicator for power transformations which might bring the data to a simpler model.

Tail behaviour. Many classical methods are strongly affected by the tail behaviour of the distribution. Sometimes, tail problems may be avoided by going to more robust methods.

In real applications, we cannot assume an i.i.d situation. Hence we have to check for dependency, trend, heteroscedasticity or other inhomogeneities as well. But for now we restrict ourselves to the i.i.d assumption.

Diagnostic plots, what are they, anyway?

The general aim of data analysis is to find interesting features in data, and to bring them to human perception. In doing this, data analysis has to avoid artifacts coming from random fluctuation, and from perception (Sawitzki (1990)). Diagnostic plots are plots tuned to serve these purposes. It depends on the context and on our intentions to say what is an interesting feature. For a general discussion, we have to ask which features can be brought to perception by a certain plot.

Full information is contained in the graph of, say, the probability density. We can easily get information about the relative location of the means, or about the standard deviations, or many other details from a plot of the probability density (Figure 1).

Although the full information is contained in this plot, it may not always come to our perception. In Figure 1, for example, one density is that of a Cauchy distribution. But the plot does not call our attention to the tail behaviour. It does not tell us that estimating the difference of the means may not be a good idea here, and it does not tell us that attempts such as studentization will run astray.

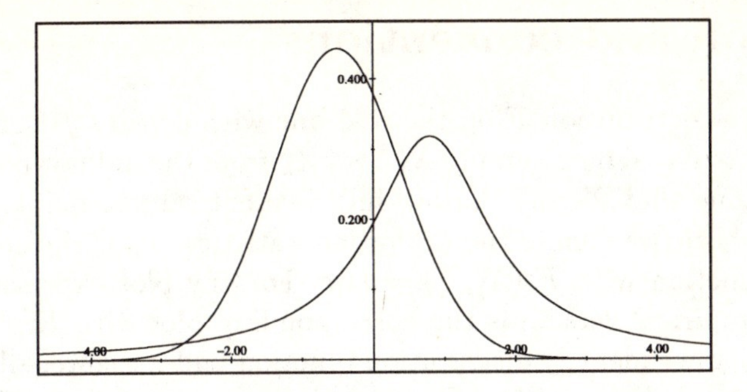

Figure 1: Densities. Differences of the means or standard deviations can be read easily from a plot of the density, but the density plot may as well be grossly misleading.

Perception is one side of the problem, and many discussions between specialists in this field and statisticians still may be necessary. There is standard literature addressing these problems: Bertin (1967, 1977, 1981) and Tufte (1990) are rich sources of possibilities. Tufte (1983) and Chambers et al. (1983) is the basic literature from a statistical point of view. Wainer (1984) is a classical article on pitfalls to be avoided.

Some of elementary lessons we have learned:

- Perception is background dependent. Avoid chart junk and background/foreground interaction.
- Visual discrimination is powerful for linear or regular structures, but weak for general curves. If you have a model case, try to represent it by a straight line or a regular structure.
- Perception knows quality and quantity. Avoid encoding quantitative information by qualitative features (such as colour).
- Perception has more dimensions than one. Make sure the information you are presenting is encoded in appropriate dimensions. In particular: avoid using 3d-effects, unless you know exactly how they are perceived.

It is left as an exercise to look for examples in which these elementary lessons are disregarded.

Here we will concentrate on the statistical side: we ask for the features (or functionals) of the distribution being represented, and for the fluctuation involved. The Cauchy example given above should be a warning: even in the absence of fluctuation, a plot may not tell the whole story. Reducing fluctuation, or only minimizing it (by some optimal choice of parameters such as a bandwidth) by itself is not a guarantee for the usability for a certain purpose.

Notation and conventions

We assume a distribution F on the real line with density f and look at diagnostic plots based on a sample $X_1, \ldots, X_n$ from the unknown distribution F. We assume that $X_1, \ldots, X_n$ are independent sample points. By $X_{(i:n)}$, or $X_{(i)}$ for short, we denote the i.th order statistics. F_n is the empirical distribution function with $F_n(X_{(i:n)}) = i/n$. For any plot expressed in terms of F, the empirical version is the corresponding plot with F_n replacing F. For simplicity, we identify distribution function and measure, allowing us to write $F((a, b]) = F(b) - F(a)$. For any plot we try to follow this sequence: We give a rough recipe how to generate the plot. Then we try to give a functional definition of the plot. Which features are preserved and which are lost by the functional? After that we study the fine points: what needs to be corrected in the rough plot? Then we turn to related tests and discuss optional choices. After each plot, we ask: how far have we got? What is the information we can gain so far, and what is still missing?

Histogram

Recipe:

Chose histogram bins. For any bin, mark the hit count of data points hitting this bin.

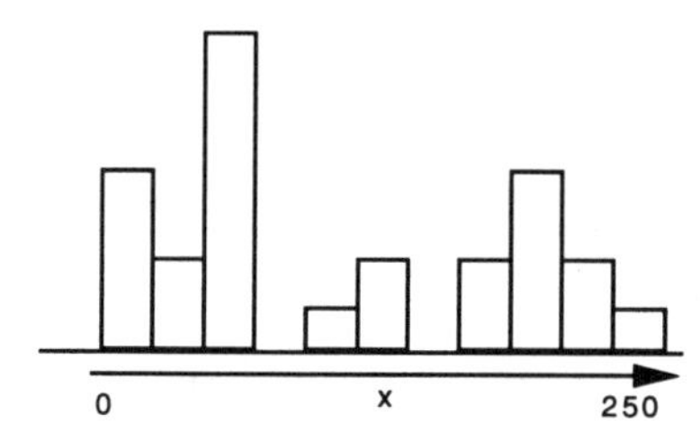

Figure 2: Histogram

The underlying functional

A histogram is "the" classical way to present a distribution. Its historical advantage is the ease of calculation - it can be reduced to putting registration notes into bins. The functional corresponding to a histogram is a discretization of the density: Given a decomposition of the real line into disjoint intervals $A_j, 0 \leq j \leq k$, we can define a histogram as the distribution, discretized to these intervals. The discretization gives a probability $p_j = F(A_j)$

for bin A_j. The vector of observed bin counts $n_j := \#\{i : X_i \in A_j\}$ has a multinomial distribution. Using Pearson's approximation (Pearson (1900))

$$P(n_1, \ldots, n_k) \approx (2\pi n)^{-1/2} (\prod p_j)^{-1/2} \exp[-1/2 \sum (n_j - np_j)^2 / np_j + \ldots]$$

we see that the χ^2-test statistics, the leading term in the exponential, is controlling the bin hit frequency for sufficiently large expected bin counts np_j.

Knowing the associated functional and its stochastic behaviour, we can tell what is to be expected from a histogram. A histogram can only show features which are preserved by the underlying functional, the discretization of the distribution. We loose all smoothness properties and local details of the distribution. The metric associated most naturally to the histogram is a χ^2-metric. So if we want to have an impression about the distributions which are compatible with our data, we should consider χ^2 confidence bands. To obtain information about the power of the histogram, we can look at the χ^2-test as a corresponding goodness of fit test.

Practical situations are slightly more complex. One complication may arise from the sampling scheme. A common case is to sample for a certain time, instead of taking a fixed sized sample. This makes the total sample size a random variable. Under independence assumptions the vector of bin counts has a multivariate poisson distribution instead of a multinomial. A similar limit applies, but we gain one degree of freedom in the limit. Another complication may arise if we define our bins in a data dependent way. If the number of bins is small compared to the number of observations, the approximation still holds with good quality even if we use the data first to estimate location and scale, and use bins based on these estimators. We still have to correct for the degrees of freedom in the goodness of fit test.

If we have fixed reference distributions, we can head for an optimal choice of bins. Common strategies are to take bins of equal probability with respect to the reference distribution, or to take bins of constant width with cut points $a_j = a_0 + j \cdot h$, $0 \le j \le k$, for some bin width h, definig bins $A_0 = (-\infty, a_0]$, $A_j = (a_{j-1}, a_j]$. This still leaves us with the problem to decide about the number of bins. The classical recommendation for equal probability bins is to let the number of bins increase as $n^{2/5}$ (Mann & Wald (1942)). A more detailed study of optimal choices of the bins with respect to the χ^2-statistics for various types of reference distributions is presented in Kallenberg et al. (1985).

Walking through our short check list, it is an easy exercise to see how histograms can be modified to compensate for known censoring or discretization. If the data recording is discretized (for example by cutting down to a small number of reported digits) and the histogram is not adapted, this discretization may interfere with the histogram discretization and lead to patterns - an effect known as "aliasing". Tuning a histogram to look for multimodality is a problem. Tuning it to identify asymmetry or tail behaviour again is comparably simple.

Perception of the histogram is a different matter. Prominent features, such as local maxima or general skewness, are perceived first: the eye does not perform the rescaling which would be necessary from a statistical point of view. Choice of bin width and bin locations are most critical to histograms (Figure 3). The rule is: if you use a histogram, don't use one. Use several.

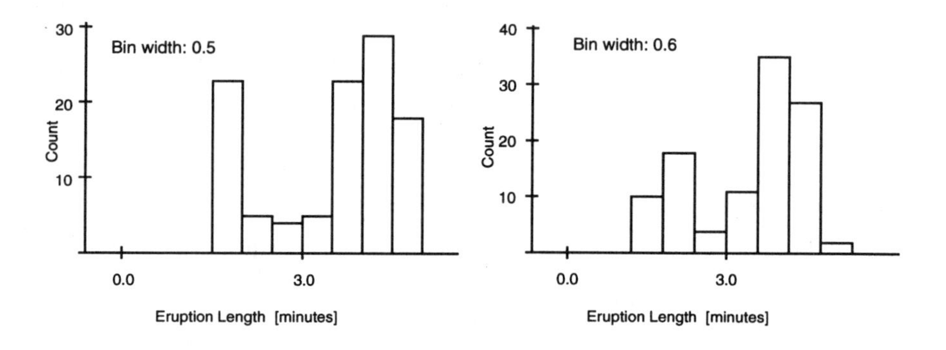

Figure 3: Histogram of eruption lengths of Old Faithful geyser. Same data set, but different bin widths. After Silverman (1986).

Using histograms to analyze for discretization effects or multimodality is inherently unsafe. A fairly safe use of histograms is for restricted purposes. Using a small number of cells to check for symmetry or tail behaviour is rather stable. But these purposes can be followed more effectively using Tukey's Box & Whisker plot, as discussed below.

Histogram density estimators

For the empirical version with bin counts $n_j := \#\{i : a_{j-1} < X_i \leq a_j\}$, the empirical histogram takes value $\widehat{p_j} = n_j/n$ on A_j. We can view a histogram as an estimator for the density, the histogram density estimator being defined by $\hat{f}(x) = \widehat{p_j}$ for $x \in A_j$. Judging the quality of this estimator by the integrated mean square error, $IMSE = \int E\{\hat{f}(x) - f(x)\}^2 dx$, at least for the case of constant bin width h asymptotic results are available (Scott (1979)). For the histogram density estimator,

$$IMSE = \frac{1}{nh} + \frac{1}{12}h^2 \int f'(x)^2 dx + O(\frac{1}{n} + h^3),$$

that is the error goes to zero in order $o(n^{-2/3})$. The integrated mean square error is minimized for $h^* = n^{-1/3}\{6/\int f'(x)^2dx\}^{1/3}$. In particular, for the Gaussian distribution, $h^* \approx 3.49 \cdot \sigma \cdot n^{-1/3}$. Choosing a suboptimal bin width $h = c \cdot h^*$ gives an error $IMSE_h \approx IMSE_{h^*}(c^3+2)/30$. For example: choosing a bin width as 50% of the optimal bin width increases the error by 42%. Under regularity assumption, the optimal bin width requires approximately at least $(2n)^{1/3}$ bins. An upper bound for the bin width is $3.55 \cdot \sigma \cdot n^{-1/3}$. The regularity assumptions are: $\int f'(x)^2dx > 0, \int f''(x)^2dx < \infty$.

But judging histograms by the integrated mean square error of the histogram density estimator is not fair. First, in constructing the histogram we deliberately restricted possible estimators to stepwise constant functions. Using the same information as contained in histograms, we can generate better density estimators by allowing piecewise linear estimators. For example we can join the bin centres of a histogram. The frequency polygon, the resulting density estimator, can achieve better error rates (Terrell & Scott (1985)). For their optimal bin width $h^* = 2[15/49n^{-1} \int\{f''(x)\}^2dx]^{1/5}$ the error decreases as $n^{-4/5}$ under regularity conditions: f'' absolutely continuous, $f''(x)^2dx > 0$, $f'''(x)^2dx < \infty$. Under these regularity conditions, we need at least $(147n/2)^{1/5}$ bins to achieve optimal bin width. An upper bound for the optimal bin width is $2.24 \cdot \sigma \cdot n^{-1/5}$. In particular, for the Gaussian distribution, $h^* \approx 2.15 \cdot \sigma \cdot n^{-1/5}$.

Thus the quality of the histogram, interpreted as a density estimator, can be greatly improved by allowing a slightly modified class of derived density estimators. But IMSE still is not an adequate criterion to judge histograms. Histograms are discretized versions of the distribution. The empirical histogram should be judged in comparison to this discretization, and competitors should be ranked on how they perform at (possible discretized) data.

Scatterplots

Recipe:

Mark the data points.

Figure 4: Scatterplot. Same data as in figure 2.

The underlying functional

The functional corresponding to the scatterplot is the density; the empirical version is $\sum \delta_{X_i}$. Problems may arise from discretizations or drawing resolutions. These effects may lead to ties in the empirical version. A simple technical solution is to use a gray pen, instead of a simple black pen. If the pen has gray level $K(y)$ at distance y from its centre, we get gray level $\sum K(x - X_i)$ at a point x: the plot corresponds to a kernel density estimator with kernel K, or to the smoothed density as a functional. Any pen (or kernel) gives rise to a whole scaled family K_h, with $K_h(y) = 1/hK(y/h)$, where h is the pen size, or bandwidth. For the functional, this means replacing the density f by the smoothed density $f * K_h$, the convolution of f and K_h. Of course you gain smoothness but loose details by using a larger bandwidth.

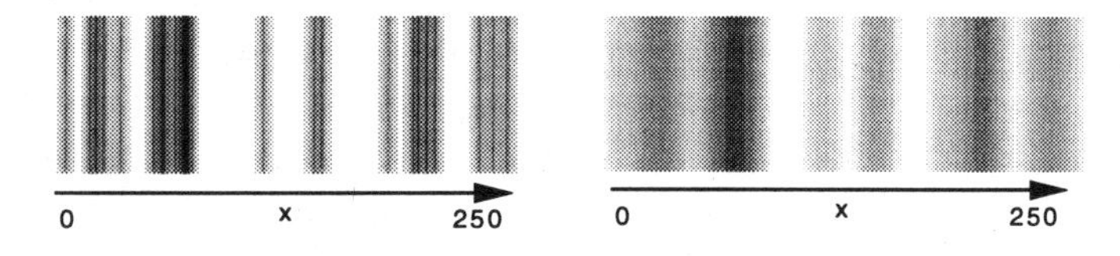

Figure 5: Scatterplot, using two different gray pens. Same data as in figure 4

Scatterplots as such can be barely considered diagnostic tools, but they may provide helpful orientation when combined with other plots: we continue to use them, but it is hard to pin down what we gain. To use them for diagnostic purposes, they must be enhanced. For example to check the symmetry behaviour, we can look at symmetry centres. An appropriate plot suggested by J.Tukey (after Wilk & Gnanadesikan (1968)) is to show $(X_{(n-i+1)} + X_{(i)})$, plotted against the distance $X_{(n-i+1)} - X_{(i)}$. To check for discretization effects, we can look at the plot of the differences $X_{(i+1)} - X_{(i)}$ against $X_{(i)}$.

Smoothed scatterplots and kernel density estimators

In principle, the complete information of a sample is represented in a scatterplot. Perception however is easily trapped by sample size effects: small sample sizes will give the impression of pattern and inhomogeneities even for uniform samples; large sample sizes will hide non-uniformities for any distribution. This problem is even more complicated for kernel density estimators: we have the choice of a pen (or a kernel, if you like). Conventionally, this problem is split into two: choosing the pen shape (or kernel type) and the pen size (or bandwidth). We meet the same problem we have encountered with histograms: what we see depends critically on these choices. But we do not know how to judge these choices. There is a mathematical hide-away. If we accept that the density is our target functional, any distance measure between the (normalized) kernel density estimate and the true density can be used as a measure of fit, and of course L_2 distance is the easiest to deal with. Call $\hat{f}_h(x) = n^{-1}h^{-1}\sum K((x - X_i)/h)$ the kernel density estimator for kernel K and bandwidth h. Let $\hat{h}_0 = \hat{h}_0(f, X)$ be the smallest minimizer of the integrated square error $\Delta(h) = \int(\hat{f}_h - f)^2dx$ and h_0 the smallest minimizer of the mean integrated square error $M(h) = \int E(\hat{f}_h - f)^2dx$. Under regularity conditions, for any (empirical) bandwidth $\hat{h}$ we have

$$\Delta(\hat{h}) - \Delta(\hat{h}_0) = 1/2(1 + o_p(1))(\hat{h} - \hat{h}_0)^2 M''(h_0)$$

(Hall & Marron (1987)). While this does not help to estimate the error, it says that minimizing the integrated square error is essentially equivalent to optimizing the bandwidth for the data at hand. But $\hat{h}_0$ can be represented as $\hat{h}_0 = A_1 + n^{1/5}A_2\int f'^2dx + o_p(n^{-3/10})$, where A_1 and A_2 are functions of the data, not depending on f (Hall & Johnstone (1992)). A_2 does not vanish asymptotically. So determining an optimal bandwidth is related to estimating $\int f'^2dx$. An optimal rate of $n^{-1/2}$ for the estimation of $\int f'^2dx$ makes the relative error of approximating $\hat{h}_0$ at best of order $n^{-1/10}$ (Hall & Johnstone (1992)). These results tell us why optimal bandwidth selection is a hard problem even for very large sample sizes and continuous distributions, let alone for real data, that is for finite sample sizes and data truncated or rounded to some finite precision.

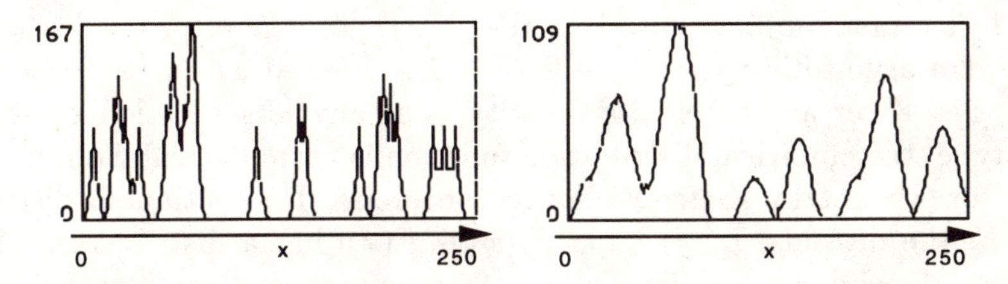

Figure 6: Pixel intensity for gray level plots of figure 5

It is possible to base goodness of fit tests on kernel density estimators (Mammen (1992, Ch. 3)). But the stochastic behavour of kernel density estimators is difficult. There is no clear notion of distance or variation associated to kernel density estimators. There are candidates, among them distances based on (penalized) square errors. These are a treatable mathematical concept, but L_2 confidence bands are not too helpful from a data analytical point of view. The information gained from scatterplots, including kernel density estimators, is doubtful. Checking the list of critical features given above, it is hard to spot a feature that is reliably detected and reported by a scatterplot. Silverman (1981) made an attempt to exploit kernel density estimators as a diagnostic tool to analyze for multimodality. A simpler approach, the densitogram (related to the excess mass test), is given below.

Distribution function and related plots

Recipe:

Sort the data points. For any point, mark the proportion covered (the frequency of data points not exceeding this point).

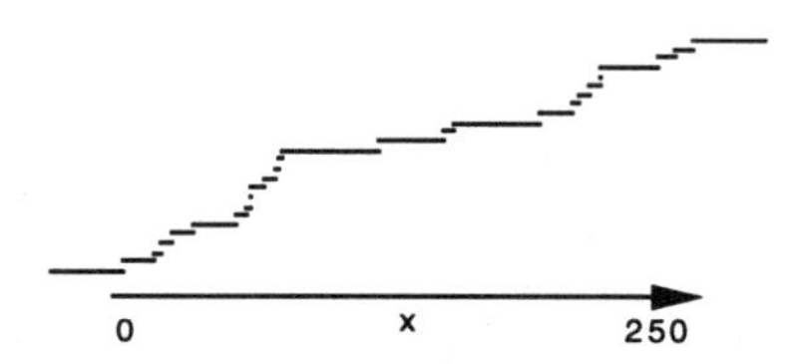

Figure 7: Distribution function

The underlying functional

The distribution function gives the probability of half-lines $F : x \mapsto F(x) = P[X \leq x]$. It can be estimated by its empirical version, F_n where $F_n(x_{(i:n)}) = i/n$. The stochastic behaviour is described by the Glivenko-Cantelli Lemma: we have $\sup_x |F_n(x) - F(x)| \overset{P}{\to} 0$. The error has a Brownian bridge asymtotics: $\sqrt{n}(F_n - F) \to Z$. Viewed as an estimator for F, F_n has a certain general optimality: For any loss function of supremum type, the empirical distribution function is asymptotically a minimax estimator (Dvoretzky-Kiefer-Wolfowitz-Theorem). For continuous distributions F, the distance $D_n = \sup_x |F_n(x) - F(x)|$ has a distribution which does not depend on F. This allows for simultaneous confidence bands: if c denotes the α-quantile of the Kolmogorov-Smirnov statistics, we have $F(x) \in [F_n(x) - c, F_n(x) + c]$ for all x with probability of at least $1 - \alpha$. The Kolmogorov distance is the metric associated most naturally to the distribution function; the distribution function is easy to reconstruct, and its statistics is well understood. Interpreting it needs some education.

Plots related to the distribution function

Comparing two distribution functions visually is quite difficult. We have to compare two graphs, both piecewise constant and monotonous. Most interesting features are hidden in details. We can help perception by using a transformation which gives a near-to linear graph for corresponding distributions. If we have a given reference distribution, our choices are to align quantiles by transforming the probability scale (the quantile-quantile-plot), or to align probabilities by transforming the data scale (the percentage-percentage plot).

Quantile-quantile-plot (Q-Q-plot)

Recipe:

Choose a reference distribution. Sort the data points. For any data point, find the proportion of observations not exceeding this data point. Plot the data point against the corresponding quantile of the reference distribution.

Q-Q-plot details

To transform the probability scale, we transform a probability to the corresponding quantile. The Q-Q-plot compares two distributions by plotting quantile against quantile. If F and G are the distributions to be compared, $X \sim F, Y \sim G$, the Q-Q-plot shows the curve $\alpha \mapsto (x_\alpha, y_\alpha)$. In terms of the probability distributions, this is the graph of $x \mapsto G^{-1} \circ F(x)$. Again, orientation has been chosen to give an easy empirical version $x \mapsto G^{-1} \circ F_n(x)$.

If F and G coincide, the Q-Q-plot is a diagonal line. If one is a linear transformation of the other, the Q-Q-plot is linear. The Q-Q-plot shows a high resolution in regions of low densities and vice versa. As a consequence, it emphasizes the tail behaviour for long-tailed distributions (Wilk & Gnanadesikan (1968)), and emphasis on the tails combines unluckily with high variation.

If G is the true distribution, $G = F$, the Q-Q-plot of F_n against F is given by $(X_{(i:n)}, x_{i/n})$ where $x_{i/n}$ is the i/n quantile. In particular, for $G = U[0,1]$ we have $x_{i/n} = i/n$; so in this case the Q-Q-plot coincides with the empirical distribution function.

If U_i are iid $\sim U[0,1]$, $U_{(i:n)}$ is distributed as $\beta(i, n-i+1)$. Hence in the general case $E(F(X_{(i:n)})) = E(U_{(i:n)}) = i/(n+1)$: the empirical version of the Q-Q-plot is biased. We can take this into account to get an "unbiased" empirical plot by using plot positions $(X_{(i:n)}, G^{-1}(i/(n+1)))$ for an empirical Q-Q-plot to compare the observations X_i with the model distribution G. This is the convention used by Weibull (1939). But getting the mean behaviour right is only one part of the difficulty. Since you will not apply a diagnostic plot to a mean situation, but to a sample, you are prone to be affected by the notorious skewness of empirical quantile distributions. This is the origin for many fine points to be considered in the actual mapping (Kimball (1960), Harter (1984)).

Direct relatives of the Q-Q-plot are goodness-of-fit tests based on the regression of order statistics on expected order statistics, like for example the Shapiro-Wilk test (Shapiro & Wilk (1965)).

Percentage-percentage plot (P-P-plot)

Recipe:

Choose a reference distribution. Sort the data points. For any data point, find the proportion of observations not exceeding this data point. Plot the proportion against the corresponding proportion of the reference distribution.

P-P-plot details

To transform the data scale for linearity, we have to transform X to the corresponding probability under the reference distribution. If F and G are the distributions to be compared, $X \sim F$, $Y \sim G$, the P-P-Plot shows the curve $X \mapsto (G(X), F(X))$. In terms of the probability distributions, this is the graph of $\alpha \mapsto F \circ G^{-1}(\alpha)$. We apply this with G in the role of a reference distribution. Orientation has been chosen here to avoid the discontinuities in F_n, that is to give an empirical version $\alpha \mapsto F_n \circ G^{-1}(\alpha)$.

If F and G are identical, the P-P-plot will be straight line. P-P-plots are not preserved under linear transformations: they are not equivariant. So usually P-P-plots will be applied only to distributions standardized for location and scale. For the empirical version, this preferably is done using robust estimators of location and scale. As for the Q-Q-plot, the skewness of the empirical quantile function should be considered in the actual mapping. But in contrast to the Q-Q-plot, for the P-P-plot, high variability is not combined with sensitivity in the tails. So choice of the proper plotting position is a fine point for the P-P-plot, whereas it is critical choice for the Q-Q-plot. Goodness-of-fit tests can be constructed based on the linearity of the P-P-plots (see Gan & Koehler (1990)).

Other plots related to the distribution function plot

The plots based on the distribution function suffer from the tail-orientation of the distribution function. It measures half infinite intervals, and local behaviour can be judged only by looking at differences. This is easy to compensate using a third dimension: you can localize the probability mass to intervals and define a probability mass plot $(a, b) \mapsto F(a, b]$, with the obvious empirical version. But readability and practical use are doubtful.

Box & Whisker-plot

Recipe:

Find the median and quartiles, and mark them. Connect the range of points which are not too far from the median (judged by the interquartile distance). Highlight all points which are out or far out.

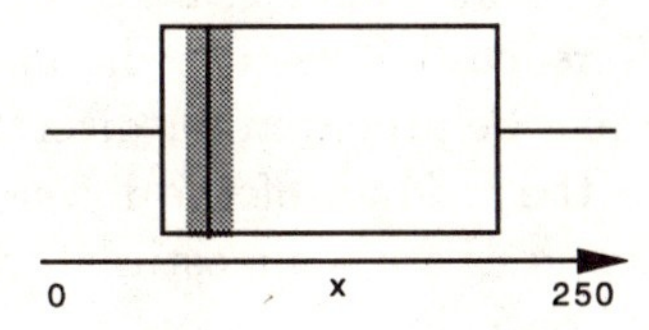

Figure 8: Box & Whisker plot, with a confidence interval for the median marked grey.

Box & Whisker plot details

In more detail, the construction is: Find the median of the data points, and mark it. Find the median of the subset above the general median, mark it, and call it the upper hinge. Find the median of the subset below the general median, mark it, and call it the lower hinge. Let Δ be the distance between the hinges. Draw a whisker from the box to the last data point not exceeding $upper\ hinge + 1.5 \cdot \Delta$. Mark all data points in the out area between $hinge + 1.5 \cdot \Delta$ and $hinge + 2.5 \cdot \Delta$. Highlight all far out points exceeding $hinge + 2.5 \cdot \Delta$. Do the corresponding for the lower hinge.

John Tukey's Box & Whisker plot is one of the gems of data analysis. Like the histogram, the Box & Whisker plot represents a discretization of the density. But where the histogram discretizes on the observation scale, the Box & Whisker plot discretizes on the probability scale. The discretization varies, from a rough 25%-discretization in the centre part, to a $1/n$ discretization for a sample size n in the tails.

The Box & Whisker plots achieve to present general information about the core of the data, with information hiding in this area. On the other hand, they highlight the exceptional. The exceptional data might be just tail effects, or it might be genuine outliers - they are worth a second look anyway.

The Box & Whisker plot is best understood by following its construction. Roughly, the Box & Whisker plot marks median and quartiles, and exceptional points. We will try to look at the ideas of Box & Whisker plots more carefully here. For the Box & Whisker plot, first we try to get an estimator for the location.The data median is used as the obvious (robust) candidate. The centre line of the Box & Whisker plot marks the median. Now we estimate the scale. Since we have already estimated the location, we can use this information. Given an estimator for the location, estimating the scale would be useful in exceptional cases: it would be meaningful only for symmetric distributions. Given the location estimator, we construct two scale estimators, a lower and an upper scale estimator. In the absence of ties, we could use the differences between median and lower/upper quartile as estimators. Since we must be prepared for discretization effects, we must be more careful. We use the median of the lower or upper half instead - Tukey's hinges. Finally, using these scale estimators, we estimate "central" areas, and mark all points outside.

Tukey's Box & Whisker plot takes into account many possibilities and pitfalls of real data sets. It is very easy to miss these fine points, as can be seen from popular software packages. The Box & Whisker plot is particularly powerful in analyzing the overall structure of a distribution, like location, scale and outliers. But it still leaves the needs to diagnose other features. Discretizations are in no way reflected in the Box & Whisker plot. The tail behaviour is made a caricature: if there are tails, outliers are identified. But if the tails are too steep, heavier tails are invented: even a uniform distribution is shown with tails. Multimodality is ruled out: the Box & Whisker plot knows about central location, but has no space for modes. It must be accompanied by other plots.

Silhouette and densitogram

Recipe:

Choose a family of sets serving as a model (e.g. sets composed of one or two intervals, if you are looking for bimodality). Choose a level λ. Mark the maximal set with average hit density exceeding level λ. Do this for a choice of levels λ.

The underlying functional

If you are looking for specific features in your data, it is possible to design diagnostic plots for these features. Silhouette and its accompanying plot, the densitogram, are plots tuned to inspect multimodality (Müller & Sawitzki (1987)). Both are based on the idea that a mode of a distribution is a location where the probability mass is concentrated. A corresponding

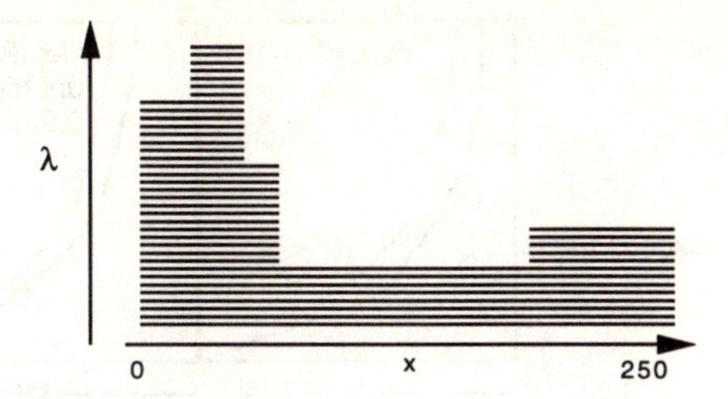

Figure 9: Silhouette. Locations of maximal excess mass $E(\lambda) := \int (f - \lambda)^+ dx$ for varying levels λ.

functional is the excess mass,

$$E(\lambda) := \int (f - \lambda)^+ dx,$$

giving the probability mass exceeding λ. Restricting the allowed sets in an appropriate way to a family $\mathcal{C}$, we define

$$E_{\mathcal{C}}(\lambda) := \sup_{C \in \mathcal{C}} \int_C (f - \lambda) dx = \sup_{C \in \mathcal{C}} (F - \lambda \mathrm{Leb})(C).$$

The silhouette marks the maximizing sets, for any level λ. The densitogram shows the excess mass, as a function of lambda. The cue lies in the freedom to choose $\mathcal{C}$. For unimodal distributions, $\mathcal{C} = \mathcal{C}_1$ should be the family of intervals and $E(\lambda) = E_{\mathcal{C}_1}(\lambda)$. For bimodal distributions, $\mathcal{C} = \mathcal{C}_2$ is made of the disjoint unions of two intervals and $E(\lambda) = E_{\mathcal{C}_2}(\lambda)$. Given a hypothesis on the modality, silhouette and densitogram can be estimated using the empirical excess mass

$$\widehat{E_{\mathcal{C}_m}(\lambda)} = \sup_{C \in \mathcal{C}_m} (F_n - \lambda \mathrm{Leb})(C), \quad m = 1, 2, \ldots.$$

Silhouette and densitogram details

As an estimator for the location of the mode, the silhouette shares a rate of order $n^{-1/3}$ with density estimation based methods. But the number of modes in the silhouette is more reliable even for small sample size. For the densitogram, the associated test is the excess mass test for multimodality (Müller & Sawitzki (1991)): $\sup_\lambda \widehat{E_{\mathcal{C}_2}(\lambda)} - \widehat{E_{\mathcal{C}_1}(\lambda)}$, the maximal difference between excess mass, estimated on the assumption of bimodalitity, and excess mass, estimated on the assumption of unimodality, can be used as a

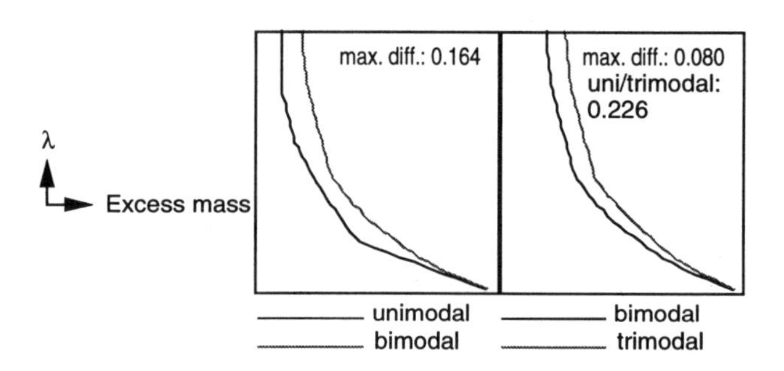

Figure 10: Densitogram, the excess mass concentration curve. Same data as figure 2. Excess mass estimated under assumption of uni- bi- and trimodality. By assuming bimodality, an additional excess mass of 16.4% of the data is covered.

test statistic for multimodality. For a bimodal distribution, the maximal excess mass difference $\sup_\lambda \widehat{E_{\mathcal{C}_2}(\lambda)} - \widehat{E_{\mathcal{C}_1}(\lambda)}$ is half the total variation distance, between F and the closest unimodal distribution. This points to the Kuiper metric as a distance measure related to excess mass.

On the unimodal distributions, the error rate of these excess mass estimates is of order $n^{-1/2}$. In more practical terms: the difference between both excess mass curves starts providing a reliable indicator for multimodality for a sample size n in the range 20 to 50.

Shorth-plot

Recipe:

Choose a coverage α. For any point, get the length of the shortest interval containing this point and covering at least an α-fraction of the data (at least $\alpha \cdot n$ data points). Do this for a selection of coverages α.

The underlying functional

The shorth is the smallest interval containing at least 50% the distribution:

$$S = \arg\min\{|I| : I = [a, b], P(X \in I) \geq 0.5\}.$$

Here $|I|$ is the length of an interval I and arg min is the minimizing argument. More generally the α-shorth is the smallest interval containing at least an α fraction of the distribution:

$$S_\alpha = \arg\min\{|I| : I = [a, b], P(X \in I) \geq \alpha\}.$$

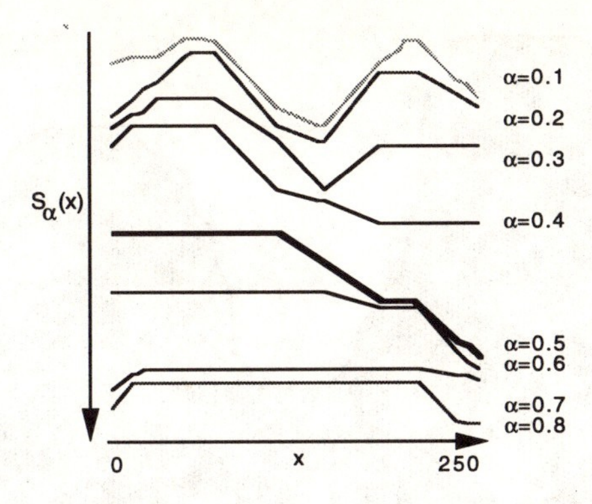

Figure 11: Shorth Plot. The shorth length axis points downwards.

For data analysis, we can localize the shorth. We define the α-shorth at x as the smallest interval at x containing at least a proportion α of the distribution

$$S_\alpha(x) = \arg\min\{|I| : I = [a,b], x \in I, P(X \in I) \geq \alpha\}.$$

In particular, the shorth at x is defined as $S(x) := S_{0.5}(x)$. The shorth plot is the graph of $x \mapsto |S_\alpha(x)|$ for a selection of coverages α.

More about shorth-plots

Andrews et al. (1972) use the (not localized) shorth to construct a robust estimator of location. The shorth procedure takes the centre of the empirical shorth as location estimator. Unfortunately this estimator of location has an asymptotic rate of only $n^{-1/3}$, with non-trivial limiting distribution. However Grübel (1988) shows that the length of the empirical shorth is a reasonable estimator of scale, converging with a rate of $n^{-1/2}$ to a Gaussian limit. This result can be carried over to the localized shorth, using a modified version of Grübel's proof (Sawitzki (1992)). Mass concentration now can be represented by the graph of $x \mapsto |S_\alpha(x)|$. A small length of the shorth signals a large mass concentration. To facilitate optical interpretation, we plot the negative of the lengths. The shorth plot was proposed as a means to investigate mass concentration by Sawitzki (1992). It is easy to compute, avoids the bandwidth selection problems, and allows scanning for local as well as for global features of the distribution. The good rate of convergence of the shorth estimator makes it useful already for moderate sample size.

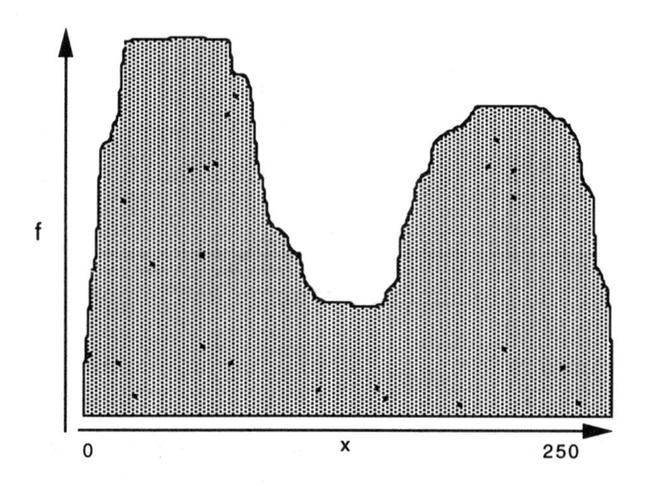

Figure 12: True distribution used for the examples; data points sampled by acceptance/rejection; sample size: 25 data points.

Summary

How far have we got? The general purpose plots (histogram, scatterplot, distribution function) can be applied, but provide doubtful information per se. They can be sufficiently restricted to provide reliable information on questions as rough symmetry or tail behaviour. But the rough information seems to be read off more readily from constructions like the Box & Whisker plot. The general purpose plots may have an advantage if we move to the one-sample problem or the two sample problem, where no immediate generalization of the Box & Whisker is available. Multimodality stays a critical feature. The classical general purpose plots tend to be misleading: random fluctuation may appear as modes, and no controlled measure of significance is available. The general purpose plots are not likely to oversee modes, but are prone to show more than should be shown. The Box & Whisker plot does not address the problem of modes at all. We can construct special plots for the detection of modes, such as silhouette and densitogram. We loose information on density and tails in these plot. The shorth plot tries to make a compromise, allowing for information about modality as well as on local density, but avoiding the fluctuation affecting (smoothed) scatterplots and other classical plots. It may be a candidate for a general purpose plot. But practical evaluation and analysis is still necessary.

The true distribution is usually hidden from our eyes. Since we were using simulated data here, we are able to look at the true distribution. In the examples shown here, we used a bimodal distribution with two strong modes. Sample size for the illustrations was 25 data points.

References

Andrews, D.F., Bickel, P.J., Hampel, F.R., Huber, P.J., Rogers, W.H. & Tukey, J.W. (1972) Robust Estimation of Location: Survey and Advances. Princeton University Press, Princeton, N.J.

Bertin, J. (1967). Semiologie Graphique. Gauthier-Villars, Paris.

Bertin, J. (1977). La Graphique et le Traitement Graphique de l'Information. Flammarion, Paris.

Bertin, J. (1981). Graphics and Graphic Information Processing. De Gruyter, Berlin.

Chambers, J.M., Cleveland, W.S., Kleiner, B. & Tukey, P.A. (1983). Graphical Methods for Data Analysis. Wadsworth Statistics/Probability Series, Wadsworth, Belmont.

Gan, F.F. & Koehler, K.J. (1990). Goodness-of-Fit Tests Based on P-P Probability Plots. Technometrics 32, 289 - 303.

Grübel, R. (1988). The Length of the Shorth. Annals of Statistics 16, 2:619-628.

Hall, P. & Marron, S. (1987). Extent to which Least Squares Cross Validation Minimises Integrated Squared Errors in Nonparametric Density Estimation. J. Probab. Theory and Related Fields 74, 567-581

Hall, P. & Johnstone, I. (1992). Empirical Functionals and Efficient Smoothing Parameter Selection. Journal of the Royal Statistical Society, Series B. 54, 475-530.

Harter, H.L. (1984). Another Look at Plotting Positions. Communications in Statistics-Theory and Methods 13, 1613-1633.

Kallenberg, W.C.M, Oosterhoff, J. & Schriever, B.F. (1985). The Number of Classes in Chi-Squared Goodness-of-Fit Tests. Journal of the American Statistical Association 80, 959 - 968.

Kimball, B.F. (1960). On the Choice of Plotting Positions on Probability Paper. Journal of the American Statistical Association 55, 546-550.

Mammen, E. (1992). When does Bootstrap Work? Lecture Notes in Statistics. Springer, Heidelberg.

Mann, H.B. & Wald, A. (1942). On the Choice of the Number of Intervals in the Application of the Chi-Squared Test. Annals of Mathematical Statistics 13, 306-317.

McGill, R., Tukey, J.W. & Larsen, W.A. (1978) Variations of Box Plots. American Statistician 32, 12-16.

Müller, D.W. & Sawitzki, G. (1987). Using Excess Mass Estimates to Investigate the Modality of a Distribution. Universität Heidelberg, Sonderforschungsbereich 123 (Stochastische Mathematische Modelle). Reprinted in: Proceedings of the ICOSCO-I Conference, (First International Conference on Statistical Computing, Çesme, Izmir 1987) Vol II. American Science Press, Syracuse 1990.

Müller, D.W. & Sawitzki, G. (1991). Excess Mass Estimates and Tests for Multimodality. Journal of the American Statistical Association 86, 738-746.

Pearson, K. (1900) On a Criterion that a Given System of Deviations from the Probable in the Case of a Correlated System of Variables is such that it can be Reasonably Supposed to Have Arisen from Random Samples. Philosophical Magazine (5th Series) 50, 157-175.

Sawitzki, G. (1990). Tools and Concepts in Data Analysis. In: Faulbaum, F., Haux, R. & Jöckel, K.-H. (eds.) SoftStat '89 Fortschritte der Statistik-Software 2. Gustav Fischer, Stuttgart. 237-248.

Sawitzki, G. (1992). The Shorth Plot. Technical Note. Heidelberg 1992.

Shapiro, S.S. & Wilk, M.B. (1965). An Analysis of Variance Test for Normality (Complete Samples). Biometrika 52, 591-611.

Scott, D.W. (1979). On Optimal and Data-based Histograms. Biometrika 66, 605 - 610.

Silverman, B.W. (1981). Using Kernel Density Estimates to Investigate Multimodality. Journal of the Royal Statistical Society, Ser. B., 43, 97-99.

Silverman, B.W. (1986) Density Estimation for Statistics and Data Analysis. London: Chapman and Hall

Terell, G.R. & Scott, D.W. (1985). Oversmoothed Nonparametric Density Estimators. Journal of the American Statistical Association 80, 209 - 214.

Tufte, E.R. (1983). The Visual Display of Quantitative Information. Grapics Press, Cheshire, Connecticut.

Tufte, E.R. (1990). Envisioning Information. Grapics Press, Cheshire, Connecticut.

Tukey, J.W. (1962). The Future of Data Analysis. Annals of Mathematical Statistics 33, 1-67.

Wainer, H. (1984). How to Display Data Badly. The American Statistician 38, 137-147

Weibull, B.A. (1939). The Phenomenon of Rupture in Solids. Ingeniors Vetenskaps Akademien Handlingar 153, 7.

Wilk, M.B. & Gnanadesikan, R. (1968). Probability Plotting Methods for the Analysis of Data. Biometrika 55, 1 - 17.

Graphical Data Analysis using LISP-STAT

Axel Benner
Biostatistik, Deutsches Krebsforschungszentrum
PF 101 949, D-69009 Heidelberg, Germany

Key Words:
object-oriented programming, dynamic graphics, Lisp-Stat.

Abstract
In this paper the object-oriented programming environment Lisp-Stat and its underlying programming concept is described and examples are presented to illustrate the potential such an environment has for graphical data analysis. Object-oriented programming can help to simplify the development of new and more sophisticated methods. It also provides more flexibility in the application of dynamic graphics tools in routine data analysis.

Introduction

Informal data analysis is an important complement to mathematical statistics and should be applied if the assumptions made by theoretical models are unjustified or at least uncertain. In this case exploratory data analysis methods make sense if used according to the maxime *"Finding the question is often more important than finding the answer"*.[1] Statistical graphics belong to the most effective tools for exploratory data analysis because they work on data in a visualized form. Eliciting the information hidden in the data is more easily done with graphical tools than with any other technique. The Lisp-Stat environment provides flexible tools for graphical data analysis through its object-oriented programming concept and its dynamic graphics system.

[1] John W. Tukey (1980): "We need both exploratory and confirmatory". Amer. Statist., 34, p.24

In the next section the Lisp-Stat environment will be introduced, followed by an overview of the object-oriented programming paradigm and a description of techniques for dynamic graphics. In the last section some Lisp-Stat examples of dynamic graphics tools are presented.

The Environment

Lisp-Stat is a graphical interactive programming environment for statistical computing. It's functionality is strongly influenced by developments in the S language (Becker et.al. (1988)). Lisp-Stat offers numerical and elementary statistical operations and an extendable graphics system which makes use of a graphical user interface and standard graphs, forming "building blocks" for more elaborate graphical tools. The standard graphs include histograms and boxplots, bivariate scatterplots, scatterplot matrices and spinning plots. Basic features are brushing and linking of different plots of the same data set.

New numerical and graphical tools can be added using Lisp programming. In most Lisp-Stat implementations extensions of the system are also possible by dynamic or static loading of user written C functions or FORTRAN subroutines.

Lisp-Stat adds to Lisp's features functions to support vectorized arithmetics, functions for linear algebra, matrix manipulation and maximization techniques as well as functions for elementary statistical computations, a graphical user interface and tools for building static and dynamic graphs. It makes use of an object-oriented programming system to support e.g. programming of graphs and organizing statistical model classes, such as generalized linear models. Lisp-Stat does not offer "presentation graphics" nor does it provide "pre-packaged" statistical methods. In addition Lisp-Stat actually does not support missing data handling. Some extensions to Lisp-Stat exist, including methods for regression diagnostics using dynamic graphics (Cook & Weisberg (1989)).

The first implementation of Lisp-Stat is XLISP-STAT, which is based on XLISP (Betz (1988)), a programming language combining features of Common Lisp with an object-oriented extension capability. XLISP is completely written in C and both, XLISP and XLISP-STAT are freely available in source form to non-commercial users. XLISP-STAT runs on different computers with different window systems including the Apple Macintosh, Personal Computers with MS Windows, and Unix workstations using the X Window System. The source code of XLISP-STAT and PostScript files for a tutorial introduction to Lisp-Stat are available e.g. by *anonymous-ftp* from *umnstat.stat.umn.edu.* Documentation is given by the book of Luke Tierney (Tierney (1990)).

Object-Oriented Programming

Object-oriented programming is not a new concept in statistical applications and it was often associated with Lisp (Hurley & Oldford (1991)). Statistical analysis environments built in Lisp exist since the early 1980's (cp. McDonald (1986), Stuetzle (1987), McDonald & Pederson (1988)).

Object-oriented programming has it's origins in the late 1960's with the simulation language Simula and the introduction of the class concept. The first real object-oriented language was Smalltalk, developed at Xerox Palo Alto Research Center since the early 1970's (cp. Goldberg & Robson (1989)). In contrast to Smalltalk most other object-oriented systems use an existing programming language, extended by object-oriented programming facilities (e.g. `C++` or Lisp-Stat). The term *object-oriented* itself is defined differently by different people. Often object-oriented is defined basically by three concepts: objects, classes, and (class) inheritance. Sometimes other requirements are added, such as polymorphism, dynamic binding or encapsulation.

Objects are the basic entities in an object-oriented system. Associated with every object is a set of procedures that define operations on that object. In Lisp-Stat an object is described as a data structure that contains the information about its properties in named locations called *slots.* The fundamental processing approach is to send a message to an object asking it to take certain actions (*methods*). A set of possible objects is defined by a construct called *class* or *type.* This is a construct for implementing a user-defined data type. The definition of a class specifies the properties that its instances (the objects of this class) should have by means of *instance variables.* One has to define an object as an instance of a class together with a set of specific methods. An alternative concept, realized e.g. in Lisp-Stat, is to use a type-based object-oriented system. Objects are now organized themselves in a hierarchy in which they inherit directly from other objects, their *ancestors. Types* can be used like classes, but their use is not obligatory. Every object has a *type* which is also an object, that describes the behaviour of the object. In Lisp-Stat types are called *prototypes.*

Prototype objects serve as templates for building new objects, instances of the prototype. In Lisp-Stat new prototypes are defined by using

```
(defproto <name> <instance slots> <shared slots>
                    <parents> <doc string>)
```

<u>*Example:*</u>
(defproto time-series-proto '(origin spacing)()data-set-proto)
defines a time series prototype having a single ancestor, a data set prototype called data-set-proto, *and two additional slots,* origin *and* spacing.

In Lisp-Stat a message is send to an object (or prototype) by

```
(send <object> <selector> <arg 1> ... <arg n>)
```

with <selector> being a symbol used to select a method. If e.g. plot is a scatterplot object, then adding the data which belong to a pair of variables (x, y) to the scatterplot plot can be achieved by

```
(send plot :add-points x y)
```

with :add-points being the <selector> belonging to the specific plotting action.

New methods are defined in Lisp-Stat by

```
(defmeth <object> <selector> <parameters> <body>)
```

Again *object* can be an existing object or a type.

<u>*Example:*</u>
Let histo *be a histogram object, then a new sample of 50 Gaussian random data is displayed through object* histo *by*

```
(send histo :new-sample 50)
```

with method new-sample *defined by*

```
(defmeth histo :new-sample (sample-size)
    (send self :clear :draw nil)
    (send self :add-points (normal-rand sample-size) :draw nil)
    (send self :adjust-to-data))
```

Alternatively the method could also be defined for the corresponding type, which in this case is the type used for histograms:

```
(defmeth histogram-proto :new-sample (sample-size) ...
```

In the first case the method new-sample *is defined exclusively for an existing histogram object* histo, *in the second situation the method is defined for all objects defined by type histogram-proto.*

Inheritance is a relation usually defined between classes or types that allows to define an implementation of one class to be based on that of other existing classes. Inheritance helps to construct software systems from reusable parts and minimizes the amount of new code needed when adding additional features. Lisp-Stat supports multiple inheritance and also instance-based inheritance.

In Lisp-Stat statistical graphs are build using prototypes which all inherit from the *graph-proto* prototype (Figure 1). The prototype graph-proto implements a scatterplot as a two-dimensional view of an m-dimensional data set ($m \geq 2$). Scatterplots can be constructed according to the scatterplot-proto prototype e.g. by use of the functions `plot-points` and `plot-lines`.

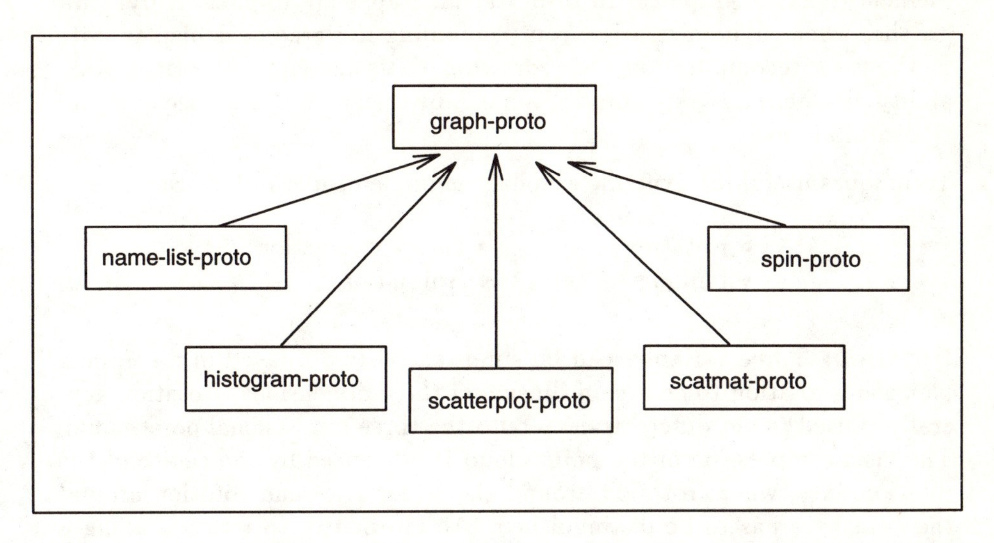

Figure 1: *Inheritance graph for statistical graphics prototypes in Lisp-Stat.*

Dynamic Graphics[2]

To provide efficient graphical data analysis the graphics display needs to be flexible with all its components being dynamically changeable. The graphical data analysis procedures should be extendable with minimal programming effort. This can be achieved e.g. by use of object-oriented programming and the use of graphical hierarchies as provided by the Lisp-Stat language.

Lisp-Stat makes use of menus and dialog windows and provides static and editable text fields, radio and push buttons, sliders, scrollable lists and check boxes. Basic graphical objects and actions are similar to those of ordinary painting software. Lisp-Stat offers special tools for graphical data analysis like parallel coordinates plots (Bolorforoush & Wegman (1988)) or projection techniques like grand tours (Asimov (1985)).

Many (commercial) statistical software packages for personal computers and workstations now contain some form of dynamic graphics, but most of them do not allow users to customize their plots or to develop functions for producing specialized graphical tools. In addition very few statisticians have

[2]The term "dynamic graphics" as used here is only a generic term to characterize methods which influence in interactive or animative way graphical displays of data.

had access to an environment that would allow them to implement dynamic graphics ideas. Lisp-Stat provides an easy way to support the user to implement dynamic graphics. In Lisp-Stat all graphs are implicitely dynamic graphs, where dynamics arise from combining the graphical display with methods for recomputation and redrawing of it's contents. A normal probability plot becomes a dynamic graph e.g. by assigning a sequence of power transformations.

Techniques useful for dynamic graphics as implemented in Lisp-Stat are:

- Rotation
- Linking
- Direct interaction
- Animation

Rotation: Three variables can be simultaneously displayed in a spinning plot using rotation to create an illusion of three dimensions. Rotation generally is used to get different views onto the three-dimensional point cloud. The visual impression of the point cloud is influenced by the choice of the rotation axes, where rotation around the screen axes and rotation around the data axes has to be distinguished. An alternative to rotation along a given axis is called *Hand Rotation* or *Pushing* (Becker et al. (1988)), which is controlled by user actions. The idea is to think of having all data points inside a ball, which can be turned (pushed) in any direction. This allows the user to easily get to a desired orientation of the point cloud. The impression of perspective depth can be increased, if the size of plot symbols varies depending on the position of the points corresponding to the plot coordinates (*Depth-Cuing*), or by adding a frame box drawn around the point cloud, where the rotation origin should be used as the center of the box.

To preserve the three-dimensional impression when the rotation stops one can use the *Rocking* technique. Rocking a point cloud means to produce views of the cloud by varying the viewing direction by small amounts back and forth around the vertical screen axis to get an impression of perspectivity without changing the main viewing direction.

Example: The method rock-plot rotates the point cloud by the current rotation angle three times in one direction, then rotates back and forth ten times by twice the angle. After this it rotates back to the original position.

```
(defmeth spin-proto :rock-plot (&optional (n 10) (k 3))
  (let ((angle (send self :angle)))
    (dotimes (i k) (send self :rotate-2 0 2 (- angle)))
    (dotimes (i n)
      (dotimes (i (* 2 k)) (send self :rotate-2 0 2 angle))
      (dotimes (i (* 2 k)) (send self :rotate-2 0 2 (- angle))))
    (dotimes (i k) (send self :rotate-2 0 2 angle))))
```

Direct interaction: To interact with plot objects two highlighting techniques called *brushing* and *selection* for working on (subsets of) points have proved to be quite useful. Selecting is done either to identify single points or to choose a group of points using a selection rectangle around the group. Brushing is done by moving a rectangle over a plot, where each observation belonging to points inside the brush changes its state corresponding to the operation mode of brushing. In Lisp-Stat the only operation mode in brushing is to select points by highlighting. Using the plot menu the user can do more on selected points, e.g. change the color or the symbol used, remove the selection or focus on the selected points. If different plots of the same data set are linked together the brushing operation or the operation chosen from the plot menu is also performed on corresponding points in the other plots.

Linking: One of the most useful and rarely found techniques for graphical analysis of multivariate data is the connected use of different views on the data corresponding to different tasks. This can be realized by *linking* of the views in an analysis system using special *constraints* (McDonald (1986)). McDonald used an approach which he called *object-viewing*, taking the windows in a window environment as views of the current state of observations. Observations are taken as objects which have state properties like color or plotting symbol and which are viewed using different graphs at one time. *Leader-follower constraints* with a leader object and follower objects are used as the linking strategy. When the leader object changes state, the followers need to update their own state to make the constraint true. The Antelope system (McDonald (1986)), Plot Windows (Stuetzle (1987)), or Data Desk (Data Description Inc.) are examples of systems using this strategy.

An easy way to look at $m \geq 3$ variables at one time is to use a scatterplot matrix of all pairwise scatterplots. In a scatterplot matrix the points in the different scatterplots are "naturally" linked together. But in Lisp-Stat it is possible to link different "views" (graphical displays) on a data set by choosing `Link View` from the menus of the corresponding plots. In the Lisp-Stat system the default linking strategy is a loose association model in which points in different linked plots are related only by their indices. This makes it impossible to use this kind of links between plots showing different subsets of the same data set. But instead of using the default linking technique it is possible to implement a more complex linking strategy, e.g. similar to the one used in Antelope, within the Lisp-Stat graphics system.

Animation: The basic idea is to show a smooth sequence of graphs fast enough to simulate motion. In an animation one starts with a graph object, characterized by a parameter vector. If the parameters change the graph is adjusted accordingly and redrawn, which leads (if done fast enough) to animated displays of the graph.

Two basic animation techniques are available in Lisp-Stat: *XOR drawing* and *double buffering.* The drawing mode XOR ("eXclusive OR") deletes a symbol on the screen by drawing it again, thus making use of being its own inversion. XOR makes sense if the aim is to move simple objects across a plain background. Disadvantages are: (i) XOR may destroy drawings on the background of the picture, (ii) using XOR drawing in animation results in flickering pictures due to permanently drawing and redrawing of individual points, and (iii) XOR is only well-defined in monochrome painting. In Lisp-Stat XOR is used for drawing the brushing rectangle and point labels.

A flicker-free animation can be achieved by drawing the picture in an invisible background buffer and then copying it to the screen as a whole. If this copying process is fast enough a smooth and flicker-free "motion" of the picture results. In general background buffers are used to prepare or preserve (parts of) pictures. In double buffering a single background buffer is used. Double buffering is also useful to preserve a coordinate system if only the contents of a plot has to be redrawn. In Lisp-Stat double buffering is used e.g. for drawing points in rotations.

Examples

This section gives some examples of applications of graphical data analysis using Lisp-Stat. The examples describe several forms of implementing dynamic graphics in Lisp-Stat, which will give an impression of the potentials of the Lisp-Stat system.

Plot Interpolation

Plot interpolation (Buja et al. (1988)) can be applied in situations where a data set with at least four continuous variables, e.g. (X_1, Y_1, X_2, Y_2), is to be analyzed. It is useful to detect clusters due to position and motion or to keep track of (subsets of) points when moving from one plot to the other.

The procedure can be implemented in Lisp-Stat as an animated scatterplot controlled by a single parameter p which regulates the interpolation between two plots (X_1, Y_1) and (X_2, Y_2). The graph object corresponding to p is the plot of (X_p, Y_p) and a possible method for constructing the artificial variables X_p and Y_p is the trigonometric interpolation defined by

$$\begin{aligned} X_p &= cos(p \cdot \pi/2) \cdot X_1 + sin(p \cdot \pi/2) \cdot X_2 \\ Y_p &= cos(p \cdot \pi/2) \cdot Y_1 + sin(p \cdot \pi/2) \cdot Y_2 \end{aligned} \qquad p \in [0,1] \qquad (1)$$

Continually plotting of (X_p, Y_p) for a sequence of values $p \in [0,1]$ then leads to animated plot interpolation.

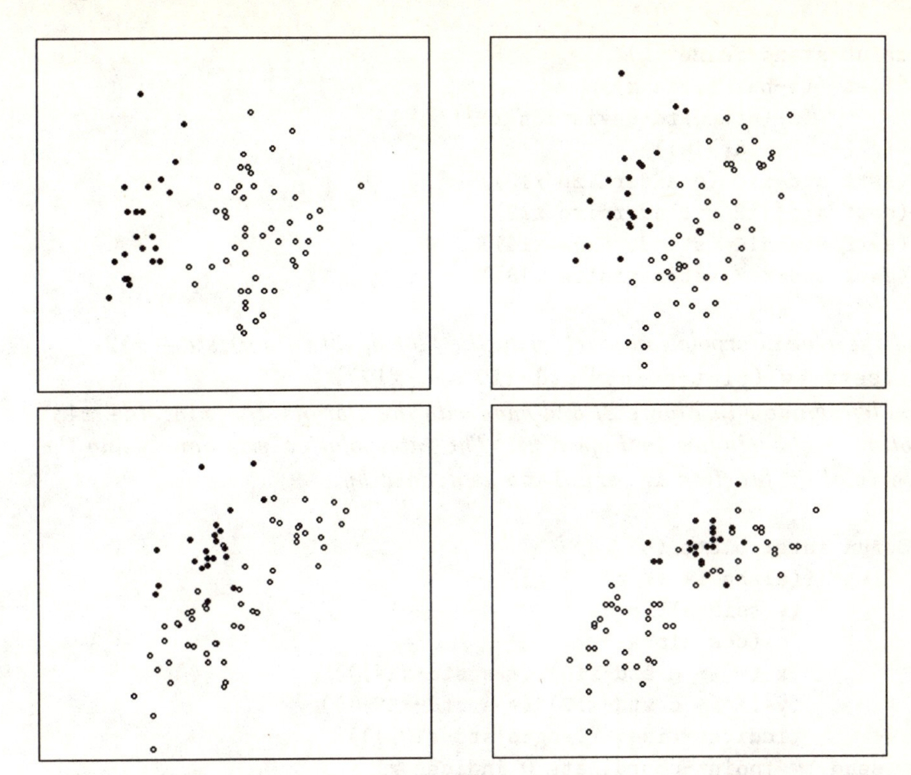

Figure 2: *Four views from plot interpolation of the Beetles data set. The plot of* $(std - x10, std - x12)$ *is given in the top left window, and moving from left to right and top to bottom the animation ends with the plot of* $(std - x14, std - x48)$. *From left to right and top to bottom the values for* p *corresponding to formula (1) are:* $p = 0,\ p = 0.4,\ p = 0.7,\ p = 1$. *In this case the results would have been the same as one would get by displaying the two scatterplots linked together.*

Example:

Plot interpolation is performed for the flea beetles data set, reported in Lubischew (1962) and often used to demonstrate projection pursuit techniques.

74 beetles of species Ch. heikertingeri (# 31), Ch. heptapotamica (# 22) and Ch. concinna (# 21) belonging to genus Chaetocnema were analyzed by plot interpolation of four out of six variables as used in Lubischew (1962):

- $x10$*: width of the first joint of the first tarsus in microns*
- $x12$*: the same for the second joint*
- $x14$*: the maximal width of the aedeagus in the fore-part in microns*
- $x48$*: the aedeagus width from the side in microns*

The variables have been standardized using function `standardize`

```
(defun standardize (x)
  (let ((x-bar (mean x))
        (s (standard-deviation x)))
  (/ (- x x-bar) s)))
 (setf std-x10 (standardize x10))
 (setf std-x12 (standardize x12))
 (setf std-x14 (standardize x14))
 (setf std-x48 (standardize x48))
```

and the plot interpolation starts with the plot of $(std - x10, std - x12)$

```
(setf bw (plot-points std-x10 std-x12))
```

(top left window in Figure 2) and ends with the plot of $(std - x14, std - x48)$ *(bottom right window in Figure 2). The interpolation was done using the interpolation function* `interpolate` *as defined by*

```
(defun interpolate (p)
  (let* ((alpha (* (/ pi 2) p))
         (s (sin alpha))
         (c (cos alpha))
         (x (+ (* c std-x10) (* s std-x14)))
         (y (+ (* c std-x12) (* s std-x48)))
         (indices (iseq (length std-x10))))
  (send bw :point-coordinate 0 indices x)
  (send bw :point-coordinate 1 indices y)
  (send bw :redraw-content)))
```

Figure 2 shows four "freeze frames" of the animation corresponding to values $p = 0$, $p = 0.4$, $p = 0.7$ *and* $p = 1$.

Grand Tour

The grand tour is a method for "viewing" high-dimensional data via projections onto a sequence of one- or two-dimensional subspaces to search for patterns or cluster. The basic idea as described in Asimov (1985) was to provide a sequence of orthogonal two-dimensional projections, chosen to be dense in the set of all projections, so that the user can view a sequence of two-dimensional scatterplots which, asymptotically, come arbitrarily close to all possible two-dimensional projections of the data set. The grand tour can be seen as an approximation of an m-dimensional data set by a time-indexed family of two-dimensional views. Grand tours can be combined with numerical projection pursuit techniques to steer the tour towards "interesting" views (Cook et.al. (1991)).

A graphical version of the grand tour is implemented in Lisp-Stat by choosing two points in the m-dimensional unit sphere describing a rotation axis. The views are generated by rotating a random number of times by a given rotation angle α before two new points are chosen and rotation starts again using the new axis.

<u>*Example:*</u>
Grand tour of the Australian rock crabs data (Campbell & Mahon (1974)). The data set consists of five measurements on each of 200 Australian rock crabs: length of front lip, width of rear end, length and width of the carapace and thickness of the body . The crabs split into two species (one with blue carapace and one with orange carapace) and the two sexes, altogether four groups with each having 50 individuals. The resulting views of a grand tour for this data set is shown in Figure 3.

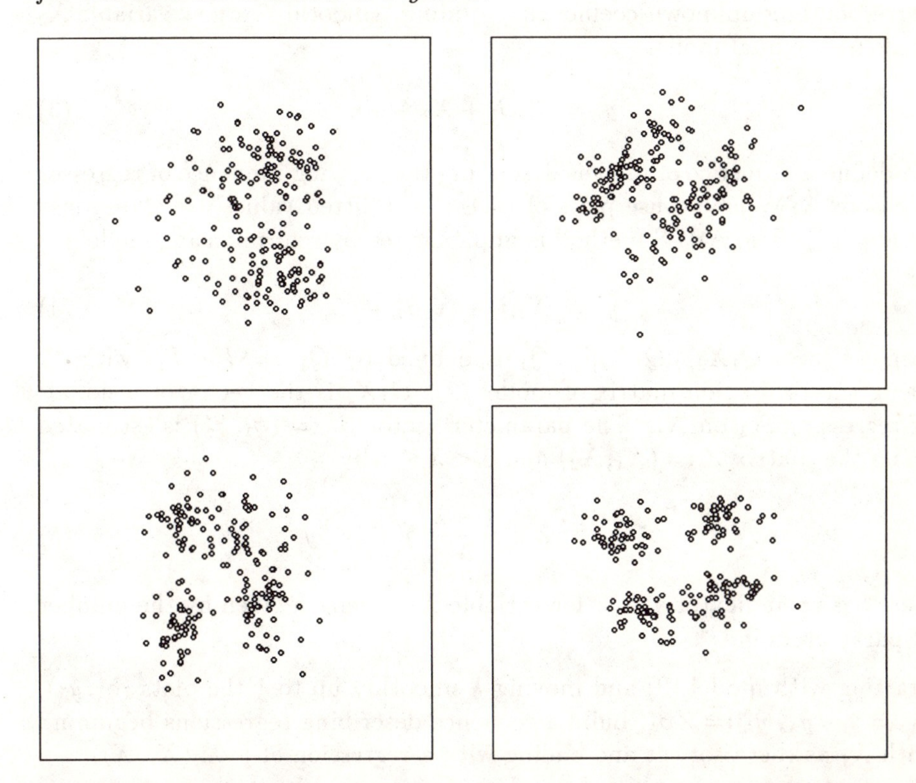

Figure 3: *Four "freeze frames" from a grand tour of the Australian rock crabs data. The grand tour was performed on the principal components of the data. The first plot (top left) shows the starting plot of the first two principal components. At the end the grand tour found the structure of the data (bottom right).*

ARES Plots

Animation can be very useful in regression diagnostics, e.g. for analyzing the effects of adding or removing variables in a linear modeling environment (which is described here only for one variable under discussion). More on the use of dynamic graphical regression diagnostics as well as a detailed description of the technique shown here is provided in Cook & Weisberg (1989). Cook & Weisberg implemented their ideas for regression diagnostics in Lisp-Stat.

To perform ARES[3] plots one starts with a basic linear model

$$y = X_1\beta_1 + \varepsilon \tag{2}$$

X_1 represents the matrix of one or more regressors and β_1 is the vector of corresponding unknown coefficients. Adding "smoothly" a new variable X_2 leads to the final model

$$y = X_1\beta_1 + X_2\beta_2 + \varepsilon \tag{3}$$

To define a sequence of models describing the growing influence of regressor X_2 Cook & Weisberg use plots of $(\hat{y}, y - \hat{y})$ ("fitted values" vs. "response residuals"). The ARES method is implemented by reformulating model (3) by

$$y = X_1\beta_1 + \hat{X}_2\beta_2^* + \varepsilon \tag{4}$$

using $\hat{X}_2 := Q_1X_2/||Q_1X_2||$. Q_1 is defined by $Q_1 := I - P_1$, with P_1 being the projection matrix of model (2). Q_1X_2 is the vector of residuals of regressing X_2 on X_1. The parameter vector $\beta^* := (\beta_1, \beta_2^*)$ is estimated using the matrix $Z := (X_1, \hat{X}_2)$ and $0 < \lambda \leq 1$ by

$$\hat{\beta}_\lambda^* = (Z^TZ + \frac{1-\lambda}{\lambda}bb^T)^{-1}Z^Ty \tag{5}$$

Here b is an indicator vector for variable X_2 of length given by the number of parameters in β^*.

Starting with model (2) and moving λ smoothly up to 1 the plots $(e_\lambda, \hat{y}_\lambda)$, $e_\lambda := y - \hat{y}_\lambda$, $\hat{y}_\lambda := Z\hat{\beta}_\lambda^*$, build a sequence describing regressions beginning with X_1 as regressor set and ending with a regression of y on (X_1, X_2).

Example:
The example shown here is taken from Cook & Weisberg (1989). The data set is a set of measures on 19 rats, randomly selected to investigate the amount of a drug retained in the liver of a rat (reported in Weisberg (1985)). Predictors are the relative dose of the drug, body weight and liver weight. The response is the percentage of the dose in the liver. Figure 4 gives an impression of the effect of influential observations (here: case 3) on regression.

[3] ARES: Adding REgressors Smoothly

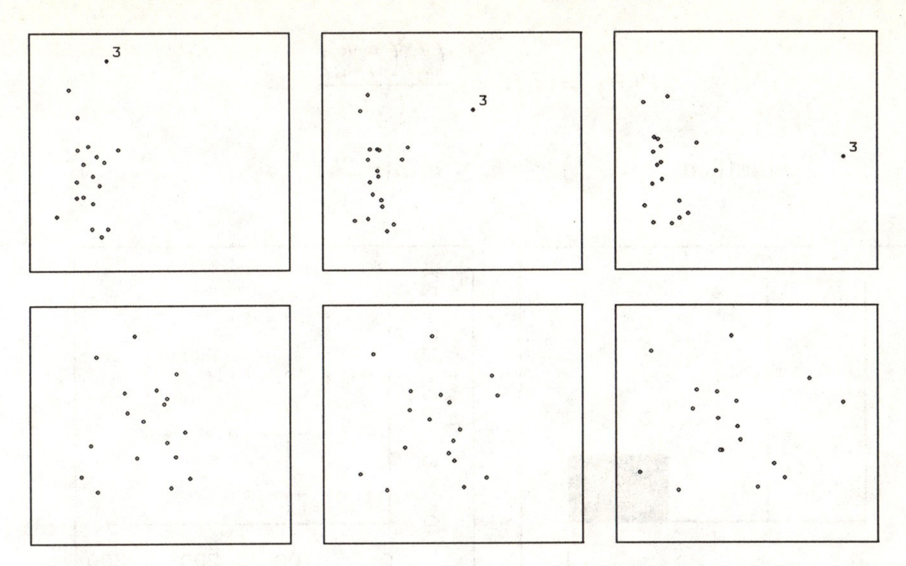

Figure 4: *ARES plots of the rats data set. The basic model uses regressors body weight and liver weight. Animation is used to describe the effect of adding dose into the model (upper sequence: left to right). The regression of the model depends fully on one single observation (case 3 as highlighted in the upper sequence). Removing case 3 and repeating the procedure (lower sequence: left to right) shows no additional effect due to variable dose. (All plots in a sequence are scaled identical).*

Odds Plots

The last example describes an implementation of odds plots (Stuetzle (1991)) in Lisp-Stat. The aim of the procedure is to find association between variables through graphical linking using leader-follower constraints. This is done by estimating and displaying the conditional probabilities for realizing a variable X inside bins B, given the realization of variable Y, $P(X \in B|Y = y)$. Points are selected (highlighted) in a leader plot and the conditional probabilities are computed by superimposing a regular grid of bins on the follower plot. For each bin the proportion of highlighted points is computed. Icons + (or −) are drawn at the center of each bin if there are more (or less) highlighted points than is expected under uniform distribution. A "dot" symbol ($\diamond$) is drawn if the observed number of highlighted points are about as many as expected. No symbol is drawn in a bin if it is empty. The size of the symbols + and − is determined depending on the magnitude of the difference between observed and expected highlighted points. If n is the total number of observations and h the total number of highlighted points, then the number of highlighted points in a bin j has a hypergeometric distribution

$$P(H_j = h_j) = \frac{\binom{h}{h_j}\binom{n-h}{n_j-h_j}}{\binom{n}{n_j}} \qquad (6)$$

$$max(0, n_j - n + h) \leq h_j \leq min(n_j, h)$$

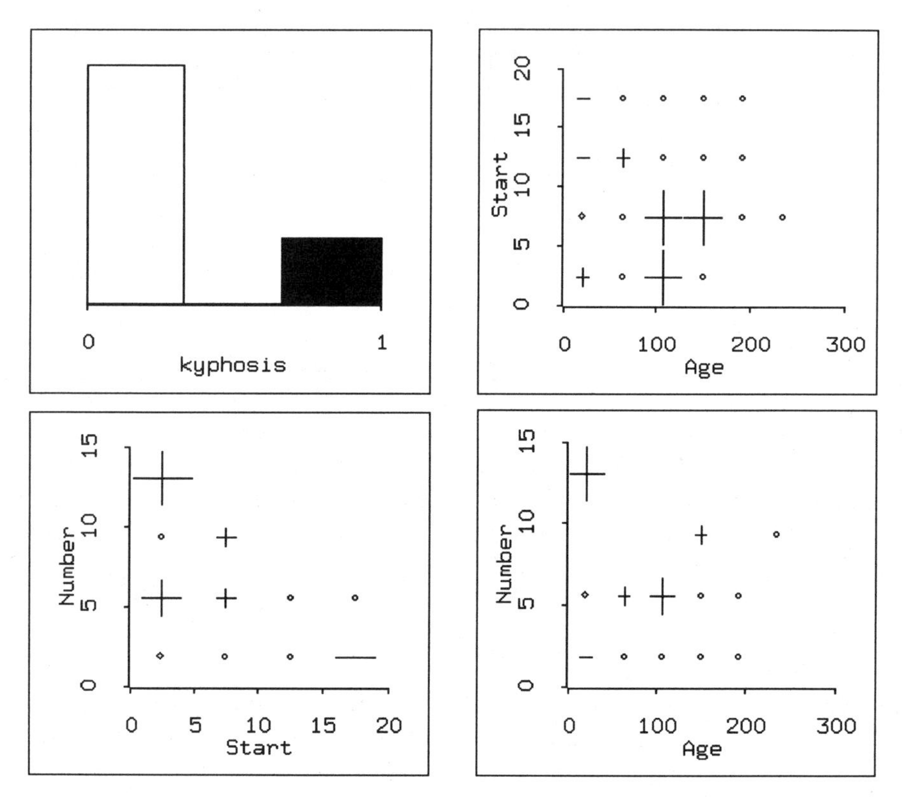

Figure 5: *Odds plots of the kyphosis data set. The leader plot is a bar chart of the kyphosis indicator variable, and the cases with kyphosis being present are selected (highlighted). Three follower odds plots describe the dependencies between kyphosis and the three predictors variables age, start and number. A change in the dependencies at about start* = 10 *is obvious and a nonlinear behaviour is shown in the plot of variables age and number.*

Using the tail probability $p_j = min(P(H_j \geq h_j), P(H_j \leq h_j))$ the relative size s_j of the symbols is given by

$$s_j = \begin{cases} 0.3 & : \quad 0.1 > p_j \geq 0.01 \\ 0.6 & : \quad 0.01 > p_j \geq 0.001 \\ 0.9 & : \quad 0.001 > p_j \end{cases} \qquad (7)$$

Example:
A study of postoperative deformities following a corrective spinal surgery was performed in 83 children (reported in Hastie & Tibshirani (1990)). The response y denotes the presence (y = 1) or absence (y = 0) of kyphosis, defined to be a forward flexion of the spine of at least 40 degrees from vertical, which was present in 18 children. Available predictors are age of the children at time of operation (in months), the starting and ending range of vertebrae levels involved, and the number of levels involved. Since `end = start + number - 1`*, only age, start and number have been used in the analysis. The results of applying odds plots to the data are shown in Figure 5.*

Summary

Lisp-Stat is a programming environment which allows the user to program statistical applications, especially for graphical data analysis, in a very efficient way by using object-oriented programming. It provides a foundation on which computational statisticians can build their own analysis environments. The ViSta system (Young et al. (1993)), a system for statistical visualization, which is written using the Lisp-Stat environment, is only one example.

In this paper the object-oriented paradigm is described and examples of some well-known graphical tools are given which illustrate the power of the Lisp-Stat environment for building data analysis applications which include dynamic graphics and graphical user interfaces.

Lisp-Stat is no standard statistical package, it is still experimental, but its advantages for experimenting with graphical applications can ideally be used if Lisp-Stat is used together with a statistical analysis environment like S-Plus (Statistical Sciences Inc (1992)).

References

Asimov, D. (1985). The grand tour: A tool for viewing multidimensional data. SIAM Journal on Scientific and Statistical Computing, Vol. 6, 128-143.

Becker, R.A., Chambers, J.M. & Wilks, A.R. (1988). The new S language: A programming environment for data analysis and graphics. Wadsworth, CA.

Becker, R.A., Cleveland, W.S. & Weil, G. (1988). The use of brushing and rotation for data analysis. In: Cleveland, W.S. & McGill, M.E. (eds.) Dynamic Graphics for Statistics. Wadsworth, CA, 247-275.

Betz, D.M. (1988). XLISP: An object-oriented Lisp, Version 2.0 . Reference Manual.

Bolorforoush, M. & Wegman, E.J. (1988). On some graphical representations of multivariate data. Interface '88: Proceedings of the 20th Symposium on the Interface, 121-126.

Buja, A., Asimov, D., Hurley, C. & McDonald, J.A. (1988). Elements of a viewing pipeline for data analysis. In: Cleveland, W.S. & McGill, M.E. (eds.) Dynamic Graphics for Statistics. Wadsworth, CA, 277-308.

Campbell, N.A. & Mahon, R.J. (1974). A multivariate study of variation in two species of Rock Crab of the genus Leptograpsus. Austral. J. Zool., Vol. 22, 417-425.

Cook, R.D. & Weisberg, S. (1989). Regression diagnostics with dynamic graphics (with discussion). Technometrics, Vol. 31, 277-311.

Cook, R.D., Buja, A. & Cabrera, J. (1991). Direction and motion control in the grand tour. Interface '91: Proceedings of the 23rd Symposium on the Interface.

Goldberg, A. & Robson, D. (1989). Smalltalk-80: The language. Addison-Wesley, MA.

Hastie, T.J. & Tibshirani, R.J. (1990). Generalized additive models. Chapman & Hall, London.

Hurley, C.B. & Oldford, R.W. (1991). A software model for statistical graphics. In: Buja, A. & Tukey, P.A. (eds.) Computing and graphics in statistics. Springer, New York. 77-94.

Korson, T. & McGregor, J.D. (1990). Understanding object-oriented: A unifying approach. Commun. ACM , Vol. 33, 40-60.

Lubischew, A.A. (1962). On the use of discriminant functions in taxonomy. Biometrics, Vol. 18, 455-477.

McDonald, J.A. (1986). Antelope: Data analysis with object-oriented programming and constraints. ASA Proceedings of Statistical Computing Section, 1-10.

McDonald, J.A. & Pedersen, J. (1988). Computing environments for data analysis III: Programming environments. SIAM Journal on Scientific and Statistical Computing, Vol. 9, 380-400.

Stuetzle, W. (1987). Plot windows. Journal of the American Statistical Assoc., Vol. 82, 466-475.

Stuetzle, W. (1991). Odds plots: A graphical aid for finding associations between views of a data set. In: Buja, A. & Tukey, P.A. (eds.) Computing and graphics in statistics. Springer, New York. 207-217.

Tierney, L. (1990). Lisp-Stat: an object oriented environment for statistical computing and dynamic graphics. John Wiley & Sons, New York.

Tukey, J.W. (1980). We need both exploratory and confirmatory. Amer. Statist., Vol. 34, 23-25.

Weisberg, S. (1985). Applied linear regression. John Wiley & Sons, New York.

Young, F.W., Faldowski, R.A. & McFarlane, M.M. (1993). Multivariate statistical visualization. In: Rao, C.R. (ed.) Handbook of Statistics, Vol. 9. Elsevier Science Publishers B.V., Amsterdam. 959-998.

Multivariate Graphics: Current Use and Implementations in the Social Sciences

Rainer Schnell[1], Herbert Matschinger[2]
[1]University of Mannheim, A5, 68131 Mannheim
[2]Central Institute for Mental Health, P.O. 122120, 68072 Mannheim

Key Words:
graphics, projection plots, principal components, statistical program packages

Abstract
The purpose of the paper is to provide an introduction to multivariate graphics for social scientists. Graphical data analysis is seldom used in the social sciences. Multivariate graphics are mainly but erroneously perceived as a tool for exploratory data analysis. Two broad classes of multivariate graphics can be distinguished: icons and projection plots. The most useful exploratory multivariate graphics are projection plots. The main types of projection techniques (PCA, biplots, MDS, NLM) are described. The recently popular projection techniques for categorical data (CA, MCA, JCA, NLPCA) are reviewed. Common usage of some techniques is criticized. Commonly available software in standard packages is evaluated.

Introduction

Multivariate graphics in the social sciences have todate been seldom used. The practice of data analysis in the social sciences is currently dominated by two techniques: ritualistic significance testing and the fitting of regression models of different kinds without any thoughts about generating mechanisms. Both kinds of "data analysis" restrict the possible outcome of an empirical research process: to *yes* or *no* in the case of a significance test and the estimated size of the parameters of a model in the case of regression (cf. Verboom (1988, p.2)). This practice of data analysis may be one cause for the hardly noticable theoretical advances in the social sciences.[1] The discontent with the current state of empirical research is growing. A lot of panaceas have been suggested. Some of them are based on graphical procedures. Multivariate graphics are mainly perceived as a tool for "exploratory data analysis" (EDA). Especially since "correspondence analysis" (CA) has gained widespread popularity, a revitalization of essentially inductivistic arguments can be observed. Contrary to widespread beliefs among dedicated users of multivariate graphics (for example Weihs & Schmidli (1990)), multivariate graphics are not suitable for the generation of theoretical ideas. At least a few ideas based on background knowledge concerning the structure of a given dataset are absolutely necessary. The reasons are twofold: logical and psychological. The logical argument in the context of justification is simple: The discussion of the problems of induction in the philosophy of science (Popper (1972), Chalmers (1980)) proves the impossibility of any logical foundation for induction. In the context of discovery, the more relevant argument is psychological. The psychology of perception and cognition strongly suggests that inductive generalizations as a result of the study of a graphical data representation are highly improbable: You predominantly see structures you expect to see ("schemata").[2] With the trivial exception of outliers in multivariate space, even in highly structured data sets (for example "iris-data")[3] you need luck and a lot of experience to see the structure. The reader is invited to sample randomly from a collection of data sets and try his favourite "induction technique" (PCA, CA, MDS or even TETRAD): If neither the experimenter nor the testperson knows the data set, we would be very surprised if the testperson discovers the "known" structure. Tukey's famous sentence "graphics helps to see the unexpected" therefore must be taken cum grano salis. The interest of a researcher must be focused on special features. Recently, Tukey (1990) distinguishes between "exploratory" and "prospecting" data analysis: A prospector knows what he is searching for. A trained data analyst has a long list of models in

[1]There has been a discussion on the use of "significance tests" in the social sciences for more than 20 years. The main result of the "significance test controversy" (Morrison & Henkel (1970), Oakes (1986)) was the realization of the practical uselessness of significance tests in the research process. A similar debate about the questionable use of regression models in the social sciences is at least as old (cf. Freedman (1991)). Unfortunately, most statistic texts do not mention these debates or their results. They are therefore widely unknown.

[2]Almost unnoticed by mathematicians (see for example Wegman & Carr (1993)), current psychology of perception is not primarily concerned with perceptual illusions as such, but with cognitive processing of expected patterns or elementary graphical units (Pinker (1990), Banks & Krajicek (1991)).

[3]The iris-data is a classical multivariate demonstration data set with four variables and 150 observations with three species of iris. The data set can be found in many texts on multivariate statistics. Furthermore, it is supplied with many program packages (CSS, SYSTAT, S-Plus).

his mind and tries to fit one of these models to explain the data. The given dataset is interesting only in relation to such a model, e.g. for a sound data analysis the examination of residuals is essential. To repeat this central point once more: The data not explain itself. The generation of theoretical explanations by "mechanical" means is dependent on a high amount of prior theoretical knowledge. [4] Therefore, the scientifically most interesting use of multivariate graphics is the display of residuals of statistically simple models. In this case, most "multivariate graphics" consist of standard two-dimensional scatterplots: Only the variables, which define the axes, are the results of "multivariate statistics". In this sense multivariate graphics is mostly the application of special techniques of residual analysis to standard multivariate techniques, like factor analysis or analysis of variance.[5]

Furthermore, for many standard techniques, there is a large collection of special plots to aid in the interpretation of results. For example, there are dozens of special plots in cluster analysis. Only very few of them are available in program packages.

Available software for multivariate graphics

During the discussion we will frequently refer to "standard program packages". In order to clarify our definition, we have to remind the reader, that "data analysis" is the analysis of real data sets by subject matter specialists in order to settle a substantive research problem. Therefore, an account of "available" implementations of statistical programs has to be restricted to standard packages like SPSS, SAS, BMDP or SYSTAT, STATGRAPHICS, STATA and CSS. That some special techniques are available as stand-alone programs, S-Plus macros, or "can be easily programmed in XLISP" or GAUSS, or ISP or whatever is totally irrelevant for practical data analysis. The publication history of every major multivariate data analysis technique (e.g.: path analysis, ML-factor analysis, latent class models, correspondence analysis) tells the same story: The widespread use of an analysis technique starts with the availability of the technique in a standard package or a user friendly stand-alone program. It's an incredible waste of human resources to expect that a subject matter specialist has to learn a programming language in order to do a conditional scatterplot with a nonparametric smoother. Furthermore, data analysis in practice is constrained by financial resources. No one can expect that a research team will buy all available packages,

[4]For a demonstration by computer simulations based on past discoverys, see Holland et al. (1986) and Langley et al. (1987). Tukey & Wilk (1970, p.385) note: "Contemplation of raw observations with an empty mind, even when it is possible, is often hardly more beneficial than not studying them at all."

[5]For examples for this kind of multivariate graphics, see Schnell (1994).

languages and hardware platforms. For example, nearly all data analysis in the social science is currently done with DOS-based microcomputers. Some interesting programs do not work in such environments (e.g. DATA DESK). Other implementations (like S-Plus) are crippled. In short, procedures that are not available in widely used standard program packages (SPSS, SAS etc.) will not be used in practice. [6]

Displays of more than three dimensions

Despite that many plots of multivariate graphics are essentially two-dimensional scatterplots which display the results of multivariate procedures, multivariate graphics is seen by most data analysts and presented in reviews[7] as a collection of methods for the display of multivariate data in two or three dimensions. In this tradition, we will start with a demonstration of the most popular displays of more than three dimensions on the "Iris" dataset. Multivariate data has more than 3 dimensions; therefore in order to display more than 3 dimensions in an essentially two-dimensional space, special tricks must be used. Two kinds of tricks can be distinguished: Icons and dimension-reducing procedures. Icons are graphical representations of multivariate data. In general, each observation is represented by one icon.

The degree of a visual feature of an icon is an indicator of one variable. The simplest icons consists of the use of different plotsymbols, plotsizes and colors. In this way, more than 3 variables may be displayed in a scatterplot (for 3D-scatterplots, see Huber (1983)). The results of such plots are in general not convincing. For nontrivial amounts of data, such displays are cluttered at best. The most famous icons are the "Chernoff-faces" (Chernoff (1973)). Here the characteristics of a cartoon-like face are determined by the variables. For example, the size and the form of the eyes, the orientation of the mouth etc. may be controlled by the variables of the data set. Naturally, the use of icons is limited to very small datasets. Furthermore, icon displays are extremely sensitive to different standardizations, the order of the observations and the correspondence between visual features and variables. In most cases, the display must be organized according to the results of a prior dimension reducing procedure or a cluster analysis. In spite of the large number of suggested different icons,[8] a convincing demonstration of

[6] In statistical graphics as in other fields of statistics, mathematicians and academic computer scientists haven't pressured the producers of standard packages too much to improve their programs (cf. Mason (1991, p.348)). It's remarkable that many procedures in standard packages have been written by psychologists.

[7] see for example the reviews of Wainer & Thissen (1981, 1993), Gabriel (1985), Verboon (1988) and Gessler (1993).

[8] for an uncritical review, see Hartung & Elpelt (1985).

the actual usefulness of icons is still missing. [9]

In our view, icons have a limited use as plotsysmbols in scatterplots of dimension reducing procedures. Interpretation of outliers is facilitated by these icon-enhanced scatterplots. For example, two or three characterizing variables may be displayed with the help of an icon in a scatterplot of the first two principal components. For the display of more than 3 dimensions, many different kinds of dimension reducing procedures have been proposed. The most common forms are multidimensional scaling procedures and different kinds of principal component analysis.

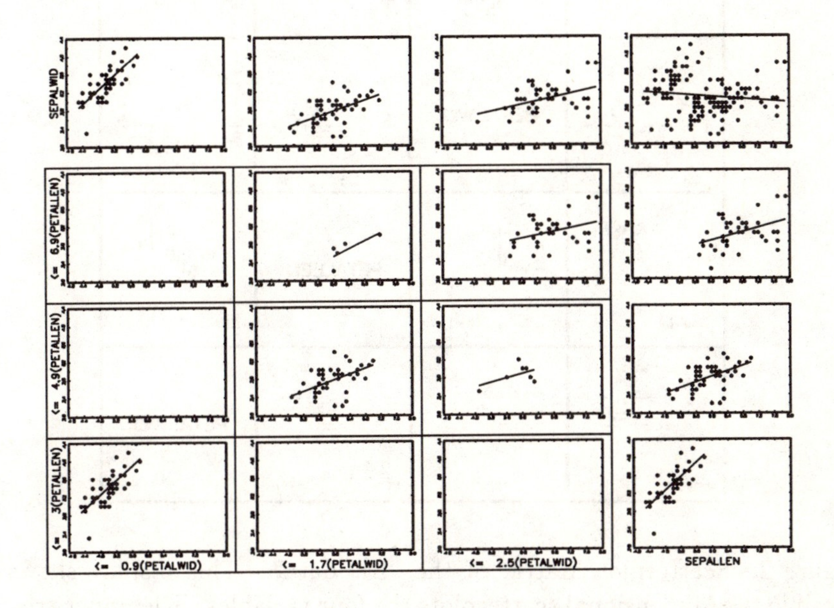

Figure 1: Multiwindow Display of the "Iris data". The unconditional scatterplot of SEPALLEN with SEPALWID is in the upper right corner. Conditioning variables are PETALLEN (rows) and PETALWID (columns)

A "truly" multivariate display for 4 Variables is the "multiwindow display" (Chambers et al. (1983)). Essentially, this is a matrix of conditional scatterplots (see fig. 1). For each combination of two (grouped) variables r and c, the scatterplot of two other variables x and y is plotted. The margins of this matrix are the unconditioned scatterplots, e.g. the top row consists of the scatterplots conditioned only on the column group variable c. The rightmost upper scatterplot is the unconditioned scatterplot x-y. After a short

[9] see for example, Everitt (1987). The most of the few empirical tests of icons are based on a comparison between the cluster recovery ability of icon sorting humans and traditional multivariate techniques.

training period, these plots are extremely useful. Such displays exhibit the relationship of two variables controlling the other variables. Nonlinear effects, clusters, problems due to outliers are easily detected with such plots. Interestingly, only a few programs (and none of the standard packages) provide such facilities (for example CSS, S+, STATGRAPHICS).

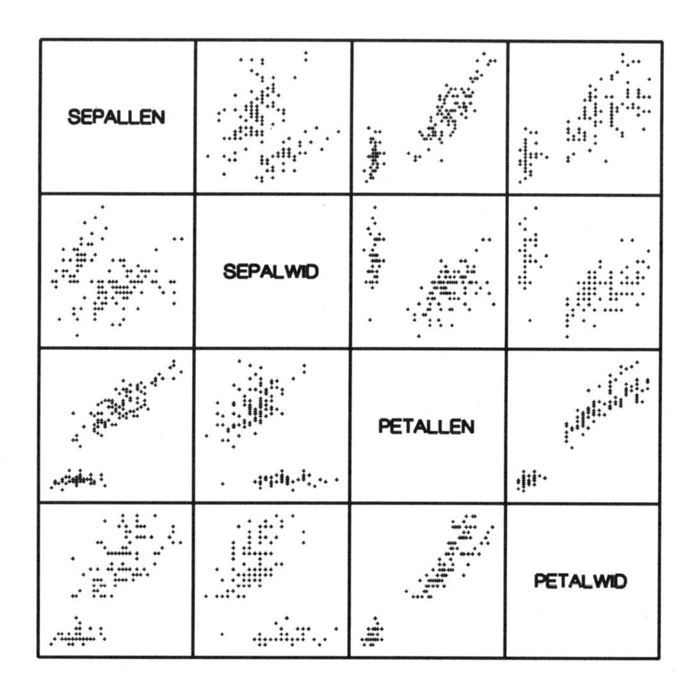

Figure 2: Scatterplot Matrix of the "Iris data". This display shows all possible two-dimensional scatterplots the four variables. The upper triangle is a transpose of the lower triangle and therefore redundant.

Up to 10-15 variables can be checked by a scatterplot matrix ("splom"). A splom is a display of all bivariate scatterplots of a set of variables in a correlation matrix like configuration (see fig. 2). In general, only the lower triangle is shown. The additional inclusion of univariate displays (especially one-dimensional scatterplots) in the diagonal is useful. The detection of outliers in a splom is greatly facilitated by the inclusion of a data ellipsoid. The data ellipsoid of radius r is the set of bivariate observations, that satisfy $[x-\mu]'\sum^{(-1)}[x-\mu] \leq r^2$ where $\sum^{(-1)}$ is the inverse covariance matrix, μ the mean vector and r the radius of the ellipsoid (Monette (1990, p.254)). Data ellipsoids can be computed for any desired proportion, e.g. a 50% data ellipse. When the data are approximately bivariate normal distributed, the ellipse will enclose the desired proportion of cases. Nonlinear relationships between the variables are easier to detect, if a scatterplotsmoother like

LOWESS is included. Scatterplotsmoothers are nonparametric regressions, which may show a linear relationship, but as an empirical description and not by assumption (Hastie & Tibshirani (1990)). Simple SPLOMS (without additions like LOWESS) are available in a lot of programs, but surprisingly missing in SAS. Enriched sploms are implemented in SYSTAT, STATA and SPSS and S+. The interactive identification of points in sploms is the most useful simple technique of multivariate graphics; this is currently possible with two Windows-programs: SYSTAT and S+.

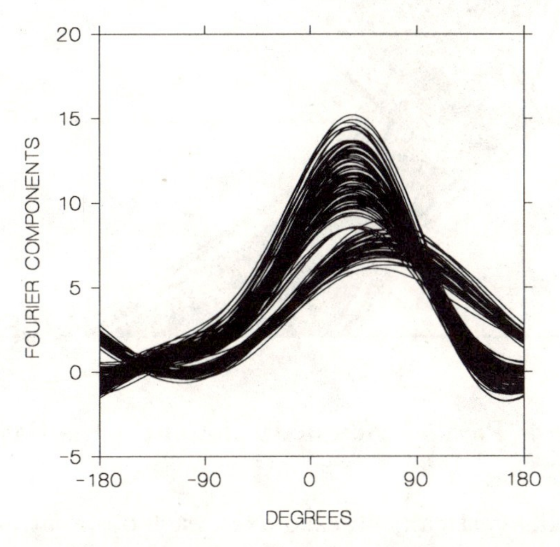

Figure 3: Andrews-plot of the Iris-data

For limited purposes, two special techniques for the display of multivariate data are available. A very popular technique among mathematicians is the "Fourier-" or "Andrews"-plot. Andrews (1972) suggests a plot of p variables of an observation $(x_1, x_2, ..., x_p)$ as function

$$f_x(t) = \frac{x_1}{\sqrt{2}} + x_2 \cdot \sin(t) + x_3 \cdot \cos(t) + x_4 \cdot \sin(2t) + x_5 \cdot \cos(2t) + \cdots$$

$(-\pi \leq t \leq \pi)$. Each observation is represented by one curve. "Similar" observations have similar curves (see fig. 3). Such plots are used for the visual identification of clusters. Unfortunately, the plot is very sensitive to the range and sequence of variables. Therefore, in general the variables have to be standardized and the sequence of variables is determined by their "importance" in a previous principal component analysis. In practice, a lot of trials are necessary to produce useful results. The plot is often unreadable already with 150 observations. The Andrews-plot is implemented in SYSTAT. It may be easily programmed with any data transformation language.

A simple and useful technique for the display of many variables is the use of parallel-coordinates (Inselberg (1985), Wegman (1990)). The variables

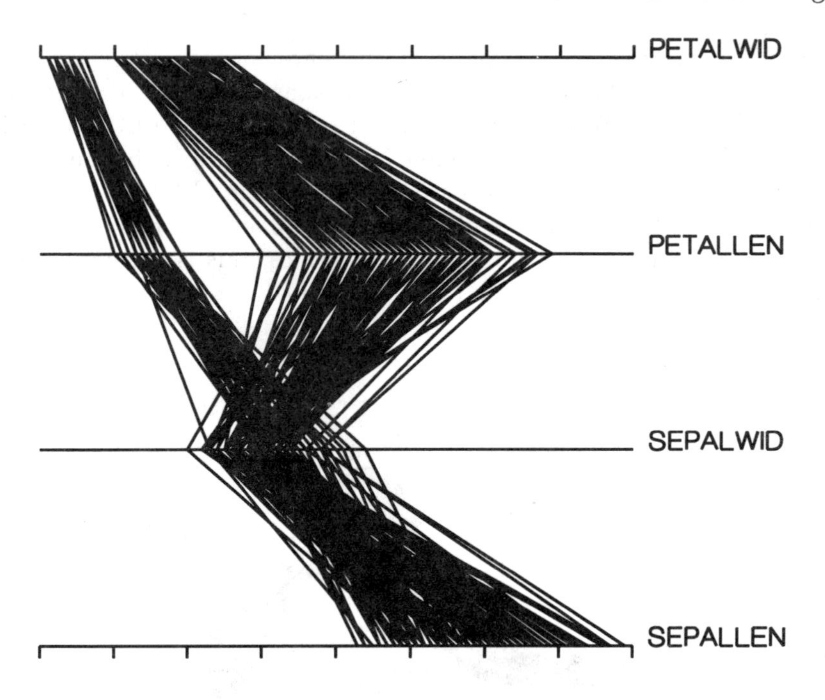

Figure 4: Parallel coordinate plot of the Iris-Data

are plotted as parallel vertically stacked axes, each observation is plotted as a line through all axes (see fig 4). Such a plot is a simultanenous display of one-dimensional scatterplots for each variable and a visual display of the correlations (the number of crossings). Clusters may be easily detected. With more than a few hundred observations, the plot becomes useless. Due to the publications of Bolorforoush & Wegman (1988, p.124-125) and Miller & Wegman (1991) the use of nonparametric density estimators for line densities and a colorcoded contourplot of these densities has gained popularity. For such plots, there is no upper limit for the number of observations. Practical experience and software for these plots are currently not available. Simple plots with parallel coordinates are implemented in SYSTAT.

Projectionplots of metric variables

More than 10-15 Variables can not be displayed simultaneously in a way, which is easily understable to humans. Therefore, a reduction of dimensionality is necessary. The most widely known techniques are principal component analysis (PCA), biplots, multidimensional preference scaling (MDPREF), multidimensional scaling (MDS), correspondence analysis (CA) and homogeneity analysis (nonlinear principal component analysis).[10] With the exception of MDS, all procedures are based on the singular value decomposition of a data matrix, only the pre- and postprocessing of the data differs (Weller & Romney (1990, p.16)).

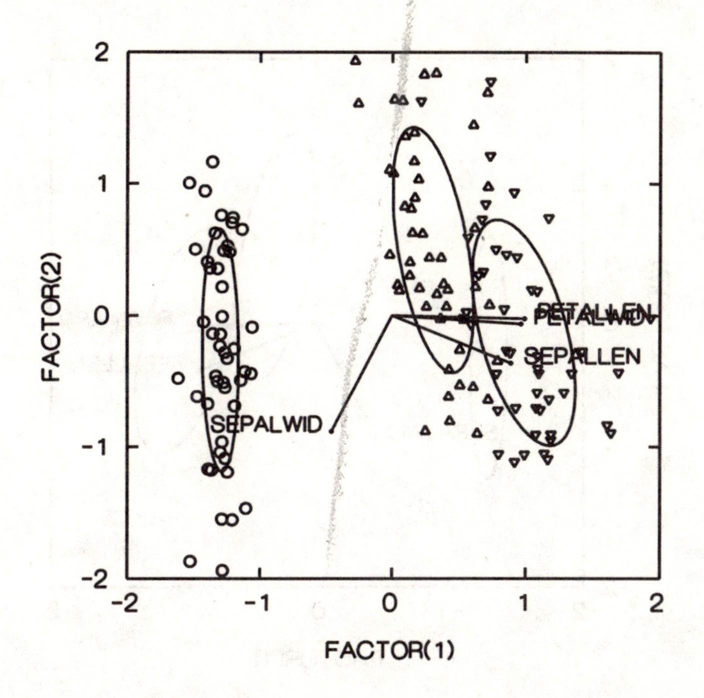

Figure 5: Biplot with separate data ellipses of each species of the iris-data. Observations are displayed as points, variables as vectors. Due to the high correlation, two vectors (PETALLEN, PETALWID) are nearly indistinguishable in the plot.)

The most important projection technique is principal component analysis (PCA). PCA is computed by means of a singular value decomposition of the correlation matrix R in two matrices: $R = ULU'$, where U are the eigenvectors and L is a diagonal matrix of eigenvalues. The principal components are computed as the product of the standardized data matrix Z and $U : PC = ZU$. Usually only the first two columns (the first two principal components) are used for the plot. Principle component plots display the projection of the objects into the space spanned by the principle components. In the case of two axes a scatterplot of the first two principle

[10]There seems to be no published experiences with the recently suggested dimension reducing technique "sliced inverse regression" (SIR) of Li (1991). Although SIR needs a specification of a dependent variable, it is also a kind of PCA. An implementation is available in XPLORE (Broich et al. (1990)).

component scores gives us the best possible linear two-dimensional representation of objects from a multidimensional space. Points close together in the plot have similar variable patterns; distances in the plot approximate Euclidean distances of the objects.

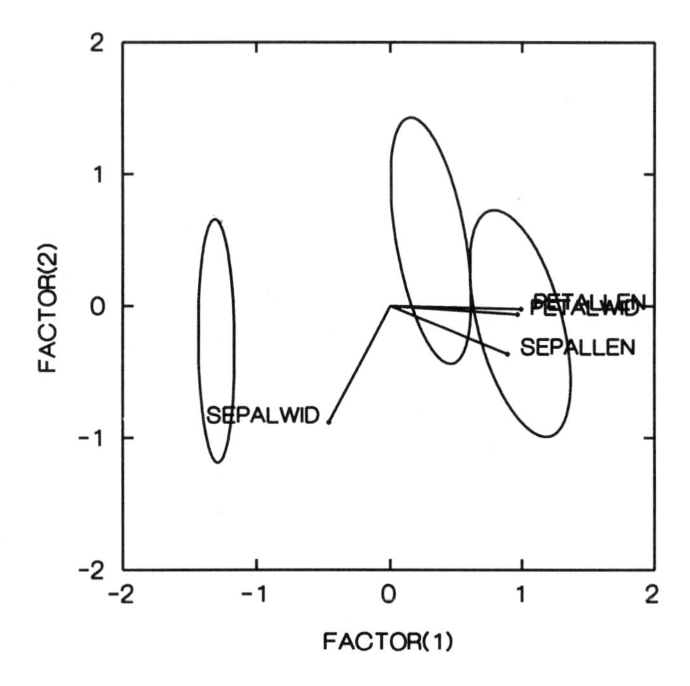

Figure 6: h-plot with separate data ellipses of each species of the iris-data. Observations are not displayed. As in the biplot the vectors of PETALLEN and PETALWID are nearly indistinguishable.)

The simultaneous mapping of both the variables and the objects of a PCA in a single plot is called "biplot".[11] The biplot is computed on the mean centered data matrix Y by a singular value decomposition: $Y = ULV'$, where U are the left singular vectors for the observations, L is the diagonal matrix of singular values, and V are the right singular vectors for the variables. Three different kinds of biplots are distinguished (SQ, GH and JK). The JK-biplot gives the same positions as the PCA for the observations (UL), the plotpositions of the variables are given by V'. The other variations of the biplot give the same configuration as the JK-biplot, but with different scaling factors of the axes. The objects are plotted as points, the variables as vectors from the origin. The distances of the objects in

[11]Gabriel (1971). Interestingly, biplots are nearly identical with MDPREF, see Jackson (1991, p.204-210)). indexMDPREF

the plot are approximations of their Euclidean distances. The cosine of the angle of two variable vectors is an approximation of their correlation. The length of the vectors is an approximation of the standard deviation of the variable. The projection of an object on a vector indicates the quantification of the observation on this variable. For an example, see fig. 5. With large datasets, biplots with individual data points are confusing. A simple way for handling large data sets in biplots is the use of density ellipsis for subgroups. Only the ellipsis and the vectors are plotted. For an example of such an h-plot, see fig. 6.

Two other projection techniques have to be mentioned: multidimensional scaling (MDS) and nonlinear mapping (NLM). Both can be used on any set of objects, for which distance measures (like Euclidean distance) can be defined.

MDS starts with a configuration of the points in the desired lower dimensional space (for example the results of a PCA). The departure from monotonicity between object distances in the plot to the distance matrix of the objects is assessed by a goodness of fit measure. The goodness of fit is maximized by standard numerical techniques. MDS is available in almost every standard program. Nonlinear mapping (Sammon (1969)) is very similar to MDS, but NLM minimizes the sum of squared differences between observed and projected distances. The second difference to MDS is that larger weights are given to small differences. The plots of MDS and NLM will therefore differ slightly in most cases. NLM is not implemented in standard programs.

Projectionplots of categorical variables

Projection plots for categorical data are not so widely known as the older projection techniques. But in the last 5 years, the old and long forgotten technique of correspondence analysis (CA) has become very popular in the social sciences. One indicator is the addition of CA-programs to all standard packages. In some journals CA-plots are the only kind of multivariate graphic ever published. The popularity of CA in social sciences can be explained by the urgent desire for nearly automatic "interpretations" of huge amounts of badly measured data. CA seems to offer a simple plot: Columns and rows of a two-dimensional contingency table are projected in the same plane. The interpretation of this "french plot" (symmetric plot) seems to be easy: Points close by each other in the space are similar. The popularity of CA is almost entirely due to this plot. Unfortunately, a lot of substantive interpretations of these plots are erroneous. Contrary to the seemingly simple surface, the geometric interpretation of CA is far from simple (Greenacre & Hastie (1987)). Therefore, we will discuss CA-plots and their refinements in more detail.

The use of CA has been obfuscated by a strange terminology (due to French statisticians); the underlying mathematics is simply a singular value decomposition of the matrix $D = ULV'$, where D is the matrix of the cell contributions to the χ^2-statistic multiplied by $\frac{1}{\sqrt{n}}$ (Friendly (1991, p.514)). The plot positions of rows and columns are given by postmultiplications of U and V with L and premultiplication with $\frac{1}{\sqrt{R}}, \frac{1}{\sqrt{C}}$ resp. R and C are the diagonal matrices containing the row and colums totals of the contingency table divided by n (the so called "profiles"). CA therefore can be viewed as a picture of the residuals of a loglinear independence plot (Van der Heijden et al. (1989, p.253)). The scores for the rows and the columns are normalized so that the column points are weighted averages of the row points and vice versa. Although it sounds pretty convincing to interprete the distances between row points and column points within the same plot such interpretations are misleading. Inter-set distances should be interpreted only according to the distances to the origin or the angle between the vectors defined by the origin and the points (for a discussion, see Greenacre (1993)). The only reasonable interpretation of the axis (the "dimensions" of the plot) has to rely on the projections of the points on the axis (of course, this are the category scores themselves). In this plot "similarity" can only be judged separately for each dimension. Nevertheless all available software provides the user with the possibility of producing the so called symmetric ("French") plot and seduce him to draw conclusions, which are simply wrong. In contrast to loglinear models, standard software does not provide procedures for statistical inference of CA. If the interaction is not significantly different from zero, CA will "decompose" the sampling error. To summarize: there are at least three problems using CA for EDA:
1. The number of dimensions of the projection must be justified on theoretical grounds or at least by the evaluation of the relative size of the eigenvalues. Habitual use of two dimensions to facilitate a plot is hardly justified.
2. The popular interpretation of the "French plot" is grossly in error.
3. The interpretation of any CA-solution is an interpretation of a sample statistic. Results are liable to sampling errors like any other statistic.

The first problem is not specific for CA and furthermore it is only a problem of documentation and careful interpretation. The solution of the second problem is easy: don't use symmetric CA-plots. The solution for the third problem is a bit more tedious: Any plot based on a sample can be evaluated by the use of the bootstrap. The application of a bootstrap to a plot is simple: Use the sample (size n) as a population and sample from this population with replacement new samples of size n. Compute the coordinates of the plot for each sample and display the resulting coordinates simultaneously. To enhance the visual impact of this plot, calculate the convex hull for each object in the plot. The smaller the convex hull in the plot, the better the stability of the graphical display (see Greenacre (1993)). For an example of a bootstrapped CA, see fig. 7.

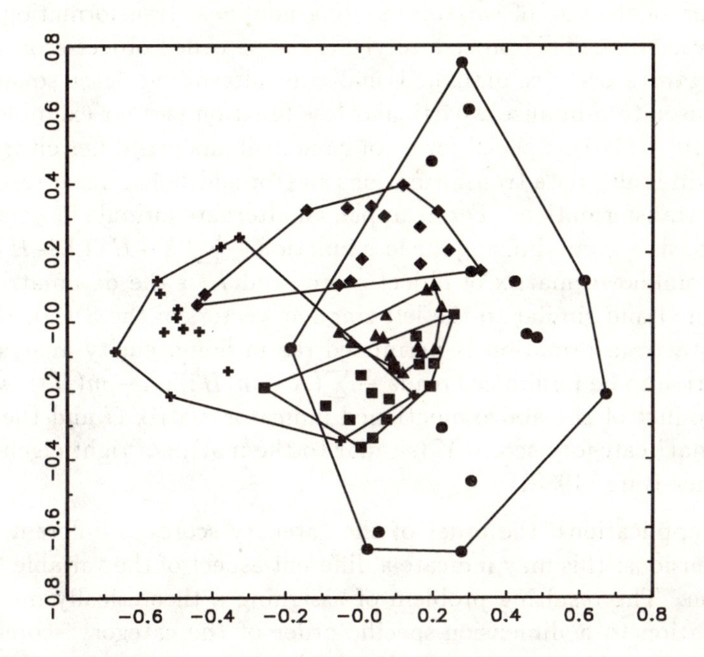

Figure 7: Bootstrapped CA-plot of Greenacre's (1993:75) "Funding Data" with 5 Categories. The 5 convex hulls of 15 replications are plotted. Most convex hulls have elements in every quadrant: Due to sample fluctuation, any interpretation of proximities between plot positions is nearly impossible.

CA is not restricted to two variables, the application of CA to more variables is called multiple correspondence analysis (MCA). The $(n \times p)$ data matrix is transformed to an indicator matrix G by the concatenation of dummy variables. Each $(n \times 1)$ column of G is a dummy variable which represents one category of one of the p original variables. Each row sums up to p and each column sums to the frequency of the corresponding category. Like the CA, the MCA is performed by a SVD: $GD^{-1/2} = ULV'$, where $D = diag(G'G)$. The geometric interpretation of a plot of the MCA suffers from the inclusion of the marginal frequencies in the analysis. Greenacre (1993, p.131-140) therefore developed an alternative form of correspondence analysis for multivariate data: "Joint Correspondence Analysis" (JCA). The interpretation of JCA is almost identical to that of the CA.[12] A lot of other techniques are identical with MCA. Dual scaling, optimal scaling, MCA and homogeneity analysis are all special cases (respectively only other names) of generalized non-linear canonical analysis with k sets of variables (van der Burg (1988), van der Burg & Dijksterhuis (1989)). Most applications use only one variable in each set. The aim of such an analysis is to maximize the canonical correlation of the sets of variables finding nonlinear transformations of the original variables. This procedure yields the so called object scores (rows) and category scores (columns). Usually an alternating least square algorithm is used to minimize a particular loss function (see for example Young (1981), Gifi (1990)). Special forms of canonical analysis differ either in the class of admissible data transformations and/or additional linear restrictions for these transformations. For example, an alternate formula for computing a PCA consists in minimizing the loss function $\frac{1}{m}\sum(X-H)'(X-H)$, where X is the unknown matrix of object scores and H is the datamatrix (X is orthonormal and similar to the left singular vectors of the SVD). If a non-linear data transformation is permitted (as in homogeneity analysis), the loss function to be minimized reads $\frac{1}{m}\sum(X-m(H))'(X-m(H))$, where m is the product of the above mentioned indicator matrix G and the matrix of "optimal" category scores Y (similar to the matrix of right eigenvectors, see van der Burg (1988)).

In most applications, the order of the category scores is different for any two dimensions: this may indicate a different aspect of the variable for each dimension. The resulting problem of assigning a theoretically meaningful interpretation to a dimension-specific order of the category scores is the most difficult and important task of a multiple correspondence or HOMALS-analysis. This interpretation problem of MCA is avoided by an additional constraint on the category quantifications: All resulting quantifications for one original variable will be on a line through the origin. This is the idea of nonlinear principal component analysis (NPCA). The resulting matrix of new category scores has rank one; the quantifications for all dimensions are identical except for a multiplication. In contrast to PCA, nonlinear canonical correlation can be applied to variables with different and unknown measurement levels; especially nominal variables. Therefore, an increasing use of these techniques in the social sciences can be expected. [13] In principle,

---

[12] Unfortunately, for actual programming Greenacre (1993) does not give enough technical details of JCA. Recently, a new release of SIMCA has been announced, which will include JCA.

[13] The discussion in the philosophy of science shows, that "measurement" and "theory" are dependent. It is therefore not astonishing, that in theoretical weak fields "measurement" in the sense of the representational theory of measurement can hardly be archieved. As a consequence, many statisticians have sought to improve the current practice of data analysis by the development of techniques for nominal or ordinal "measurement levels"

the problems of the interpretation of resulting row and column scores are completely analogous to simple CA. In practice, most applications of these techniques use only a plot of the category quantifications. Therefore, the problem of interset distance interpretation is avoided.

Currently very little is known about the consequences of a choice between MCA and NPCA. Despite the large number of stand-alone programs mentioned in literature, very few applications of NPCA exist. Even the inclusion of MCA and NPCA in two standard packages (SPSS and SAS, both with incomprehensive documentation) has only slightly increased the number of published applications.

Other projection techniques

If the number of interesting variables exeeds trivial limits, the number of possible relationships will be so large, that the available time will not even permit the examination of all bi- and trivariate plots. A number of solutions for this problem has been suggested. The "small tour" is a sequential view of 3d-rotating scatterplots (Stuetzle (1984, p.67)); the user controls the selection and sequence of the variables. The "grand tour" is a sequence of 2d-projections with changing directions of projection (Asimov (1985)). Even a small number of dimensions will lead to such a large number of projections so that only a subset can be examinated. Stuetzle (1984, p.74) considers a maximum of 6. Most of the projections will be uninteresting; for example, projections to a point or an evenly spaced projection. The basic idea of projection pursuit is the computation of an index of "interestingness" of a projection and the display of only the most "interesting" projections (Huber (1985), Friedman (1987)). Whereas "small tour" and "grand tour" need powerful dynamic graphics subsystems, but can in principle be done with any PC, "projection pursuit" need only static displays but very high computing power. The results of a projection pursuit algorithm are simple two-dimensional plots, which exhibit more or less interesting configurations of points. Substantive interpretation of these plots is certainly not easy; examples on real data sets are unknown to us. Implementations in standard packages are not available. [14] Fortunately, two factors will prevent these techniques from widespread misuse, especially in the social sciences: The necessary mathematical sophistication for the interpretation will restrict

(for a critique of this classification, see Duncan (1984)). In contrast to this practice, history of science suggests that better measurement will be a consequence of better theories (c.f. Kuhn (1961)). As the now declining popularity of log-linear models has shown, new analysis techniques will not help.

[14]To our knowledge, there is only an implementation in ISP, widely unknown in the social sciences.

the application to a very small subset of social scientists and the coarse measurement procedures will doom most plots to be dull.

Dynamic graphics

"Dynamic graphics" is the label for a collection of techniques which require a high amount of interaction between the user and the display. Some authors therefore suggest that "high interaction graphics" is a more useful label. The following techniques are examples for high interaction graphics (see Becker et al. (1987)): Interactive identification of points by labeling or highlighting the same observation in sploms ("scatterplot brushing", see Becker & Cleveland (1987)), rapid change of different observation subsets in the display ("alternagraphics"), linking the same observations in different plots and 3D-rotation. Some applications of multivariate interactive graphics are simple, but very convenient. A nice example is the choice of the bandwidth of a multivariate nonparametric density estimator (see Scott (1992)) by an onscreen slider. Really interesting features of interactive graphics like conditioned plots ("slicing") or linked plots (different plots of the same data, selected observations highlighted, see Young et al. (1993)) are unavailable for nonprogrammers. The practical use of such techniques is severely limited by the available software. None of the major packages have such features beyond the relativc unimportant 3D-rotation. [15] A few standard programs have scatterplot brushing (e.g. SYGRAPH) or interactive outlier identification (e.g. STATGRAPHICS). Many of the current implementations of such features are done with Lisp-Stat (Tierney (1990)). Unfortunately, the complexity of such environments is too high for the average user.

Conclusion

Even the simple "descriptive" techniques of multivariate graphics like SPLOMS or conditional scatterplots are not implemented in most current "statistical analysis systems". Simple projection techniques like MDS, PCA or CA are implemented, but even the biplot is still missing in most programs. Aids for the interpretation of projection plots like smoothers, convex hulls, density ellipsoids of subgroups, rotation, brushing and interactive point identification are not available in standard programs, at least not for technically unexperienced users. The same is true for techniques of dynamic graphics.

[15] For an interesting discussion of the use and limitations of dynamic graphics for a special application, see Cook & Weisberg (1989).

If despite these obstacles a user succeeds in producing an interesting projection plot, the problem of validating the plot remains. Internal and external validation is necessary. The internal validation consists of the comparison with other projections; for example a MDS-plot with a PCA-plot or a nonlinear mapping plot. The only formal method for the comparison of graphical displays of multivariate data is procrustes analysis (Sibson (1978)). Standard software for procrustes analysis simply does not exist. The only possible external validation of the results of any plot is a replication of the plot on other samples. In practice, other samples are not immediately available. To prevent exaggerated interpretations of projection plots, bootstrapping techniques must be used (Stine (1990)). Such techniques are widely unknown and almost never used in practice. They are not supported by any analysis package.

Graphical techniques in multivariate analysis are widely misunderstood as tools of a purely "exploratory data analysis". Tukey (1990, p.332) calls this the "tabula-rasa fallacy of display": The illusion of an automatic and unbiased way to analyze the data by graphical display. Multivariate EDA suffers from logical and cognitive limitations: Without the use of theoretical models, researchers will be lost in multivariate space. A useful multivariate EDA is mainly an aid for the interpretation of results of computational techniques. Very often this will be a kind of residual analysis. Most of the available programs do not support graphical techniques as aids for the analysis of residuals of standard techniques like analysis of variance (Emerson (1991)), logistic regression (Hamilton (1992)), factor analysis (Bollen & Arminger (1991)) or event history analysis. Techniques with a large graphical component like nonparametric density estimation (Härdle (1991)) or plots as an aid to the interpretation or the discovery of clusters (Chambers & Kleiner (1982)) have been largely ignored in textbooks and programs. Even a simple scatterplot smoother (Hastie & Tibshirani (1990)) like LOWESS is not available in current releases of large standard packages. Many of the mentioned techniques exists only as humble FORTRAN programs, S-Plus functions, or SAS-IML-macros. Ironically, those techniques, which are most easily understood, are available only for technically inclined statistical experts.

References

Andrews, D.F. (1972) Plots of High-Dimensional Data; Biometrics, 28, 125-136

Asimov, D. (1985) The Grand Tour: A tool for viewing multivariate data; SIAM Journal on Scientific and Statistical Computing, 6, 1, 128-143

Banks, W.P.& Krajicek, D. (1991) Perception; Annual Review of Psychology, 42, 305-331

Becker, R.A. & Cleveland, W.S. (1987) Brushing Scatterplots; in: Technometrics, 29, 2, 127-142

Becker, R.A., Cleveland, W.S. & Wilks, A.R. (1987) Dynamic Graphics for Data Analysis; Statistical Science, 2, 4, 355-395 (with discussion)

Bollen, K.A. & Arminger, G.(1991) Observational Residuals in Factor Analysis and Structural Equation Models; Sociological Methodology, 21, 235-262

Bolorforoush, M. & Wegman, E.J. (1988) On Some Graphical Representations of Multivariate Data; in: Wegman, E.J., Gantz, D.T. & Miller, J.J. (eds.) Computing Science and Statistics: Proceedings of the 20th Symposium on the Interface, Alexandria, VA (ASA), 121-126

Broich, T., Härdle,W. & Krause,A. (1990) XploRe. A computing envinronment for eXploratory Regression and data analysis, Institut für Gesellschafts- und Wirtschaftswissenschaften, University of Bonn.

Chalmers, A.F.(1987) What is this thing called science? London, Open University Press

Chambers, J.M. & Kleiner, B. (1982) Graphical Techniques for Multivariate Data and for Clustering; in: Krishnaiah, R. & Kanal, L.N. (eds.) Handbook of Statistics, Amsterdam, New York, Oxford, Vol. 2, 209-244

Chambers, J.M., Cleveland, W.S., Kleiner, B. & Tukey, P.A. (1983) Graphical Methods for Data Analysis, Belmont (Wadsworth)

Chernoff, H. (1973) The use of faces to represent points in k-dimensional space graphically; Journal of the American Statistical Association, 68, 361-368

Cook, R.D. & Weisberg, S. (1989) Regression Diagnostics With Dynamic Graphics; Technometrics, 31, 3, 277-311 (with Discussion)

Duncan, O.D. (1984) Notes on Social Measurement. Historical and Critical, New York, Russel Sage Foundation

Emerson, J.D. (1991) Graphical Display as an Aid to Analysis; in: Hoaglin, D.C., Mosteller,F. & Tukey, J.W. (eds.) (1991) Fundamentals of Exploratory Analysis of Variance, New York (Wiley), 165-192

Everitt, B.S. (1987) Graphical Displays of Complex Data - Scientific Tools or Simply Art for Art's sake; Bulletin of the International Statistical Institute, Proceedings of the 46th Session, 353-367

Freedman, D.A.(1991) Statistical Models and Shoe Leather; in: Maerdsen, P.V. (ed.) Sociological Methodology 1991, vol. 21, 291-313

Friedman, J.H. (1987) Exploratory Projection Pursuit; Journal of the American Statistial Association, 82, 397, 249-266

Friendly, M. (1991) SAS System for Statistical Graphics, Cary/NC (SAS Institute)

Gabriel, K.R. (1971) The biplot graphic display of matrices with application to principal component analysis; Biometrika, 58, 453-467

Gabriel, K.R. (1985) Multivariate Graphics; Encylopedia of Statistical Science, Vol. 6, 66-79

Gessler, J.R. (1993) Statistische Graphik, Basel (Birkhäuser)

Gifi, A. (1990) Nonlinear multivariate analysis, New York (Wiley)

Greenacre, M. & Hastie, T. (1987) The Geometric Interpretation of Correspondence Analysis; Journal of the American Statistical Association, 82, 437-447

Greenacre, M.J. (1993) Correspondence Analysis in Practice, London (Academic Press)

Härdle, W. (1991) Smoothing Techniques. With Implementation in S, New York (Springer)

Hamilton, L.C. (1992) Regression with Graphics. Pacific Grove (Brooks)

Hastie, T.J. & Tibshirani, R.J. (1990) Generalized Additive Models, London (Chapman and Hall)

Hartung, J. & Elpelt, B. (1985) Multivariate Statistik, München (Oldenbourg), 4. edition

Holland, J.H., Holyoak, K.J, Nisbett, R.E. & Thagard, P.R. (1986) Induction Processes of Inference, Learning and Discovery, London, Mit-Press

Huber, P.J. (1983) Experience with Three Dimensional Scatterplots; Journal of the American Statistical Association, 82, 448-453

Huber, P.J. (1985) Projection Pursuit; The Annals of Statistics, 13, 435-475

Inselberg, A. (1985) The plane with parallel coordinates; The Visual Computer, 1, 69-91

Jackson, J.E. (1991) A User's Guide to Principal Components, New York

Kuhn, T.S. (1961) The function of measurement in modern physical science; Isis, 52, 161-190

Langley, P., Simon, H.A., Bradshaw, G.L. & Zytkow, J.M. (1987) Scientific Discovery. Computational Explorations of the Creative Processes, Cambridge, Mass, Mit-Press

Li, K.-C. (1991) Sliced Inverse Regression for Dimension Reduction; Journal of the American Statistical Association, 86, 414, 316-342

Mason, W.M. (1991) Freedman is right as far as he goes, but there is more, and it's worse; in: Marsden, P.V. (ed.) Sociological Methodology 1991, vol. 21, 337-351

Miller, J.J. & Wegman, E.J. (1991) Construction of Line Densities for Parallel Coordinate Plots; in: Buja, A. & Tukey, A. (eds.) Computing and Graphics in Statistics, New York (Springer), 107-123

Monette, G. (1990) Geometry of Multiple Regression and Interactive 3-D-Graphics; in: Fox, J. & Long, J.S. (eds.) Modern Methods of Data Analysis, Newbury Park, 209-256

Morrison, D.E. & Henkel, R.E. (eds.) (1970) The Significance Test Controversy, London, Aldine

Oakes, M.(1986) Statistical Inference: A Commentary for the Social and Behavioral Sciences, Chichester (Wiley)

Pinker, S. (1990) A Theory of Graph Comprehension; in: Freedle, R. (ed.) Artificial Intelligence and the Future of Testing, Hillsdale (Erlbaum), 73-126

Popper, K.R. (1972) Objective Knowledge, Oxford, Oxford University Press

Sammon, J.W. (1969) A Nonlinear Mapping for Data Structure Analysis; IEEE Transactions on Computers, C-18, 5, 401-409

Schnell, R. (1994) Graphisch gestützte Datenanalyse, München, Oldenbourg, forthcoming

Scott, D.W. (1992) Multivariate Density Estimation, New York, Wiley

Sibson, R. (1978) Studies in the Robustness of Multidimensional Scaling: Procrustes Statistics; Journal of the Royal Statistical Society, B, 40, 234-238

Stine, R. (1990) An Introduction to Bootstrap Methods; in: Fox, J. & Long, J.S. (eds) Modern Methods of Data Analysis, Newbury Park, 325-373

Stuetzle, W. (1984) Graphische Exploration multivariater Daten am Computer; in: Allgemeines Statistisches Archiv, 68, 63-80.

Tierney, L. (1990): Lisp-Stat: An object-oriented environment for statistical computing and dynamic graphics, New York (Wiley).

Tukey, J.W. & Wilk, M.B. (1970): Data Analysis and Statistics: Techniques and Approaches; in: Tufte, E.R. (ed.): The Quantitative Analysis of Social Problems, Reading/Mass.,370-390.

Tukey, J.W. (1990) Data-Based Graphics: Visual Display in the Decades to Come; in: Statistical Science, 5, 3, 327-339.

van der Burg, E. (1988) Nonlinear canonical correlation and some related techniques, Leiden (DSWO-Press).

van der Burg, E. & Dijksterhuis, G. (1989) Nonlinear Canonical Correlation Analysis of Multiway Data; in: Coppi, R. & Bolasco, (eds.) Multiway Data Analysis, Amsterdam (North Holland), 245-255.

Van der Heijden, G.M., Falguerolles, A.D. & DeLeeuw, J.D. (1989) A combined approach to contingency table analysis using correspondence analysis and log-linear analysis; in: Applied Statistics, 38, 2, 249-292.

Verboon, P. (1988) Graphical Tools in Multivariate Analysis, University of Leiden.

Wainer, H. & Thissen, D.(1981) Graphical Data Analysis; in: Annual Review of Psychology, 32, 191-241.

Wainer, H. & Thissen, D.(1993) Graphical Data Analysis; in: Keren, G. & Lewis, C. (eds.) A Handbook for Data Analysis in the Behavioral Sciences: Statistical Issues, Hillsdale, New Yersey, (Erlbaum), 391-457.

Wegman, E.J. & Carr, D.B. (1993) Statistical Graphics and Visualization; in: Rao, C.R. (ed.): Handbook of Statistics, Vol. 9, New York, 857-958.

Wegman, E.J. (1990) Hyperdimensional Data Analysis Using Parallel Coordinates; in: Journal of the American Statistical Association, 85, 664-675.

Weihs, C. & Schmidli, H. (1990) OMEGA (Online Multivariate Exploratory Graphical Analysis) Routine Searching for Structure; in: Statistical Science, 5, 2, 175-226 (with discussion).

Weller, C. & Romney, C.(1990) Metric Scaling. Correspondence Analysis. Newbury Park (Sage).

Young, F.W. (1981) Quantative Analysis of Qualitative Data; in: Psychometrika, 46, 4, 357-388.

Young, F.W., Faldowski, R.A. & McFarlane, M.M. (1993): Multivariate Statistical Visualization; in: Rao, C.R. (ed.): Handbook of Statistics, Vol. 9, New York, 959-998.

Interactive Analysis of Spatial Data

Matthias Nagel
ZEG Zentrum für Epidemiologie und Gesundheitsforschung Berlin GmbH, Außenstelle Zwickau, Marchlewskistr. 2, D – 08062 Zwickau

Key Words:
classification, data analysis languages, disease mapping, EDA, environmental analysis, interactive graphics, large data sets, spatial data, special models on lattices, thematical maps

Abstract
Since a large proportion of environmental and environmetal health data is held in spatial form geographical information systems appear to be the natural choice for the storage and processing of data. The complexity of environmental data requires special concepts and solutions and a comprehensive array of facilities. Aim of the paper is gathering of the geographical object descriptions and environmental/disease informations in the open structure of the ISP language to provide an easier access to the geographical and (the large set of) relevant data. We illustrate the powertools of dynamical graphics and interactive data management used in the system in application to pollution data of heavy metals in rivers, pesticide contamination of soil and some forms of disease mapping of prevalence data of the former GDR.

The Problem

Most data sets in ecology, environmental protection, epidemiology, etc. have a geographic context which needs to be considered in the analysis of the data. Aside from the numerical analysis of this influencing factor (via a suitable categorization based on relevant regional or demographic features), one will find a visualization of the data within their geographical structure very helpful. It seems only natural to combine data analytical concepts such as EDA with methods of computer-aided cartography and regional- or geostatistics to achieve a rapid and uncomplicated visualization of this complex information.

Geographic Information Systems (GIS) represent the most elaborate, yet not the only way to tackle data analytical problems with a geographical context (Schilcher (1989), De Root (1992)). It turns out that some data analysis languages show particularly strong suitability to address such questions due to their great flexibility.

This applies in particular to ISP (DOS and soon UNIX), S-Plus (DOS, UNIX and MS Windows), Lisp-Stat (DOS, MS-Windows and Macintosh; see Tierney (1990)), Datadesk, and REGARD (see Unwin (1993), for Macintosh) may also be useable for such applications. However, with the latter two systems one is restricted to the features available, Datadesk offers only methods for EDA but none for spatial data, the exact opposite applies to REGARD. Since the Mac operating system allows for easy data exchange, an excellent combination of the two may make analyses similar to those addressed in this paper possible.

The author maintains that the usability of any such system is determined by its ability to handle frequently changing problems with large amouts of data: The systems should offer good response times for interactive work (graphics) even with several thousands of observations and a few dozens of variables. S-Plus must be excluded under MS-DOS because of the poor garbage collection, the constant disk access and lack of interactive graphics. The author has no experience with the performance of the Lisp-Stat system with large amounts of data.

A precondition for analysis systems for spatial data to be suitable is a powerful, easy to use graphics facility and easy programmability, i.e. expandability, so that new methods can be incorporated into the system anytime. The following requirements must be fulfilled by a data analysis system suitable for the analysis of geographical data:

- Interface to external data bases: Ease of access to external data bases and efficient management of subject data
- Geographic data: Ability to combine these with subject data
- Graphics: Interactive generation and analysis of thematic maps. In the analysis one must consistently apply the principles and methods of EDA as adapted for the analysis of data with a territorial context
- Availability of commonly used map types: Point maps, isopleth or choropleth maps and gradient graphs should be available or producible
- Data analysis and management: Should be possible within the graphical user interface of a thematic map

- Hardcopy output: Either as "quick-and-dirty" screen dumps or as high-quality presentation maps

- Digitizing of maps: Should be possible directly in the analysis system to produce relevant maps of sufficient accuracy quickly and with the necessary degree of detail

- Implementation of complex algorithms: To adapt mathematical and statistical models and to add generated or modified data (eg. via a transformation) to the data already available

- Easy handling of the system: The system should offer both a customizable, easy to use menu interface as well as the flexibility provided by a high-level command language.

To achieve fast response time, one will find the workspace concept useful: It is advantageous to keep important pieces of data in a workspace so that they and also intermediate results of calculations are always available immediately. Thus one can avoid or reduce the need for repeating analyses steps. Furthermore, it should be possible to document the analysis in the form of a re-executable protocol or script.

Maps are Special Graphics!

When constructing maps one needs to consider that the human eye only can grasp three visual variables simultaneously: two dimensions are needed for the position coordinates of the graphics objects, leaving a third dimension, such as brightness, size, or colour of an object as a simultaneously useable parameter. If a map contains more visual parameters, the eye resorts to a sequential inspection, i.e. the dimension "time" is used.

In contrast to other graphics, maps already use two dimensions for the representation of the geographic component, the y-axis of the plane therefore is no longer available to separate of individual factors of the subject data (see Bertin (1974, 1981, 1982)). For thematical maps one, therefore, needs to:

- Draw one map per factor. The map then answers the questions "where is this factor?" and "what can be found at this location?"

- Superimpose all factors on a single map. In this case there is no longer a visual answer to the question "where is this factor?"

Thematical maps of the former type are maps to look at, which corresponds to exploratory analysis of the data. They show the distribution of a factor as well as structural groupings, which arise due to spatial proximity as well as numerical proximity of values found in a certain region. The latter type represents maps to read. For analyses purposes they have only limited use, however, have a justification for specific presentation of relationships and information already known to the analyst.

Interactive Analysis - Definition

The rules of graphical semiology mentioned above (see Bertin (1982)) apply to static graphics. Using methods of dynamic graphics one has technical means to break through these visualization barriers by rotating a scatterplot or by linking or brushing within scatterplot matrices (see Cleveland (1988), Buja & Tukey (1991), Huber & Nagel (1993), Ostermann (1992), Ostermann & Nagel (1993), Fleischer et.al. (1992)).

Dynamic graphics become a vital component of interactive analysis of spatial data. It is possible to properly adjust or position a three dimensional view, to identify an object through its label, display a variable value, or to mask selected portions or layers of the map to emphasize or suppress information. One can select interesting regions (and thus subsets of the data) interactively, which then can be analyzed separately, for example to study heavy metal contamination along a river. Another important aspect of dynamic graphics is visual classification: data showing a similarity can be grouped in subsets which can then be numerically analyzed or one can mark selected objects to determine whether they also qualify as outliers in other variables.

The entire **data management** is also **interactive**: During analysis one selects variables, forms groups, fits models, and transforms data. More generally, on may see interactive analysis as the enlargement of the system though the addition of new variables (which may be derived by transformation of existing variables), new observations, or incorporation of further analytical methods which may seem appropriate. It is of considerable advantage, if one can use one and the same analysis system for all tasks, avoiding the need to switch between data base systems and/or programming language.

Finally, one has the ability to **interactively develop maps** whenever several correction iterations are needed and useful.

Sample Applications

The following four case studies will present different types of analytical maps and explain what techniques are indispensable. Only some of the possible analysis methods are mentioned here. With ISP most of the typical uni- and multivariate methods of EDA are available or can be created very easily using the array-oriented language (nearest neighbor estimators, splines, smoothers, projections, etc., are all available as ISP building blocks). For example, the join-count-test with a convenient graphical representation was added as a simple macro to the system as a useful method for investigating spatial auto-correlations. Other methods commonly used in geostatistics, such as variogram-analysis and Kriging were not used, since they were not applicable to our examples.

Exploration of Data with Point Maps: Heavy Metals in Rivers

In 1988, after a lengthy dry period, several sediment and water samples were taken from river and lake beds in teh South of the then still existing GDR. The locations of the 52 sampling points are shown in figure 1 (Nagel & Lippold (1992)).

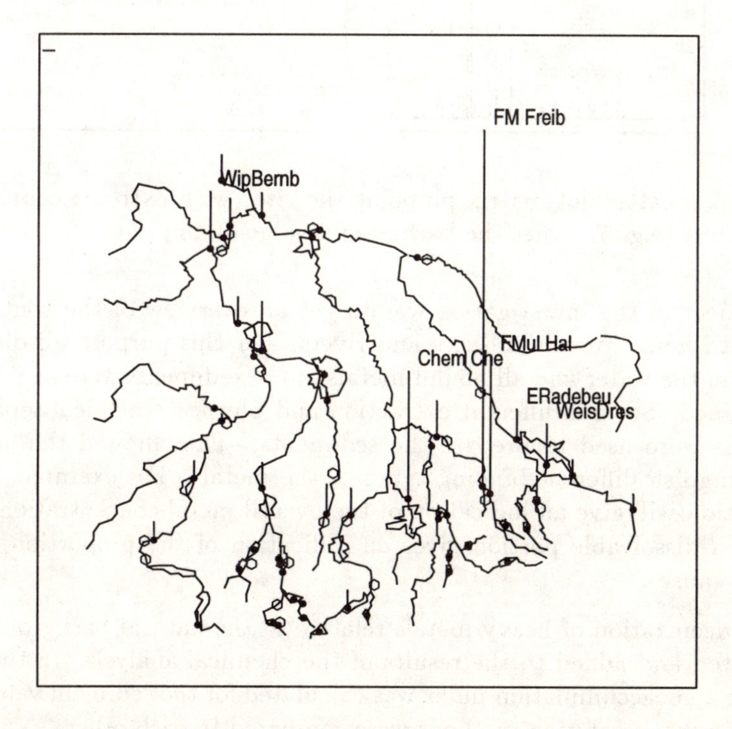

Figure 1: The locations of the 52 sampling points and the concentration of copper (displayed as columns over the sample points)

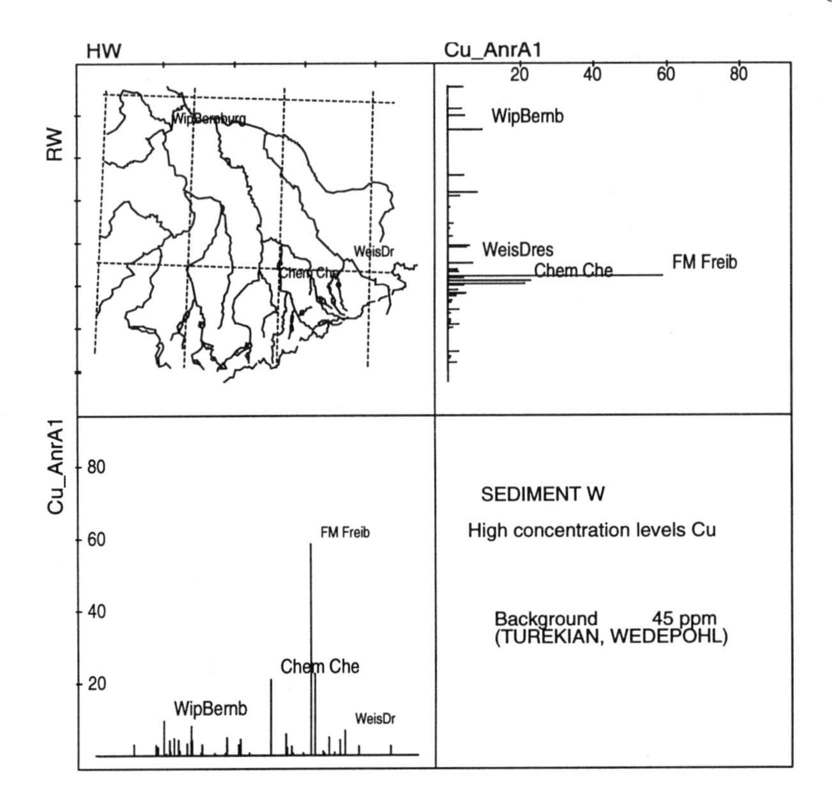

Figure 2: Scatterplot-matrix pinpoint the sites with extreme copper concentrations (e.g. 57 times the background value of 45 ppm).

The object of the investigation was to get an overview of the contamination with heavy metals in lakes and rivers. For this purpose all dissolved metals in the water and all bound metals in the sediments were analytically determined. Several different extraction and physical/chemical separation methods were used to prepare the sediments. This allowed the analysts to distinguish different binding types of the metals. For example, a total dissolution will give an indication of the overall metal concentration, while the HCl-dissolvable portion gives an indication of the proportion due to human causes.

The concentration of heavy metals relative to the natural background was calculated and added to the results of the chemical analysis. In the same manner a geoaccumulation-index was calculated for the sediment values and the different dissolution methods were compared to each other.

Geographic latitude and longitude of the sample points were added to the data values to enable a display in a map. These values were digitized carefully, so that they could be combined with various available maps. The rivers and lakes, as well as major towns represent additional information which is not of critical importance for the analysis, however, facilitates orientation. Accordingly, not as much effort was invested into the preparation here.

The environmental researcher can now use the dynamic graphics system of PC-ISP (the analysis software used) to subject the area of interest to a thorough investigation. Figure 1 shows a snapshot of the map rotated slightly in the 3D-mode: the concentration of copper is displayed as columns over the sample points. By pointing the cursor at the tip of a column one can now read off the variable values (also for those variables not displayed at the moment, for example 2640 ppm Cu on sample point Freiberg). This is essentially a visualization of data base entries enhanced with the results of numerical analyses.

In the scatterplot-matrix display (figure 2) of the same data one can immediately pinpoint the sites with extreme copper concentrations (e.g. 57 times the background value of 45 ppm). In similar fashion one can quickly shift through each of the variables and localize them in the region. Most of the high concentration levels can be explained by the settlement of heavy industry in the region (Saxony, East Germany).

Using scatterplot matrices, one can quickly analyze possible connections and develop hypotheses for modelling. (For example, the relationship between *ph*-value and dissolved aluminum can be described through a simple regression model with a correlation coefficient of 0.91.)

A principal components analysis of the heavy metals reveals correlation structures between the variables and amplifies similarities or dissimilarities amongst the sampling points. By adding the principal component values to the display in the dynamic graphics, the environmental researcher gets additional indications on the origin of contaminations. The Freiberg valley (copper industry) sticks out through the second principal component, while the lakes have high values on the third component.

One can see that the first principal component shows a strong loading through Pb, Cd, and Zn, while the second one has a heavy loading of Ni and Cu, and the third one is determined by Cr and Al. The first three principal components describe 76% of the variance in the data.

Spatial Model on Lattices: Pesticide Contamination of Soils

The ministry of agriculture of the former GDR built up an extensive data base to monitor and develop strategies for pesticide control. It included all regions used agriculturally as well as detailed information on the following data:

- for each plot of land the type of pesticide and quantity used
- date and mode of application
- crop planted on the field
- exact location of plot with plot number

Detailed information on the main active ingredient, the water pollution index, and a toxicological classification is available for the total of 193 different pesticides and 81 different tank mixtures. It was our task to make an assessment of the contamination risk for drinking water through pesticides and fertilizers to determine the

- effective contamination of soil due to agricultural activity and
- "optimal" location of sample points for a measuring and monitoring program to investigate the influence of pesticides on drinking water.

The amount of data is considerable. In the course of a pilot study for the two districts of Hagenow and Ludwigslust in Mecklenburg-Vorpommern, we extracted over 13 MB of data (over 500'000 records for the period of 1985 to 1989, just for these two districts alone!).

Given this amount of data, a traditional data base or tabulation solution becomes impractical. Therefore, we suggested to the owner of the data, the *Agrarbörse Deutschland Ost e.V.*, to make a cartographic analysis of the data in which applied quantities of pesticides would be displayed together with the drinking water protection zones and water generation facilities (see Bericht an die FKST (1993)).

For this purpose we digitized 45 survey maps in which the exact locations of water generation facilities and water protection zones were marked by the *Institute of Hygiene in Schwerin.* While preparing data and maps, we discovered that despite the very exact documentation of the location of each plot, the survey maps were not detailed enough to reconstruct the exact positions of the plots.

For this reason we decided to use a (rectangular) grid within which the individual plots could be allocated. The pesticide data were aggregated at the grid points in such a way, that all relevant information about pesticides, the amount and date of application, as well as soil quality and water protection class were maintained. By packing this data in a binary direct access file, we stored it in a very compact yet easily retrievable form. To ease the analysis, several summary statistics were also coded as variables, so that they could be viewed just like any of the other variables. These 24 basic variables can be displayed in the dynamic graphics as **isosurfaces**, i.e. surfaces of equal levels of pesticide application as shown in figure 3.

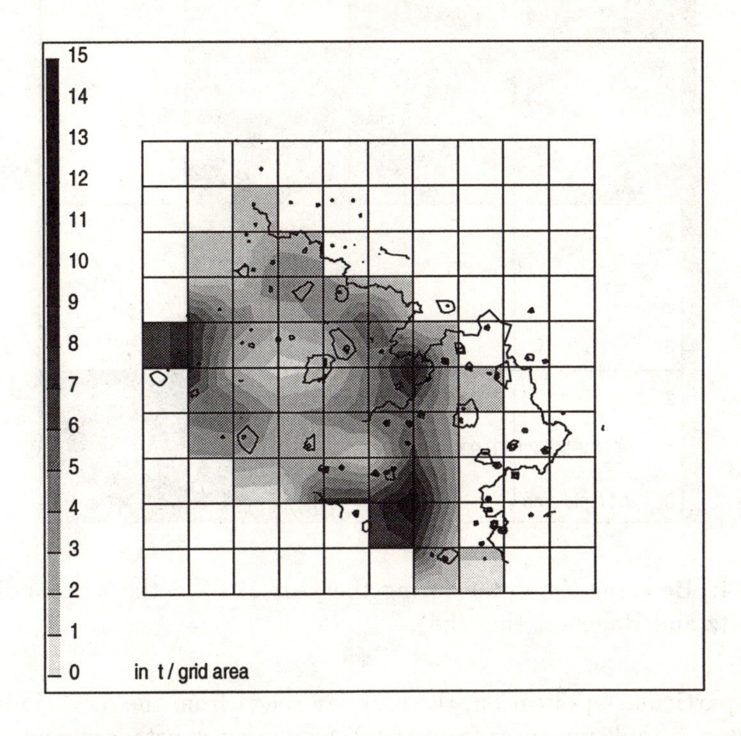

Figure 3: The figure shows the distribution levels of applied pesticides of water quality index 3 during the year 1986 in the districts of Hagenow and Ludwigslust (in t/grid area).

Obviously, maps like these create immediate hints in which areas of drinking water protection zones the land was treated particularly heavily with pesticides. The hygienist has the choice of either a global overview over the applied quantities or can also probe for detailed information such as year of application, crop type and characteristics of the pesticides, which might allow conclusions about the ability of the soils to absorb pesticides.

A comparison of the quantities applied over a period of years is easily possible using this system. One just selects a variable and the desired map, into which one can zoom and add labels to get details, e.g. in the surrounding of a water generation facility. The maps can be saved and viewed later as a slide show or can be printed as high-quality color postscript files directly from the application.

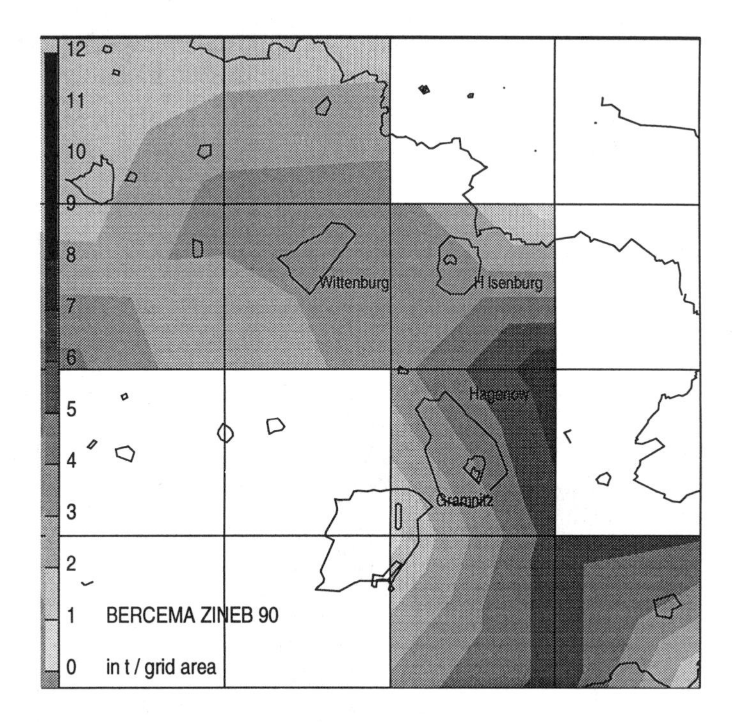

Figure 4: Bercema Zineb 90 during the year 1987 in the surroundings of Gramnitz and Hagenow (in t/ha).

In this particular application, the user can select from the menu individual pesticides or tank mixtures from the 274 different substances contained in the data set. The relevant data are extracted for the selected substances from among the 35'744 different records pertinent to these two districts, split by year of application and allocated to the area of application on the map. Figure 4 shows the amount of pesticide Bercema Zineb 90 applied during the year 1987 in the surroundings of Gramnitz and Hagenow. The map was zoomed and the scaling of the isosurfaces was modified interactively.

This software application represents a prototype for the graphical processing of large data bases. The wishes and needs of the end user were incorporated directly during the development. The user received a "personal" analysis system, whose realization was strongly influenced by himself. He can easily use it by use of complete and application specific menus and the intuitive actions of the dynamic graphics.

Considering the high costs of a soil analysis (approx. DM 1.000 to DM 1.700 per analysis) it is sensible to select sampling points carefully based on this data (if one cannot find anything at a site with high pesticide application, it does not make any sense to measure at sites with low applications). These analyses are being conducted right now and the results will also be evaluated within this system, i.e. compared to the quantities applied. The aim is to be able to model absorption processes.

Disease Mapping: Prevalence Data of the Former GDR

After the previous examples of point and grid maps, this example illustrates an analysis of disease incidences for regional structures such as districts or countries. In this particular case the data represent the complete hospitalization data of the years 1985 through 1989 of the former GDR. The data set contains almost 1000 different diagnoses according to the International Classification of Diseases (see ICD-9, 1975).

After an initial processing of the raw data - the ID-numbers had to be made anonymous and multiple identical diagnoses on one and the same person had to be eliminated during the prevalence period - the big problem remained how to analyze the 14.923.300 records.

As a first step, individual files were generated for certain diagnostic groups. Redundant information was removed and the data stored in compressed-binary form (for example district, age, and sex were packed into a single word, since initially one would look for these together). Further, the data compressed by summing up cases for identical diagnoses, domicile, age groups and sex.

For this task the workspace-concept of PC-ISP was used to load as much data as possible into memory, sort, aggregate, and store it again in a direct access file. This whole process of re-coding, compression, splitting the 15 million individual records into (a four-way-table) 216 districts, 900 diagnoses, 18 age groups and sex took 508 minutes on a 33 MHz PC 486.

The resulting data set can now be analyzed interactively (i.e. in real time). From a menu one can select sex and diagnosis. This information is then provided in several tables as standardized (by age and sex) prevalence rates per 100.000 population for arbitrary regions and/or maps to be viewed on screen or printed in Postscript quality. In addition to diagnoses, one can investigate the age structure of any region. Figure 5 shows the distribution of the male congenital anomalies (ICD-9: 740-759) in the New German countries as a choropleth-map.

Figure 6 shows the age structure of these diagnostic-groups for Brandenburg. With scatterplot-matrices various prevalence rates are plot against each other. Figure 7 and figure 8 show the prevalence of Acute Myocardial

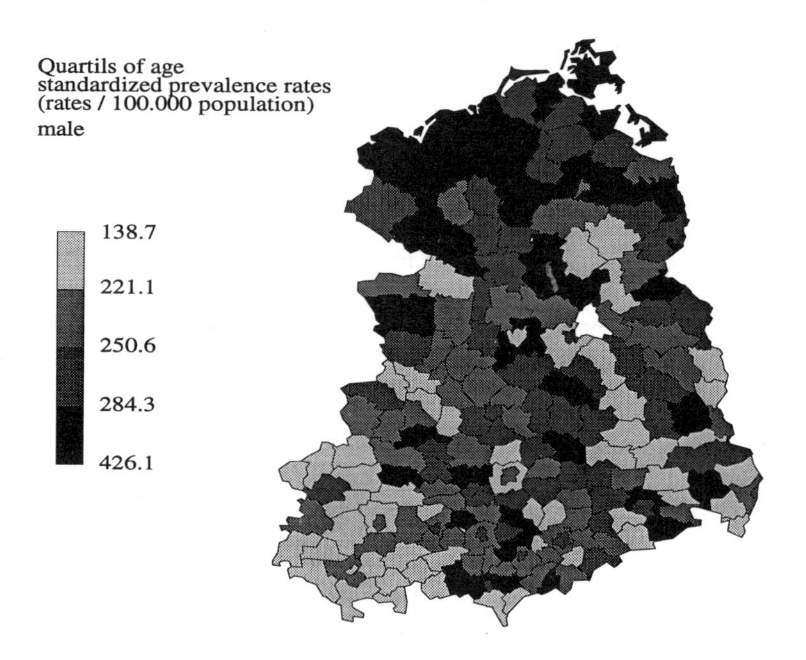

Figure 5: Choropleth map of the distribution of the male congenital anomalies (ICD-9: 740-759) in the new German countries.

Infarction (AMI) (ICD 9: 410) a well as the exogenous variable showing the proportion of graduates of technical schools in the population.

Both maps show remarkable structure in nearly identical regions. The map in figure 8 represents the typical population structure found in the former GDR: skilled people stay in the surroundings of the educational institutions. A notable concentration of highly qualified people can be found in and around Jena (the center of optical and electronical industry around Carl-Zeiss Jena). Aside from a few areas of industrial concentration we find a predominantly rural structure with only marginal differences in education. Just in these regions one also finds strong concentration of heart attacks, which may be a result of differing lifestyles. These correlations between certain illnesses and exogenous variables can be seen in the scatterplot matrices, too. The analysis of connections with other exogenous

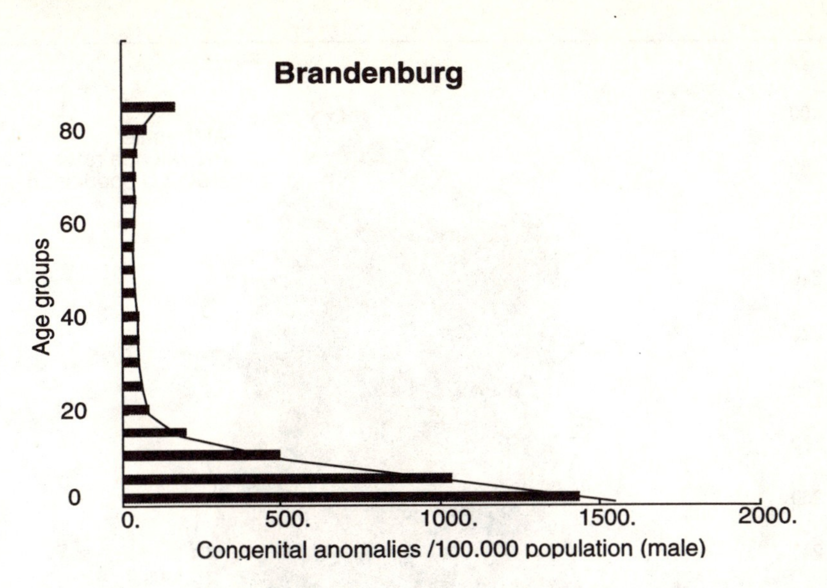

Figure 6: Age structure of the male congenital anomalies (ICD-9: 740-759) for Brandenburg. The overlaying line shows the prevalence rates for all five new countries.

information, such as pollution or environmental variables or indicators for socio-economic status is of similar interest, if the information is available for the same region.

Map Pattern and Spatial Autocorrelations

The previous examples merely represent a description of a situation from which one may or may not see something. We now would like to address the subject of spatial autocorrelations. This problem can be attacked in several different ways using the available system (see also Cliff (1988), Cressie (1991), Moran (1948) and Nagel et.al. (1993)):

- Each district can be described as a point D_i (for example, the center of the district). A grid is placed over these points and the cases summed up for each grid cell and the rates computed. Next, one applies a median polish to the so defined matrix (with the prevalence data D_i) and the result is displayed on the map:

$$D_i = a + r_{k(i)} + c_{l(i)} + R_i,$$

where the ith district is located at grid node $(k(i), l(i))$, a is the overall mean, r_k is the kth row effect and c_l is the lth column effect. R_i are

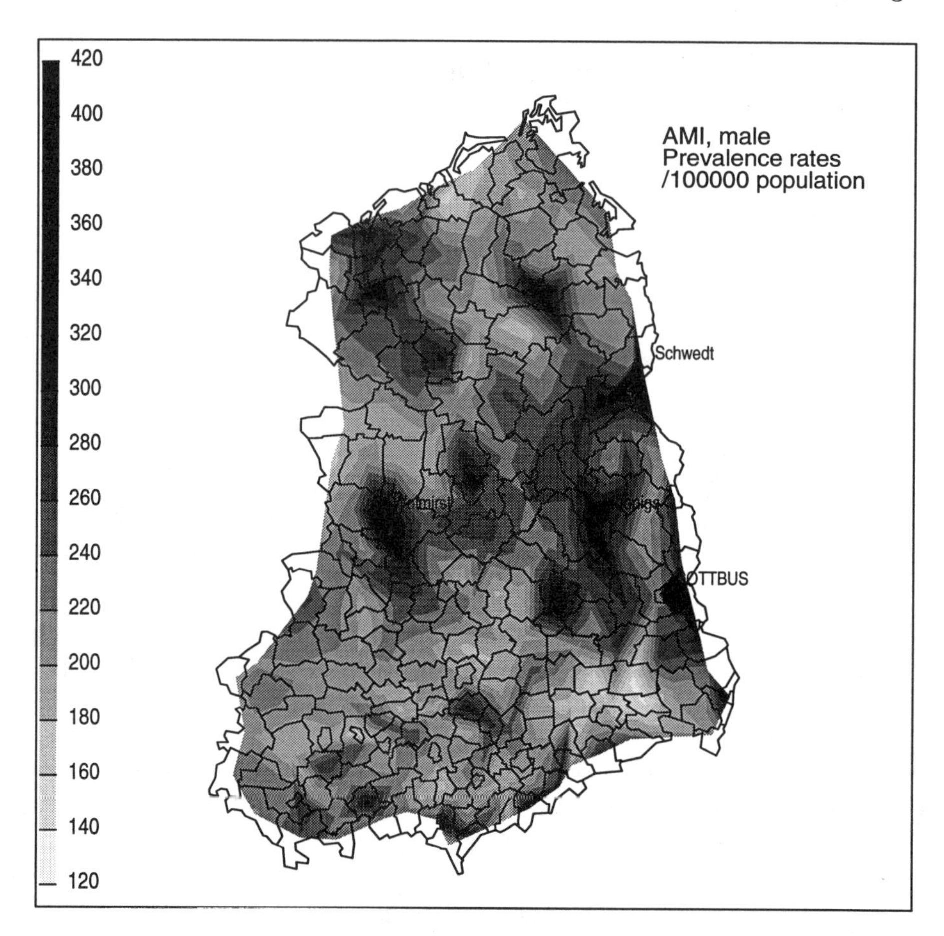

Figure 7: Prevalence rates of Acute Myocardial Infarction (AMI) (ICD 9: 410)

the random errors with $var(R_i) = const$, and $cov(R_i, R_j)$ a function of the Eucliden distance between the ith and jth counties.

In this particular case the problem arises, that districts in the south of the former GDR are very small and densely populated, while the opposite is true in the north. When using such additive models one must achieve a stability of the variance through a transformation (heavily populated regions show a smaller variance than sparsely populated ones) (see Freeman & Tukey (1950)).

- Since diseases are sporadic, we may assume a Poisson distribution. One can try to fit generalized linear models to the data with a link

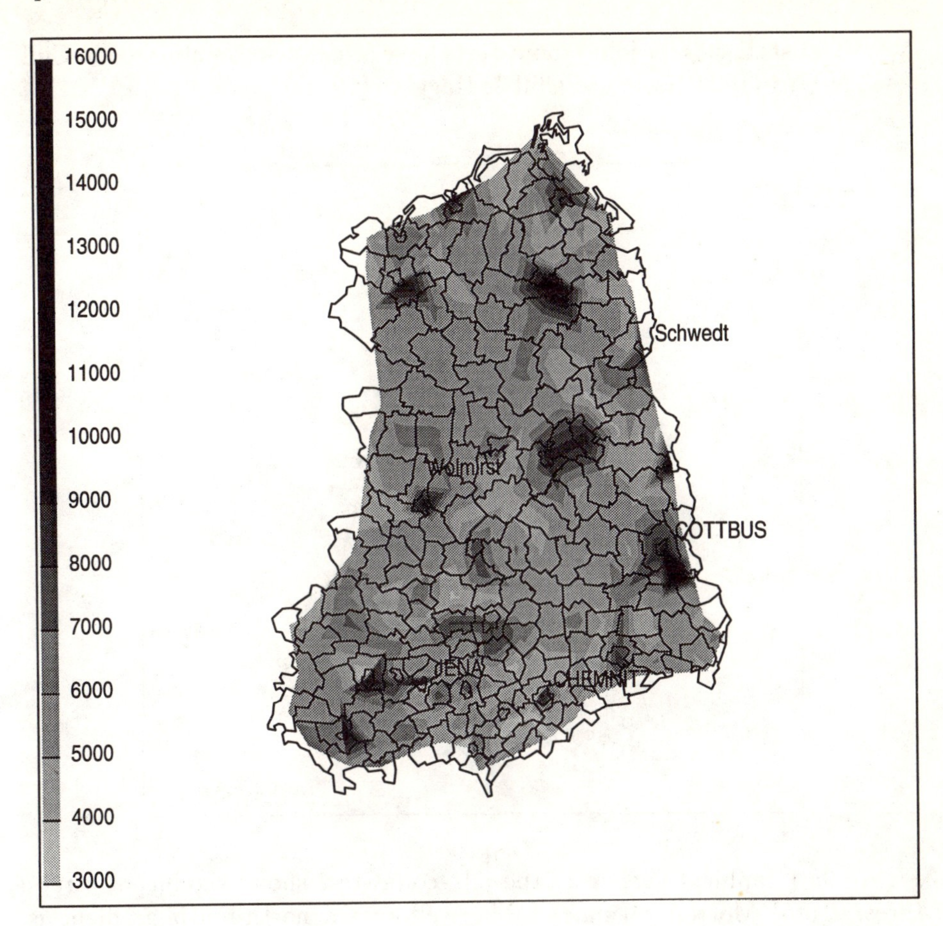

Figure 8: Concentration of highly qualified people (proportion of graduates of technical schools in the population of the former GDR)

function $g(x) = (.)$ and analyze the influence of positions of districts via a discussion of models (analysis of deviance and comparison of deviations under varying model assumptions). In this case one should also incorporate other variables into the analysis, if these are available.

- Finally one can connect all districts with a common border by lines. Two neighboring districts with a common border are connected along the edges, if their rates are from opposite sides of the median. Thus areas above or below the median will not be separated by borders. Under the hypothesis of spatial non-correlation the "black" and "white" areas should be randomly distributed and therefore also the borders.

Test statistics for joint-count-tests have already been published in Cliff & Ord (1973) (see also Cliff & Haggett (1988)).

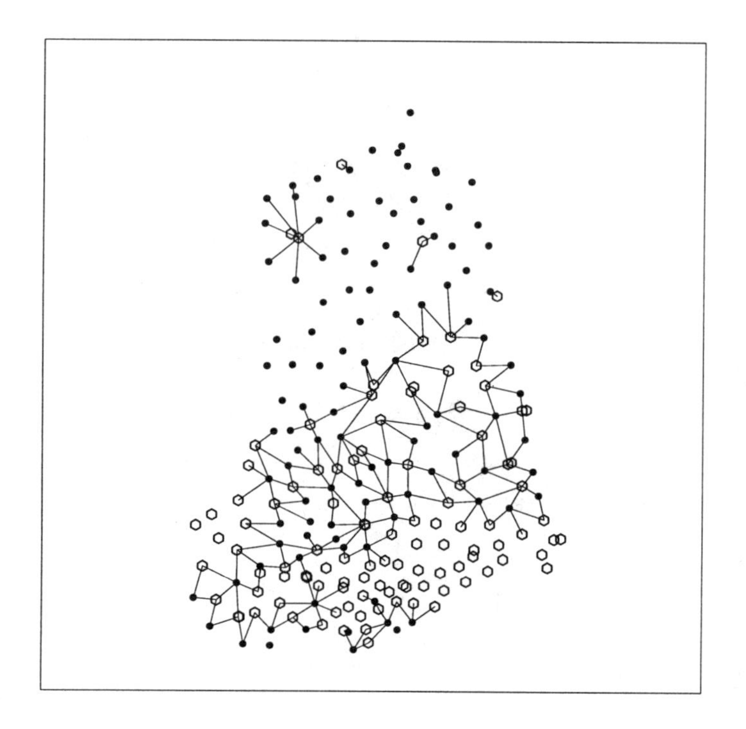

Figure 9: Graphical version of the join-count-test shows significant structures: Total Mortality (male). The well-known north-south gradient is shown.

When applied to the prevalence data, this shows significant structures in figure 9. Interesting are the clusters due to congenital anomalies or the distribution of chronic bronchitis (ICD-9: 490-496) in figure 10 (compare this to the SO_2-emmission in figure 11).

One should note, that this approach reveals interesting phenomena, which should be subjected to a more detailed analysis incorporating other variables such as smoking and nutrition. Unfortunately, such data are often available only for small samples of the population.

Conclusions

The case studies presented here represent a portion of the many applications of spatial data analysis which I have developed on 386 and 486 PCs using

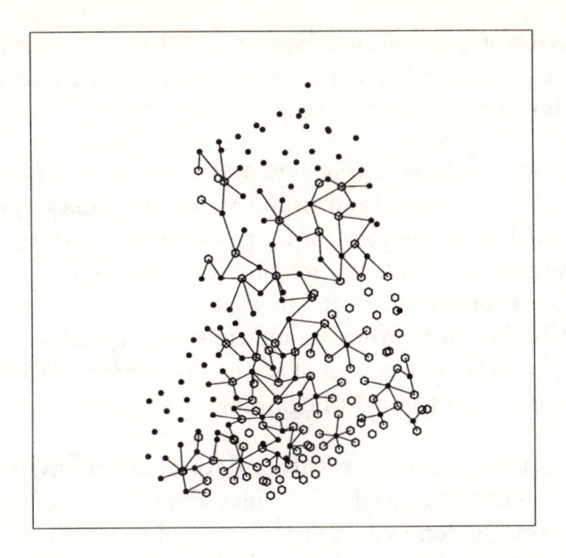

Figure 10: Join count statistics of chronic bronchitis (ICD-9: 490-496) (male).

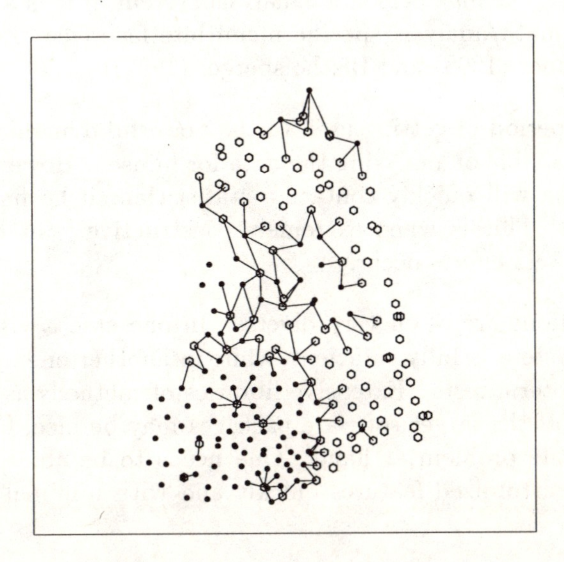

Figure 11: Spatial autocorrelation of the SO_2-emissions due to home furnaces.

the ISP-data analysis language (see Nagel et.al. (1992) *for Nitrate in drinking water*, Nagel et.al. (1992) *for PCB's in fishes*, Fuchs & Wernecke (1993) *for milk monitoring in Austria*). Based on the experience gained through the constructive working relationships with subject researchers and in the development of specialized spatial evaluation systems, I would like to emphasize the following points:

- The user always needs his system and his analysis "right away". Usually, he gets significant and useful ideas for the analysis from a preceding solution in the form of a prototype with which he can see the data in their full complexity for the first time. This is a common communication problem found in any system specification. Therefore, one needs to budget for several, sometimes considerable, revisions of the system. For this reason it is of critical importance to involve the end user directly from the very beginning.

- Development of prototypes with a data analysis environment such as PC-ISP is very fast and efficient, since a major part of the development of the system is done while analyzing real data. (The author needed about one month for completion of the pesticide project.)

- The user typically likes to analyze his data on his own and, therefore, is grateful for his "personal" analysis system, if it is accessible via a suitable data/analysis specific menu interface (see "rapid prototyping", Huber (1993) and Bischofsberger (1992)).

- After a period of getting used to the powerful dynamic graphics, the user is capable of analyzing the data for himself. However, experience shows, he will quickly contact a statistician, if he finds interesting features. This is when extremely constructive joint modelling and analysis sets efforts occur.

- Applications are often very diverse: In one case a visual data base access system is fully sufficient, while in another one may need some fairly esoteric tests. Extensive libraries of methods as can be found in some of the larger software packages may be nice, but rarely help solving the problem at hand. One needs to be able to incorporate special/customized features quickly and with minimal programming effort!

- Maps ought to be a useful vehicle to help understanding spatial data. However, it is very difficult to find digitized data in a useable and simple format. Often one has considerable and superfluous information, while the relevant information is well hidden. Therefore, it is recommended to digitize the maps based on the needs and not on availability. (In the meantime it is possible to digitize directly data in a special version of the dynamic graphics of PC-ISP.)

- Large data sets have their own particular problems: On medium sized data sets one can quickly assess the validity of the data using graphics (erroneous values are usually quickly picked out in the dynamic graphics). With huge data sets this no longer works. For these more than half the time is consumed for sorting, selection, and aggregation algorithms which are far too slow on generally available data base systems.

Within the workspace structure of PC-ISP one gets very good response times: On a PC the selection of an arbitrary diagnostic group, selection of an age group, splitting into regions, classification and display on screen as a map will take between 13 sec, and a maximum of 400 sec, when searching for 200 ICD-codes, 18 age groups, 216 counties and sex in 389.039 records.

References

Bericht an die FKST (1993), *Software-Lösung zur Einschätzung des Gefährdungspotentials des Grund- und Trinkwassers in den Kreisen Hageow und Ludwigslust durch Pflanzenschutzmittel,* ZEG Zwickau

Bertin, J. (1974), *Graphische Semiologie, Diagramme-Netze-Karten,* Walter de Gruyter, Berlin

Bertin, J. (1981), *Graphics and Graphic Information Processing,* Walter de Gruyter, Berlin

Bertin, J. (1982), *Graphische Darstellung und die graphische Weiterverarbeitung der Information,* Walter de Gruyter, Berlin

Bischofsberger, W. & Pomberger, G. (1992) *Prototyping Oriented Software Development. Concepts and Tools,* Springer-Verlag, Berlin

Buja, A. & Tukey, P.A. (Eds.) (1991) *Computing and Graphics in Statistics,* Springer-Verlag, New York

Cleveland, W.S, & McGill, M.E. (Eds.) (1988) *Dynamic Graphics for Statistics,* Wadsworth & Brooks/Cole, Belmont (CA)

Cleveland, W.S. (1985) *The Elements of Graphing Data,* Wadsworth, Monterey (CA)

Cliff, A.D & Haggett, P. (1988), *Atlas of Disease Distributions,* Blackwell References, Oxford

Cliff, A.D. & Ord, J.K. (1973), *Spatial Autokorrelation,* Psion, London

Cressie, N. (1991), *Statistics for Spatial Data,*, J. Wiley, New York

De Roo, A.P.J. & Burrough, P.A. (1992) *Using GIS for Handling Data from Surveys of Monitoring and Pollution,* Soil Monitoring Workshop, Monte Verità, Ascona, Oktober 1992

Fleischer, W., Nagel, M. & Ostermann, R. (Eds.) (1992) *Interaktive Datenanalyse mit ISP,* Westarp Wissenschaften, Essen

Freeman, M.F. & Tukey, J.W. (1950), *Transformations related to angular and the square root.* In: Annals of Mathematical Statistics, 21, 607-611

Fuchs, K. & Wernecke, K.-D. (1993) *Milch-Monitoring in Österreich,* Biometrisches Kolloquium Berlin'93, 16-19 March 1993

ICD-9 (1975) *Manual of the International Statistical Classification of Diseases, Injuries and Causes of Death,* ninth edition, World Health Organization, Geneva

Höring, H., Englert, N., Krause, C., Meyer, E., Nagel, M. & Schiller, F. (1992), *Lead in Drinking Water and Blood. A Study in School Children from Southern Saxonia, Germany,* In: Proc. WHO/UNEP GEMS/HEAL Meeting, Bangkok, Thailand, 16-19. November 1992

Huber, T. M. (1993) *Statistics software versus problem solving,* Proc. SoftStat'93, Heidelberg, 14-18 March 1993 (submitted)

Huber, T.M. & Nagel, M. (1993) *Understanding The Data: An initial exploratory investigation using interactive graphics of PC-ISP,* In: Computational Statistics (submitted)

Moran, P.A.P (1948), *The implementation of statistical maps,* In: Journal of the Royal Statistical Society B, 10, pp.243-251

Nagel, M. & Lippold, U. (1992), *Exploratorische Analyse raumbezogener Daten,* (p. 291-312) In: Festschrift for F. Eicker, J. Eul Verlag, Düsseldorf

Nagel, M., Dahlke, D., Claßen, E & Heinemann, L. (1993), *Die Analyse von Massendaten mit dem PC,* Proc. SoftStat'93, Heidelberg, 14-18 March 1993 (submitted)

Nagel, M., Huber, T & Höring, H. (1992) *Hochinteraktive Graphik und ihre Anwendung auf Daten mit territorialem Bezug,* In: Faulbaum (Ed.) SoftStat'91, G. Fischer, Stuttgart, pp. 177-184

Nagel, M., Hothorn, L & Hartmann, P, (1992) *Hochinteraktive Datenanalyse – Werkzeuge und Prinzipien,* In: H. Enke, J. Gölles, R. Haux and K.-D. Wernecke, (Eds.): Methoden und Werkzeuge für die exploratorische Datenanalyse in den Biowissenschaften, G. Fischer, Stuttgart, pp. 75-92

Nagel, M. & Höring, H. (1993) Agrarbörse Deutschland (Ost), *Kartographische und graphische Analyse von Umweltrisiken,* Proc. SoftStat'93, Heidelberg, 14-18 March 1993 (submitted)

Nagel, M. & Ostermann, R. (1993) *Dynamic Graphics for Discrete Data,* in: Computational Statistics, 8, 197-205

Ostermann, R. (1992) *Exploratory-Graphical Methods for Time Series Analysis,* In: Schader, M. (Ed.): Analyzing and modeling data and knowledge. Springer-Verlag, Heidelberg, pp. 79 -84

Schilcher, M. & Fritsch, D. (Eds.) (1989) *Geo-Informationssysteme,* Wichmann, Karlsruhe

Tierney, L. (1990) *LISP-STAT. An Object-Oriented Environment for Statistical Computing and Dynamic Graphics,* John Wiley, New York

Unwin, A. (1994) *REGARDing Geographic Data,* In: Dirschedl, P. & Ostermann, R. (Eds.): Computational Statistics, Papers Collected on the Occasion of the 25th Conference on Statistical Computing at Schloß Reisensburg, Physica-Verlag, Wien (in this volume)

REGARDing Geographic Data

Antony Unwin
Department of Statistics, Trinity College Dublin,
Dublin 2, Ireland

Key Words:
spatial data, interactive graphics

Abstract
Geographers seek to explain patterns in space. Standard statistical techniques are little help and so new methods have been sought. REGARD is a software package which provides high interaction graphics tools for spatial data. This paper describes geographic applications of the current version of REGARD, discusses how well it meets geographers' needs and outlines additional features which could be added.

Introduction

There are various examples of geographic data: locations and populations of towns, distributions of types of trees, breeding sites of birds, clusters of occurrences of diseases, layouts of electricity supply systems and many more. Geographic Information Systems have been developed in recent years to store these data in accessible formats but they provide only cumbersome means of exploring the data to turn them into information. Geographers need to define regions and boundaries, to identify patterns in space, to seek explanations for those patterns through links with economic or other factors and to study the processes that have led to those patterns. Classical statistical tools cannot cope with the variety, interdependence and multiplicity of variables that characterise real spatial data sets. Interactive graphics can.

Exploring data

Data exploration has had a bad press in the statistical literature. It has been associated with data mining and hunting for p- values. This is partly because few tools have been available for exploration and partly because, although exploration was common practice in real analyses, it was rarely reported. The situation is similar to that in the philosophy of science where many texts exist offering descriptions of how scientists act which bear little relation to how they actually work (Beveridge (1950) and Holton (1978) may be consulted for a more realistic view). The denigration of exploratory work has meant that exploratory tools and procedures have been slow to develop while some contorted and intricate mathematical tools have been advanced whose sophistication and complexity must be marvelled at but whose practical value is negligible.

The term 'Exploratory Data Analysis' has become so inseparably linked with Tukey's EDA book (Tukey (1977)) that it is frozen in time. Even the panel of eminent statisticians who reported to the International Statistical Institute recently (Moriguti et al (1992)) seem to be unaware of recent research developments, giving as their recommended reference the book by Velleman and Hoaglin (Velleman & Hoaglin (1981)). The first author has said that the book is so out of date they have refused the publishers suggestion of preparing a second edition (Velleman, private communication 1992). Although historically of great value, Tukey's book is misleading as a current text because it has no knowledge of modern computing tools (after all, it was published before the first Apple II appeared and must have been written round about the time Ken Olsen, ex-President of Digital, said he saw no reason why anyone would ever want a personal computer). Nevertheless, it does emphasise the thinking behind exploratory analysis.

Now effective exploratory tools are available in the form of interactive graphics. It is no longer necessary to make innumerable assumptions, accept limiting restrictions or insist on particular data structures before any analysis can begin. The benefits of these interactive tools are not only seen in initial exploration of data but complement more traditional tools at all stages of data analysis. The benefits are especially clear with geographic data where data points are definitely not independent, where commonly there are many variables of many different types and where contextual knowledge of diverse kinds may contribute substantially to analyses.

Interactive tools are still in the early stages of development and are not as widely used as might be expected. A parallel may be seen in the introduction and growth of personal computers. Much could be achieved with even the earliest PC, but with continually increasing power and improvements in interfaces more and more use is made of them and more and more people are convinced of their value. This paper therefore presents an intermediate view of the current state of what exists and what is possible by describing some applications of interactive graphics with real data sets. New tools will be developed (some possibilities are outlined at the end of the paper), existing tools will be made more efficient and the set of tools will be combined more effectively into a coherent whole.

Stages of exploratory analysis

Exploring data is valuable in several ways which may be considered as different stages of an analysis:

- Investigate data quality
- Indicate exceptions
- Identify patterns
- Inform analyses
- Illuminate conclusions

The suggestion is not that these form a strictly ordered list, rather that this is a natural order to start with. An analysis might cycle round through a mixture of the stages several times.

Initially displaying data quickly and flexibly allows extensive investigation of data quality with little effort. Discovering errors through residual analysis is a waste of modelling effort if they can be found directly. A non-spatial example is the geyser data where some data were shifted in time. Unwin and Wills (1988) describe how this may be found naturally and directly by interactive graphical exploration as distinct from the more roundabout ways previously suggested.

Data points, which are not errors but are exceptional are indicated and possibly explained through linking several variables. In the MMR vaccination example discussed below there was one area of Dublin with only old people living in it. Linking to the map showed that this was the Phoenix Park, a large public park with an old people's home.

Having satisfactorily established data quality and ensured that outliers have been accounted for, structural analysis can start. Good statisticians have always checked that their data sets are clean before pursuing elaborate analyses, but textbooks assume this and journal articles imply it so that routine analyses often ignore it. Patterns of a wide variety of types can be identified and studied with graphical tools. Statistical models are restrictive in the structures they recognise but are valuable when the analyst is confident a structure of a particular, analysable kind exists which may be obscured by noise. These models are not useful when the reason for poorness of fit is not noise but the fact that the wrong relationship is being assumed. Graphical methods are excellent for identifying complex relationships provided that there is relatively little noise. In the Irish elections data set used later on, the constituencies where the Workers Party were strongest were selected. Most of these were in Dublin (which was well known) but what was interesting (and not well known) was that they formed a contiguous band round the North and West of the city rather than being in the centre.

Should a standard statistical analysis be carried out, its results have to be informed by graphical displays. The German election results data set described below has previously been analysed by Gebhardt (1988). He identified a regional Bundesland effect using linear models. To assess the practical significance of statistically significant results the data have to be inspected. Was one region distinct from all the rest or were there two separate groups or was there a gradual change across the range?

Finally, in presenting results, it is always helpful to illuminate conclusions with good graphics. When those graphics can be interrogated, linked and explored themselves, communication is much more effective.

REGARD - exploring geographic data

REGARD (Radical Effective Graphical Analysis of Regional Data) runs on colour Macintosh computers and has been developed specifically to provide interactive graphical tools for exploring spatial data. There are four types of layer: point, region, line and picture. Data may be stored in all but picture layers which are used for background context and for loading in maps to be converted into region layers. There may be more than one layer of any kind, for instance if data have been collected at different sets of locations. In a study of oil imports into the EC over a number of years (Unwin et al. (1992)), the oil flows were recorded in a single line layer, enabling comparisons across years. They were also recorded in individual line layers for each year (in work not reported in the paper) in attempts to obtain clearer map displays of the flows.

The central REGARD display is the map window in which the layer locations are drawn. A map may be loaded as one picture in a picture layer or as several pictures in several layers. The advantage of the latter approach is that different aspects of a map (such as the road network or regional boundaries or factory locations) may be turned on or off by turning on or off the appropriate layers. Maps may also be loaded as regions in a regional layer using a standard GIS polygon format. All other displays, whether statistical ones like boxplots, scatterplots and rotating plots or control ones like the layers window and the tools palette, are secondary in importance. The map window and the statistical displays are linked and may be interrogated, highlighted, resized and rescaled interactively. Advanced features include zooming into submaps, animation across ordered variables, cross-layer linking, network analysis tools and interactive query tools across all graphical displays.

The principal features of REGARD which distinguish it from other interactive statistical packages as a tool for analysing spatial data are its map window with the facility to load maps as pictures to give context, its multiple layers and cross-layer linking permitting the loading of several data sets of different support, and its ability to handle points, lines and regions as data objects. The linking of statistical displays to point locations could be managed in Data Desk but no maps could be loaded. In ISP (Nagel et al (1992)) and S-Plus it is possible to load maps but only limited linking is possible between displays. Spatial data sets have many variables of many different types and REGARD's strengths lie in providing the user with tools to explore such data sets quickly and effectively through multiple statistical displays linked to the key map window.

REGARD is under development and currently allows up to 8000 objects in each layer, each having values on up to 60 variables. The software has been used on a number of collaborative research projects and articles discussing applications of REGARD include Unwin et al. (1990), Wills et al. (1991) and Unwin et al. (1992).

Examples of applying REGARD

Geographical data sets commonly have many variables of different kinds. Initial uninformed analyses demand reviewing a large number of variables quickly. Even informed analyses benefit from scanning many variables, as data errors may be highlighted or unexpected relationships made apparent. Figure 1 shows data on bird-breeding collected in part of Nort-East Scotland. In each of some 400 small areas there are data on over 100 birds and about fifteen environmental variables. The selected bar in the bar chart for red grouse highlights the definite breeding areas for the species. The map window shows that these are predominantly in the South-West of the region round the borders with a few sites on the Eastern sea coast. Boxplots for many of the environmental variables have been drawn. The standard box and whiskers boxplot represents data for the selected areas, the background box plot represents data for all the areas. At a glance it is clear that red grouse breed in areas of higher altitude with less grass. (It is also clear that several of the variables have very skew distributions which suggests they should be transformed before further analyses) This example shows how REGARD can handle many variables at once, enabling quick decisions

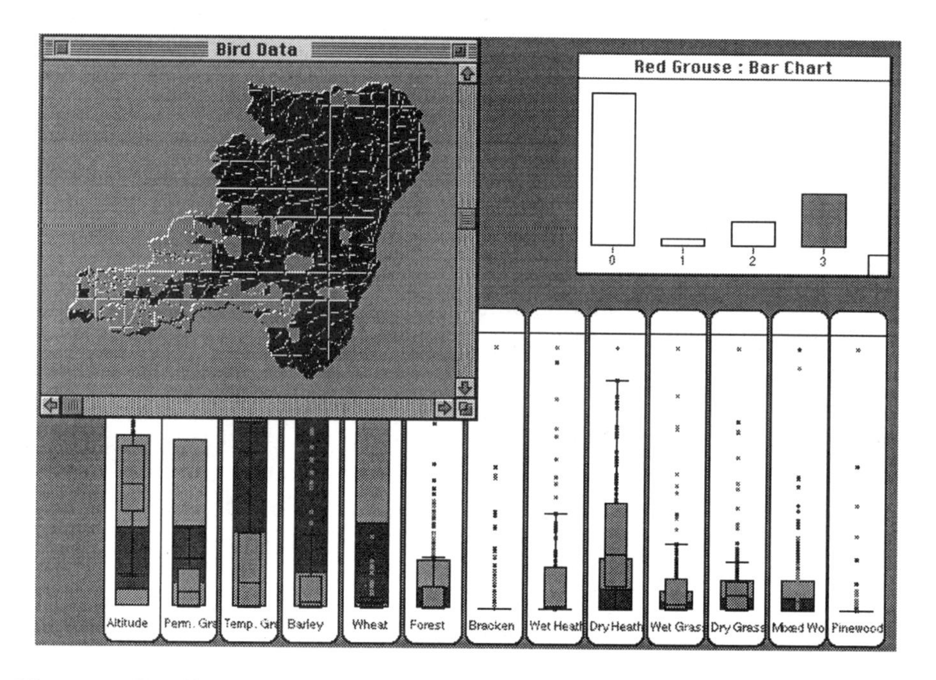

Figure 1: Profile of confirmed Red Grouse breeding in North-East Scotland

to be made about which variables to follow up for more detailed analyses. Linked boxplots are very efficient in space terms (contrast the screen space required for the 13 boxplots with how much would be needed for 13 histograms). The interpretation of multiple plots, particularly interdependent ones like those for the groundcover variables, is tricky. Results from graphical analyses should always be checked by more traditional analytic methods, but then results from analytic methods should always be checked graphically.

Ideally any spatial analysis program of this kind should provide a wide variety of transformation tools as it may be necessary to transform or recode the data. In general REGARD provides no tools as under the Macintosh operating system it is as easy to copy the variables to be transformed, paste them into a package running in parallel (a spreadsheet or a statistics program like Data Desk), transform them and paste them back. Thus variables may be logged, grouped, combined or conditioned. For one particular transformation REGARD does include a tool. After drawing a histogram it is possible to create a new variable based on the histogram classes. This is useful in forming groups especially after using the interactive histogram rescaling feature. The transformation was included within REGARD to prepare continuous variables for controlling animation.

The second example illustrates REGARD's innovative scatterplot displays.

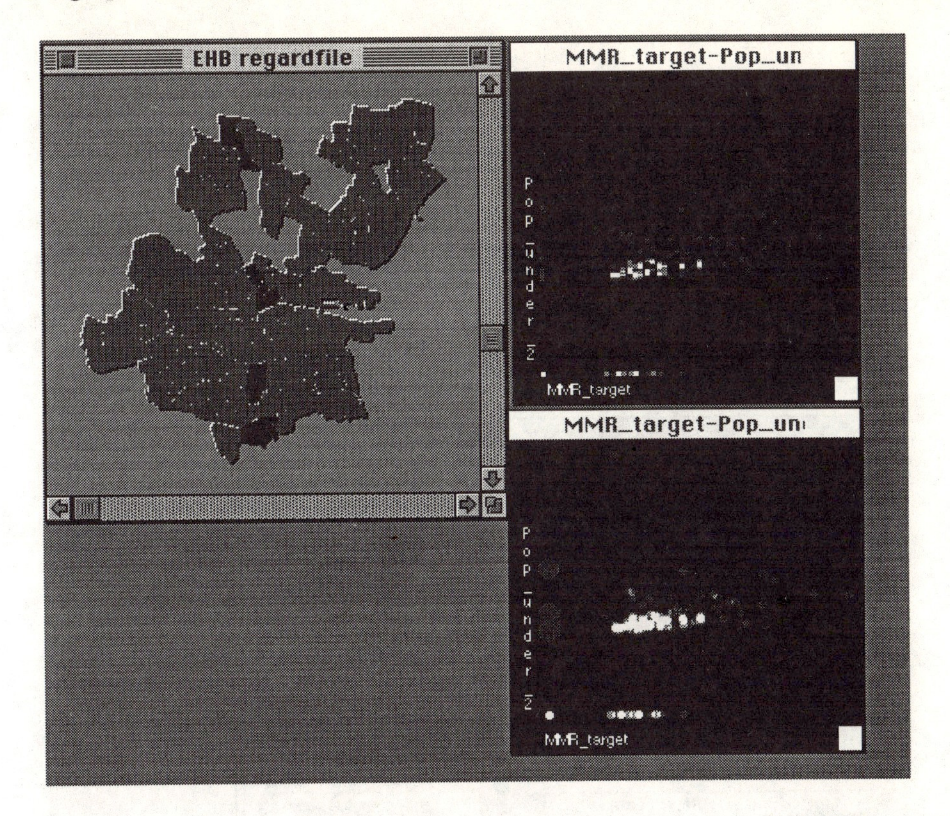

Figure 2: MMR vaccination targets and numbers of young children in Dublin

Figure 2 shows displays from the first look at a sample data set partly from the Eastern Health Board in Dublin and partly from the Central Statistics Office census. The data were based on District Electoral Divisions (DED's), the smallest areas for which data were collected, and contain population numbers by age and class and information on an MMR vaccination campaign for children. At that stage it was known that the map information was incomplete and there were concerns about how much of the data had been included. The main window shows gaps where DED's had not been mapped. The top scatterplot is a standard display of the population under 2 from the census plotted against the MMR vaccination target numbers. The lower scatterplot is the same one with larger point sizes for all points and even bigger ones for highlighted points. (This is a feature of REGARD which allows many different sizes and also a choice of tonal levels for displaying overlapping points with increased brightness. It enables the effective display of scatterplots with very large numbers of points, especially as point size can be changed interactively allowing fast scans of a range of different views of the same plot.) From the lower plot the DED's with non-zero population but apparently no MMR target have been selected (the large circles on the left). From the map these are all in the same vertical strip that includes some of the missing DED's so there could have been a systematic error. The scatterplot also shows DED's with MMR targets but no population under 2. This was due to the census data being several years older and not including housing estates developed in the meantime. A further point emphasised by the second plot is the single DED with a very high population under 2. Initially this was thought to be an error but the figure was later confirmed.

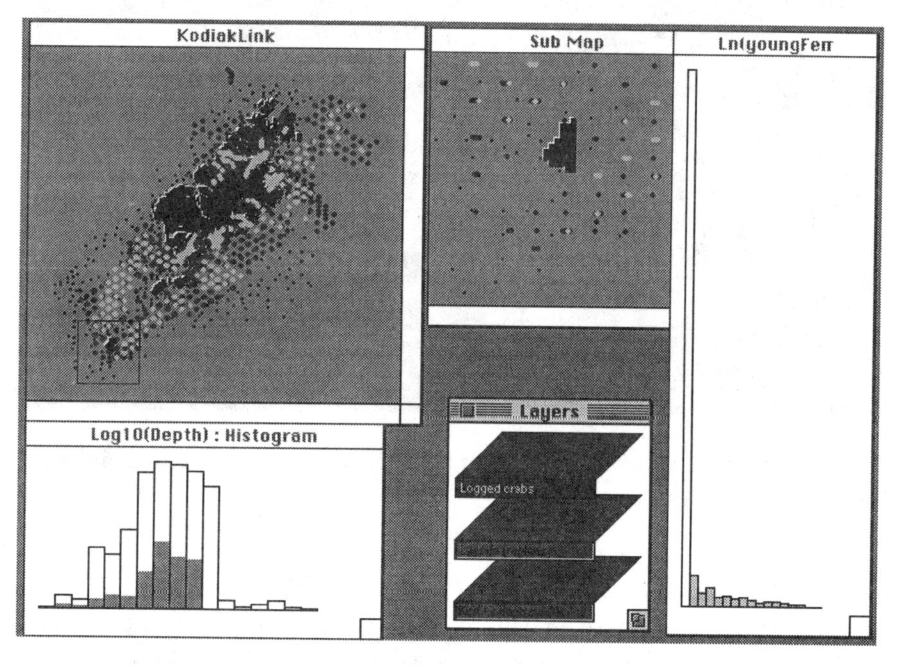

Figure 3: Cross-linking of depths and numbers of crabs found around the Kodiak Islands

The next example illustrates two advanced features of the REGARD software, how it deals with zooming and how it handles data sets with variables recorded at different locations. The main map window in Figure 3 shows the Kodiak Islands of Alaska. The original data set concerned the numbers of king crabs of different ages counted at a variety of locations over several years. The data were extremely skewed so the numbers were logged, which is why the point layer containing the locations and numbers found is labelled 'Logged crabs' in the Layers window. The second layer is a region layer containing the islands (the only data stored in this layer are the islands' names) and the third layer is another point layer. It was thought that depth might influence the numbers of crabs found and so admiralty maps were used to add this information. The main map window displays points from the crabs layer with the larger size and depth points with the smaller size. It can readily be seen that there were many points with depth information far beyond where crab sample counts were available. To see the inshore locations at which depths were recorded it is necessary to turn off the crabs layer or to zoom in. The submap window shows a zooming in around Chiroff island and this area is shown in the main map by the superimposed rectangular border. Zooming can be to any size but is simple magnification and not logical zooming, as the submap shows clearly with the obvious approximation to the island's shape and the location of one point on top of the island instead of in the sea. Were the submap the active window its scroll bars would be visible. It is possible to use the submap like a magnifying glass, moving round the main map window. Retaining the main map is important for it provides the essential frame of reference.

Variables stored in the two point layers cannot be directly linked because they are recorded at different locations. In this case a special REGARD function has been run to fix a link between each crab sampling point and the closest point at which depth is recorded. This is, of course, not a 1-1 link and many of the depth points are not linked to any crab sampling point. Bearing this in mind the other displays in Figure 3 can be examined. The histogram on the far right is of the logged numbers of young female crabs found and the locations where most have been found have been selected. The histogram to the bottom left of the figure is for the logged depths and the corresponding nearest depth points have been highlighted. The main point here is that young females appear to be found at greater than the shallowest depths but this could be because there were not depths recorded nearby and the linked closest depths are not representative. The fact that there are shallow depths which are not highlighted must also be treated with care. These could be points in areas which were not sampled or points which are not linked to crab sample points because there are depth points which were closer. Both these issues can be investigated in detail by zooming and then scrolling round the key areas.

Simple selection and linking is limiting. It is often necessary to ask more complicated questions than where are the high values on a variable (selecting from a histogram) or where do a cluster of points in two variables occur (selecting from a scatterplot). REGARD provides addition and intersection selection tools which are on the right-hand side of the Tools palette shown in Figure 4. Intersection or subselection is used here to pick out the constituencies in which the largest Irish political party, Fianna Fail, did well in all five parliamentary elections in the 1980's. The map displays the Irish political constituencies with the Dublin Constituencies magnified on

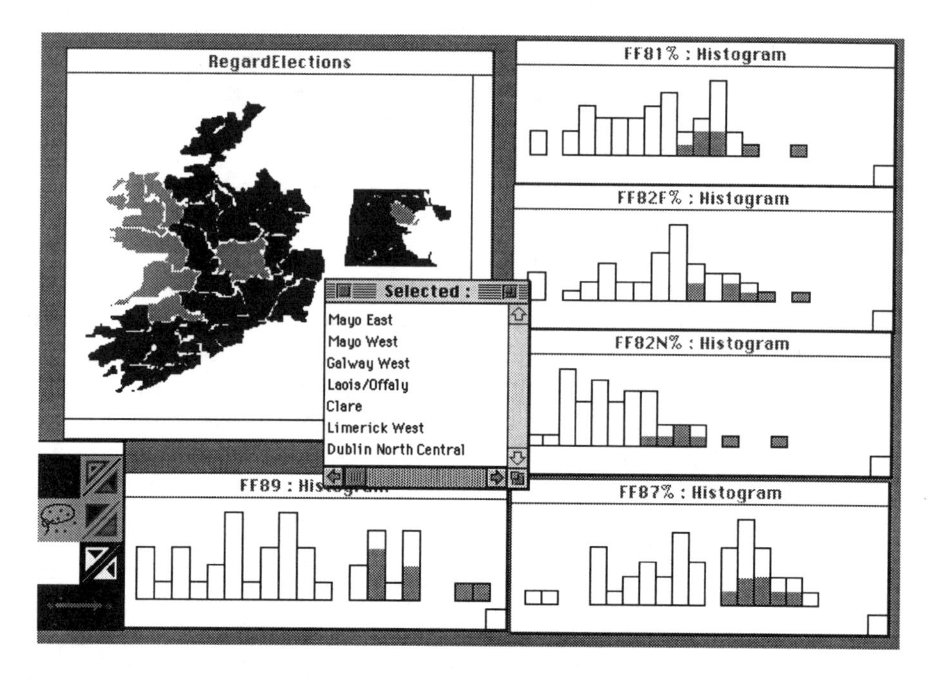

Figure 4: Fianna Fail strongholds, identified by subselection across the histograms

the right. The five histograms are for the first preference percentages obtained by Fianna Fail in the five elections. They have been selected starting with those where Fianna Fail did best in 1989, then selecting from within those the ones where they did best in 1987 and so on. The map shows a strong contiguous base on the Western seaboard, a constituency in the Midlands and the constitency of the party leader in Dublin. A useful additional capability is of listing the names of the selected regions.

Research developments

This paper has used a variety of examples from geographic data sets to illustrate the need for new approaches to the analysis of spatial data. The examples have been chosen to demonstrate the features of the REGARD software package and to make clear the philosophy underlying it and the concepts it is built on.

Geographers would like to determine spatial order. Tools for identifying, adjusting and defining boundaries would be valuable. This would include looking for large differences across short distances (which could be set up as a special function with a distance parameter) and assessing homogeneity across contiguous regions (which would need a facility to interactively aggregate and compare at several levels).

Geographers would like to study the processes which may be responsible for patterns they observe in space. The animation feature in REGARD is a start in that direction but will have to have greater flexibility and control. Some ways of identifying boundaries in time (presumeably related to those for identifying boundaries in space) will be needed.

In investigating multiple interrelationships it is essential to be able to query the data in complex, yet intuitive ways. Both Data Desk (Velleman (1992)) and REGARD implement limited multivariable graphical selection tools, but these only emphasise the need for faster and more powerful interfaces. Graham Wills' thesis reviews the problems of linking in general data structures and suggests how they might be tackled (Wills (1992)).

Points closer together tend to be more similar, but not always. Sometimes it is useful to identify groups of similar points or regions even when they are not contiguous. Interactive defining of similarity and clustering will require flexible tools to encourage intelligent exploration of the possibilities. Extensions of Data Desk's selector and group tools and of its slider facility look promising.

Interactive statistical graphics is an exciting field of research which is continually stimulated by the demands of applications areas. Geographic data pose especially complex and interesting problems which should lead to important further developments.

Acknowledgements

Thanks to Dr.F.Gebhardt of the GMD for providing the German data, to Dr.S.Buckland for the Scottish data, to Dr.Z.Johnston and the CSO for the Dublin data, to the ASA for the Kodiak data and to Suzanne Wills for map preparation.

References

Beveridge, W.I.B., (1950) The Art of Scientific Investigation, Heinemann, London

Buckland, S.T., Bell, M.V. & Picozzi, N., (1990) The Birds of North-East Scotland, North-East Scotland Bird Club, Aberdeen

Gebhardt, F. (1988) Prospects for Expert Systems to Analyse Election Data, Classification and Related Methods of Data Analysis (ed. H.H. Bock), pp 691-696, Elsevier

Holton, G. (1978) The Scientific Imagination: Case Studies, Cambridge University Press, Cambridge

Moriguti, S., Diggle, P.J., Gower, J.C., Wallman, K.K. & Wang Shou Ren (1992) The Role of Statisticians, International Statistical Review, vol 60, pp 227-246

Nagel, M., Huber, T.M. & Hoering, H. (1992) Hochinteraktive Graphik und ihre Anwendung auf Daten mit territorialem Bezug, in: Faulbaum, F. (ed) Softstat 91, G. Fischer, Stuttgart, pp 177-184

Tukey, J.W., (1977) Exploratory Data Analysis, Addison-Wesley, London

Unwin, A.R. & Wills, G. (1988) Eyeballing Time Series. Proceedings of 1988 ASA Statistical Computing Section pp 263-268

Unwin, A.R., Sloan, B.J. & Wills, G. (1992) Interactive Graphical Methods for Trade Flows, Proceedings of Conference on New Techniques and Technologies, Bonn, Office for Official Publications of the EC, pp 295-303

Unwin, A.R., Wills, G. & Haslett, J. (1990) REGARD - Graphical Analysis of Regional Data, ASA Proceedings of the Section on Statistical Graphics, pp 36-41

Velleman, P.F. (1992) Data Desk, Data Description, Ithaca NY

Velleman, P.F. & Hoaglin, D.C. (1981) The Applications, Basics and Computing of Exploratory Data Analysis, Duxbury Press, Boston

Wills, G. (1992) Spatial Data: Exploration and Modelling via Distance-Based and Interactive Graphics Methods, PhD thesis, Trinity College Dublin

Wills, G., Unwin, A.R. & Haslett, J. (1991) Spatial Interactive Graphics Applied to Irish Socio-economic Data, Proceedings of 1991 ASA Statistical Graphics Section pp 37-41

Applied Nonparametric Smoothing Techniques

Wolfgang Härdle[1], Sigbert Klinke[2] & Marlene Müller[1]
[1]Humboldt-Universität zu Berlin, Wirtschaftswissenschaftliche Fakultät, Spandauer Str. 1, D-10178 Berlin, Germany
[2]Université Catholique de Louvain, C.O.R.E. & Institut de Statistique, Voie du Roman Pays 34,
B-1348 Louvain-La-Neuve, Belgium

Key Words:
nonparametric estimation, kernel density estimation, projection pursuit, implementation of kernel estimates, confidence intervals, kernel regression, multivariate regression, single index models, average derivative estimation, generalized additive models

Abstract
Nonparametric smoothing methods are applied in statistics as a flexible tool in finding structure and connections within data. We present here kernel estimators, which are easy to handle in all dimensions, in various situations and applications. We first consider density estimators and show how these are used as an exploratory tool in univariate and multivariate situations. Some theory is provided for inferential issues. Next we give a short overview on smoothing techniques in univariate regression. The last chapter deals with multivariate regression, where we present two semiparametric applications. In the appendix we present our computer implementations which are done entirely in *XploRe 3.1 - an interactive statistical computing environment.*

The Need for Fast Density Estimates

Nonparametric estimation plays an important role in modern statistical research and practice. A variety of methods has been proposed by various authors and there is a considerable amount of literature. This paper concentrates on using and implementing kernel density and kernel regression estimation. To illustrate these ideas, we consider throughout this paper the following data sets.

Example 1
The Swiss Bank Notes data (see Flury & Riedwyl (1988, page 5 ff.)) consist of 200 measurements of old Swiss bank notes. It is known that the first 100 bank notes of the sample are genuine and the second 100 are forged. The data contain the measured values of length (X_1), left height (X_2), right height (X_3), distance of the inner frame to lower border (X_4) and to the upper border (X_5), and the length of the diagonal (X_6). One is interested in discriminating between the two groups.

Example 2
The nuclear sclerotic cataract data stem from the Beaver Dam Eye Study (see Mares-Perlman et al. (1993), Klein et al. (1992)). The data set contains 1136 observations of age minus 60 (X_1), logarithm of zinc concentration in diet (X_2), and the level of the nuclear sclerotic cataract (Z) given in four categories labelled 1 to 4 (lowest to highest). To simplify the presentation we transform Z to a binary variable Y by pooling together levels 1, 2 ($Y = 0$) and 3, 4 ($Y = 1$). The interest lies in modelling the cataract as a function of the other two variables.

All computations and graphical presentations in this paper are done in XploRe 3.1. This system is an interactive, open statistical computing environment, which allows the user to implement complicated nonparametric algorithms in an easy way, to combine them into libraries and therefore to tailor the computing environment according to his own interests. The application of multivariate nonparametric methods is supported by predefined macros and libraries. The XploRe language is matrix oriented and offers various possibilities for interactive graphical representation, see XploRe (1993).

One important feature of kernel density estimates is that they provide easily an impression of the data distribution. Consider *Example 2*: how can we immediately check for modes or skewness in the distribution of the variables X_1 or X_2? Dividing the data range into bins and looking at histograms is often informative, but histograms change their shape with the origin of the bin sequence, see Scott (1992). Kernel estimates overcome this problem. Denote the observations of the variable X by $x_i, i = 1, \ldots, n$. The kernel density estimate $\hat{f}_h$ of the density f is constructed by averaging over scaled kernel functions centered in the points x_i, i.e.

$$\hat{f}_h(x) = \frac{1}{nh} \sum_{i=1}^{n} K\left(\frac{x - x_i}{h}\right). \tag{1}$$

Table 1 lists some commonly used kernel functions.

Figure 1 shows such a kernel density estimate for the zinc data of *Example 2*.

The choice of the bandwidth h is the essential problem in kernel density estimation. One simple method is the "rule of thumb" for a Gaussian kernel

$K(\bullet)$	Kernel
$K(u) = \frac{1}{2}\,\mathrm{I}(\lvert u\rvert \le \frac{1}{2})$	Uniform
$K(u) = (1 - \lvert u\rvert)\,\mathrm{I}(\lvert u\rvert \le 1)$	Triangle
$K(u) = \frac{3}{4}(1 - u^2)\,\mathrm{I}(\lvert u\rvert \le 1)$	Epanechnikov
$K(u) = \frac{15}{16}(1 - u^2)^2\,\mathrm{I}(\lvert u\rvert \le 1)$	Quartic (Biweight)
$K(u) = \frac{1}{\sqrt{2\pi}}\exp(-\frac{u^2}{2}) = \varphi(u)$	Gaussian

Table 1: Kernel functions.

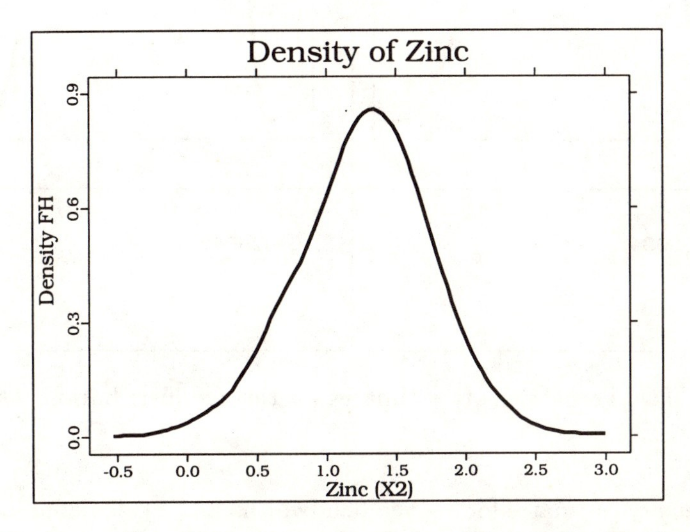

Figure 1: Kernel density estimate for the zinc distribution.

proposed by Silverman (1986) which assumes the true underlying density to be normal with variance σ^2. This yields

$$\hat{h} = 1.06\hat{\sigma}n^{-1/5} \tag{2}$$

with $\hat{\sigma}$ the usual standard deviation estimator for σ. This method was applied for the density estimate in Figure 1 using the XploRe macro `denauto`.

The same method applied to the age variable X_1 of *Example 2* gives us the curve on the left of Figure 2. The calculated bandwidth is given below the figure. The estimated density shows a slight second mode and a pronounced skewness to the right. From the assumption for the "rule of thumb" we know that the density estimation tends to oversmooth for multimodal data sets. In the right part of Figure 2 we have calculated $\hat{f}_h$ for $h = \frac{1}{2}\hat{h}$. We see that more modes appear and that this rough density estimate visualizes much better the underlying structure of the age data. The modes are of course functions of the bandwidth.

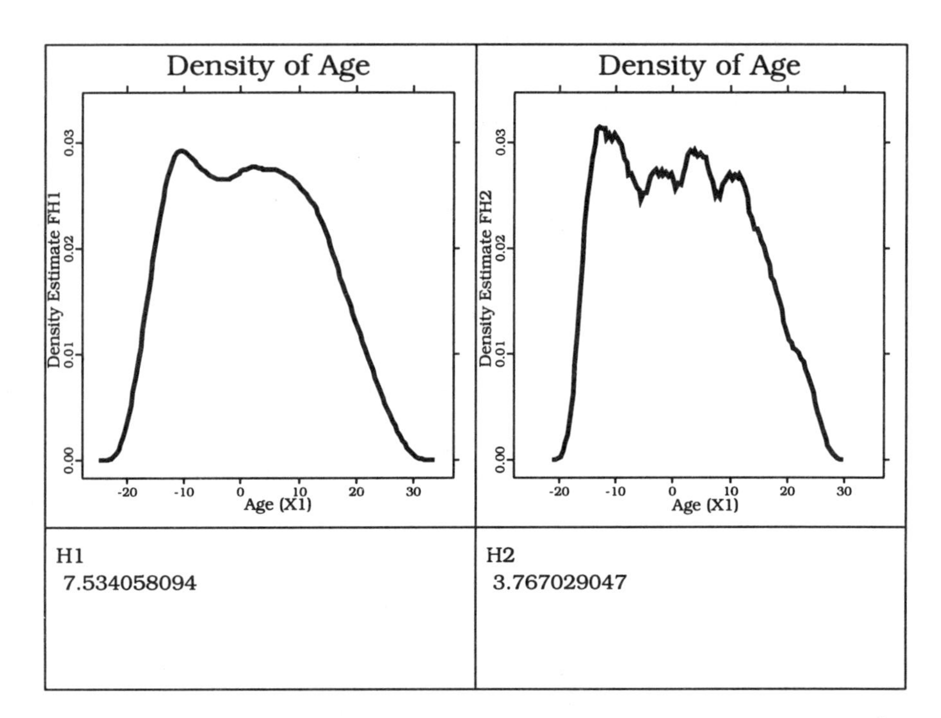

Figure 2: Two kernel density estimates for the age distribution. Output of program A.1.

Kernel density estimates for a given bandwidth can be calculated in XploRe via the macro `denest`. The whole XploRe code for Figure 2 is given in program A.1 (see Appendix).

More information on data can be provided by two dimensional density estimates, calculated in Figure 3 (XploRe macro `denest2`). One sees immediately the unimodal structure in the zinc direction (Y axis) as well as the different modes in the age direction (X axis). It is clear that the exploratory character of density estimates is restricted to low-dimensional data; the limit is for three dimensional data. Scott (1992) uses contour shells to reveal structure in three and four dimensional data. Let us here review a method for finding non-normal structures in high-dimensional data sets by combining projection and density estimation techniques.

Projection Pursuit techniques cover a wide field of interesting topics in data analysis (density estimation, regression, exploratory data analysis). The idea is to find an informative low-dimensional projection of a high-dimensional data set which help to describe the non-normal structure of the data. We recall that each projection of a normal distribution has also a normal distribution. Since there is an infinite number of possible projections

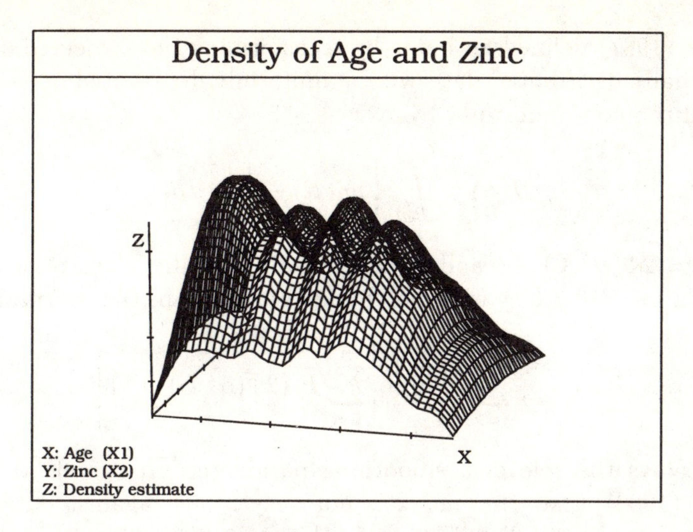

Figure 3: Two dimensional kernel density estimate for age and zinc.

there is a need for an automatic choice. Let $I(\alpha) : I\!R^n \to I\!R$ be a *projection index*, where α describes a projection vector with $|\alpha| = 1$. The aim is to maximize this index in order to detect non-normality. If we project the data points x_i, we obtain 1-dimensional data points $z_i = \alpha^T x_i$. The density estimation of z_i as a function of α should look non-normal for projections α revealing non-normality. The task is thus to find such projections.

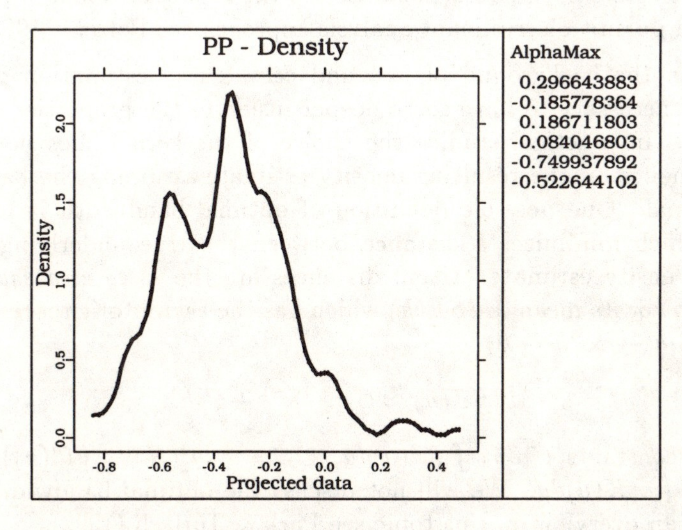

Figure 4: Estimated density of projected data in exploratory projection pursuit.

Friedman (1987) defined an index by transforming the projected data such that normally distributed data will be uniformly distributed on $[-1, 1]$ and measured the non-uniformity by

$$I(\alpha) = \int_{-1}^{1} \left\{ p_R(R) - \frac{1}{2} \right\}^2 dR \tag{3}$$

where $R = 2\Phi(\alpha^T X) - 1$ and $p_R(R)$ is the probability density of R. After expanding $p_R(R)$ by Legendre-polynomials, we obtain the estimation

$$\hat{I}(\alpha) = \frac{1}{2} \sum_{j=1}^{J} \frac{2j+1}{n^2} \left[\sum_{i=1}^{n} P_j\{2\Phi(\alpha^T x_i) - 1\} \right]^2 . \tag{4}$$

Here, J plays the role of a smoothing parameter. It is well known (see e.g. Hall (1989)) that this index is not very robust against outliers. We obtain usually skewed distributions in this case. Nevertheless we present it here to demonstrate the interplay between density estimation and projection pursuit techniques.

To demonstrate the performance of this method we apply it to the Swiss Bank Notes data set (dimension 6) described in *Example 1.* Figure 4 (obtained by the `ppexpl` macro of XploRe) shows the projection of the data maximizing $\hat{I}(\alpha)$ for $J = 3$, which is interactively selected by the user. For maximization the Nelder & Mead (1965) algorithm is used. The coordinates of the projection are displayed in the right window. We can distinguish at least two clusters in the data, separated in the Figure at about -0.4. For a projection pursuit discriminant analysis approach see Polzehl (1993).

To explain the background of fast and easy kernel estimation programming it is necessary to have some deeper inside to the properties of kernel estimates. In practical studies the choice of the kernel does not have a great influence on the resulting density estimate assuming the bandwidth h is optimal. One possible definition of optimal bandwidth is to choose that h which minimizes a distance between the true underlying density and the density estimate. Usual distances are the *integrated squared error* $ISE(h)$ or its mean $MISE(h)$ which has the asymptotic representation $(n \to \infty, nh \to \infty, h \to 0)$

$$MISE(h) \approx C_1 n^{-1} h^{-1} + C_2 h^4, \tag{5}$$

with the constants $C_1 = \int K^2(u)du$, $C_2 = \frac{1}{4}\ \mu_2^2(K)\ \int \{f''(x)\}^2 dx$ and $\mu_2(K) = \int u^2 K(u)du$. We will not discuss the optimal bandwidth choice here. For an overview on this topic see Park & Turlach (1992).

The concept of canonical kernels introduced by Marron & Nolan (1989) scales the kernels such that they are equivalent in view of $MISE$. More exactly, for two kernels K_1, K_2,

$$MISE_{K_1}(h_1) \approx c_{K_1,K_2} MISE_{K_2}(h_2) \tag{6}$$

with a constant c_{K_1,K_2} independent of h_1, h_2, if the bandwidths h_1, h_2 fulfill

$$h_2 = h_1 \frac{\delta_2^*}{\delta_1^*}, \quad \delta_i^* = \left\{ \frac{\int K_i^2(u)du}{\mu_2^2(K_i)} \right\}^{1/5}. \tag{7}$$

Table 2 shows the transformation factors for the bandwidths if we change from one kernel of Table 1 to another.

δ_j^*/δ_i^*	Uniform	Triangle	Epanechnikov	Quartic	Gaussian
Uniform	1.000	0.715	0.786	0.663	1.740
Triangle	1.398	1.000	1.099	0.927	2.432
Epanechnikov	1.272	0.910	1.000	0.844	2.214
Quartic	1.507	1.078	1.185	1.000	2.623
Gaussian	0.575	0.411	0.452	0.381	1.000

Table 2: Canonical kernel transformations

In general, the implementation of kernel density estimation is not an easy computational task. Some packages are very specialized in the sense that they can do only calculations related to kernels (e.g. N-kernel). On the other hand some are so general (e.g. GAUSS) that everything must be programmed (often with do-loops). Others have not enough built-in flexibility (e.g. Splus supports only four kernels: Cosine, Gaussian, Uniform and Triangle) which will cause difficulties if density estimation is only a part of the problem.

XploRe offers a great variety of kernels as well as predefined macros which realize the computation of kernel estimates. The user may add his own kernels of course. The fast computing algorithms are based on the WARPing technique, described e.g. in Härdle (1991). The basic idea is the "binning" of the data, i.e. dividing the real axis into small bins of length δ starting at an origin x_0 and evaluating the kernel estimate only at the bincenters. Instead of computing the kernel for all differences $(x_i - x_k), i, k = 1, ..., n$ the kernel function needs now to be evaluated only at $i\delta h^{-1}, i = 1, ..., \ell$, where ℓ is number of bins which contains the support of the kernel function. The calculation of the estimated density reduces from $\hat{f}_h(x_j)$ to

$$\bar{f}_h(\bar{x}_j) = \frac{1}{nh} \sum_{i=1}^{N_b} n_i K \left\{ (i-j)\delta h^{-1} \right\} \tag{8}$$

computed on the grid $\bar{x}_j = x_0 + \left(j + \frac{1}{2}\right)\delta$ with N_b denoting the number of non-empty bins, n_i the number of observations in the i-th bin.

The XploRe density estimation macros `denauto`, `denest` and `denest2` use the fact of the asymptotic equivalence of the kernels described above and calculate therefore the density estimates based on the Quartic kernel. This kernel has the advantage of a compact support, is almost as efficient as the Epanechnikov kernel (which is the optimal kernel of order 2, see Silverman (1986, Section 3.3)), but gives smoother estimates. In program A.2 we provide the XploRe code for the `denauto` macro, `denest` and `denest2` are programmed in an analogous way.

Since we have an asymptotic normal distribution for the kernel density estimate at fixed points x if $n \to \infty$, we can also construct confidence intervals for $\hat{f}_h(x)$. For a bandwidth $h = cn^{-1/5}$ the following formula (Silverman (1986); Härdle (1991)) holds

$$n^{2/5}\{\hat{f}_h(x) - f(x)\} \xrightarrow{\mathcal{L}} N\left(\frac{c^2}{2}f''(x)\mu_2(K)\ ,\ c^{-1}f(x)C_1\right). \tag{9}$$

For small h (in relation to $n^{-1/5}$) the mean in (9) is negligible. This yields the asymptotic confidence interval

$$\left[\hat{f}_h(x) - u_{1-\alpha/2}\sqrt{\frac{\hat{f}_h(x)C_1}{nh}}\ ,\ \hat{f}_h(x) + u_{1-\alpha/2}\sqrt{\frac{\hat{f}_h(x)C_1}{nh}}\right] \tag{10}$$

with $u_{1-\alpha/2}$ denoting the $\alpha/2$ quantile of the standard normal distribution.

The XploRe code in program A.3 calculates asymptotic 0.95-confidence intervals for the density of the age data of *Example 2*. We show these confidence intervals in Figure 5. One sees that the modality structure remains inside the confidence intervals (due to the neglection of the mean of the asymptotic normal distribution).

Formulas and computer algorithms for true confidence bands can be found in Bickel & Rosenblatt (1973), Härdle (1991). They are slightly more complicated but they have the same underlying idea to exploit the asymptotic limit distribution. For a bootstrap approach see Hall (1992, page 220 ff.).

Smooth Regression in One Dimension

The aim of this chapter is to give a very short overview and to recall the main ideas of univariate nonparametric regression methods, in particular kernel regression smoothing.

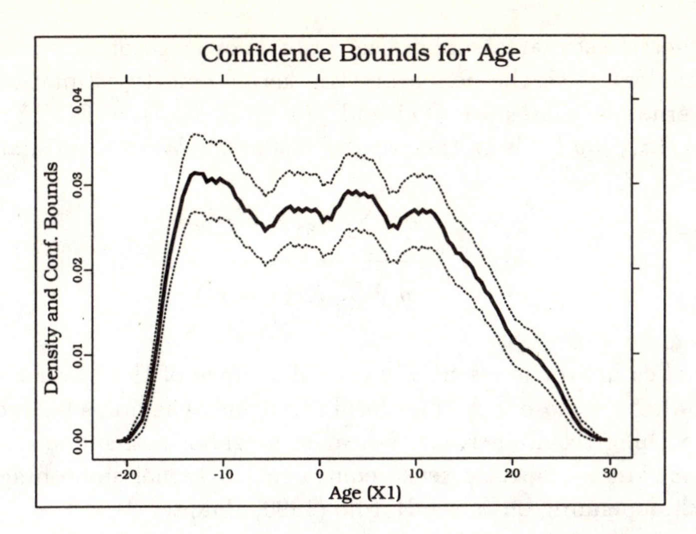

Figure 5: Density estimate and confidence intervals for the age data. Output from program A.3.

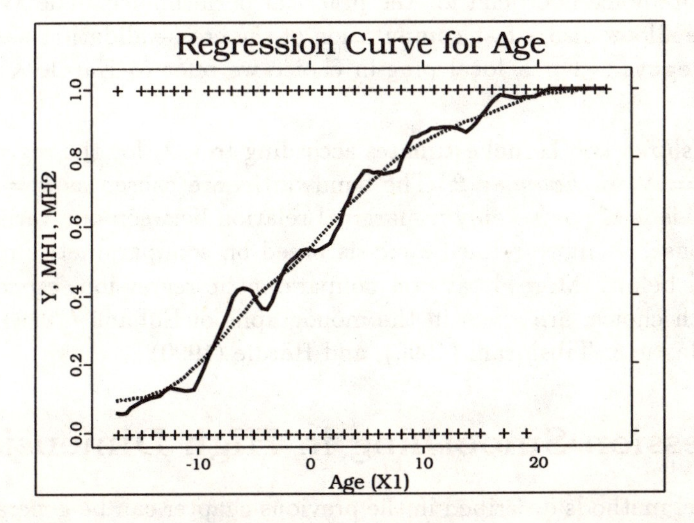

Figure 6: Two one-dimensional kernel estimates for the bandwidths $h = 2.5$ and $h = 7$. Data labelled by "+".

The univariate regression model assumes observations of two variables X and Y, i.e. data of the form $(x_i, y_i), i = 1, \ldots, n$ which are connected via an unknown regression function $m(\bullet)$ as follows:

$$y_i = m(x_i) + \epsilon_i \ , \ i = 1, \ldots, n, \tag{11}$$

ϵ_i denoting the error variables. The problem is now to estimate $m(\bullet)$.

Nonparametric estimates suppose no prior knowledge of m. There is an obvious analogy with the nonparametric kernel density estimation. Plugging in kernel estimates for $f(x)$ and $f(x,y)$ in $m(x) = \mathbf{E}(Y|X = x) = f(x)^{-1}\int yf(x,y)dy$ leads to the popular Nadaraya–Watson estimate

$$\hat{m}_h(x) = \frac{n^{-1}\sum_{i=1}^{n} K_h(x - x_i)y_i}{n^{-1}\sum_{j=1}^{n} K_h(x - x_j)} . \tag{12}$$

This kernel estimator is essentially a local average of the y_i variables with corresponding x_i's close to x. This local averaging behavior is behind several other smoothing techniques, e.g. k-nearest-neighbor and spline smoothing. They are in an asymptotic sense equivalent to kernel smoothing with a bandwidth depending on x, see Härdle (1990, Chapter 3).

In XploRe the fast computing of the kernel smoothers is again performed via the WARPing technique (macro `regest`). As in density estimation the bandwidth choice is crucial for the practical performance. The WARPing technique allows also a fast computation of the cross-validation bandwidth (macro `regcvl`). For a local plug-in choice we refer to Härdle & Marron (1991).

Figure 6 shows two kernel estimates according to (12) for the regression of Y on $X = X_1$ in *Example 2*. The bandwidths are chosen as $h = 2.5$ and $h = 7$. This is of course only a marginal relation between one variable and the response. A more refined analysis based on semiparametric models is presented below. More details on nonparametric regression, especially on bandwidth choice, are given in the monographs of Eubank (1988), Müller (1988), Hastie & Tibshirani (1990), and Härdle (1990).

Regression Smoothing in High Dimensions

The kernel methods described in the previous chapter can be generalized to the multivariate case. In most practical applications however problems will arise due to the sparseness of data. Projection based methods or additive modelling avoid this data sparseness. We will present *Single Index Models* and *Generalized Additive Models*, both generalizations of *Generalized Linear Models* (GLM), see McCullagh & Nelder (1989). The GLM generalizes the linear regression model with systematic component $\eta = X^T\beta$ to $\mathbf{E}Y = G(\eta)$ with a known (inverse) link function G. In our *Example 2*, where we have binary responses, we model

$$P(Y = 1|X = x) = G(x^T\beta). \tag{13}$$

The Single Index Model idea is to generalize (13) to arbitrary smooth link functions g. This is what is called in the statistical literature a one term *Projection Pursuit Model.* Friedman & Stuetzle (1981) proposed an iterative method for estimating β. Härdle & Stoker (1989) derived a direct non-iterative method, the so-called *Average Derivative Estimation* (ADE).

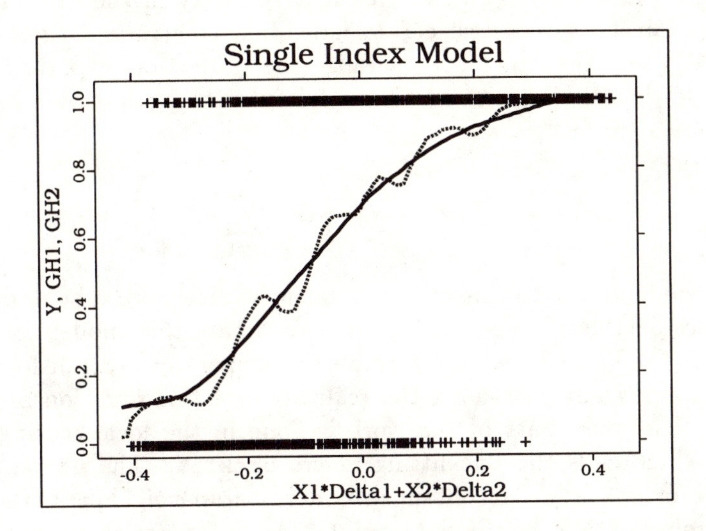

Figure 7: Single Index Model fit with two estimated link functions. ($h = 0.05, 0.15$). Projected data labelled by "+". Output of program A.4.

The ADE idea is as follows. For $m(x) = g(x^T\beta)$ the average derivative

$$\delta = \mathbf{E}m'(X) = \mathbf{E}\left[\frac{dg}{d(x^T\beta)}(X^T\beta)\right]\beta \tag{14}$$

determines β up to a scale factor. Since δ equals $\mathbf{E}\ell Y$, ℓ denoting the *score function* $-\partial \log f/\partial x = -f'/f$, it can be estimated by $\hat{\delta} = n^{-1}\sum_{i=1}^{n}\hat{\ell}_h(x_i)y_i$ with $\hat{\ell}$ based on a kernel density estimate with bandwidth h.

Figure 7 shows for *Example 2* the projected observations $x_i^T\hat{\delta}$ vs. the responses y_i as well as two link functions $\hat{g}$ (computed with `regest`) with bandwidths $h = 0.05$ and $h = 0.15$, respectively. The corresponding XploRe code (Program A.4) utilizes the macro `adeind` which calculates the ADE.

Since the two estimates of Figure 6 and Figure 7 look a bit similar it would be interesting to test whether $\delta_2 = 0$, i.e. whether there is no influence of zinc. We have here $\hat{\delta} = \binom{0.018}{-0.058}$ and its estimated covariance is $\hat{\Sigma}_\delta = \begin{pmatrix} 0.00037 & -0.00056 \\ -0.00056 & 0.14288 \end{pmatrix}$. Härdle & Turlach (1992) describe a test using a Wald statistic which goes back to Stoker. To test the hypothesis $R\delta = r$ we have to compare the test statistic $W = n(R\hat{\delta} - r)^T(R\hat{\Sigma}R^T)^{-1}(R\hat{\delta} - r)$ to the $\chi^2(\text{rank } R)$ value. For our running example this leads to $W = 26.678$ and $P(\chi_1^2 > W) \approx 0$, thus we reject the hypothesis. This result is not surprising and of course highly significant since we have a very large sample size ($n = 1136$).

In Figures 6 and 7 one can see a clear asymmetry in the link functions, speaking against a symmetric, e.g. logistic, link. *Generalized Additive Models* (GAM) keep the link but generalize the projection $x^T\beta$ to a sum of nonparametric transformations. These fall into the class of *Additive Models*, i.e. one assumes

$$P(Y = 1|X = x) = G\left(\alpha + \sum_{j=1}^{d} g_j(x_j)\right). \tag{15}$$

For an introduction into this class of models we refer to the book of Hastie & Tibshirani (1990). The algorithm to estimate this model consists of *local scoring* and *backfitting* to determine the nonparametric transformations $g_1, \ldots, g_d$. Program A.5 shows the realization of this iteration process in XploRe. The main part of the work is done in the local scoring macro `lscore`, which calls the backfitting macro `backfit`. The nonparametric estimates for $g_1, \ldots, g_d$ are obtained by a k-nearest-neighbor method with k the number of 30% of the data points. As link function we have taken the logistic link.

The output of program A.5 is displayed in Figure 8. The upper left picture shows the y_i vs. $\hat{\eta}_i = \hat{\alpha} + \sum_{j=1}^{d} \hat{g}_j(x_j)$ and the fit $G(\hat{\eta}_i)$. The lower left and upper right pictures show the estimated nonparametric components $\hat{g}_1(x_{1i})$ vs. x_{1i} and $\hat{g}_2(x_{2i})$ vs. x_{2i}. The lower right picture displays the 3-dimensional surface $(x_{1i}, x_{2i}, \hat{\eta}_i)$. We see that the nonparametric $\hat{g}_1$ for the age is almost linear. However $\hat{g}_2$ for zinc is nonlinear and has a negative slope. Recall that the ADE estimate was negative in the second component, too. But one should pay attention that $\hat{g}_1$ varies in $[-4, 6]$ whereas $\hat{g}_2$ takes values in $[-1, 1]$. So it turns out that the zinc variable has a smaller influence than the age variable.

To assess the quality of this GAM estimate we provide a brushed residual plot. Figure 9 shows in the upper left a plot of the age and zinc data, masked by a point "•" if the corresponding residual lies in the lower quartile, by a "+" if the residual lies in the interquartile range and masked by a "o" if it is in the upper quartile. The three other plots show these residual groups separately. The masking can be achieved directly by giving the corresponding commands in the program or brushing interactively on the screen since the four displays are linked. It is easy to see that the residual plots underline our conclusion from the GAM fit that the zinc has less

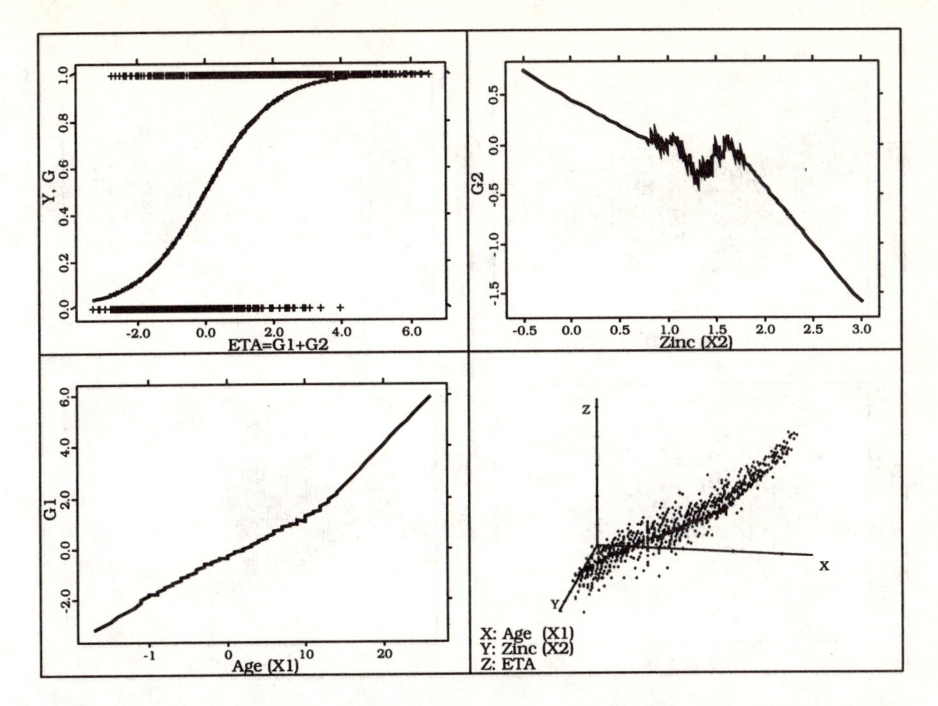

Figure 8: Generalized Additive Model fit. Output from program A.5.

influence on the results. Program A.6 gives the XploRe code for Figure 9. Inferential issues are still open for GAMs as has been pointed out by Härdle & Müller (1993b).

There are many more approaches in the analysis of high dimensional data. We would like to mention among others *Alternating Conditional Expectations* (ACE), Breiman & Friedman (1985), and *Sliced Inverse Regression* (SIR), a method introduced by Li (1991). Both methods are available in XploRe by the `acefit` and `sir1`, `sir2` macros.

Acknowledgements

We are grateful to Julie Mares-Perlman, Barbara and Ron Klein for the permission to use their data on Nuclear Sclerotic Cataract (*Example 2*). We would also like to thank Christian Ritter (Louvain-la-Neuve) for attracting our interest to this data set.

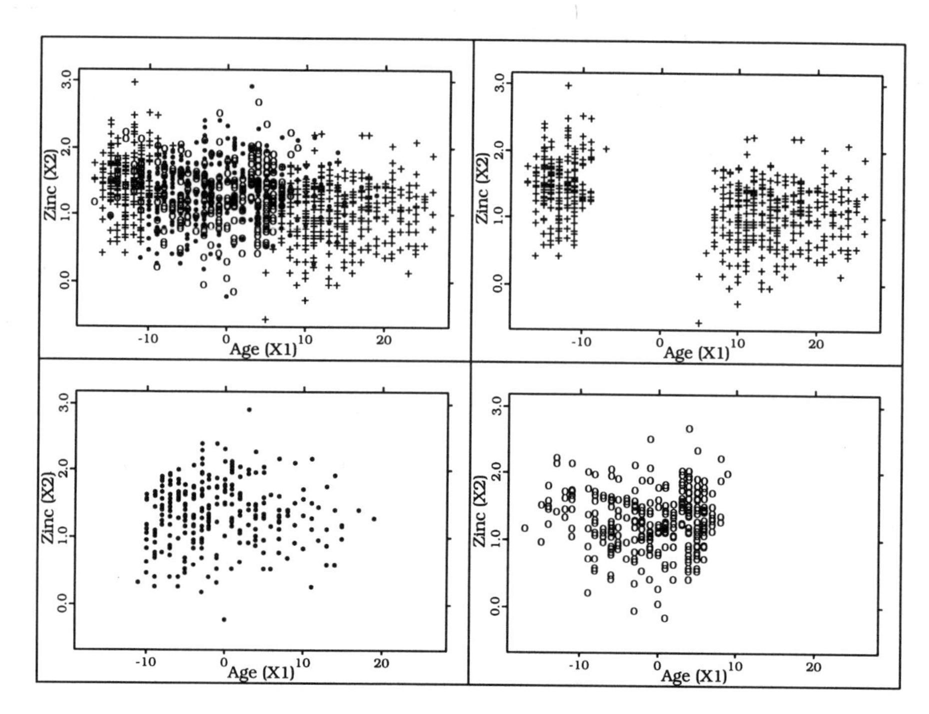

Figure 9: Brushed residual plot for the GAM fit. Output from program A.6.

References

Bickel, P.J. & Rosenblatt, M. (1973). On some global measures of the deviations of density function estimates. Ann. Statist. 1, 1071-1095.

Breiman, L., Friedman, J.H. (1985). Estimating optimal transformations for multiple regression and correlation (with discussion). J. Amer. Statist. Assoc. 80, 580-619.

Eubank, R.L. (1988). Spline smoothing and nonparametric regression. Marcel Dekker, New York.

Flury, B., Riedwyl, H. (1988). Multivariate statistics. A practical approach. Chapman and Hall, London.

Friedman, J.H. (1987). Exploratory Projection Pursuit. J. Amer. Statist. Assoc. 82, 249-266.

Friedman, J.H. & Stuetzle, W. (1981). Projection Pursuit Regression. J. Amer. Statist. Assoc. 76, 817-823.

Hall, P. (1989). Polynomial Projection Pursuit. Ann. Statist. 17, 589-605

Härdle, W. (1990). Applied nonparametric regression. Econometric Society Monographs No. 19, Cambridge University Press, Cambridge.

Härdle, W. (1991). Smoothing Techniques. With Applications in S. Springer, New York.

Härdle, W. & Marron, J.S. (1991). Fast and Simple Scatterplot Smoothing. CORE Discussion Paper No. 9143, Université Catholique de Louvain.

Härdle, W. & Müller, M. (1993a). Nichtparametrische Glättungsmethoden in der alltäglichen statistischen Praxis. Allg. Statist. Archiv 77, 9-31.

Härdle, W. & Müller, M. (1993b). Discussion of: Hastie, T.J.; Tibshirani, R.J. (1993). Varying-Coefficient Models. J. Roy. Statist. Soc. B 55, 757-796.

Härdle, W. & Stoker, T.M. (1989). Investigating Smooth Multiple Regression by the Method of Average Derivatives. J. Amer. Statist. Assoc. 84, 986-995.

Härdle, W. & Turlach, B. (1992). Nonparametric Approaches to Generalized Linear Models. In: Fahrmeir, L.; Francis, B.; Gilchrist, R.; Tutz, G. (Eds.): Advances in GLIM and Statistical Modelling. Lecture Notes in Statistics No. 78, Springer, New York, 213-225.

Hastie, T.J. & Tibshirani, R.J. (1990). Generalized Additive Models. Chapman and Hall, London.

Hall, P. (1992). The Bootstrap and the Edgeworth Expansion. Springer, New York.

Klein, B.E.K., Klein, R. & Linton, K.L.P. (1992). Prevalence of Age-related Lens Opacities in a Population. The Beaver Dam Eye Study. Ophthalmology 99, 546-552.

Li, K.-C. (1991). Sliced Inverse Regression for Dimension Reduction (with discussion). J. Amer. Statist. Assoc. 86, 316-342.

Mares-Perlman, J.A.; Klein, B.E.K.; Klein, R.; Ritter, L.L.; Freudenheim, J.L. & Luby, M.H. (1993). Nutrient Supplements Contribute to the Dietary Intake of Middle-aged and Older-aged Adult Residents of Beaver Dam, Wisconsin. J. Nutr. 123, 176-188.

Marron, S. & Nolan, D. (1989). Canonical kernels for density estimation. Stat. Prob. Letters 7, 191-195.

McCullagh, P. & Nelder, J.A. (1989). Generalized Linear Models, 2nd. Ed., Chapman and Hall, London.

Müller, H.-G. (1988). Nonparametric regression analysis of longitudinal data. Lecture Notes in Statistics No. 46, Springer, Berlin.

Nelder, J.A. & Mead, R. (1965). A Simplex Method for Function Minimization. Computer Journal 7, 308-313.

Park, B. & Turlach, B. (1992). Practical performance of several data driven bandwidth selectors (with discussion). Comp. Statist. 7, 251-271.

Polzehl, J. (1993). Projection Pursuit Discriminant Analysis. CORE Discussion Paper No.9320, Université Catholique de Louvain.

Scott, D.W. (1992). Multivariate density estimation. Wiley, New York.

Silverman, B.W. (1986). Density estimation for statistics and data analysis. Chapman and Hall, London.

XPLORE (1993). XploRe 3.1 - an interactive statistical computing environment. Available from XploRe Systems, Institut für Statistik und Ökonometrie, Wirtschaftswissenschaftliche Fakultät der Humboldt-Universität Berlin.
(XploRe 3.0 is available via anonymous ftp from the directory pub/xplore of amadeus.wiwi.hu-berlin.de = 141.20.100.2)

Appendix

```
proc()=main()
 x=read(nuc)                      ; read the data file "nuc.dat"
 library(smoother)                ; load the necessary library
 fh1=denauto(x[,1])               ; estimate fh1 by the rule of thumb
 n=rows(x)
 sigma=sqrt(cov(x))               ; estimate standard deviation
 h1=2.62*1.06*sigma*n^(-0.2)      ; bandwidth h1 used in "denauto"
 h2=h1./2                         ; bandwidth h2
 fh2=denest(x[,1] h2)             ; density estimate with h2
                                  ;
 createdisplay(d1, 2 (-2), s2d text s2d text)
                                  ; create 2 text & 2 graphics
                                  ; displays (asymmetric)
                                  ;
 show(fh1 s2d1, fh2 s2d2, h1 text1, h2 text2)
                                  ; show fh1, fh2 and h1, h2
endp
```

Program A.1: XploRe code for Figure 2.

```
proc(fh)=denauto(x)
 d=(max(x)-min(x))./100                  ; make 100 bins
 (xb yb)=bindata(x d)                    ; bin the data, binwidth d
 sigma=sqrt(cov(x))                      ; estimate standard deviation
 h=2.62*1.06*sigma*(rows(x))^(-0.2)      ; determine h by rule of thumb
                                         ; use transformation constant
                                         ; for change to quartic kernel
 wy=symweigh(0 d/h h/d &qua)             ; create weights for the
                                         ; symmetric quartic kernel
 wx=aseq(0 rows(wy))
 (xc yc or)=conv(xb yb wx wy)            ; calculate density function
                                         ; with these weights
 fh=((xc+0.5)*d)~(yc/(n*d))              ; retransform to get density
                                         ; estimate at the bincenters
endp
```

Program A.2: Automatic density estimation, XploRe macro **denauto**.

```
proc()=main()
 x=read(nuc)                                  ; read the data
 library(smoother)                            ; load necessary library
 h=3.767                                      ; used bandwidth
 fh=denest(x[,2] h)                           ; density estimate
 ci=1.96*sqrt((5/7).*fh[,2]./(rows(x)*h))
 cup=fh[,1]~(fh[,2].+ci)                      ; upper confidence bound
 clo=fh[,1]~(fh[,2].-ci)                      ; lower confidence bound
 show(fh s2d)                                 ; display the result
endp
```

Program A.3: Density estimation confidence intervals. XploRe code for Figure 5.

```
proc()=main()
 x=read(nuc)                          ; read the data
 library(smoother)                    ; load the necessary libraries
 library(addmod)
 y=x[,3]                              ; the response variable
 x=x[,1:2]                            ; the age and zinc variables
 d=(max(x)-min(x))/20                 ; make 20 bins
 (xb yb)=bindata(x d 0 y)             ; bin the data, binwidth d
 (del vdel) = adeind(xb yb d 3)       ; the ADE del and its asymptotic
                                      ; covariance matrix  vdel
 est=(x*del)~y                        ; the projected data
 gh1=regest(est 0.05)                 ; estimate g, h=0.05
 gh2=regest(est 0.15)                 ; estimate g, h=0.15
 show(est gh1 gh2 s2d)                ; display the results
endp
```

Program A.4: The Single Index Model in XploRe. Code for Figure 7.

```
proc()=main()
 x=read(nuc)                      ; read the data
 library(smoother)                ; load the necessary libraries
 library(addmod)
 y=x[,3]                          ; the response variable
 x=x[,1:2]                        ; the age and zinc variables
 (fx alpha dev)=lscore(x y 0.3)   ; run the local scoring algorithm
                                  ; which includes the backfitting
 eta=alpha+sumr(fx)
 mu=exp(eta)./(1+exp(eta))        ; calculate the link function
 createdisplay(h1, 2 2, s2d s2d s2d d3d)
                                  ; create the output display
 dat11 = eta~y
 dat12 = sort(eta~p)
 dat2  = sort(x[,1]~fx[,1])
 dat3  = sort(x[,2]~fx[,2])
 dat4  = x~eta
 show(dat11 dat12 s2d1, dat2 s2d2, dat3 s2d3, dat4 d3d1)
 res=(y-mu)./(sqrt(mu.*(1-mu)))   ; calculate the residuals
 xres=x~res                       ; combine res with x
 write(xres xres)                 ; save xres to a file "xres.dat"
endp
```

Program A.5: The Generalized Additive Model in XploRe. Code for Figure 8.

```
proc()=main()
 xres=read(xres)                   ; load file "xres.dat"
 xres=sort(xres 3)                 ; sort data after residuals
 xr=xres[,1:2]                     ; sorted age and zinc values
                                   ;
 createdisplay(h2, 2 2, s2d s2d s2d s2d)
                                   ; create output display
                                   ;
 xr1=xr[1:284,]                    ; data with lower quartile res.
 xr2=xr[284:852,]                  ; data with interquartile res.
 xr1=xr[853:1136,]                 ; data with upper quartile res.
                                   ;
 show(xr1 xr2 xr3 s2d1, xr1 s2d2, xr2 s2d3, xr3 s2d4)
                                   ;
 link("s2d1\data_1" "s2d2\data_1"); link the 4 displays
 link("s2d1\data_2" "s2d3\data_1")
 link("s2d1\data_3" "s2d4\data_1")
 link("s2d2\data_1" "s2d1\data_1")
 link("s2d3\data_1" "s2d1\data_2")
 link("s2d4\data_1" "s2d1\data_3")
 display(h2)                       ; show the linked displays
endp
```

Program A.6: Residual plot. XploRe code for Figure 9.

Missing Values: Statistical Theory and Computational Practice

Werner Vach
University of Freiburg, Institute of Medical Biometry and Medical Informatics, Stefan Meier Str. 26,
D-79104 Freiburg, Germany

Key Words:
complete case analysis, maximum likelihood principle, EM algorithm, selection bias

Abstract
There are two different paradigms for the development of methods to handle missing values: The first is the manipulation of the data in order to achieve an artificially completed data set, the second is the use of established statistical principles. The first approach aims to use standard statistical software, the second often requires additional computational efforts. Statistical theory may guarantee good statistical properties for the second approach, while the first may result in biased or inefficient estimates. This interdependence of statistical theory and computational practice is discussed in this paper.

Introduction

Statistical methods are usually developed for complete data. Statistical methods are usually applied to incomplete data. The necessity to solve the latter task with computer programs written for the first caused a rather widespread technique to evolve: the complete case analysis; i.e., all units with a missing value in at least one of the relevant variables are removed from the data set before the computation starts. To overcome the obvious inefficiency of this approach, several strategies for completing the data are used. Missing values are regarded as an additional category, they are substituted by a mean or another guess, or variables with missing values are omitted. For methods based on an empirical covariance matrix, estimation of single covariances can be based on all available pairs. Obviously, the development of these approaches aimed at an improvement of the input for existing computer programs. On the other hand there are approaches developed by the use of statistical principles. Rubin (1976) showed the general applicability of the maximum likelihood principle in missing value problems, which has been applied in a lot of different settings (cf. Little & Rubin (1987)). He also developed a theoretical foundation for multiple imputation techniques, allowing to distinguish between proper and improper ones (Rubin (1987)). Hence with respect to the field of "Computational Statistics" a double task appears: The methods representing today's computational practice must be evaluated with respect to their statistical properties, such that recommendations for their use or warnings against their use can be given. The methods representing today's statistical theory must be checked on their practical applicability and appropriate implementations should be included in statistical software packages. Both tasks will be illustrated in this paper by examples.

The handling of missing values in a univariate setting is more or less trivial; hence we focus our attention on the multivariate case. For each unit we consider the p-dimensional random variable $X = (X_1, \ldots, X_p)$. Observability is indicated by $R = (R_1, \ldots, R_p)$ with

$$R_i := \begin{cases} 1 & \text{if } X_i \text{ can be observed} \\ 0 & \text{otherwise} \end{cases} \tag{1}$$

such that we observe instead of X the random variable $Z = (Z_1, \ldots, Z_p)$ with

$$Z_i := \begin{cases} X_i & \text{if } R_i = 1 \\ ? & \text{if } R_i = 0 \end{cases} \tag{2}$$

Complete Case Analysis

Let C be a random variable indicating the observability of all variables, i.e.

$$C := \begin{cases} 1 & \text{if } R_i = 1 \quad \forall\, i = 1, \ldots, p \\ 0 & \text{if } R_i = 0 \text{ for at least one } i \in \{1, \ldots, p\} \end{cases} \tag{3}$$

In a complete case analysis all units with $C = 0$ are removed. The implicit assumption is hence

$$F(X) = F(X \mid C = 1) \tag{4}$$

or at least

$$T(F(X)) = T(F(X \mid C = 1)) \tag{5}$$

where T is the functional of interest. In principle conditions implying the validity of the assumptions above are widely discussed in the literature; however, they are usually discussed in the more general setting of sample selection (e.g. Kleinbaum et al. (1981)). There the selection of a unit from a population is indicated by the variable S. Assuming $X \sim F$ in the population, any violation of the assumption $T(F(X)) = T(F(X \mid S = 1))$ is called a selection bias. Although selection bias is known to be a serious problem for many data selection procedures, the immediate implications with respect to the bias of a complete case analysis are mostly ignored. Moreover the selection procedure given by removing units with $C = 0$ is obvious, and the validity or violation of necessary assumptions can be easily checked.

A simple example is the estimation of the odds ratio for two binary variables X_1 and X_2. The odds ratio is defined by

$$OR := \frac{p_{11}p_{00}}{p_{01}p_{10}} \quad \text{with} \quad p_{ij} := P(X_1 = i, X_2 = j) \tag{6}$$

The selection procedure due to a complete case analysis is determined by

$$q_{ij} := P(C = 1 \mid X_1 = i, X_2 = j) \quad \text{for} \quad i = 0,1; \quad j = 0,1 \tag{7}$$

such that

$$P(X_1 = i, X_2 = j \mid C = 1) = q_{ij}p_{ij} / \sum_{i,j} q_{ij}p_{ij} \tag{8}$$

Hence estimates of the odds ratio based only on units with completely observed variables are estimates of the quantity

$$Q \times OR \quad \text{with} \quad Q := \frac{q_{11}q_{00}}{q_{01}q_{10}} \tag{9}$$

i.e. if $Q \neq 1$, a complete case analysis results in biased estimates of the odds ratio. A sufficient condition for an unbiased estimation is $q_{ij} = q_i$ or $q_{ij} = q_j$, i.e. the selection due to missing values depends either only on the true value of X_1 or only on the true value of X_2. This corresponds to the well-known fact, that case-control studies allow the unbiased estimation of an odds ratio, if the selection probabilities depend only on the disease status.

If missing values only occur in one variable, say X_2, the bias vanishes if the probability for the occurrence of a missing value depends only on the value of X_2 itself, or only on the value of X_1. In a case-control study, where X_1 is the disease status and X_2 is a binary exposure affected with missing values, such an assumption is highly questionable. Different data collection procedures for cases and controls imply $q_{1j} \neq q_{0j}$. In addition an exposure like a radiation therapy is often better documented if it is present than if it is absent. This implies $q_{i0} \neq q_{i1}$. Hence in the analysis of case-control studies the complete case approach may often be associated with considerable bias (cf. Vach & Blettner (1991)).

If both variables are affected with missing values, there exists at least one non-trivial situation of practical interest, where the bias due to removing units with incomplete information can be denied. Let us assume, that the two variables are collected by direct questioning of one individual. The probability to refuse to answer the first question concerning X_1 should only depend on the true value of X_1, i.e. we can denote it by $1 - p_{1,i}$. Given answering of the first question was not refused, the probability to refuse to answer the second question concerning X_2 should only depend on the true value of X_2, i.e. we can denote it by $1-p_{2,j}$. There may be a high correlation between R_1 and R_2, such that $P(R_2 = 1 \mid X_1 = i, X_2 = j)$ depends on i and j, but nevertheless we have $Q = 1$, because $q_{ij} = p_{1,i}p_{2,j}$.

As a second example, where examination of the possible bias is quite simple, we consider regression models with missing values in the covariates. If X_1 is the outcome variable, the essential assumption is

$$F_\beta(X_1 \mid X_2, \ldots, X_p) = F_\beta(X_1 \mid X_2, \ldots, X_p, C = 1) \tag{10}$$

where β denotes the regression parameters of interest. If the probability for missing values in the covariates depends only on the covariates themselves, the conditional distribution of the outcome variable does not change by additionally conditioning on the completeness of the covariates. In a prospective study, where the outcome variable is measured after finishing measurement of the covariates, and where the outcome variable measures an event really happening after the events measured by the covariates, the validity of the assumption above can be assumed. In retrospective studies, the occurrence of missing values in the covariates may depend on the outcome variable, and hence a complete case analysis may result in biased estimates.

Maximum Likelihood Estimation

The maximum likelihood (ML) principle is a general, widely accepted estimation procedure in statistics. To apply it in missing value problems, the starting point is the density of the distribution of Z.
If for any $(r_1, \ldots, r_p) \in \{0,1\}^p$

$$\begin{aligned} P(R_i = r_i \quad \forall i = 1, \ldots p \mid X_i = x_i \quad \forall i = 1, \ldots, p) \equiv \\ P(R_i = r_i \quad \forall i = 1, \ldots, p \mid X_i = x_i \quad \forall i \text{ with } r_i = 1) \end{aligned} \tag{11}$$

then the density $g_\theta(z)$ of Z satisfies

$$g_\theta(z) \propto \int f_\theta(x)\, dx_{M(z)} \tag{12}$$

where $f_\theta(x)$ denotes the density of X and $M(z) := \{i \mid z_i = ?\}$. Hence (12) allows for units with missing values in some variables to base the likelihood on the marginal distribution of the observed variables. Due to its importance condition (11) is called *Missing At Random* (MAR) by Rubin (1976). It roughly requires, that there is no relation between the true values of one or several variables and their observability. Hence if the MAR assumption is satisfied, statistical inference of θ based on the ML principle is possible.

Maximization of the likelihood based on $g_\theta(z)$ can be tackled by the Newton Raphson method or one of its variants. However, the use of the Expectation Maximization (EM) algorithm (Dempster et al. (1977)) is widespread in this setting. The EM algorithm is an iterative method, where each iteration consists of two steps. In the first step we have to compute the expected loglikelihood of X given Z, where the expectation is taken at the current value of θ_t. In the second step, the resulting function is maximized with respect to θ leading to a new value θ_{t+1}. Hence if $l(\theta; x) := \log f_\theta(x)$ denotes the loglikelihood for θ given $X = x$, then θ_{t+1} is the solution of the maximization of

$$\sum_u l^t(\theta; z_u) \text{ with } l^t(\theta; z) = E_{\theta_t}[l(\theta; X) \mid Z = z] \tag{13}$$

where summation is over all units. This maximization task can often be solved using existing software for computing ML estimates in the complete data case. For example if all variables are categorical, we have

$$\sum_u l^t(\theta; z_u) = \sum_c l(\theta; c)\, n_c^t \tag{14}$$

where summation is taken over all cells of the contingency table built by all variables. The first step requires then to update the contingency table n^t, and the maximization task of the second step can be solved by applying an existing program for ML estimation to the updated contingency table. Similarly, it is possible to relate the Fisher information of the incomplete data likelihood to the Fisher information of the complete data likelihood, facilitating the computation of variance estimates (Louis (1982)). However, implementation requires additional efforts, whereas Newton Raphson methods yield a variance estimate directly.

In the statistical literature we find many applications of the ML principle to achieve parameter estimates in the presence of missing values. Together with the EM algorithm it has become a standard technique for sophisticated applications. However, it has not (yet) become a standard technique for applied statisticians, struggling with missing values in their daily work. There seem to be several reasons for this:

1. Application of the ML principle requires the MAR assumption which is often highly questionable. This nearly prevents the use of ML estimation in any application where incomplete data is caused by non-response of individuals or gaps in the sources of retrospective studies.

2. In many publications the use of the EM algorithm is stressed. It is often not mentioned, that the resulting estimates obey the traditional ML principle. Hence it is often thought, that estimates computed with the EM algorithm require their own statistical theory.

3. The EM algorithm has a slow rate of convergence (cf. for example Wu (1983)). On the other hand, the EM algorithm is often used in settings like latent class analysis, where maximization of the likelihood is difficult due to the existence of several local maxima or flatness. With respect to the difficulties with the EM algorithm a clear distinction between poor properties of the algorithm and poor properties of the maximization task to be solved seems to be missing.

4. The implementation of the EM algorithm requires a loop around an (available) module to solve the maximization task. This is no problem in modern languages like S. It is a much more complicate task in traditional systems like SAS or BMDP, which are designed according to sequential batch processing.

A fifth reason is discussed in the following chapter.

Missing Values in the Covariates of a Regression Model

Many statistical applications are based on regression models, because they provide a flexible tool to investigate the relation between several covariates and one outcome variable. They only require to specify the conditional distribution of the outcome variable given the covariates. However, if we want to handle missing values in the covariates using ML estimation as described in the last chapter, the joint distribution of all variables is required. To illustrate this point, let us consider the case where X_1 is the outcome variable, X_2 and X_3 are covariates, X_3 is affected by missing values, and $f_\beta(x_1 \mid x_2, x_3)$ defines the regression model. Then under the MAR assumption

$$g_\beta(z_1, z_2, z_3) \propto \begin{cases} f_\beta(z_1 \mid z_2, z_3)\, h(z_3 \mid z_2) & \text{if } z_3 \neq ? \\ \int f_\beta(z_1 \mid z_2, x)\, h(x \mid z_2)\, dx & \text{if } z_3 = ? \end{cases} \tag{15}$$

where $h(\cdot\,|\,\cdot)$ denotes the density of the conditional distribution of X_3 given X_2. Now ML estimation requires a finite parametrization for h. A reasonable parametrization seems to be out of reach in most applications if continuous covariates are involved. Only if both covariates are categorical, the conditional cell probabilities can be used as parameters without any restrictions on the distribution of the covariates (Ibrahim (1990), Vach & Schumacher (1993)).

To consider the case of continuous covariates, a straightforward idea is to replace the unknown h by a nonparametric estimate, based on the units with complete covariate information. Then it remains to maximize over β, and the resulting estimate can be regarded as a semiparametric ML estimate. It requires some efforts to establish consistency and asymptotic normality of these estimates and to provide estimates of the asymptotic variance. Special cases are solved by Pepe & Fleming (1991) and Carroll & Wand (1991). Since they open the way to solve a problem of high practical interest, it seems to be desirable to implement this solution into statistical software. However, such a program should be able to handle more than two covariates. Then it becomes necessary to estimate conditional distributions with multidimensional conditions in a nonparametric manner. This requires an appropriate distance measure, and the choice of an appropriate distance measure is heavily influenced by possible knowledge about the distribution of the covariates. Hence also with this semiparametric approach we are far away from a pure automatic handling of missing values in the covariates. Moreover this approach is still based on the MAR assumption, and consistent estimation of h may also require additional assumptions on the missing value mechanism (cf. Vach & Schumacher (1993)).

An alternative approach to handle missing values in the covariates is due to Reilly (1991). Her basic idea is to estimate the contributions to the score function of the units with incomplete data from the contributions of the units with complete data. This approach is also applicable for generalized linear models, where only estimation equations, but no likelihood, are specified.

Some General Considerations about the Handling of Missing Values in Statistical Computing

Missing values are a genuine part of data, hence any data-based statistical computing system should be able to represent them. However, missing values are a contradiction to any statistical method assuming complete data, hence a program realising a statistical method should refuse to produce a numerical result, if a missing value occurs in the necessary computations. There exists no natural way to handle missing values, especially complete case analysis is not the natural solution. The decision of how to handle missing values should be left to the user. The decision should be presented in such a way as to make it apparent to the user that he or she would be responsible for the choice made, and that this choice would be no trivial matter.

This raises the question of how one can support this decision. The validity of any approach to handle missing values is based on assumptions on the missing value mechanism. It is a straightforward idea to offer tools to check such assumptions. However, it can easily be shown, that an assumption like MAR usually cannot be checked by the data alone; hence this idea is of limited value. An alternative is to provide tools to check sensitivity of a single estimate (or other results) against violation of the MAR assumption, but this requires some knowledge about the direction of likely violations (cf. Vach (1994, chapter 8), Phillips (1993)). It seems to be unavoidable to leave a large responsibility to the user if we offer strategies to handle missing values, and the danger of misuse increases with the severity of the necessary assumptions on the missing value mechanism.

Besides the handling of missing values in statistical computations the handling in transformations of the data is of importance. The extension of arithmetic and logical operations and order comparisons causes a lot of trouble. The widespread rule, that any operation on a missing value should result in a missing value seems to be convincing at the first sight. However, it implies some curiosities like the fact, that the subtraction of a variable from itself need not result in a zero value. The behaviour of testing equality is a crucial point; sometimes it is desired that the comparison of two missing values should result in the value true, sometimes it is preferable to achieve a missing value. Allowing logical values to be missing requires to extend the definition of conditional statements appropriately, as it is provided in MATHEMATICA for example. Summarizing, there seems to be no unique optimal solution for many problems.

Finally it should be mentioned, that good statistical software should be able to distinguish different types of missing values. First, missing values in the original data should be distinguished from missing values resulting from undefined operations like division by 0. Often the latter should be better represented as ∞ or $-\infty$. Missing values in the original data should be distinguishable with respect to the validity of the MAR assumption. Also, there should be a distinction between values completely missing and values only observed fuzzily. The latter include the special cases of censored and grouped data.

Conclusions

Computational aspects have a strong impact on the development of methods to handle missing values. On the one hand, the availability of computer programs for the analysis of complete data has forced the appearance of strategies to create artificially completed data sets, which have often poor statistical properties. E.g. Vach & Blettner (1991) criticize the practice of regarding missing values as an additional category. On the other hand, theoretically well-based approaches like ML estimation, semiparametric ML estimation or multiple imputation need to be checked on their suitability to include an appropriate implementation in standard statistical software packages. One may argue, that the latter holds for any development of statistical methods. However, for other methods it may suffice to offer a highly specialized program. Handling missing values is a task of the daily work in applied statistics, and hence it seems to be unavoidable that new methods for handling missing values will occupy standard software.

Acknowledgements

This work was supported by the *Deutsche Forschungsgemeinschaft.* The author is grateful to Ina El-Kadhi for preparing the manuscript and improving language and style.

References

Carroll, R.J. & Wand, M.P. (1991), Semiparametric estimation in logistic measurement error models, Journal of the Royal Statistical Society B, 53, 573-585

Dempster, A.P.,Laird, N.M. & Rubin, D.B. (1977), Maximum likelihood estimation from incomplete data via EM algorithm (with discussion), Journal of the Royal Statistical Society B 39, 1-38

Ibrahim, J.G. (1990) Incomplete data in generalized linear models, Journal of the American Statistical Association, 85, 765-769

Kleinbaum, D.G., Morgenstern, H. & Kupper, L.L. (1981), Selection Bias in Epidemiological Studies, American Journal of Epidemiology, 113, 452-463

Little, R.J.A. & Rubin, D.B. (1987), Statistical analysis with missing data, Wiley, New York

Louis, T.A. (1982), Finding the observed information when using the EM algorithm, Journal of the Royal Statistical Society B, 44, 226-233

Pepe, M.S. & Fleming, T.R. (1991), A nonparametric method for dealing with missing covariate data, Journal of the American Statistical Association, 86, 108-113

Phillips, M.J. (1993), Contingency tables with missing data, The Statistician, 42, 9-18

Reilly, M. (1991), Semi-parametric methods of dealing with missing or surrogate covariate data, Ph.D. thesis, Dept. of Biostatistics, University of Washington, Seattle

Rubin, D.B. (1976), Inference and missing data, Biometrika, 63, 581-592

Rubin, D.B. (1987), Multiple imputation for nonresponse in surveys, Wiley, New York

Vach, W. (1994), Logistic Regression with Missing Values in the Covariates, Lecture Notes in Statistics 86, Springer

Vach, W. & Blettner, M. (1991), Biased estimation of the odds ratio in case-control studies due to the use of ad-hoc methods of correcting for missing values in confounding variables, American Journal of Epidemiology, 134, 895-907

Vach, W. & Schumacher, M. (1993), Logistic regression with incomplete observed categorical covariates - a comparison of three approaches, Biometrika, 80, 353-362

Wu, C.F.J. (1983), On the convergence properties of the EM algorithm, Annals of Statistics, 11, 95-103

A Permutation Approach to Configural Frequency Analysis (CFA) and the Iterated Hypergeometric Distribution

Joachim Röhmel[1], Bernd Streitberg, Christian Tismer[2]
[1] Bundesgesundheitsamt, Institut für Arzneimittel, Seestraße 10, 13353 Berlin, Germany
[2] AFB-Parexel, Europacenter, 10789 Berlin, Germany

Key Words:
exact distribution, CFA, hypergeometric distribution, incidence algebra, posets, binomial coefficients, contingency table.

Abstract
Configural frequency analysis (CFA) has been developed as a means to detect cells in a contingency table which deviate from the independence assumption. It can also be considered as a generalization of Fisher's exact test for 2^k-contingency tables. The exact permutation distribution of CFA can be derived from the investigation of algebraic properties of certain matrices, connected with the hypergeometric distribution. This leads to an algorithm of linear space and quadratic time complexity. The close connection of these hypergeometric matrices to the incidence algebra over the subset lattice allows an elegant proof of the main theorems. APL, TURBO PASCAL and MAPLE programs are given.

Introduction

Configural Frequency Analysis (CFA) was developed by Lienert (1969) and Krauth & Lienert (1973). In the majority of cases CFA has been applied and still is being applied on data gathered in medical or psychological research. It may for instance be used to investigate in a multivariate setting whether a certain symptom or symptom pattern occurs more frequently in the observed data than chance alone could explain. Cells in the contingency table with a higher frequency than expected are called types. One could of course search in a contingency table for anti-types, i.e. for cells with a frequency lower than expected. Up to 1984, only asymptotic approximations of the frequency distribution under the assumption of complete independence were known for 3- (or higher) dimensional tables. The correct asymptotic distribution was derived by Lehmacher (1982). Lindner (1984) published the first algorithm for the computation of exact distribution of 3-dimensional tables and gave an outlook, how this algorithm might be generalized to calculate the distribution for 4-dimensional tables. The time complexity of this algorithm, however, depends on the dimension of the table. Therefore, computations become soon prohibitive for higher dimensional tables.

Here an algorithm is presented whose time complexity does not increase with dimension, but only depends on the smallest marginal frequency. We believe that both, the results and the method of derivation are of interest, but possibly to different audiences. Therefore, the main theorems and their proofs are given in different sections of this paper. The algorithm derived here was a present for Lienert's 65th birthday. It should have been published soon afterwards, but due to numerical instabilities of the algorithms caused by the limited precision of computers we hesitated to release the paper[16]. G. Lienert was motivating us to try to remove the numerical weakness of the algorithm. After the Colloquium held in memory of Bernd Streitberg in January 1993, we took a fresh look to the algorithm and succeeded in creating a version which is not as elegant as the former, but which now is numerically stable.

CFA is a multiple test procedure in the sense that a certain portion or all cells in the contingency table are investigated, whether their observed frequency is significantly higher (or lower) than the expected frequency under the model of complete independence. The test statistic T_k, then, is the observed frequency of a given cell in the table. The task is to compute the exact distribution of T_k under the model of complete independence of the considered variables. Depending on the primary research interest one has to distinguish between the confirmatory CFA where an a priori defined cell (or a certain portion of cells) is tested at a given multiple level α and the exploratory CFA where all cells are tested without making any adjustment for multiple testing. We will not discuss the problem connected with multiple tests where some powerful methods have been developed nor the adequacy of the complete independence assumption (cf. Sonnemann (1982); Perli et al. (1987); Victor (1989)), but concentrate on the individual test for an arbitrary but fixed cell.

[16] Our friend, Bernd Streitberg, has died meanwhile (29.9.1990) and we miss his advice and friendship very much. The longer we are not able to ask him the more it becomes clear to us, how important Bernd had been in our lifes.

A Permutation Approach

The k-dimensional contingency table originates from a data matrix

$$X = (x_{ij}), \qquad i = 1, ..., n; \; j = 1, ..., k$$

of size (n, k), giving the observations of k variables on n units. Let X be the set of all data matrices of size (n, k). In the permutation approach, the usual stochastic assumptions of independence are represented by the action of a group G on the sample space X. Here, G is the k-fold direct product

$$G = (S_n)^k = S_n \cdot S_n \cdots S_n$$

of k copies of the symmetric group S_n of all permutations σ on the set $\{1, 2, ..., n\}$. The (right) action of G on X is defined by the set of all k-tupels $g = (\sigma_1, \sigma_2, ..., \sigma_k) \in G$ (one for each column of X), which permute the columns of X independently. If f is the density of X, the assumption of complete independence (under H_0) implies that

$$f(X) = f(X^g) \quad \forall g \in G \; . \tag{1}$$

It follows that all (columnwise permuted) matrices X^g in the orbit of the observed matrix X are equiprobal. Note, that this property is weaker than the usual complete independence assumption. It differs however from the pointwise independence hypotheses investigated by Perli et al. (1987).

A configuration c in the contingency table is defined by a k-tuple $c = (c_1, c_2, ..., c_k)$ of values, one for each variable. The observed frequency $T_k = T_k(X)$ of this configuration is the number of rows in X which coincide with c. The exact permutation distribution of T_k can be obtained by counting the rows of X^g, which coincide with c for each of the $(n!)^k$ permuted X^g, $g \in G$. If we use in addition (1) we can write:

$$prob(T_k = s) = \left(\frac{1}{n!}\right)^k \cdot \sum_{g \in G} \{T_k(X^g) = s\} \; ; \; s = 0, 1, 2, ... \tag{2}$$

(Here we use the notation that for a logical expression ... the symbol {...} counts as 1, if the expression is true and as 0, if it is false.)

The task, of course, is to calculate $p_s = p_s^{(k)} = prob(T_k = s)$ without generating all $(n!)^k$ matrices X^g. This has been achieved by Lindner (1984), whose methods proceeds by induction in the number of columns. He derives a formula for the conditional probability $P(T_j = s | T_{j-1} = t)$. The transition from T_{j-1} to T_j is given by

$$p_s^{(j)} = prob(T_j = s) = \sum_t prob(T_j = s | T_{j-1} = t) \cdot p_t^{(j-1)} \tag{3}$$

where the sum extends over all t with non vanishing $p_t^{(j-1)}$. We interpret this formula as a matrix product and study the algebra of the corresponding matrices. The full significance of this algebra will only become clear if certain combinatorial (lattice) constructions are introduced in the latter part of this paper. We believe, however, that the following results can be transparent to those readers also, who do not want to be deeply involved in combinatorial details.

Hypergeometric Matrices

For $u = 0, 1, 2, ..., n$ we define matrices $H(u)$, called hypergeometric matrices, by $H(u) = (h_{s,t}(u))$; $s, t = 0, ..., n$ and

$$h_{s,t} = h_{s,t}(u) = \frac{b_{s,t} \cdot b_{u-s,n-t}}{b_{u,n}} \quad . \tag{4}$$

Here, B is the usual binomial matrix:

$$B = (b_{s,t}) = \frac{t!}{s!(t-s)!} \; ; \; s, t = 0, 1, ..., n \quad . \tag{5}$$

$H(u)$ is the transition matrix of the following experiment: an urn contains n balls, some of which are marked with a certain color. From this urn, u balls are drawn randomly and without replacement, marked with a second color and put back into the urn. Then $h_{s,t}(u)$ is the conditional probability to have s balls marked with both colors, if initially t balls were marked. We will derive a formula for the probability to have s balls marked with all colors, after performing the experiment k times with k different colors and $u = u_1, u_2, ..., u_k$ on an urn with n balls, where initially all balls are marked. We call this the 'iterated hypergeometric' distribution.

Lindner (1984), has shown that the exact conditional probability for obtaining a certain frequency for configuration $c = (c_1, c_2, ..., c_j)$ of the contingency table (based on the first j variables), given that frequency t was observed for configuration $(c_1, c_2, ..., c_{j-1})$ of the contingency table (based on the first $j-1$ variables) is equal to $h_{s,t}(u_j)$, where u_j is the (marginal) frequency of c_j in the jth column of X. Hence, (3) can be rewritten as

$$p_s^{(j)} = \sum_t h_{s,t}(u_j) \cdot p_t^{(j-1)} \; , \tag{6}$$

or in matrix notation:

$$p^{(j)} = H(u_j) \cdot p^{(j-1)} \quad . \tag{6a}$$

The final permutation distribution of T_k depends only on n, the number of rows in X, and on the marginal frequencies $u_1, u_2, ..., u_k$ of $c_1, c_2, ..., c_k$ in the respective columns of X:

$$p^{(k)} = H(u_k) \cdot H(u_{k-1}) \cdots H(u_2) \cdot H(u_1) \cdot p^{(0)} \tag{7}$$

where $p^{(0)}$ is defined by $p_n^{(0)} = 1$ and $p_t^{(0)} = 0$ for $t < n$.

This is a compact representation of Lindner's algorithm. From this representation immediately follows that the time complexity is $O(u^{2(k-2)})$, where $u = min(u_1, u_2, ..., u_k)$, if the complexity of the computation of the binomial coefficients is considered to be negligible (which isn't). $k-2$ instead of k in the complexity estimation is correct, because $H(u_1) \cdot p^{(0)}$ is a unit vector $e(u_1)$ with the 1 in position u_1, and therefore, $H(u_2) \cdot H(u_1) \cdot p^{(0)}$ can be considered to be a given vector. We will use (7) as the definition of the 'iterated hypergeometric' distribution:

Definition: The iterated hypergeometric distribution with size parameters n and marginal parameters $(u_1, ..., u_k)$ has support $s = 0, 1, ..., u$ with $u = min(u_1, u_2, ..., u_k)$, and the probability density function

$$p^{(k)} = H(u_k) \cdot H(u_{k-1}) \cdots H(u_2) \cdot e(u_1). \tag{8}$$

We will present an algorithm of complexity $O(u^2)$, including the computation of the binomial coefficients. This algorithm has been found by experimentation with concrete matrices $H(u)$ calling to help APL as a mathematical laboratory. We urge the reader to re-do these experiments, e.g. with $H(u)$ matrices for $n = 4$:

$$H(0) = \begin{pmatrix} 1 & 1 & 1 & 1 & 1 \\ 0 & 0 & 0 & 0 & 0 \\ 0 & 0 & 0 & 0 & 0 \\ 0 & 0 & 0 & 0 & 0 \\ 0 & 0 & 0 & 0 & 0 \end{pmatrix} \qquad H(1) = 1/4 \cdot \begin{pmatrix} 4 & 3 & 2 & 1 & 0 \\ 0 & 1 & 2 & 3 & 4 \\ 0 & 0 & 0 & 0 & 0 \\ 0 & 0 & 0 & 0 & 0 \\ 0 & 0 & 0 & 0 & 0 \end{pmatrix}$$

$$H(2) = 1/6 \cdot \begin{pmatrix} 6 & 3 & 1 & 0 & 0 \\ 0 & 3 & 4 & 3 & 0 \\ 0 & 0 & 1 & 3 & 6 \\ 0 & 0 & 0 & 0 & 0 \\ 0 & 0 & 0 & 0 & 0 \end{pmatrix} \qquad H(3) = 1/4 \cdot \begin{pmatrix} 4 & 1 & 0 & 0 & 0 \\ 0 & 3 & 2 & 0 & 0 \\ 0 & 0 & 2 & 3 & 0 \\ 0 & 0 & 0 & 1 & 4 \\ 0 & 0 & 0 & 0 & 0 \end{pmatrix}$$

$$H(4) = \begin{pmatrix} 1 & 0 & 0 & 0 & 0 \\ 0 & 1 & 0 & 0 & 0 \\ 0 & 0 & 1 & 0 & 0 \\ 0 & 0 & 0 & 1 & 0 \\ 0 & 0 & 0 & 0 & 1 \end{pmatrix}$$

The Algebra of Hypergeometric Matrices: Statement of the Theorems

Theorem 1: The hypergeometric matrices commute:

$$H(u) \cdot H(v) = H(v) \cdot H(u). \tag{9}$$

Theorem 2: The product of two hypergeometric matrices is a linear combination of hypergeometric matrices:

$$H(u) \cdot H(v) = \sum_{w=0}^{n} \phi_{u,v}^{w} \cdot H(w), \tag{10}$$

with

$$\phi_{u,v}^{w} = \frac{b_{u-w,n-w} \cdot b_{v-w,n-u} \cdot b_{w,n}}{b_{u,n} \cdot b_{v,n}} \quad . \tag{11}$$

The properties stated in Theorem 1 and 2 are usually required to be satisfied in a so called 'algebra'. Theorem 1 reflects the fact that the order of the k variables in the contingency table is of no importance. Theorem 2 is not used in the following derivation of the algorithm, but is only mentioned because of it's theoretical importance. What follows from Theorem 1 is the fact that there exists a matrix, which simultaneously diagonalizes all hypergeometric matrices for a given n. This matrix is found to be the binomial matrix $B = (b_{s,t})$. The inverse of B is given by

$$B^{-1} = ((-1)^{s+t} \cdot b_{s,t}); \; s,t = 0,1,...,n \quad . \tag{12}$$

Theorem 3: The diagonal form of $H(u)$ is given by

$$B \cdot H(u) \cdot B^{-1} = D(u), \tag{13}$$

where

$$D(u) = diag\left(\frac{b_{u-s,n-s}}{b_{u,n}}\right); \; s = 0,1,...,n \quad . \tag{14}$$

$D(u)$ is a diagonal matrix with the nonvanishing elements in the main diagonal.

Proofs of the Theorems can be found in appendix I.

A CFA Algorithm

By application of Theorem 3, formula (8) can be transformed into

$$p^{(k)} = B^{-1} \cdot D(u_k) \cdot D(u_{k-1}) \cdots D(u_2) \cdot B \cdot e(u_1), \tag{15}$$

$$p^{(k)} = B^{-1} \cdot q, \tag{15a}$$

where

$$q = D(u_k) \cdot D(u_{k-1}) \cdots D(u_2) \cdot B \cdot e(u_1) \quad . \tag{16}$$

Having in mind that $e(u_1)$ is a unit vector, B is the binomial matrix and $D(u_k) \cdot D(u_{k-1}) \cdots D(u_2)$ is also a diagonal matrix, the vector q can easily be calculated. This leads immediately to closed formulas for $p_s = p_s^{(k)}$ either in matrix form (15a) or equivalently more detailed for the single probabilities:

$$p_s^{(k)} = \sum_{i=s}^{u} (-1)^{s+i} \cdot b_{s,i} \cdot q_i, \tag{17}$$

with

$$q_i = b_{i,u} \cdot \prod_{t=2}^{k} \frac{b_{u_t - i, n-i}}{b_{u_t, n}} \quad . \tag{18}$$

Here we have, using Theorem 1, recorded the marginal frequencies such that $u = u_1 \leq u_2 \leq \cdots \leq u_k$. The following recursive algorithm is based on (17) and (18) and allows the simultaneous computation of the q_i and the columns c_i of B^{-1}:

(A) Initialization

$$q_0 = 1, \;\; c_0 = (1), \;\; \text{(a vector of length 1)}, \; p_0 = (1) \tag{19}$$

(B) Recursion for i=0, ..., u-1

$$q_{i+1} = q_i \cdot \frac{u-i}{i+1} \cdot \frac{u_2 - i}{n-i} \cdots \frac{u_k - i}{n-i} \tag{20}$$

$$c_{i+1} = (0, c_i) - (c_i, 0) \tag{21}$$

$$p_{i+1} = (p_i, 0) + q_{i+1} \cdot c_{i+1} \tag{22}$$

(C) Result

$$p^{(k)} = p_u$$

It is easy to see from (19) that the q_i can be calculated in the recursive way indicated in (21). If we replaced the '-' in (22) by a '+', we would have the usual recursive formula for the calculation of the binomial coefficients. We notice that the binomial matrix is identical with it's inverse apart from the fact that the sign is reversed for all cells with exactly one odd index. Hence, (21) gives the recursive formula for calculating the inverse of the binomial matrix. (22) is another way to express that $p^{(k)}$ is the sum of the columns of B^{-1} multiplied with the coefficients q_i. However, the vectors p_i $(1 < i < u)$

which are the results of the recursion steps do not form a probability density nor need their components be positive. This makes it difficult to control the intermediate calculations for rounding errors which occur due to the limited precision of computers. Our observation is that the components of the vectors p_i and $q_{i+1} \cdot c_{i+1}$ may become very large and may coincide in all significant digits apart from the sign. The subsequent addition in (22) then may cancel the informative part of the entries and may leave an unpredictable error term which will subsequently be used in all further calculations. This biases the results to an unacceptable degree. Hence, we looked for alternatives to circumvent the cancelation problem. However, we will use formulas (19) to (22) for estimating the time complexity of the algorithm.

(20) consists of $(2+2\cdot k)$ multiplications, (21) has $(i+1)$ additions, (22) has $(i+1)$ multiplications and $(i+1)$ additions. This makes $(i+3+2\cdot k)$ multiplications and $(2+2\cdot i)$ additions in the ith step. Thus, in total we have $0.5\cdot u\cdot(u+1)+3\cdot u+2\cdot k\cdot u$ multiplications and $2\cdot u+u\cdot(u+1)$ additions. Hence, we have established that the time complexity of the algorithm is $O(u\cdot(k+u))$. Usually, k is small compared to u, which shows that the time complexity is $O(u^2)$.

To establish an alternative algorithm, we take (15a) in the form

$$B \cdot p^{(k)} = q. \tag{23}$$

The explicite formula (18) for the components q_i of q can also be expressed as

$$q_i = \frac{\prod_{t=1}^{k} b_{i,u_t}}{\left(b_{i,n}\right)^{k-1}} \quad . \tag{24}$$

For better reading, we replace $p_i^{(k)}$ by p_i, and therefore (24) reads as

$$\sum_{i=0}^{u} b_{s,i} \cdot p_i = \frac{\prod_{t=1}^{k} b_{s,u_t}}{\left(b_{s,n}\right)^{k-1}} \;;\; s = 0, ..., u \quad . \tag{25}$$

The triangular form of the binomial matrix B suggests to compute the p_i in the reverse order. Let us therefore substitute the p_i's by the a_i's which are related with each other by the equations

$$p_{u-i} = (i!) \cdot q_u \cdot b_{i,u} \cdot a_i \quad .$$

Once the a_i's are known, the p_{u-i} can easily be calculated. The defining equations for the a_i are as follows, as can be derived from (25):

$$\sum_{i=0}^{s} \frac{a_i}{(s-i)!} = \prod_{i=0}^{s} Q_i; \quad s = 0, ..., u \tag{26}$$

where Q_i is defined by

$$Q_i = \frac{(n-u+i)^{k-1}}{\prod_{j=1}^{k} u_j - u + i}, \quad i > 0; \; Q_0 = 1$$

(26) has the advantage that only one additional term of the a_i (and thus of the p_i) is involved in the calculations when s increases from $s = 0$ to $s = u$. The a_i have to be positive. Hence, we can stop, if $a_i < 0$ occurs the first time due to the limited precision of the computer. However, this algorithm does not resolve the general problem caused by the precision limits of available computers. The best advice we can give at present is to use a software which is able to deal with long integers, and to reformulate the algorithm in a way that it only uses integers instead of ratios in their decimal representation. We recommend to use our algorithms only, if the total number of observations is below $N = 200$, and to switch to Lehmacher's approximation in all other cases. This is in line with the fact that exact algorithms are of particular value for small numbers of observations, while for larger numbers the approximations are quite satisfactory.

The programs will be found in Appendix II. What remains is to prove the Theorems (in Appendix I). In principle, these proofs could be performed by tedious manipulations with binomial coefficients. The following method of derivation, however, is more general and insightful. The method relies on the modern concept of finite partially ordered sets, cf. Aigner (1975), as a brilliant textbook, and the original paper by Rota (1964).

Example

The data for CFA are given in a 5 dimensional 2^5- contingency table. The example is discussed by Lienert (1978 Volume II, page 791). Five dichotomous variables out of 50 items of the Hamburg depression scale, called Q (qualvolles Erleben), G (Grübelsucht), A (Arbeitsunfähigkeit), N (Nicht-Aufstehen-Wollen) und D (Denkstörung), i.e. five symptoms of affective disorder, are observed in 150 depressed out-patients. The question was whether these 5 items reflect 5 independent aspects of the patient's experience of grief. The observed frequencies are as follows:

The observed frequency in cell + + + + + is 12, the marginal frequencies are 46, 63, 70, 76, and 111, the total number of observations is 150. Hence, the discrete conditional probability distribution ranges from 0 to 46.

As one can see from below, the differences between the exact distribution and the approximative version of Lehmacher are not very large. These

QGAND	f	QGAND	f	QGAND	f	QGAND	f
+ + + + +	12	+ - + + +	0	- + + + +	7	- - + + +	2
+ + + + -	4	+ - + + -	2	- + + + -	4	- - + + -	1
+ + + - +	7	+ - + - +	1	- + + - +	11	- - + - +	2
+ + + - -	1	+ - + - -	0	- + + - -	7	- - + - -	9
+ + - + +	7	+ - - + +	0	- + - + +	7	- - - + +	0
+ + - + -	2	+ - - + -	2	- + - + -	8	- - - + -	5
+ + - - +	7	+ - - - +	0	- + - - +	9	- - - - +	4
+ + - - -	1	+ - - - -	0	- + - - -	17	- - - - -	11

differences, however, may become important, if the CFA is used in a confirmative way, where all cells are tested, and where an appropriate adjustment for multiple testing has to take place.

In the table below, probabilities for $X = 18$ to $X = 46$ have been omitted because they are below $0.5 \cdot 10^{-10}$ in both distributions.

X	exact	Lehmacher's approximation	X	exact	Lehmacher's approximation
0	0.0249481967	0.0335114240	9	0.0027189676	0.0010719480
1	0.1011444530	0.0895879124	10	0.0006551351	0.0001322718
2	0.1943590785	0.1702025976	11	0.0001346027	0.0000115774
3	0.2357961490	0.2298473531	12	0.0000237433	0.0000007184
4	0.2029752961	0.2206559682	13	0.0000036154	0.0000000315
5	0.1320952800	0.1505873062	14	0.0000004773	0.0000000010
6	0.0676257866	0.0730464242	15	0.0000000548	0.0000000000
7	0.0279799000	0.0251790804	16	0.0000000055	0.0000000000
8	0.0095392578	0.0061653858	17	0.0000000005	0.0000000000

Table 1: A one-sided p-value is given by Prob $(X \geq 12)$:
$p_{exact} = 0.0000278968$ $p_{approx} = 0.0000007509$

References

Aigner, M. (1975) Kombinatorik I, Springer Verlag,

Bortz, J., Lienert, G.A. & Boehnke, K. (1990) Verteilungsfreie Methoden in der Biostatistik. Springer Verlag, Berlin, Heidelberg, New York.

Krauth, J. & Lienert, G.A. (1973) Die Konfigurationsfrequenzanalyse und ihre Anwendung in Psychologie und Medizin. Freiburg, Alber Verlag.

Lehmacher, W. (1981) A More Powerful Simultaneous Test Procedure in Configural Frequency Analysis. Biometrical Journal, Vol 23, 429- 436.

Lienert G.A. (1969) Die 'Konfigurationsfrequenzanalyse' als Klassifikationsmittel in der klinischen Psychologie. In: Irle, M. (Hrsg.) Bericht über den 26. Kongress der Deutschen Gesellschaft für Psychologie in Tübingen 1968, 244- 253, Göttingen, Hogrefe Verlag.

Lienert, G.A. (1978) Verteilungsfreie Methoden in der Biostatistik, 2. völlig überarbeitete Auflage, Band II, Verlag Anton Hain, Meisenheim am Glan.

Lindner, K. (1984) Eine exakte Auswertemethode zur Konfigurationsfrequenzanalyse, Psychologische Beiträge, Band 26, 393- 451.

Perli, H.G., Hommel, G. & Lehmacher, W. (1987) Test Procedures in Configural Frequency Analysis (CFA) Controlling the Local and the Multiple Level. Biometrical Journal 29, 255 - 269

Rota, G.C. (1964) On the Foundation of Combinatorial Theory I: Theory of the Möbius Functions. Zeitschrift für Wahrscheinlichkeitsrechnung 2.

Sonnemann, E. (1982) Allgemeine Lösungen multipler Testprobleme. EDV in Medizin und Biologie 13, 120- 128.

Victor, N. (1989) An Alternative Approach to Configural Frequency Analysis. Methodika Vol III, 61- 73.

APPENDIX I: Proof of the Theorems

The Incidence Algebra

Let $(\Omega, \leq)$ be a finite partially ordered set. The incidence algebra $J(\Omega)$ is the set of all matrices $A = \{a_{S,T} | S, T \in \Omega\}, a_{S,T} \in R$, with $a_{S,T}$ unequal to 0 only if $S \leq T$. Note that $J(\Omega)$ indeed is an algebra (with identity), since the axioms:

(i)	$A, B \in J(\Omega)$	implies $A + B \in J(\Omega)$
(ii)	$A \in J(\Omega)$ and $\alpha \in R$	implies $\alpha \cdot A \in J(\Omega)$,
(iii)	$A, B \in J(\Omega)$	implies $A \cdot B \in J(\Omega)$
(iv)	$I \in J(\Omega)$,	where I is the unit matrix

can be verified immediately.

A basic example is the incidence algebra of $\Omega = 2^{[n]}$ = set of all subsets of $[n] = \{1, 2, ..., n\}$, ordered by set inclusion. The zeta matrix $Z \in J(\Omega)$ is, as usual, the matrix of the order relation, i.e. $Z_{S,T} = 1$ iff S is contained in T $(S \leq T)$. For $n = 2$, we have the Hasse diagram and the zeta matrix

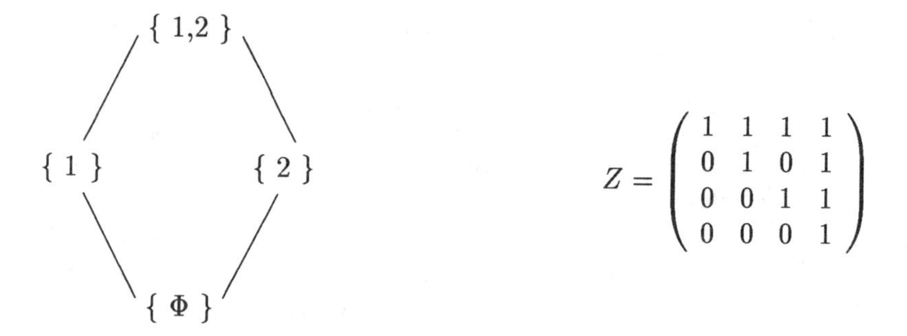

where rows and colums of Z are indexed by $\Phi, \{1\}, \{2\}, \{1,2\}$. All matrices in $J(\Omega)$ are obtained by replacing the 1's in Z by arbitrary real numbers. We write $z_{S,T} = \{S \leq T\}$, again using the convention that for a logical expression ... the symbol $\{...\}$ counts as 1 if the expression is true, and is 0 if the expression is false.

Let Ω admit a rank function, i.e. a function $r : \Omega \rightarrow N$ with the following properties

$$\begin{array}{ll} (1) & r(S) = 0 \text{ iff } S \text{ is minimal in } \Omega \\ (2) & S < \cdot T \text{ implies } r(T) = 1 + r(S) \end{array}$$

where $S < \cdot T$ is the covering relation $S < \cdot T$ if $S < T$, and no U exists with $S < U < T$. The rank function in $2^{[n]}$ is given by $r(S) = |S|$, the number of elements in S.

The Rank Algebra

Binomial identies are often easily deduced by using simple matrix operations in the incidence algebra. The basic reason for this is that the binomial coefficient $b_{s,n}$ is obviously equal to the number of subsets in $[n]$ in the rank level set $\Omega(s) = \{S | r(S) = s\}$. Matrices of binomial coefficients are obtained by considering 'small' versions of certain matrices in the incidence algebra over $2^{[n]}$. An example is the binomial matrix $B = (b_{s,t})$, which is obtained from Z by first summing all those rows in Z which belong to the same rank level set $\Omega(s)$, and then selecting in the result an arbitrary column from each rank level set $\Omega(t)$:

$$b_{s,t} = \sum_{S:r(S)=s} z_{S,T} \text{ , for an arbitrary } T \text{ with } r(T) = t \ . \tag{27}$$

The inverse $B^{-1} = ((-1)^{s+t} \cdot b_{s,t})$ is obtained in the same way from the Möbius matrix Z^{-1}. An analogous construction is possible for all matrices

$A \in J(\Omega)$ with the rank preserving property. Here, a matrix A is called rank preserving if

$$\sum_{S:r(S)=s} (a_{S,T} - a_{S,U})) = 0 \quad \forall\, s \text{ and } T, U \text{ with } r(T) = r(U) \quad . \tag{28}$$

For rank preserving matrices A, the following diminution map is well defined

$$\delta(A) = (\delta(A)_{s,t}) = \sum_{S:r(S)=s} a_{S,T} \quad , \tag{29}$$

where T is an arbitrary element with $r(T) = t$. $\delta(A)$ is a $(n+1, n+1)$-matrix, where n is the maximal rank of an element in Ω. As shown above, we have

Lemma 1: $\delta(Z) = B$.

Our purpose is the derivation of identities between matrices of binomial coefficients, which can be shown to be images under δ. As it is often true in mathematics, the larger tasks can more easily be solved than smaller ones, we will derive these identities by simple computation with the preimages. In order to do this, an algebraist would ask for two natural properties, which are indeed true.

Lemma 2: The set $R(\Omega)$ of all rank preserving matrices in $J(\Omega)$ is an algebra with identity. We call $R(\Omega)$ the rank algebra.

Proof: Axioms (i), (ii) and (iv) are immediate. For proving (iii), let $A, B \in R(\Omega)$ be rank preserving. Then for T with $r(T) = t$ we have

$$\begin{aligned}\sum_{S:r(S)=s} \sum_{U} a_{S,U} \cdot b_{U,T} &= \sum_{S:r(S)=s} \sum_{u=0} \sum_{U:r(U)=u} a_{S,U} \cdot b_{U,T} \\ &= \sum_{u=0} \sum_{U:r(U)=u} \left(\sum_{S:r(S)=s} a_{S,U} \right) \cdot b_{U,T} \quad .\end{aligned}$$

The inner sum is constant in U with $r(U) = u$. Since B is rank preserving, therefore, the total expression is also constant in T with $r(T) = t$, i.e. $A \cdot B$ is rank preserving.

Lemma 3: The diminuition is an identity-preserving algebra homomorphism from $R(\Omega)$ into the algebra $R(n)$ of all upper triangular (n+1,n+1)-matrices:

$$\begin{array}{llcl} (i) & \delta(A+B) & = & \delta(A)+\delta(B) \\ (ii) & \delta(\alpha\cdot A) & = & \alpha\cdot\delta(A) \\ (iii) & \delta(A\cdot B) & = & \delta(A)\cdot\delta(B) \\ (iv) & \delta(I) & = & I,\text{ for the corresponding unit matrices.} \end{array}$$

Proof: Again, only (iii) needs proof. This, however, is straight forward and uses the same arguments as in the proof of Lemma 2.

Application to Hypergeometric Matrices

We work in $\Omega = 2^{[n]}$, where we have $r(T) = |T|$. We introduce 'large' versions $K(a)$ of the hypergeometric matrices $H(a)$ by defining an intersection matrix $K(A) \in J(\Omega)$ by

$$\begin{array}{lcl} K(A) & := & (k_{S,T}(A)|S,T \in 2^{[n]}) := (\{S = T \cap A\}) \text{ and set} \\ K(a) & := & \sum\limits_{A:|A|=a} K(A). \end{array}$$

Example for n = 2:

$$K(\{1\}) = \begin{pmatrix} 1 & 0 & 1 & 0 \\ 0 & 1 & 0 & 1 \\ 0 & 0 & 0 & 0 \\ 0 & 0 & 0 & 0 \end{pmatrix}, K(1) = \begin{pmatrix} 2 & 1 & 1 & 0 \\ 0 & 1 & 1 & 2 \\ 0 & 0 & 0 & 0 \\ 0 & 0 & 0 & 0 \end{pmatrix}, H(1) = \begin{pmatrix} 2 & 1 & 0 \\ 0 & 1 & 2 \\ 0 & 0 & 0 \end{pmatrix},$$

(the K-matrices are indexed by Φ, {1}, {2}, {1,2}).

Lemma 4: $K(a)$ is rank preserving, and $\delta(K(a)) = b_{a,n} \cdot H(a)$.

Proof: Select T with $|T| = t$. Then

$$\begin{array}{rcl} (\delta(K(a)))_{S,T} & = & \sum\limits_{S:|S|=s} \sum\limits_{A:|A|=a} \{S = T \cap A\} \\ & = & \sum\limits_{A:|A|=a} \sum\limits_{S:|S|=s} \{S = T \cap A\} \\ & = & \sum\limits_{A:|A|=a} \{s = |T \cap A|\} \end{array}$$

All sets A with $|A| = a$ for which $s = |S \cap A|$ can be constructed by selecting s elements from T and, independently, $a - s$ elements from $[n]\backslash T$.

The number of these sets does depend on T only via its rank function $t = |T|$ and is equal to $b_{s,t} \cdot b_{n-t,a-s}$. Hence we have

$$\delta(K(a)) = b_{a,n} \cdot \frac{b_{s,t} \cdot b_{n-t,a-s}}{b_{a,n}} = b_{a,n} \cdot H(a).$$

Lemma 5: The intersection matrices provide a matrix representation of the intersection operation in $J(\Omega)$: $K(V) \cdot K(W) = K(V \cap W)$ and the matrices $K(u)$ are the basis of a commutative linear associative algebra

$$\begin{aligned} K(u) \cdot K(v) &= \sum_w f^w_{u,v} \cdot K(w) \text{ with structure coefficients} \\ f^w_{u,v} &= b_{u-v,n-w} \cdot b_{v-w,n-u} \end{aligned}$$

Proof:

$$\begin{aligned} (K(U) \cdot K(V))_{S,T} &= \sum_W k_{S,W}(U) \cdot k_{W,T}(V) \\ &= \sum_W \{S = W \cap U\} \cdot \{W = T \cap V\} \\ &= \{S = T \cap U \cap V\} \\ &= k_{S,T}(U \cap V). \end{aligned}$$

Therefore,

$$\begin{aligned} K(u) \cdot K(v) &= \sum_{U:|U|=u} \sum_{V:|V|=v} K(U \cap V) \\ &= \sum_W (\sum_{U:|U|=u} \sum_{V:|V|=v} \{W = U \cap V\}) \cdot K(W) \end{aligned}$$

All sets U, V with $|U| = u$ and $|V| = v$ and a given intersection W can be contructed by first selecting $u - |W|$ elements from $[n] \backslash W$ to give $\frac{U}{W}$ and then selecting $v - |W|$ elements from $[n] \backslash U$ to give $V \backslash W$. Hence it follows that

$$\begin{aligned} K(u) \cdot K(v) &= \sum_W (b_{u-|W|,n-|W|} \cdot b_{v-|W|,n-u}) \cdot K(W) \\ &= \sum_w b_{u-w,n-w} \cdot b_{v-w,n-u} \cdot K(w). \end{aligned}$$

The last formula also shows that the matrices $K(u)$ and $K(v)$ commute

$$b_{u-w,n-w} \cdot b_{v-w,n-u} = \frac{n - w!}{(u - w!) \cdot (v - w!) \cdot (n + w - u - v!)}.$$

Theorem 1 is an immediate consequence of this. Theorem 2 follows from lemma 5 together with $H(u) = \frac{\delta(K(u))}{b_{u,n}}$ and lemma 3(iii). Theorem 3 will be proven by

Lemma 6: The zeta matrix Z simultaneously diagonalizes all intersection matrices $K(U), U \in 2^{[n]}$:

$$Z \cdot K(U) = D(U) \cdot Z,$$

where $D(U) = diag(\{S \leq U | S \in 2^{[n]}\})$

$$Z \cdot K(u) = D(u) \cdot Z,$$

where $D(u) = diag(\{b_{u-|S|,n-|S|} | S \in 2^{[n]}\})$

Proof:

$$\begin{aligned} \sum_S z_{R,S} \cdot k_{S,T}(U) &= \sum_S \{R \leq S\} \cdot \{S = T \cap U\} \\ &= \{R \leq T \cap U\} \\ &= \{R \leq U\} \cdot \{R \leq T\} \\ &= D(U)_{R,R} \cdot z_{R,T}. \end{aligned}$$

Next,

$$\begin{aligned} Z \cdot K(u) &= Z \cdot \sum_{U:|U|=u} K(U) = \sum_{U:|U|=u} D(U) \cdot Z \\ &= D(u) \cdot Z \quad \text{, by the definition of } D(u). \end{aligned}$$

APPENDIX II: Programs

APL Programs

```
    ∇P←N CFAexact1 A;C;Q;I;U;K
[1]  ⍝The left argument N is the total number of observations.
[2]  ⍝The right argument A is the vector of the marginal frequencies
[3]  ⍝for a cell in the contingency table. If no lef t argument is
[4]  ⍝supplied, the program assumes that the first entry in A is N
[5]  ⍝(the total number of observations), and the remaining values
[6]  ⍝in A correspond to the marginal frequencies.
[7]  ⍝The result P is the discrete probability density.
[8]  ⍝Program CELLMARGIN can be used to create the right argument
[9]  ⍝A in the form: N, marginal frequencies
[10]
```

```
[11]  ⍎(2≠ ⎕NC'N')/'N←A[1] ◇ A←1↓A'
[12]  C←P←,Q←2+I←¯1
[13]  U←1↑A←A[⍋A]
[14]  K←ρA←1↓A
[15]  LOOP:→(U≤I←I+1)/0
[16]  Q←Q×( (U-I)÷I+1)×(×/A-I)÷(N-I)*K
[17]  C←(0,C)-C,0
[18]  P←(P,0)+C×Q
[19]    →LOOP
      ∇

      ∇ P←N CFAexact2 A;I;ALPHA;a;b;K;NULL;COUNT
[1]   ⍝the same syntax as for CFAexact1
[11]  ⍎(2≠ ⎕NC'N')/'N←A[1] ◇ A←1↓A'
[12] NULL←a←P←1+I←0 ◇ COUNT← ⍳0
[13] N←N-ALPHA←⌊/A
[14] A←A-ALPHA
[15] K←¯1 + ρ A
[16] LOOP:→(ALPHA<I←I+l)/END
[17] a←a×((N+I)*K)÷× /A+I
[18] P←P÷ COUNT←I , COUNT
[19] →(∨/0≥NULL,b←a-+/P)/ZERO
[20] P←P,b ◇ →LOOP
[21] ZERO:P←P,NULL←0 ◇ →LOOP
[22] END: P← ⌽÷ a
      ∇

      ∇ Z←C CELLMARGIN T;I;N;M
[1] ⍝C is the index set of a cell with entries from 1 or 2
[2] ⍝T is a (2 × 2 × ··· × 2) - table
[3] ⍝Z is a vector with N , the total number of observations
[4] ⍝as the first entry, followed by the marginal frequencies
[5] ⍝which correspond to cell with index C
[6] Z←+/,T ◇ C←l+C=2
[7] I←0 ◇ N←ρρT ◇ M←2*N-1
[8] LOOP:→(N<I←I+l)/0
[9] Z←Z,+/((2,M)ρT)[C[I];]
[10] T←(1⌽⍳⍴⍉T
[11] →LOOP
      ∇
```

PASCAL Programs

```
PROGRAM CFAdemo (input, output) ;
IMPORT PasDos (ParamStr) ;

CONST
  Kmax = 100 ;  { maximum number of marginal sums }
  Nmax = 400 ;  { maximum number of observations }
TYPE
  Marg = ARRAY[1..Kmax] OF INTEGER ;
  Prob = ARRAY[0..Nmax] OF REAL ;
  Coef = ARRAY[0..Nmax] OF REAL ; { INTEGER too small }

{ Version 1, 07.07.87 }
PROCEDURE CFA1 (
 {i}  N: INTEGER ; { total number of observations }
 {i}  K: INTEGER ; { size of ARRAY A              }
 {i}  A: Marg    ; { marginal sums A[1..K]        }
 VAR {o} L: INTEGER ; { top index of P    }
 VAR {o} P: Prob ; {the exact distribution on [0,min(A)]}
 VAR {o} R: INTEGER   { return code.
     R=0  on normal termination
     R=-1 on wrong sideconditions,
     R=-2 on arithmetic overflow } ) ;
 VAR S, SUM, MIN, MAX, I, J: INTEGER ; Q: REAL ; C: Coef ;
 LABEL 999;
BEGIN
  SUM := 0 ; FOR i:=1 TO K DO SUM := SUM+A[I] ;
  {-- find the index S of the first observation
      of nonzero probability }
  S := SUM - N * (K-1) ; IF S<0 THEN BEGIN  S := 0  END ;
  {-- if S>0, find an equivalent vector that has S=0 }
  N := N-S ; MIN := N+1 ; MAX := -1 ;
  FOR I:=1 TO K DO BEGIN
    A[I] := A[I]-S ;
    IF A[I] < MIN THEN MIN := A[I] ;
    IF A[I] > MAX THEN MAX := A[I] ;
  END ;
  {-- does the vector exist? }
  IF (N<0) OR (MIN<0) OR (MAX>N) THEN
     BEGIN R := -1 ; GOTO 999 END ;
  {-- initialize }
  C[0] := 1.0 ; Q := 1.0 ; P[0] := 1.0 ;  R := 0 ;
```

```
  {-- compute }
  FOR I:=0 TO MIN-1 DO BEGIN
    Q := Q * (A[1]-I) / (I+1) ;
    FOR J:=2 TO K DO Q := Q * (A[J]-I) / (N-I)
    C[I+1] := C[I] ;
    FOR J:=I DOWNTO 1 DO C[J] := C[J-1] - C[J]
    C[0]   := -C[0] ;
    P[I+1] := 0 ;
    FOR J:=0 TO I+1 DO P[J] := P[J] + Q * C[J]
  END ;
  {-- retransformation }
  FOR I:=MIN DOWNTO 0 DO P[I+S] := P[I] ;
  FOR I:=0 TO S-1     DO P[I] := 0.0 ;
  L := MIN + S ;
  {-- check for negative results }
  FOR I:=0 TO L DO BEGIN
      IF P[I] < 0 THEN R := -2 ;
  END ;
  999:
END ; { CFA1 }

{ Version 2, 29.05.93 }
PROCEDURE CFA2 (
  {i}  N: INTEGER ; { total number of observations }
  {i}  K: INTEGER ; { size of ARRAY A              }
  {i}  A: Marg    ; { marginal sums A[1..K]        }
  VAR {o}  L: INTEGER ; { top index of P            }
  VAR {o}  P: Prob ; { the exact distribution on [0,min(A)] }
  VAR {o}  R: INTEGER   { return code.
  R=0  on normal termination
  R=-1 on wrong sideconditions,
  R=-2 on arithmetic overflow } ) ;
  VAR S, SUM, MIN, MAX, I, J: INTEGER ; xa, xb: REAL ;
      ok : BOOLEAN ;
  LABEL 999;
  BEGIN
     SUM := 0 ; FOR i:=1 TO K DO SUM := SUM+A[I] ;
     {-- find the index S of the first observation
         of nonzero probability }
     S := SUM - N * (K-1) ; IF S<0 THEN S := 0 ;
     {-- if S>0, find an equivalent vector that has S=0 }
     N := N-S ; MIN := N+1 ; MAX := -1 ;
     FOR I:=1 TO K DO BEGIN
```

```
          A[I] := A[I]-S ;
          IF A[I] < MIN THEN MIN := A[I] ;
          IF A[I] > MAX THEN MAX := A[I] ;
      END ;
      {-- does the vector exist? }
      IF (N<0) OR (MIN<0) OR (MAX>N) THEN
         BEGIN R := -1 ; GOTO 999 END ;
      {-- initialize }
      xa := 1.0 ; P[MIN] := 1.0 ;  ok := TRUE ; R := 0 ;
      N := N - MIN ;
      FOR I:=1 TO K DO A[I] := A[I] - MIN ;
      {-- compute }
      FOR I:=1 TO MIN DO BEGIN
          xa := xa / (A[1]+I)  ;
          FOR J:=2 TO K DO xa := xa * (N+I) / (A[J]+I) ;
          FOR J:=I DOWNTO 2 DO P[MIN-I+J] := P[MIN-I+J] / J ;
          xb := xa ;
          FOR J:=I DOWNTO 1 DO xb := xb - P[MIN-J+1] ;
          IF xb < 0 THEN ok := FALSE ;
          IF NOT ok THEN xb := 0 ;
          P[MIN-I+1-1] := xb ;
      END ;
      {-- divide by xa }
      FOR I:=0 TO MIN DO P[I] := P[I] / xa ;
      {-- retransformation }
      FOR I:=MIN DOWNTO 0 DO P[I+S] := P[I] ;
      FOR I:=0 TO S-1 DO P[I] := 0.0 ;
      L := MIN + S ;
      {-- check for negative results }
      IF NOT ok THEN BEGIN  R := -2  END ;
      999:
   END ; { CFA2 }
VAR A      : Marg ;
    K      : INTEGER ; { Number of marginal sums }
    Q      : REAL ;
    C      : Coef ;    { Coefficients }
    P1, P2 : Prob ;    { Probabilities }
    L1, L2 : INTEGER ; { length of Prob-Vector }
    R1, R2 : INTEGER ; { return codes }
    N      : INTEGER ; { Observations }
    S      : INTEGER ;
    x, i : INTEGER ;
    OUT  : TEXT ;
```

```
PROCEDURE Abort (Msg: maxstring) ;
   BEGIN
      writeln(Msg) ; writeln('Program aborted') ; HALT ;
   END ;

   BEGIN {----- Data setup -----}
    N := 0 ;
    write ('Give number of observations (1..',Nmax,') ') ;
    readln(N) ;
    IF (N<1) OR (N>Nmax) THEN Abort ('Value out of range');
    {-- enter marginal sums with zero to stop }
    K := 1 ;
    x := 0 ;
    writeln ('Give marginal sums, one per line.') ;
    writeln ('A value of Zero ends the list.') ;
    REPEAT
    IF x<0 THEN
       BEGIN
       Abort('negative marginal sum impossible')
       END ;
    write ('A[',K,'] = ') ;
    x := 0 ;
    readln (x) ;
    A[K] := x ;
    K := K+1 ;
    UNTIL (x=0) OR (K>Kmax) ;
    K := K-1 ;
    IF A[K] = 0 THEN BEGIN  K := K-1  END ;
    {----- Observe the side conditions -----}
    IF K=0 THEN Abort('I expexted at least one input');
    writeln ('Running CFA1') ;
    CFA1 (N, K, A, L1, P1, R1) ;
    IF R1=-1 THEN Abort('Sideconditions not satisfied');
    IF R1=-2 THEN writeln('Arithmetic Overflow in CFA1');
    writeln ('Running CFA2') ;
    CFA2 (N, K, A, L2, P2, R2) ;
    IF R1=-1 THEN Abort('Sideconditions not satisfied');
    IF R2=-2 THEN writeln('Arithmetic Overflow in CFA2');
    IF L1<>L2 THEN Abort('L1 <> L2: Programm Error');
    {----- Print results for both versions -----}
    writeln ('Printing Results') ;
    close(output) ;
```

```
    assign(output, paramstr(1)) ;
    { assign output to file, if supplied }
    rewrite(output) ;
    FOR i:=0 TO L1 DO BEGIN
        writeln('P1(',i:3,') = ',
        P1[i], '
        P2(',i:3,') = ',
        P2[i]);
    END ;
    close(output) ;
END.
```

Appendix III

Michael Kaplan
Mathematisches Institut Technische Universität München

The algorithm given in (20)–(23) causes problems with most computer languages because of the limited precision offered by those languages. Computer Algebra languages like Maple, Mathematica, Macsyma etc. by-pass the cancellation effects in (23) by offering multiprecision integers (only limited by the size of the computer) and exact rational arithmetic.

The following Maple-program is an implementation of the cfa-algorithm presented above. The data-file contains the data of the 5-dimensional 2^5-contingency table example by Lienert, that is discussed in the text. Since all numbers in the data-file are integers, Maple uses automatically exact arithmetic. Floating point numbers are only used for the final output. Therefore one can be sure, that the printed results are precise.

Of course this program doesn't use the full power of computer algebra packages, but only the exact rational arithmetic. For other applications computer algebra programs turn out to be even more valuable tools.

Finally we should point out, that the estimation of the time complexity changes, if we use multiprecision integers, since the time for every calculation is no more fixed, but depends on the size of the numbers involved. A very simple looking computer algebra program may cause intermediate expressions well and increase calculation time considerably.

```
########################## cfa-algorithm ###################
#
#
# Read data from the file data
```

```
line:=sscanf(readline(data),`%d %d`):
n:=line[1];
k:=line[2];
U:=sscanf(readline(data),cat(`%d`$`i`=1..k));

# Initialization (20)
u:=min(op(U)); # op=Operands of ...
with(linalg):  # read in Maple-package for doing
               # Linear Algebra (array, add, augment..)
q:=array(0..u):
q[0]:=1:
c:=array(1..1,[1]):
p:=array(1..1,[1]):
null:=array(1..1,[0]):
Digits:=15:    # Set precision for floating point operations
# Recursion (21)-(23)
for i from 0 to u-1 do
 q[i+1]:=q[i]*(u-i)/(i+1)*product((U[j]-i),j=2..k)/(n-i)^(k-1):
 c:=add(augment(null,c),augment(c,null),1,-1):
 p:=add(augment(p,null),c,1,q[i+1]):
od:
# augment=join matrices together
# Output
j:=`j`:
[[j-1,'evalf(p[1,j])']$j=1..u+1]; # evalf=Evaluate
                                  # as a floating point
Sum(`Prob(X)`,`X`=12..u)=sum("[`j`][2],`j`=13..u+1);
# Sum=print pretty sum-symbol
# sum=calculate sum
# "=last evaluated expression
###############################################################
?endgroup
```

The output produced by this program on an ASCII text terminal is the following. (By use of the X Window user interface of Maple, output looking much prettier is produced).

```
[[0,.0249481966826436],[1,.101144453035155],[2,.194359078497952],
[3,.235796149034118],[4,.202975296144438],[5,.132095279980054],
[6,.0676257866490939],[7,.0279798999776058],[8,.0095392578291431],
  [9, .00271896757499038], [10, .000655135098219975],
  [11, .000134602712931590], [12, .0000237433202097744],
                          -5                            -6
```

```
[13, .361537238621015*10  ], [14, .477270631809488*10  ],
                      -7                          -8
[15, .548100059633264*10  ], [16, .549028705163231*10  ],
                      -9                          -10
[17, .480672219601712*10  ], [18, .368348647647574*10   ],
                      -11                          -12
[19, .247313862347742*10   ], [20, .145561524145536*10   ],
                      -14                           -15
[21, .751110485586444*10   ], [22, .339704726927607*10   ],
                      -16                           -18
[23, .134573257822846*10   ], [24, .466481688593928*10   ],
                      -19                           -21
[25, .141295546986602*10   ], [26, .373315965772135*10   ],
                      -23                           -24
[27, .858509852660880*10   ], [28, .171404120631033*10   ],
                      -26                           -28
[29, .296208368644652*10   ], [30, .441523517863665*10   ],
                      -30                           -32
[31, .565371931470991*10   ], [32, .619032275069477*10   ],
                      -34                           -36
[33, .576442264159914*10   ], [34, .453693634439834*10   ],
                      -38                           -40
[35, .299639620524641*10   ], [36, .164664034047199*10   ],
                      -43                           -45
[37, .745468736918177*10   ], [38, .274729964567833*10   ],
                      -48                           -50
[39, .812294880805062*10   ], [40, .189228302848469*10   ],
                      -53                           -56
[41, .339331975661182*10   ], [42, .454071317716200*10   ],
                      -59                           -62
[43, .433853276833264*10   ], [44, .276492451440547*10   ],
                       -65                           -69
[45, .104051308127000*10   ], [46, .172644520172876*10   ]]

                    46
                   -----
                    \
                     )    Prob(X) = .0000278967836544583
                    /
                   -----
                   X = 12

?endgroup
```

Dynamic Modelling of Discrete Data

Ludwig Fahrmeir
Universität München, Seminar für Statistik,
Ludwigstr.33, D-80539 München, Germany

Key Words:
discrete data, dynamic models, time series, longitudinal data, grouped survival data, filtering and smoothing

Abstract
This paper surveys dynamic or state space modelling approaches for time series or longitudinal data where the variables of primary interest are discrete. Observations of this type may be count data, where the number of events within successive time periods are recorded, or qualitative, with ordered or unordered categories. For approximately normal data, dynamic linear models and the famous Kalman filter have found numerous applications in various fields. Extensions to non–Gaussian data started with robustifying these models, while work on discrete data began only more recently. While the formulation of discrete dynamic or state space models is easy, the estimation problem ('filtering' and 'smoothing') generally becomes harder. We report on current research in this area, and illustrate methods with some applications. Software has been implemented in GAUSS and in SAS/IML.

Introduction

As a basis for the discrete case let us shortly review Gaussian linear dynamic models. In standard state space form, uni– or multivariate time series observations y_t are related to unobserved state vectors α_t by a Gaussian linear observation model

$$y_t = Z_t \alpha_t + \varepsilon_t, \quad t = 1, \ldots, T \tag{1}$$

where Z_t is an *observation* or *design matrix* of appropriate dimension, and $\{\varepsilon_t\}$ is Gaussian white noise with $\varepsilon_t \sim N(0, \Sigma_t)$. The sequence of states obeys a *linear transition equation*

$$\alpha_t = F_t \alpha_{t-1} + \xi_t, \quad t = 1, \ldots, T \tag{2}$$

with *transition matrix* F_t, Gaussian white noise $\{\xi_t\}$ with $\xi_t \sim N(0, Q_t)$, and initial state $\alpha_0 \sim N(a_0, Q_0)$. In addition, $\{\varepsilon_t\}, \{\xi_t\}$ and α_0 are mutually independent.

A major field of application of linear dynamic models are structural time series models

$$y_t = \tau_t + \gamma_t + x_t'\beta_t, \quad t = 1, \ldots, T,$$

where the states are unobserved stochastic trend and seasonal components τ_t, γ_t, and possibly time–varying effects of covariates x_t. Simple nonstationary models for trend or time–varying effects are first or second order random walk models, e.g.

$$\tau_t = \tau_{t-1} + u_t \quad \text{resp. } \tau_t = 2\tau_{t-1} - \tau_{t-2} + u_t, \; u_t \sim N(0, \sigma_u^2)$$

By appropriate definition of Z_t and F_t they can be put into state space form, see e.g. Harvey (1989, ch.4).

Given the observations $y_1, \ldots, y_T$, estimation of states α_t is of primary interest. This is termed *filtering* for $t = T$ and *smoothing* for $t < T$. Optimal estimates are conditional or posterior means

$$\alpha_{t|T} := E(\alpha_t | y_1, \ldots, y_T), \quad t = 0, \ldots, T.$$

Due to the linearity and normality assumptions in (1),(2), the posterior distribution of α_t is also normal

$$\alpha_t | y_1, \ldots, y_T \sim N(\alpha_{t|T}, V_{t|T}),$$

with posterior covariance matrix $E((\alpha_t - \alpha_{t|T})(\alpha_t - \alpha_{t|T})')$. Linear Kalman filtering and smoothing (e.g. Anderson & Moore (1979)) provides $\alpha_{t|T}, V_{t|T}$ in a computationally efficient, recursive way. Very short proofs are based on Bayesian conjugate prior–posterior properties of normal distributions. A different approach, which will be of importance for discrete dynamic models, is to derive $\{\alpha_{t|T}\}$ as posterior modes, i.e. maximizers of the posterior

$$p(\alpha_0, \alpha_1, \ldots, \alpha_T | y_1, \ldots, y_T).$$

Due to normality, posterior modes and means coincide. Using the model assumptions and taking logarithms, this is equivalent to maximizing the *penalized log–likelihood*

$$\begin{aligned} PL(\alpha) = & - \frac{1}{2}\sum_{t=1}^{T}(y_t - Z_t\alpha_t)'\Sigma_t^{-1}(y_t - Z_t\alpha_t) - \\ & - \frac{1}{2}(\alpha_0 - a_0)'Q_0^{-1}(\alpha_0 - a_0) - \\ & - \frac{1}{2}\sum_{t=1}^{T}(\alpha_t - F_t\alpha_{t-1})'Q_t^{-1}(\alpha_t - F_t\alpha_{t-1}). \end{aligned} \tag{3}$$

with respect to $\alpha = (\alpha_0, \alpha_1, \ldots, \alpha_T)$.

Defining $y_0 := a_0$, $Z_0 := I$, and introducing $y = (a_0', y_1', \ldots, y_T')'$, $Z = \text{blockdiag}(Z_0, Z_1, \ldots, Z_T)$, $W = \text{blockdiag}(Q_0^{-1}, \Sigma_1^{-1}, \ldots, \Sigma_T^{-1})$, this can be written as

$$L = -\frac{1}{2}(y - Z\alpha)'W(y - Z\alpha) - \frac{1}{2}\alpha' K\alpha \rightarrow \max_{\alpha}. \tag{4}$$

The penalty matrix K is symmetric and block–tridiagonal, and formulae for the blocks are easily obtained from (3). The first term in (3) or (4) measures closeness between fit and data, while the second term measures smoothness of the fit. The maximizer of (4) is

$$\hat{\alpha} = (Z'WZ + K)^{-1} Z'Wy, \tag{5}$$

and $\hat{\alpha}$ is obtained by the linear Kalman filter and smoother without explicitly solvings for $\hat{\alpha}$, but making efficient use of the block–banded structure of $Z'WZ + K$. The penalized likelihood approach sketched above indicates close relationship to non– and semiparametric smoothing in additive models (Hastie & Tibshirani (1990)). It should be noted that the assumption of nonsingular Σ_t, Q_t can be avoided, e.g. by working with appropriate pseudo–inverses.

Up to now it was tacitly assumed that the system matrices Z_t, F_t, Q_t and Σ_t are known. In practice, Q_t and Σ_t as well as initial priors α_0, Q_0 will often contain unknown hyperparameters, which should be estimated from the data. Usually this is done by likelihood methods (e.g. Harvey (1989, ch.4)), but cross–validation may also be considered (Kohn & Ansley (1989)).

In the next section, we consider dynamic models, mainly by replacing the normal linear observation model (1) by an observation model for discrete data, such as a logit model or a Poisson model. Filtering and smoothing based on posterior means or on posterior modes is considered in the third section. Due to non–normality of the posterior, means and modes do not coincide, and the problem becomes harder. The fourth section reports on extensions to longitudinal data.

Dynamic models for discrete time series

Let us rewrite the Gaussian linear observation equation (1) as

$$y_t|\alpha_t \sim N(\eta_t = Z_t\alpha_t, \Sigma_t).$$

The obvious modification for discrete data is to specify the observation model by *discrete observation densities*

$$y_t|\alpha_t \sim p(y_t|\eta_t = Z_t\alpha_t), \quad t = 1, \ldots, T, \tag{6}$$

which are appropriate for counts or qualitative observations. In (6) we retain the assumption that states α_t influence observations in form of a linear predictor $\eta_t = Z_t\alpha_t$, where Z_t is appropriately defined and may be a function of covariates x_t or also of past responses. The conditional density for $y_t|\alpha_t$ may also contain unknown hyperparameters, which are suppressed notationally in (6).

Let us consider some typical examples. For *counts* univariate loglinear Poisson models are a standard choice:

$$y_t|\alpha_t \sim Po(\lambda_t), \quad \lambda_t = exp(\eta_t). \tag{7}$$

The linear predictor may be chosen as in simple structural time series models:

$$\eta_t = \tau_t + x_t'\beta_t = z_t'\alpha_t$$

where $\alpha_t = (\tau_t, \beta_t')'$, and the design matrix Z_t reduces to the row vector $z_t' = (1, x_t')$. Of course, a loglinear Poisson model will not always be appropriate, and other choices such as a negative binomial may also be considered. If the number of counts in t is limited by n_t, say, binomial regression models, such as logit or probit models, are often appropriate:

$$y_t|\alpha_t \sim B(n_t, \pi_t), \quad \pi_t = h(\eta_t), \tag{8}$$

where $h(\cdot)$ is a response function, linking π_t to the predictor $\eta_t = z_t'\alpha_t$. For $h(\cdot) = exp(\cdot)/(1 + exp(\cdot))$ one obtains the logit model, for $h(\cdot) = \Phi(\cdot)$ the probit model. For $n_t = 1$, this is the most common way of modelling *binary time series.*

The observation model has to be supplemented by a stochastic mechanism governing the transition from α_{t-1} to α_t. *Explicit transition models* are defined by a transition density $p(\alpha_t|\alpha_{t-1})$. Most work on non–Gaussian dynamic modelling assumes that this transition density is defined by a *Gaussian linear transition equation* as in (2). Though the transition model remains simple, posterior mode or mean smoothing becomes more difficult since non–conjugate distributions are linked together.

To retain the relative simplicity of a conjugate prior–posterior analysis, a number of authors defined an *implicit transition mechanism* . However, this approach is restricted to more limited situations. To illustrate this, let us consider a dynamic Poisson model with stochastically varying rate $\lambda_t(=$ state $\alpha_t)$ as proposed by Harvey & Fernandes (1988). The observation density (6) is given by

$$p(y_t|\lambda_t) = \lambda_t^{y_t}\frac{exp(-\lambda_t)}{y_t!}$$

The conjugate prior for a Poisson distribution is the gamma distribution. Suppose that the filtering prior $p(\lambda|y^{t-1})$ for λ given $y^{t-1} = (y_1, \ldots, y_{t-1})$ is gamma with

$$p(\lambda; a, b) = \frac{exp(-b\lambda)\lambda^{a-1}}{\Gamma(a)b^{-a}},$$

where $a = a_{t-1}$ and $b = b_{t-1}$ are computed from the first $t-1$ observations y^{t-1}. To imitate the effect of prediction steps in linear Kalman filtering with a random walk model, Harvey and Fernandes (1988) assume a gamma distribution for $\lambda_t|y^{t-1}$ with parameters $a_{t|t-1}$ and $b_{t|t-1}$ such that

$$a_{t|t-1} = wa_{t-1}, \; b_{t|t-1} = wb_{t-1}, \quad 0 < w \leq 1.$$

This implies

$$E(\lambda_t|y^{t-1}) = E(\lambda_{t-1}|y^{t-1}), \; var(\lambda_t|y^{t-1}) = w^{-1}var(\lambda_{t-1}|y^{t-1}),$$

so that means are kept constant in prediction steps while variances are inflated. (In this special case, it is possible to show that this is formally equivalent to a multiplicative transition equation, $\lambda_t = w^{-1}\lambda_{t-1}\eta_t$, where η_t has a beta distribution.) Once the observation y_t becomes available, the posterior ('filtering') distribution is again gamma with updated parameters

$$a_t = a_{t|t-1} + y_t, \quad b_t = b_{t|t-1} + 1,$$

providing an easy solution to the filtering problem. Harvey & Fernandes (1988) extend this to Poisson models with rate $\lambda_t^+ = \lambda_t exp(x_t\beta)$, and *time constant* effects as well as similar simple forms of dynamic models for qualitative observations. For more general dynamic models, where other components than the level are allowed to change over time, West et al. (1985), see also West & Harrison (1989), define analogous implicit transition mechanisms. However, various approximations have to be made to arrive at a reasonable filtering algorithm.

Extensions to *time series of multicategorical or multinomial responses* proceed along similar lines (see e.g. Fahrmeir (1992a,b)): If k is the number of categories, responses y_t can be described by a vector $y_t = (y_{t1}, \ldots, y_{tq})$, with $q = k-1$ components. If only one multicategorical observation is made for each t, then $y_{tj} = 1$ if category j has been observed, and $y_{tj} = 0$ otherwise, $j = 1, \ldots, q$. Corresponding categorical response models are completely determined by response probabilities $\pi_t = (\pi_{t1}, \ldots, \pi_{tq})$, with $\pi_{tj} = P(y_{tj} = 1)$, $j = 1, \ldots, q$. If there are n_t independent repeated responses at t, then $y_t = (y_{t1}, \ldots, y_{tq})$ is multinomial with parameters n_t, π_t, and y_{tj} is the absolute frequency for observations in category j.

Dynamic versions of categorical response models are obtained from corresponding static models by defining linear predictors $\eta_t = (\eta_{t1}, \ldots, \eta_{tq})' = Z_t\alpha_t$ and relating it to π_t by a multivariate response function h:

$$\pi_t = h(\eta_t), \quad \eta_t = Z_t\alpha_t. \tag{9}$$

This observation model will be supplemented again by an explicit Gaussian linear transition model. For example, a dynamic multivariate logistic model with trend and covariates is specified by

$$\pi_{tj} = \frac{exp(\eta_{tj})}{1 + \sum_{r=1}^{q} exp(\eta_{tr})}, \ \eta_{tj} = \tau_{tj} + x_t'\beta_{tj}, \ j = 1, \ldots, q, \tag{10}$$

together with a transition model for trend and covariate components.

The simplest models for ordered categories are dynamic cumulative models. They can be derived from a threshold mechanism for an underlying linear dynamic model. The resulting (conditional) response probabilities are

$$\pi_{tj} = F(\eta_{tj}) - F(\eta_{t,j-1}), \quad j = 1, \ldots, q, \tag{11}$$

with linear predictors $\eta_{tj} = \tau_{tj} - x_t'\beta_t$, ordered threshold parameters $-\infty = \tau_{t0} < \ldots < \tau_{tq} < \infty$, a global covariate effect β_t, and a known distribution function F, e.g. the logistic one.

Dynamic cumulative models can be written in state space form along the previous lines. In the simplest case thresholds and covariate effects obey a first-order random walk or are partly constant in time. Then

$$\begin{aligned} \alpha_t &= (\tau_{t1}, \ldots, \tau_{tq}, \beta_t')' = \alpha_{t-1} + \xi_t, \\ Z_t &= \begin{bmatrix} 1 & & & -x_t' \\ & \ddots & & \vdots \\ & & 1 & -x_t' \end{bmatrix}, \end{aligned}$$

and the response function can be appropriately defined. Dynamic versions of other models for ordered categories discussed e.g. in Fahrmeir & Tutz (1993, ch.3.4), such as sequential models, can be designed with analogous reasoning.

Conjugate prior–posterior approaches for multicategorical observations are more restricted and far less developed: Harvey & Fernandes (1988) extend their analysis to a simple multinomial model while it seems at least quite difficult to adapt the methods of West et al. (1985) to the multivariate case.

Filtering and smoothing

In the following let $y^t = (y_1, \ldots, y_t)$ and $\alpha^t = (\alpha_0, \ldots, \alpha_t)$ denote 'histories' of observations and states up to t. Filtering and smoothing is based on posterior densities like $p(\alpha_t|y)$ or $p(\alpha|y)$ for smoothing, or $p(\alpha_t|y^t)$ for filtering. One may distinguish three approaches: (i) conjugate prior-posterior analyses, trying to solve necessary integrations in Bayes' theorem analytically, perhaps making additional approximations, (ii) full Bayes or at least posterior mean analyses based on numerical integration or Monte Carlo methods, (iii) posterior mode estimation, avoiding integration at all. Type (i) has already been discussed in Section 2 for a special case. More detailed exposition can be found in the literature cited there. It should be noted that in smoothing there is no merit in assuming a conjugate prior. For example with a loglinear Poisson observation model, calculation of the smoothing density cannot be simplified, even if a gamma prior is assumed for λ_t.

We will consider the other two estimation methods for dynamic models with discrete observation densities $p(y_t|\eta_t)$, $\eta_t = Z_t\alpha_t$ of the form (6) and Gaussian linear transition models of the form (2). The focus will be on dynamic Poisson, binomial and multinomial models with observation densities defined by (7), (8),(9). To define the models in terms of genuine likelihoods, we have to introduce conditional independence assumptions which replace the assumptions of mutually independent error processes in linear dynamic models. For simplicity, covariates $\{x_t\}$ are assumed to be deterministic; for stochastic covariates an additional assumption is necessary (see e.g. Fahrmeir & Tutz (1994, Section 8.2.1)).

(A1) Conditional on α_t and y^{t-1}, current observations y_t are independent of α^{t-1}, i.e. $p(y_t|\alpha^t, y^{t-1}) = p(y_t|\alpha_t, y^{t-1})$.

Combined with (6), this implies that the design matrix Z_t can be a function of covariates and past responses y^{t-1}, but not of past states α^{t-1}.

(A2) Errors ξ_t in the transition equation (2) are independent of y^{t-1}.

Posterior mode smoothing and penalized likelihood estimation

Posterior mode smoothers and filters are defined as maximizers of posterior densities. Using the model assumptions and taking logarithms, it turns out that maximizing $p(\alpha_0, \ldots, \alpha_T, |y_1, \ldots, y_t)$ is equivalent to maximizing the penalized log–likelihood

$$
\begin{aligned}
PL(\alpha) = \sum_{t=1}^{T} l_t(\eta_t(\alpha_t)|y_t) &- \frac{1}{2}(\alpha_0 - a_0)'Q_0^{-1}(\alpha_0 - a_0) \\
&- \frac{1}{2}\sum_{t=1}^{T}(\alpha_t - F_t\alpha_{t-1})'Q_t^{-1}(\alpha_t - F_t\alpha_{t-1}),
\end{aligned}
\tag{12}
$$

where $l_t(\eta_t(\alpha_t)|y_t) = \log p(y_t|\eta_t = Z_t\alpha_t)$ are observation log–densities. Compared to (3), least squares distances implied by the Gaussian observation model, are replaced by Kullback–Leibler distances. Posterior mode smoothers

$$\hat{\alpha} = (\alpha_{0|T}, \ldots, \alpha_{t|T}, \ldots, \alpha_{T|T})$$

are maximizers of (12). Numerical maximization of $PL(\alpha)$ can be achieved by various algorithms, differing in approximation quality and computational effort. Fahrmeir (1992a) suggests a generalized extended Kalman filter and smoother, which works satisfactorily for common dynamic models such as (7) to (11). Fahrmeir & Kaufmann (1991) develop iterative forward–backward Gauss–Newton (Fisher scoring) algorithms. Gauss–Newton smoothers can also be obtained by iterative application of linear Kalman filtering and smoothing to a 'working' model, similarly as Fisher scoring in static generalized linear models can be performed by iteratively weighted least squares applied to 'working' observations. To see this, let us first rewrite $PL(\alpha)$ in compact notation as

$$PL(\alpha) = l(\alpha) - \frac{1}{2}\alpha' K\alpha,$$

where

$$l(\alpha) = \sum_{t=0}^{T} l_t(\eta_t(\alpha_t)|y_t),$$

with $l_0(\eta_0(\alpha_0)) = -(\alpha_0 - a_0)'Q_0^{-1}(\alpha_0 - a_0)$, $\eta_0 = \alpha_0$.
Defining $r_0 = \partial l_0/\partial\eta_0$, $r_t(\alpha) = \partial l_t/\partial\eta_t$, $r(\alpha) = (r_0, \ldots, r_t(\alpha), \ldots)$ and $Z = blockdiag(Z_0, \ldots, Z_t, \ldots)$, the score function $u(\alpha) = \partial PL(\alpha)/\partial\alpha$ is

$$u(\alpha) = Z'r(\alpha) - K\alpha.$$

(Expected) information $U(\alpha) = -(\mathrm{E})\partial^2 PL(\alpha)/\partial\alpha\partial\alpha'$ is then

$$U(\alpha) = Z'W(\alpha)Z + K,$$

with $W(\alpha) = -blockdiag[(\mathrm{E})\partial^2 l_t/\partial\eta_t\partial\eta_t']$. A Gauss–Newton (Fisher scoring) step from α^0 to α^1 is then

$$(Z'W(\alpha^0)Z + K)(\alpha^1 - \alpha^0) = Z'r(\alpha^0) - K\alpha^0.$$

This can be rewritten as

$$\alpha^1 = (Z'W(\alpha^0)Z + K)^{-1}Z'W(\alpha^0)\tilde{y}^0,$$

with 'working' observation

$$\tilde{y}^0 = (W(\alpha^0))^{-1}r(\alpha^0) + Z\alpha^0.$$

Comparing with (5), it can be seen that one iteration step can be performed by applying the linear Kalman smoother with 'working' weight $W = W(\alpha^0)$ to the 'working' observation $\tilde{y}^0$. A closely related algorithm has been obtained recently by Durbin & Koopman (1993). It is recommended to initialize iterations with the generalized extended Kalman smoother of Fahrmeir (1992a). Iterations will often stop after very few steps. Estimation of unknown hyperparameters in α_0, Q_0, Q_t can be carried out by an EM–type algorithm, see e.g. Fahrmeir & Tutz (1994, ch.8). An alternative would be the use of cross–validation techniques.

As a very simple example let us consider rainfall data (Kitagawa (1987)), analyzed previously by several authors and methods. Figure 1 displays the number of occurences of rainfall in the Tokio area for each calendar day during the years 1983-1984 together with smooth estimators for the probability π_t of rainfall on calendar day $t = 1, \ldots, 366$. As in Kitagawa (1987), the following simple dynamic logit model has been chosen:

$$y_t \sim \begin{cases} B(1, \pi_t) & t = 60(\text{February } 29) \\ B(2, \pi_t) & t \neq 60, \end{cases}$$

$$\pi_t = h(\alpha_t) = \exp(\alpha_t)/(1 + \exp(\alpha_t)),$$

$$\alpha_{t+1} = \alpha_t + \xi_t, \quad \xi_t \sim \mathrm{N}(0, \sigma^2),$$

so that $\pi_t = P(\text{rain on day } t)$ is reparameterized by α_t. Figure 1 shows posterior mode smoothers $\hat{\pi}_t = h(\alpha_{t|366})$, together with approximate pointwise confidence bands (obtained from curvatures evaluated at the mode). Gauss–Newton smoothing (——), was initialized by generalized Kalman smoothing (– – –) and stopped after two iterations. Starting values and the random walk variance were estimated by an EM–type algorithm. In this example Gauss–Newton smoothers and generalized Kalman smoothers are more or less indistinguishable.

Filtering and smoothing based on integration or simulation

This section shortly discusses some approaches which have been suggested to calculate posterior densities like $p(\alpha_t|y)$, $p(\alpha_t|y^t)$, $p(\alpha_t|y^{t-1})$ or at least

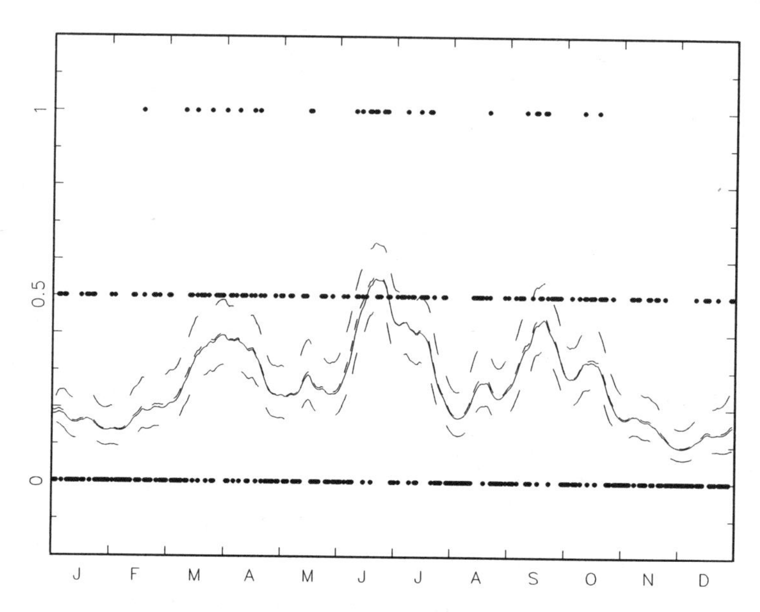

Figure 1: Smoothed probabilities $\hat{\pi}_t$ of daily rainfall, obtained by generalized Kalman (- - -) and Gauss–Newton smoothing (——), and observed relative frequencies (••).

posterior means or variances. Let $p(y_t|\alpha_t)$ and $p(\alpha_t|\alpha_{t-1})$ denote observation and transition densities. Then the following recursions can be derived (Kitagawa (1987)): The prediction step is

$$p(\alpha_t|y^{t-1}) = \int p(\alpha_t|\alpha_{t-1})p(\alpha_{t-1}|y^{t-1})d\alpha_{t-1}, \tag{13}$$

the correction step for filtering is

$$p(\alpha_t|y^t) = p(y_t|\alpha_t)p(\alpha_t|y^{t-1})/p(y_t|y^{t-1}) \tag{14}$$

with the one–step prediction density

$$p(y_t|y^{t-1}) = \int p(y_t|\alpha_t)p(\alpha_t|y^{t-1})d\alpha_t, \tag{15}$$

and backward smoothing steps are

$$p(\alpha_t|y) = p(\alpha_t|y^t)\int p(\alpha_{t+1}|y)p(\alpha_{t+1}|\alpha_t)/p(\alpha_{t+1}|y^t)d\alpha_{t+1}.$$

Integration based approaches approximate analytically intractable integrals in the recursions above or in similar expressions, e.g. for posterior means and variances, by numerical or Monte Carlo integration. For scalar observations y_t and scalar states α_t, Kitagawa presents algorithmic solutions by recursively approximating posterior densities by piecewise linear functions and applying simple numerical integration techniques. He also applies his method to the rainfall data example above. Compared to posterior mode smoothing, there are only minor departures for the initial time period. Kashiwagi & Yanagimoto (1992) propose a closely related method for a loglinear Poisson observation model with a Gaussian random walk model for the lograte $\alpha_t = \log \lambda_t$ and apply it to three sets of weekly desease incidence data for illustration. A similar integration based approach has been proposed by West & Harrison (1989). They suggest the crude application of Gauss–Hermite quadrature to solve integrals in posterior moments. However, all these methods increase exponentially with time and dimension of states. Therefore, they are mainly restricted to shorter time series of low dimension. A more practicable solution to the prediction and filtering problem has been given by Schnatter (1992) for dynamic generalized linear models with linear Gaussian transition equation. She approximates the predictive density $p(\alpha|y^{t-1})$ in (13), (14) and (15) by a normal density $\mathrm{N}(\alpha_{t|t-1}, \mathrm{V}_{t|t-1})$, where $\alpha_{t|t-1}$ and $\mathrm{V}_{t|t-1}$ are evaluated as in the linear Kalman prediction step. Then the integrals in (14),(15) can be calculated by Gauss–Hermite quadrature without exponentially increasing effort in time. Alternatively these integrals can be calculated by Monte Carlo integration as in Hennevogl (1991). Experience with artificial and real data has shown that Monte Carlo integration becomes more efficient for higher (> 3) dimensions of the state vector.

A Gibbs sampling approach. A promising solution to obtain smoothing densities and moments is the application of the Gibbs sampler, which is a general simulation based tool for estimating marginal posterior densities and moments. A Gibbs sampling approach to dynamic models with normal mixture error structure has been proposed by Carlin et al. (1992), and Fahrmeir et al. (1992) as well as Knorr–Held (1993) adopted their method to dynamic models for discrete data.

To obtain estimates for $p(\alpha_t|y)$ or corresponding moments by Gibbs sampling, it is required that conditional posterior densities $p(\alpha_t|\alpha_{s\neq t}, y)$ of α_t given all other states α_s, $s \neq t$, are available for sampling. Applying the model assumptions (A1), (A2) and Bayes' theorem it can be shown that the following proportionality holds:

$$p(\alpha_t|\alpha_{s\neq t}, y) \propto \begin{cases} p(\alpha_{t+1}|\alpha_t)p(\alpha_t) , & \text{if } t = 0 \\ p(y_t|\alpha_t)p(\alpha_{t+1}|\alpha_t)p(\alpha_t|\alpha_{t-1}), & \text{if } t = 1, \ldots, T-1 \\ p(y_t|\alpha_t)p(\alpha_t|\alpha_{t-1}) , & \text{if } t = T \end{cases}$$

Since observation densities $p(y_t|\alpha_t)$ and transition densities $p(\alpha_t|\alpha_{t-1})$ are available in closed form, the random numbers from $p(\alpha_t|\alpha_{s\neq t}, y)$ can be drawn by rejection sampling techniques, see references above for details.

With this rejection sampling technique, the Gibbs sampling procedure runs as follows: Given a set of arbitrary starting values $(\alpha_t^{(0)})$, $t = 0, \ldots, T$ one has to draw $\alpha_0^{(1)}$ from the conditional density $p(\alpha_0|\alpha_1^{(0)}, \ldots, \alpha_T^{(0)}, y)$, then $\alpha_1^{(1)}$ from $p(\alpha_1|\alpha_0^{(1)}, \alpha_2^{(0)}, \alpha_3^{(0)}, \ldots, \alpha_T^{(0)}, y)$ and so on up to $\alpha_T^{(1)}$ from $p(\alpha_T|\alpha_0^{(1)}, \alpha_1^{(1)}, \ldots, \alpha_{T-1}^{(1)}, y)$ to complete one iteration. After a large number K of iterations which define one Gibbs run, the $(T+1)$–tuple $(\alpha_0^{(K)}, \alpha_1^{(K)}, \ldots, \alpha_T^{(K)})$ is obtained. Under mild assumptions, the joint density of this $(T+1)$–tuple converges to the joint posterior density $p(\alpha_0, \alpha_1, \ldots, \alpha_T|y)$, as $K \to \infty$. Carrying out G Gibbs–runs yields $g = 1, \ldots, G$ i.i.d. $(T+1)$–tuples $(\alpha_0^{(K,g)} \alpha_1^{(K,g)}, \ldots, \alpha_T^{(K,g)})$ from the joint posterior density $p(\alpha_0, \alpha_1, \ldots, \alpha_T|y)$. These can be used to estimate the marginal posterior densities $p(\alpha_t|y)$ and moments. Estimation of unknown variances in Gaussian linear transition models can be incorporated as in Knorr-Held (1993). Application to the rainfall data gives posterior mean smoothers which are again very close to posterior mode smoothers. However, experience with simulated data shows that generally posterior modes may differ from means more distinctly. It is again recommended, to use simple posterior mode smoothers as starting values $(\alpha_t^{(0)})$ for speeding up convergence of Gibbs–runs.

Dynamic modelling of discrete longitudinal data

In many applications, more than one individual or object is observed sequentially over time. Let us first consider longitudinal or panel data which consist of observations

$$(y_{it}, x_{it}), \quad i = 1, \ldots, n, \; t = 1, \ldots, T,$$

with discrete y_{it}, for a population of n units observed across time. The state space modelling approach to longitudinal data allows, in principle, to deal with random effects ('states') across units and across time, like stochastic trend and seasonal components. We will confine attention to the case where states are constant across units.

In analogy to the second and third sections it is assumed that the individual observation densities are of the form

$$p(y_{it}|\eta_{it}), \quad \eta_{it} = Z_{it}\alpha_t, \tag{16}$$

where the design matrix or vector Z_{it} may be a function of covariates x_t and past responses y^{t-1}.

In the sequel, individual observations are collected in 'panel wave' vectors

$$y_t = (y_{1t}, \ldots, y_{nt}), \quad x_t = (x_{1t}, \ldots, x_{nt}), \quad t = 1, \ldots, T,$$

and histories up to t are denoted by y^t, x^t as before. The observation model has to be supplemented by a transition model, e.g. a linear Gaussian transition equation. To specify the model completely, a further assumption is added to (A1),(A2):

(A3) Given α_t, y^{t-1}, x^t, individual responses y_{it} within y_t are conditionally independent, i.e.

$$p(y_t|\alpha_t, y^{t-1}, x^t) = \prod_{i=1}^{n} p(y_{it}|\eta_{it} = Z_{it}\alpha_t)$$

This corresponds to independence assumptions in purely cross–sectional situations. It allows for interaction among units via the common history y^{t-1}, x^t.

Estimation of unknown states can in principle be achieved by applying filtering and smoothing approaches of the previous section to the time series of panel waves $\{y_t, x_t\}$. However, only posterior mode smoothing has been adopted to the longitudinal data situation in this way at the moment. Using assumption (A3) above, the log–likelihood contributions in the penalized log–likelihood (12) have to be modified to the sum

$$\sum_{t=1}^{T} \sum_{i=1}^{n} l_{it}(\eta_{it}(\alpha_t)|y_{it}), \quad \eta_{it} = Z_{it}\alpha_t \tag{17}$$

of individual contributions, and score functions and information matrices are sums of of individual contributions as well. Fahrmeir & Goss (1992) apply a dynamic cumulative logit model and a generalized Kalman smoother to monthly business test data collected for an industrial branch of 55 companies over the period January 1980 to December 1990, revealing nonstationary trend, seasonal components of the three–categorical response 'productivity' and partly time–varying effects of covariates.

Another longitudinal data situation, where dynamic modelling has been applied, are survival data. Gamerman (1991,1992) studied a dynamic version of the piecewise exponential model and an extension to more general point processes, adopting the approximate Bayes methods of West et al. (1985). Posterior mode smoothing for grouped survival data, where survival times are only known to lie within certain intervals $I_1, \ldots, I_t, \ldots, I_d$, is considered in Fahrmeir (1994). Survival time T is then discrete, and $T = t$ denotes failure in I_t. For individual i with survival time T_i and possibly time–varying covariate x_{it}, the discrete hazard function is defined by

$$\lambda(t|x_{it}) = P(T_i = t|T_i \geq t, x_{it}).$$

Dynamic grouped survival models are obtained by assuming

$$\lambda(t|x_{it}) = h(\theta_t + x'_{it}\beta_t), \tag{18}$$

where θ_t is the baseline hazard parameter and β_t a time-varying covariate effect, e.g. a time-varying therapy effect. Important models are the grouped Cox model with $h(\cdot) = 1 - \exp(-\exp(\cdot))$ and the logistic model. The observation model (18) has again to be supplemented by a transition model for (θ_t, β_t).

Let R_t be the set of individuals under risk in I_t, and define failure indicators y_{it}, $i \in R_t$ by $y_{it} = 1$, if individual i fails in I_t and $y_{it} = 0$, if individual i survives in I_t. Assuming a non-informative random censoring mechanism, posterior mode smoothing can be applied to the data $\{y_{it}, x_{it}\}$,$i \in R_t$, $t = I_1, \ldots, I_d$, in the same way as for panel data. Compared to (17), the second sum runs only over $i \in R_t$.

A number of real data applications is contained in Aydemir (1992). As an example, let us consider a reanalysis of gastric cancer data of 90 patients divided into two treatment groups (Gamerman (1991)). After grouping the data into one-month intervals a dynamic logit model

$$\lambda(t, x) = h(\theta_t + x\beta_t)$$

with first order random walks for θ_t, β_t was applied. The treatment covariate is 0 for chemotherapy and 1 for a combined therapy. Figure 2 shows the smoothed estimate of the treatment effect β_t. For shorter survival time up to about 21 months, the combined treatment has a positive influence on the hazard function, that is a negative influence on survival probabilities. For survival times longer than 21 months however, a combined treatment is reasonable. Figure 3 displays corresponding hazard functions for both groups. Although information has been lost by grouping the data into months, the results are in good agreement with Gamerman's analysis.

References

Anderson, B. & Moore J. (1979). Optimal Filtering. Englewood Cliffs: Prentice Hall.

Aydemir, S. (1992). Dynamische Modelle zur Ereignisanalyse. Diplomarbeit, Universität München.

Carlin, B.P., Polson, N.G. & Stoffer, D.S. (1992). A Monte Carlo Approach to Nonnormal and Nonlinear State-Space Modelling. JASA, 87, 493–500.

Durbin, J. & Koopman, S. (1993). Filtering, Smoothing and Estimation for Time Series Models when the Observations come from Exponential Family Distributions. London School of Economics, Discussion paper.

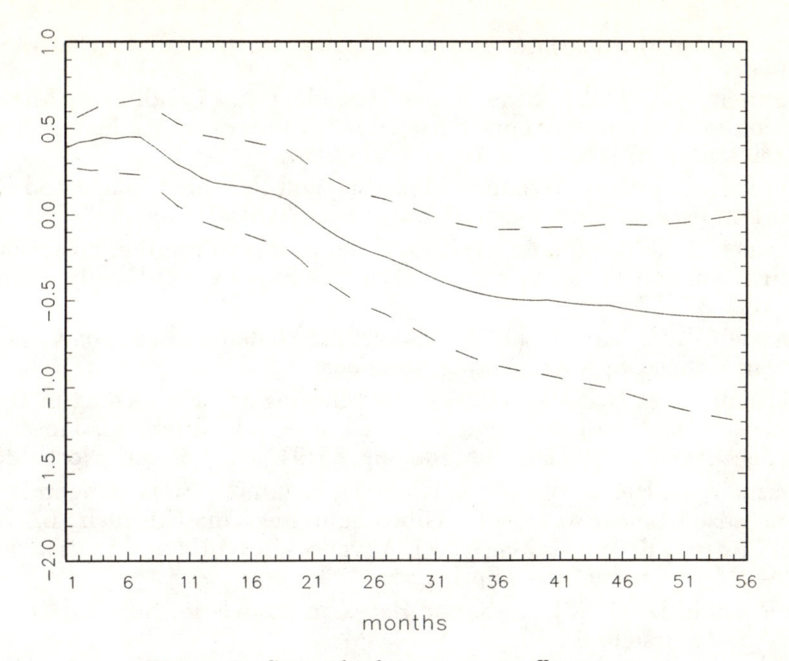

Figure 2: Smoothed treatment effect.

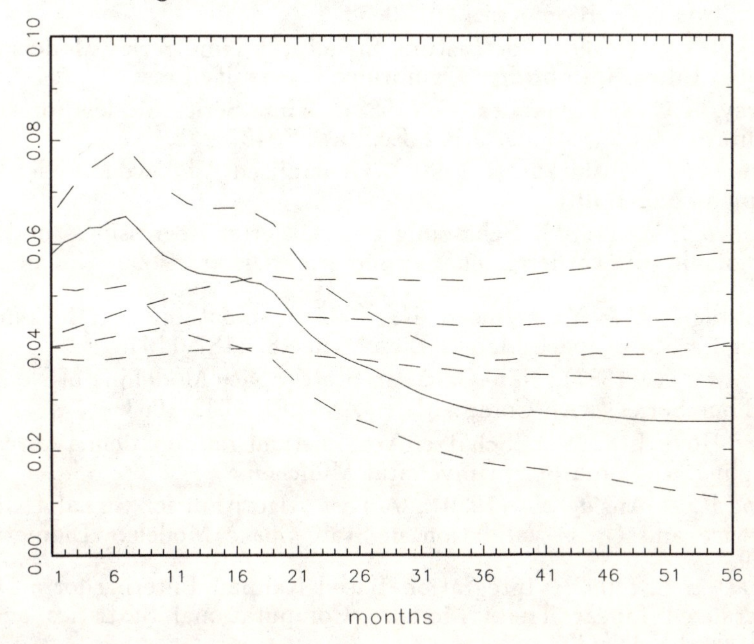

Figure 3: Smoothed hazard functions for chemotherapy (– – –) and combined therapy (——).

Fahrmeir, L. (1992a). Posterior Mode Estimation by Extended Kalman

Filtering for Multivariate Dynamic Generalized Linear Models. JASA, 87, 501–509.

Fahrmeir, L. (1992b). State Space Modelling and Conditional Mode Estimation for Categorical Time Series. IMA Volumes on Mathematics and its Applications, 87–110, New York: Springer.

Fahrmeir, L. (1994). Dynamic Modelling and Penalized Likelihood Estimation for Discrete Time Survival Data. Biomatrika, to appear.

Fahrmeir, L. & Kaufmann, H. (1991). On Kalman Filtering, Posterior Mode Estimation and Fisher Scoring in Dynamic Exponential Family Regression. Metrika, 38, 37–60.

Fahrmeir, L. & Tutz, G. (1994). Statistical Modelling Based on Generalized Linear Models. Springer Verlag, to appear.

Fahrmeir, L. & Goss, M. (1992). On Filtering and Smoothing in Dynamic Models for Categorical Longitudinal Data. In: V.d.Heijden, Jansen, Francis, Seeber (eds.), Statistical Modelling, 85–94. Amsterdam: North Holland.

Fahrmeir, L., Hennevogl, W. & Klemme, K. (1992). Smoothing in Dynamic Generalized Linear Models by Gibbs Sampling. In: Fahrmeir, L., Francis, B., Gilchrist, R. & Tutz, G. (eds.), Advances in GLIM and Statistical Modelling, 85–90. Heidelberg: Springer.

Gamerman, D. (1991). Dynamic Bayesian Models for Survival Data. Applied Statistics 40, 63–79.

Gamerman, D. (1992). A Dynamic Approach to the Statistical Analysis of Point Processes. Biometrika 79, 39–50.

Harvey, A.C. (1989). Forecasting, Structural Time Series Models and the Kalman Filter. Cambridge: Cambridge University Press.

Harvey, A.C. & Fernandes, C. (1988). Time Series Models for Count or Qualitative Observations. J.Bus.Ec.Stat., 7, 407–422.

Hastie, T. & Tibshirani, R. (1990). Generalized Additive Models. London: Chapman and Hall.

Hennevogl, W. (1991). Schätzung generalisierter Regressions- und Zeitreihenmodelle mit variierenden Parametern. Dissertation. Universität Regensburg.

Kashiwagi, N. & Yanagimoto, N. (1992). Smoothing Serial Count Data through a State–Space Model. Biometrics 48, 1187–1194.

Kitagawa, G. (1987). Non–Gaussian State–Space Modelling of Nonstationary Time Series (with Comments). JASA, 82, 1032–1063.

Knorr–Held, L. (1993). Schätzen von Zustandsraummodellen durch Gibbs Sampling. Diplomarbeit, Universität München.

Kohn, R. & Ansley, C. (1989). A Fast Algorithm for Signal Extraction, Influence and Cross–Validation in State–Space Models. Biometrika 76, 65–79.

Schnatter, S. (1992). Integration–Based Kalman–Filtering for a Dynamic Generalized Linear Trend Model. Computational Statistics and Data Analysis, 13, 447–459.

West, M., Harrison, J. & Migon, M. (1985). Dynamic Generalized Linear Models and Bayesian Forecasting. JASA, 80, 73–97.

West, M. & Harrison, P.J. (1989). Bayesian Forecasting and Dynamic Models. New York: Springer.

Evaluating the Significance Level of Goodness-of-Fit Statistics for Large Discrete Data

Gerhard Osius [1]
Universität Bremen, Institut für Statistik, Fachbereich 3
Postfach 330440, D-28334 Bremen, Germany

Key Words:
binomial data, bootstrap, edgeworth expansion, multinomial data, power-divergence statistic, saddlepoint approximation, sparse data

Abstract
Goodness-of-fit tests for multinomial models with parameters to be estimated are usually based on Pearson's X^2 or the deviance (likelihood ratio) D, both belonging to the family of power-divergence statistics SD_λ. Our aim is to evaluate the significance level of SD_λ for large samples and models with large degrees of freedom, but not necessarily large expected counts in each cell, allowing sparse data too. We consider approximations based on limiting distributions, Edgeworth and saddlepoint expansions, and the parametric bootstrap. Computational details are given, and two larger studies serve as numerical examples.

Introduction

Two devices are in common use to judge the fit of a chosen model for discrete data: the more informal analysis of suitably defined residuals and formal gooodness-of-fit tests based on summary statistics. Our focus will be on goodness-of-fit statistics, both from a theoretical and computational point of view. The most common goodness-of-fit statistics are Pearson's X^2 and the deviance D (i.e. the likelihood ratio statistic). Cressie & Read (1984) have embedded these and other familiar statistics in the family of power-divergence statistics SD_λ with a real parameter $\lambda \in \mathbb{R}$, which are defined as follows. For each cell $i = 1, ..., I$ the deviation of the observed count

[1] The author thanks the referees for helpful comments.

Y_i from its fitted value (expected count) $\hat{Y}_i$ is measured using a "distance" function

$$a_\lambda(y,\hat{y}) = \frac{2y}{\lambda(\lambda+1)}\left[\left(\frac{y}{\hat{y}}\right)^\lambda - 1\right] - \frac{2}{\lambda+1}(y-\hat{y}) ,$$

where the second term is introduced to make a_λ non-negative. The cases $\lambda = 0, -1$ are defined by continuity as $\lambda \to 0, -1$. The function a_λ is not symmetrical, but the arguments are reversed in passing from λ to $-(\lambda+1)$

$$a_{-(\lambda+1)}(y,\hat{y}) = a_\lambda(\hat{y},y) .$$

Since we consider here only models with positive fitted values $\hat{y} > 0$ but allow zero observed counts $y = 0$, which are typically encountered in sparse data, we restrict ourselves to values $\lambda > -1$.

The power-divergence statistic is the sum of all deviations

$$SD_\lambda = \sum_i a_\lambda(Y_i,\hat{Y}_i) .$$

In almost all applications, the observed total Y_+ equals the fitted total $\hat{Y}_+$ (the index "+" indicates summation over the index it replaces) and hence the sum of all second terms in a_λ vanishes, thus reducing SD_λ to the original definition of Cressie & Read (1984).

For $\lambda = 1$ we get Pearson's statistic $X^2 = SD_1$, with

$$a_1(y,\hat{y}) = (y-\hat{y})^2/\hat{y} ,$$

the deviance $D = SD_0$ is obtained for $\lambda = 0$ with

$$a_0(y,\hat{y}) = 2\left[y\log(y/\hat{y}) - (y-\hat{y})\right] ,$$

and $\lambda = -1/2$ yields Freeman-Tukey's statistic $FT = SD_{-1/2}$ with

$$a_{-1/2}(y,\hat{y}) = 4(y^{1/2} - \hat{y}^{1/2})^2 .$$

Read & Cressie (1988) suggested the value $\lambda = 2/3$ as a compromise between the rival values 0 and 1, and we denote their statistic by $CR = SD_{2/3}$. Despite the richness of the family SD_λ only the traditional values $\lambda = 0, 1$ are widely used.

The classical goodness-of-fit test is based on an asymptotic χ^2-distribution for SD_λ, which is appropriate for an increasing sample size $n := Y_+$, provided the number I of cells remains *constant* as $n \to \infty$. This so-called *fixed-cells asymptotic* implies that all fitted counts increase, i.e. $\hat{Y}_i \to \infty$.

Hence the approximation of the significance level $P\{SD_\lambda \geq sd_\lambda\}$ for the observed value sd_λ based on the asymptotic χ^2-distribution can be expected to be sufficiently accurate, only if all fitted counts $\hat{Y}_i$ are reasonably "large". According to a popular rule of thumb, going back to Cochran (1952), $\hat{Y}_i \geq 5$ seems large enough, but we will not pursue this topic.

Although very important, the fixed-cells asymptotic is not generally reliable in practice, in particular not for sparse data, where sample size n is large, but many (or even all) fitted values $\hat{Y}_i$ remain small, because the number of cells I is large, too. For such a situation an *increasing-cells asymptotic* is appropriate which only requires $I \to \infty$ as $n \to \infty$. Several authors have investigated the asymptotic behaviour of D and X^2 under increasing cells, specific sampling distributions for the observed counts (e.g. multinomial, binomial or Poisson) and particular types of models (e.g. log-linear), cf. McCullagh (1985a,b, 1986), Dale (1986) and Koehler (1986). For general $\lambda > -1$, the asymptotic normality of SD_λ was derived in Osius (1985) and generalized in Rojek (1989). The increasing-cells approach thus provides an alternative approximation of the significance level based on the normal distribution of SD_λ, provided that both n and I are large.

Although the limiting distribution of SD_λ differs for the two asymptotics, it would be desirable for practical purposes, to have approximations for the tail probabilites $P\{SD_\lambda \geq sd_\lambda\}$ which apply for large n, no matter whether the fitted counts $\hat{Y}_i$ are small, moderate or large. Our aim is to propose approximate significance levels based on different approaches

- limiting distributions for SD_λ (chi-squared and normal),
- Edgeworth and saddlepoint expansions,
- the parametric bootstrap.

The computational aspects for these methods will be given detailed enough to enable implementation in common statistical software packages (e.g. GLIM) or programming languages (e.g. S-Plus). For illustration purposes the methods are applied to two sets of sparse data, namely a study on infant mortality and on cancer.

The basic ideas are presented here in an expository and informal way, giving references to detailed work. We choose the convenient setup of multinomial models for contingency tables with no attempts to achieve a maximum of generality.

Multinomial Models

To fix the setup, we suppose that the observed counts form a $J \times K$ contingency table $\mathbf{Y} = (Y_{jk})$ with $I = JK$ cells. Let the columns represent a categorical response with K levels (e.g. a binary response for $K = 2$), and the rows are interpreted as groups j which are usually characterized by an additional S-dimensional vector $\mathbf{x}_j = (x_1, ..., x_S) \in \mathbb{R}^S$ of observed covariables. Thus the random variable Y_{jk} is the number of items in group j with response k, with an assumed positive expectation $\mu_{jk} = E(Y_{jk}) > 0$. Our primary interest focuses on the (conditional) probability $\pi_{jk} = \mu_{jk}/\mu_{j+} \in (0,1)$ for response k in group j, for which we consider models of a generalized linear type

$$\pi_{jk} = \pi_{jk}(\boldsymbol{\theta}) := G_k(\mathbf{x}_j^T \boldsymbol{\theta}). \tag{1}$$

Here $\boldsymbol{\theta} = (\theta_1, ..., \theta_S)$ is an unknown parameter vector, and G_k are known smooth functions with values in $(0,1)$.

An adequate *sampling model* for this situation is the *product-multinomial*, in which the vectors of row counts $\mathbf{Y}_j = (Y_{j1}, ..., Y_{jK})$ are independent for $j = 1, ..., J$, each having a multinomial distribution $M_K(N_j, \boldsymbol{\pi}_j)$ of size $N_j = Y_{j+}$ with probability vector $\boldsymbol{\pi}_j = (\pi_{j1}, ..., \pi_{jK})$. The product-multinomial model, assumed throughout, also arises from other popular sampling models for $\mathbf{Y}$ (e.g. Poisson or single-multinomial) by conditioning on the observed row totals Y_{j+}, which contain no information about the probabilities $\boldsymbol{\pi}$ of interest (Haberman (1974)).

The log-likelihood function for the product-multinomial model is up to a constant given by

$$l(\boldsymbol{\theta}) = \sum_j \sum_k Y_{jk} \log \pi_{jk}(\boldsymbol{\theta}) \ .$$

To fit the model, the maximum likelihood estimator $\hat{\boldsymbol{\theta}}$ (or some asymptotically equivalent estimate) has to be determined – typically by an iterative procedure – to obtain the fitted probabilities $\hat{\pi}_{jk} = \pi_{jk}(\hat{\boldsymbol{\theta}})$ and fitted cell counts $\hat{Y}_{jk} = N_j \hat{\pi}_{jk}$, both being positive.

The power-divergence statistic may now be obtained by summing up first the deviations within each group and then over all groups

$$SD_\lambda(\hat{\boldsymbol{\theta}}) = \sum_j \sum_k a_\lambda(Y_{jk}, \hat{Y}_{jk}) = \sum_j A_\lambda(\mathbf{Y}_j, \hat{\mathbf{Y}}_j) \ , \tag{2}$$

where A_λ serves as a "distance measure" between K-dimensional vectors

$$A_\lambda(\mathbf{y}, \hat{\mathbf{y}}) := \sum_k a_\lambda(y_k, \hat{y}_k) \ . \tag{3}$$

In view of $a_\lambda(c\mathbf{y}, c\hat{\mathbf{y}}) = c \ a_\lambda(\mathbf{y}, \hat{\mathbf{y}})$ for any real $c > 0$, the statistic may also be written in terms of the observed and fitted frequencies $\hat{P}_{jk} := Y_{jk}/N_j$ and $\hat{\pi}_{jk}$ as

$$SD_\lambda(\hat{\boldsymbol{\theta}}) = \sum_j N_j \ A_\lambda(\hat{\mathbf{P}}_j, \hat{\boldsymbol{\pi}}_j) \ .$$

Usually large values of the statistic $SD_\lambda(\hat{\boldsymbol{\theta}})$ indicate a lack of fit, at least if its expectation does not depend upon the parameter $\boldsymbol{\theta}$. Since the exact distribution of $SD_\lambda(\hat{\boldsymbol{\theta}})$ under the model is not tractable in practice (except in very simple situations), we have to rely on approximations for the significance level $P\{SD_\lambda(\hat{\boldsymbol{\theta}}) \geq sd_\lambda\}$ to perform a goodness-of-fit test.

Limiting Distributions

The most popular and simple approximation of the significance level is based on the limiting χ^2-distribution of SD_λ (the argument $\hat{\boldsymbol{\theta}}$, being fixed here, is now omitted) with $df = J(K-1) - S$ degrees of freedom (cf. Read & Cressie (1988))

$$P\{SD_\lambda \geq sd_\lambda\} \approx P\{\chi^2_{df} \geq sd_\lambda\} , \qquad \text{(CCA)}$$

which will be referred to as the *classical* χ^2*-approximation.* The accuracy of this approximation (derived for increasing-cells asymptotics) increases as *all* fitted counts $\hat{y}_{jk}$ resp. all N_j tend to infinity, but may be extremely poor if a relevant fraction of fitted counts are small.

In contrast, the increasing-cells asymptotic requires the number J of groups to tend to infinity as $n \to \infty$, while the number K of responses categories and the number S of parameters remain *fixed.* No restrictions are imposed here on the fitted counts $\hat{Y}_{jk}$ resp. the group sizes N_j, some or even all of which may be low. Only in the extreme case of individual groups (i.e. $N_j = 1$ for *all* j) the statistic SD_λ may have no diagnostic power for particular model-dependent values of λ (cf. McCullagh (1985a), Osius & Rojek (1992)).

Under increasing-cells asymptotic, the power divergence statistic SD_λ has an asymptotic *normal* distribution (Rojek (1989)) with expectation μ_λ and variance σ^2_λ, given by (16) and (17) in the next section. This is not surprising, since SD_λ is an increasing sum of components $A_\lambda(\mathbf{Y}_j, \hat{\mathbf{Y}}_j)$, which are almost independent except for their dependence through the common estimate $\hat{\boldsymbol{\theta}}$ of fixed dimension. The asymptotic normality of the *standardized* power-divergence statistic

$$T_\lambda = (SD_\lambda - \mu_\lambda)/\sigma_\lambda \qquad (4)$$

leeds to the *normal approximation*

$$P\{T_\lambda \geq t_\lambda\} \approx P\{N(0,1) \geq t_\lambda\} \ . \qquad \text{(NA)}$$

for the significance level of the observed value $t_\lambda = (sd_\lambda - \mu_\lambda)/\sigma_\lambda$. The accuracy of (NA) increases with J, its error being $O(J^{-1/2})$.

Before deriving the moments μ_λ and σ_λ^2 in general, we note a remarkable approximation, whose accuracy depends on the harmonic mean

$$HM = \left[(N_1^{-1} + \cdots + N_J^{-1})/J\right]^{-1}$$

of the group sizes. For large J we get (cf. Osius and Rojek 1992)

$$\sigma_\lambda^2 \approx 2J(K-1) = 2(df+S), \qquad \text{if HM is large,} \tag{5}$$
$$\mu_\lambda \approx J(K-1) = df + S, \qquad \text{if } HM/\sqrt{J} \text{ is large,} \tag{6}$$

In view of $df + S \approx df$ (for large df) theses approximate moments agree with the moments of the classical χ^2-limit, and lead to comparable results for (NA) and (CCA), provided $HM/\sqrt{J}$ is large, too. The main disadvantage of the classical χ^2-approximation for large J stems from the fact that the expectation of SD_λ can differ markably from that of the limiting χ^2-distribution (i.e. df) if $HM/\sqrt{J}$ is not large enough, and this may result (see example 1 later) in completely misleading significance values based on (CCA).

For practical purposes, however, a smooth transition between the two approximations (CCA) and (NA) is available, which is due to one referee of the paper by Osius & Rojek (1992). The idea is to approximate the normal distribution $N(\mu, \sigma^2)$ by a scaled χ^2-distribution $\beta \cdot \chi_\nu^2$ with the same expectation $\mu = \beta\nu$ and variance $\sigma^2 = 2\beta^2\nu$. Hence the *rescaled power-divergence statistic* SD_λ/β_λ with

$$\beta_\lambda = \sigma_\lambda^2/2\mu_\lambda \tag{7}$$

has an approximate χ^2-distribution with a real-valued degree of freedom

$$\nu_\lambda = 2\mu_\lambda^2/\sigma_\lambda^2 \, , \tag{8}$$

provided ν_λ is large.

For fixed-cells asymptotic on the other hand, the approximations (5) and (6) yield $\beta_\lambda \approx 1$ and $\nu_\lambda - S \approx df$. Hence both approximations (CCA) and (NA) may be incorporated in the *rescaled* χ^2-*approximation*

$$P\{T_\lambda \geq t_\lambda\} \approx P\{\chi^2_{df_\lambda} \geq sd_\lambda/\beta_\lambda\} \tag{RCA}$$

with $df_\lambda = \nu_\lambda$ or preferably $df_\lambda = \nu_\lambda - S$ degrees of freedom, provided both are large. The approximation (RCA) will give comparable results to (CCA) resp. (NA) for large J and large $HM/\sqrt{J}$.

Expectation and Variance of the Statistics

We now derive the asymptotic expectation μ_λ and variance σ_λ^2 of the power-divergence statistic $SD_\lambda(\hat{\boldsymbol{\theta}})$ for large J, which are needed for the normal and rescaled χ^2-approximation. The derivation is only sketched here to give the general idea, referring the interested reader to Rojek (1989) or Osius & Rojek (1992) for details. We first look at the sum $SD_\lambda(\boldsymbol{\theta})$ with the true parameter $\boldsymbol{\theta}$ instead of its estimate, and denote its expectation and variance (under the model) by

$$\mu_\lambda(\boldsymbol{\theta}) \quad := \quad E_\theta\{SD_\lambda(\boldsymbol{\theta})\}\,, \tag{9}$$

$$v_\lambda^2(\boldsymbol{\theta}) \quad := \quad \mathrm{var}_\theta\{SD_\lambda(\boldsymbol{\theta})\}\,. \tag{10}$$

A first order expansion of the centered sum $Z_\lambda(\boldsymbol{\theta}) = SD_\lambda(\boldsymbol{\theta}) - \mu_\lambda(\boldsymbol{\theta})$ gives

$$Z_\lambda(\hat{\boldsymbol{\theta}}) = Z_\lambda(\boldsymbol{\theta}) + \mathbf{D}Z_\lambda(\boldsymbol{\theta})\cdot(\hat{\boldsymbol{\theta}} - \boldsymbol{\theta}) + O_p(1),$$

where $\mathbf{D}$ is the differential operator. Using the score vector $\mathbf{U}(\boldsymbol{\theta}) = \mathbf{D}^T l(\boldsymbol{\theta})$ and the information matrix $\mathbf{I}(\boldsymbol{\theta}) := \mathrm{cov}_\theta\{\mathbf{U}(\boldsymbol{\theta})\}$, we get the familiar expansion

$$\hat{\boldsymbol{\theta}} - \boldsymbol{\theta} = \mathbf{I}^{-1}(\boldsymbol{\theta})\cdot\mathbf{U}(\boldsymbol{\theta}) + O_p(n^{-1})\,.$$

The derivative of $Z_\lambda(\boldsymbol{\theta})$ may be written as $\mathbf{D}Z_\lambda(\boldsymbol{\theta}) = -\mathbf{c}_\lambda^T(\boldsymbol{\theta}) + O_p(n^{1/2})$ with the covariance vector

$$\mathbf{c}_\lambda(\boldsymbol{\theta}) := \mathrm{cov}_\theta\{SD_\lambda(\boldsymbol{\theta}), \mathbf{U}(\boldsymbol{\theta})\}\,. \tag{11}$$

This leads to the fundamental representation

$$Z_\lambda(\hat{\boldsymbol{\theta}}) = \psi_\lambda(\boldsymbol{\theta}) + O_p(1)\,. \tag{12}$$

with

$$\psi_\lambda(\boldsymbol{\theta}) := Z_\lambda(\boldsymbol{\theta}) - \mathbf{c}_\lambda^T(\boldsymbol{\theta})\cdot\mathbf{I}^{-1}(\boldsymbol{\theta})\cdot\mathbf{U}(\boldsymbol{\theta}) \tag{13}$$

Simple calculations give

$$E_\theta\{\psi_\lambda(\boldsymbol{\theta})\} = 0\,,$$

$$\sigma_\lambda^2(\boldsymbol{\theta}) := \mathrm{var}_\theta\{\psi_\lambda(\boldsymbol{\theta})\} = v_\lambda^2(\boldsymbol{\theta}) - Q_\lambda(\boldsymbol{\theta}), \tag{14}$$

with the quadratic form

$$Q_\lambda(\boldsymbol{\theta}) := \mathbf{c}_\lambda^T(\boldsymbol{\theta})\cdot\mathbf{I}^{-1}(\boldsymbol{\theta})\cdot\mathbf{c}_\lambda(\boldsymbol{\theta})\,. \tag{15}$$

The variance $\sigma_\lambda^2(\boldsymbol{\theta})$ typically increases with J and, in fact, the following *variance condition*

$$J/\sigma_\lambda^2(\boldsymbol{\theta}) \quad \text{is bounded as } J \to \infty \tag{VC}$$

is assumed in Osius & Rojek (1992) to derive the asymptotic normality of T_λ. The asymptotic expectation and variance of SD_λ are now taken from (9) and (14) evaluated at the estimate $\hat{\boldsymbol{\theta}}$

$$\mu_\lambda := \mu_\lambda(\hat{\boldsymbol{\theta}}) \tag{16}$$

$$\sigma_\lambda^2 := \sigma_\lambda^2(\hat{\boldsymbol{\theta}}) \ . \tag{17}$$

Using (VC) and the consistency of $\hat{\boldsymbol{\theta}}$, we get from (12) the following important repesentation of the standardized power-divergence statistic

$$T_\lambda = \psi_\lambda(\boldsymbol{\theta})/\sigma_\lambda(\boldsymbol{\theta}) + O_p(J^{-1/2}) \ . \tag{18}$$

We now turn to the computation of $\mu_\lambda(\boldsymbol{\theta}), v_\lambda^2(\boldsymbol{\theta})$ and $\mathbf{c}_\lambda(\boldsymbol{\theta})$. For Pearson's statistic X^2 (i.e. $\lambda = 1$) the expectation (9) is constant

$$\mu_1(\boldsymbol{\theta}) = J(K-1) \ .$$

The variance (10) is given by (cf. McCullagh & Nelder (1989, p.169))

$$v_1^2(\boldsymbol{\theta}) = 2J(K-1) + \sum_j \frac{1}{N_j}\left[\sum_k \frac{1}{\pi_{jk}(\boldsymbol{\theta})} - K^2 - 2(K-1)\right] ,$$

and the covariance vector (11) does not depend on the group sizes N_j:

$$c_{1s}(\boldsymbol{\theta}) = \sum_j \sum_k \frac{1}{\pi_{jk}(\boldsymbol{\theta})} \cdot \frac{\partial}{\partial \theta_s} \pi_{jk}(\boldsymbol{\theta}) \ , \qquad s = 1, ..., S \ .$$

Note in particular, how small expected counts $N_j \pi_{jk}(\boldsymbol{\theta})$ can blow up the variance $v_1^2(\boldsymbol{\theta})$.

Unfortunately, explicit expressions like the ones given above are not available for general λ. Nevertheless, the moments $\mu_\lambda(\boldsymbol{\theta})$, $v_\lambda^2(\boldsymbol{\theta})$ and $\mathbf{c}_\lambda(\boldsymbol{\theta})$ can be computed for any λ and $\boldsymbol{\theta}$ in a straightforward way. Introducing for a multinomial $M_K(N, \boldsymbol{\pi})$ random vector $\mathbf{Y}$ the notations

$$e_\lambda(N, \boldsymbol{\pi}) := E_\pi\{A_\lambda(\mathbf{Y}, N\boldsymbol{\pi})\} \tag{19}$$

$$v_\lambda^2(N, \boldsymbol{\pi}) := var_\pi\{A_\lambda(\mathbf{Y}, N\boldsymbol{\pi})\} \tag{20}$$

$$c_{\lambda k}(N, \boldsymbol{\pi}) := cov_\pi\{Y_k, A_\lambda(\mathbf{Y}, N\boldsymbol{\pi})\} \qquad \text{for } k = 1, ..., K, \tag{21}$$

the expectation (9) may be written as

$$\mu_\lambda(\boldsymbol{\theta}) = E_\theta\left\{\sum_j A_\lambda(\mathbf{Y}_j, N_j \boldsymbol{\pi}_j(\boldsymbol{\theta}))\right\} = \sum_j e_\lambda(N_j, \boldsymbol{\pi}_j(\boldsymbol{\theta})) \ . \tag{22}$$

And similiarly the independence of all rows $\mathbf{Y}_j$ yields

$$v_\lambda^2(\boldsymbol{\theta}) = \sum_j v_\lambda^2(N_j, \boldsymbol{\pi}_j(\boldsymbol{\theta})) , \quad (23)$$

$$\mathbf{c}_\lambda(\boldsymbol{\theta}) = \left(\sum_j \sum_k \mathbf{c}_{\lambda k}(N_j, \boldsymbol{\pi}_j(\boldsymbol{\theta})) \cdot \frac{\partial}{\partial \theta_s} \log \pi_{jk}(\boldsymbol{\theta})\right)_{s=1,\ldots,S} . \quad (24)$$

The moments (19) to (21) can be computed using the definition of an expectation

$$E\{f(\mathbf{Y})\} = \sum_{\mathbf{y}} P\{\mathbf{Y} = \mathbf{y}\} \cdot f(\mathbf{y}) , \quad (25)$$

where the sum extends over all outcomes of the underlying multinomial distribution. The computational burden increases heavily with the number K of response categories, but is acceptable for moderate K and in particular for the binomial case $K = 2$, later to be treated in detail.

Edgeworth and Saddlepoint Expansions

Let us now turn to a different approach to approximate the tail probabilities of T_λ, which is based on the first four cumulants only and not on a particular limiting distribution (like χ^2 or normal) for the power-divergence statistic. For notational convenience, the parameter λ – being fixed in the general discussion to follow – is omitted as an index. The third and fourth standardized cumulants ρ_3 and ρ_4 of T are measures of *skewness* and *kurtosis* of T (and of SD).

A direct Edgeworth expansion leads to the *Edgeworth approximation* (cf. Barndorff-Nielsen & Cox (1989, Sec. 4.2))

$$P\{T \geq t\} \approx 1 - \Phi(t) + \varphi(t)\left[\frac{1}{6}\rho_3 H_2(t) + \frac{1}{24}\rho_4 H_3(t) + \frac{1}{72}\rho_3^2 H_5(t)\right] . \quad \text{(EA)}$$

Here Φ and φ denote the distribution and density function of the standard normal distribution $N(0,1)$, and $H_2(x) = x^2 - 1$, $H_3(x) = x^3 - 3x$ and $H_5(x) = x^5 - 10x^3 + 15x$ are Hermite polynomials.

For fixed t, the Edgeworth expansion is of order $O(J^{-3/2})$ and typically quite accurate around the center $t = 0$ but less reliable in the tails, where $H_3(t)$ and $H_5(t)$ will be appreciable and may even cause negative values for (EA), cf. example 1 later. This deficiency can be tempered by using a tilted Edgeworth or saddlepoint expansion, which is basically an Edgeworth expansion for a modified distribution centered around the argument of interest, see Barndorff-Nielsen & Cox (1989), Sec. 4.3, for details on the following discussion. More precisely, if $M(s)$ resp. $K(s) = \log M(s)$ are the moment resp. cumulant generating function of the standardized variable $Z(\hat{\boldsymbol{\theta}})/\sigma$, we first seek a root $\hat{\tau}$ of the equation

$$K'(\hat{\tau}) = t \tag{26}$$

in order to obtain

$$\begin{aligned} r &= \operatorname{sgn}(\hat{\tau}) \cdot [2\hat{\tau} \cdot K'(\hat{\tau}) - 2K(\hat{\tau})]^{1/2} \\ v &= \hat{\tau} \cdot [K''(\hat{\tau})]^{1/2} . \end{aligned} \tag{27}$$

The *saddlepoint approximation* of the significance level is now given by

$$P\{T \geq t\} \approx 1 - \Phi(r) - \varphi(r) \left[\tfrac{1}{r} - \tfrac{1}{v}\right] \tag{SA}$$

with error $O(J^{-1})$ uniformly in the observed value t of T. The computation of $\hat{\tau}$, r and v requires knowledge of the cumulant generation function $K(s)$, which in turn characterizes the unknown distribution. However, Taylor expansions are available

$$\begin{aligned} K(s) &\approx \frac{1}{2}s^2 + \frac{1}{6}\rho_3 s^3 + \frac{1}{24}\rho_4 s^4 , \\ K'(s) &\approx s + \frac{1}{2}\rho_3 s^2 + \frac{1}{6}\rho_4 s^3 , \\ K''(s) &\approx 1 + \rho_3 s + \frac{1}{2}\rho_4 s^2 , \end{aligned} \tag{28}$$

which can be used to approximate $\hat{\tau}$, r and v.

The Edgeworth and saddlepoint approximations and (28) depend on the cumulants ρ_3 and ρ_4 of T, which are difficult to obtain. In view of (18) these cumulants can be approximated with an error $O(J^{-1/2})$ by the corresponding *standardized* cumulants $\rho_3(\boldsymbol{\theta})$ and $\rho_4(\boldsymbol{\theta})$ of $\psi(\boldsymbol{\theta})$, which are given in the next section, cf. (30), (33) and (34). Evaluating the cumulants at the estimate $\hat{\boldsymbol{\theta}}$ leads to the approximation

$$\rho_k \approx \rho_k(\hat{\boldsymbol{\theta}}) \quad \text{for } k = 3, 4 \tag{29}$$

with error $O_p(J^{-1/2})$. Using these approximations in the Edgeworth expansion increases the error in (EA) to the order $O_p(J^{-1})$. The additional error in (SA) due to (28) and (29) is not so easily tractable.

Skewness and Kurtosis of the Statistics

In this section we derive the cumulants $\kappa_{\lambda 3}(\boldsymbol{\theta})$ and $\kappa_{\lambda 4}(\boldsymbol{\theta})$ of $\psi_\lambda(\boldsymbol{\theta})$ – with λ reintroduced as an index – from which the skewness and kurtosis are easily obtained as the standardized cumulants

$$\rho_{\lambda k}(\boldsymbol{\theta}) = \kappa_{\lambda k}(\boldsymbol{\theta})/\sigma_\lambda^k(\boldsymbol{\theta}) \quad \text{for } k = 3, 4 \; . \tag{30}$$

The random variable $\psi_\lambda(\boldsymbol{\theta})$ may be written as a sum

$$\psi_\lambda(\boldsymbol{\theta}) = \sum_j \psi_{\lambda j}(\boldsymbol{\theta})$$

with a contribution $\psi_{\lambda j}(\boldsymbol{\theta}) := Z_{\lambda j}(\boldsymbol{\theta}) + V_{\lambda j}(\boldsymbol{\theta})$ from each group j with expectation 0, consisting of the sum of the centered derivation

$$Z_{\lambda j}(\boldsymbol{\theta}) := A_\lambda(\mathbf{Y}_j, N_j \boldsymbol{\pi}_j(\boldsymbol{\theta})) - e_\lambda(N_j, \boldsymbol{\pi}_j(\boldsymbol{\theta}))$$

and a linear combination $V_{\lambda j}(\boldsymbol{\theta}) := \mathbf{d}_{\lambda j}^T(\boldsymbol{\theta}) \cdot \mathbf{Y}_j$ given by the vector

$$\mathbf{d}_{\lambda j}^T(\boldsymbol{\theta}) := -\mathbf{c}_\lambda^T(\boldsymbol{\theta}) \cdot \mathbf{I}^{-1}(\boldsymbol{\theta}) \cdot \mathbf{D}^T \log \boldsymbol{\pi}_j(\boldsymbol{\theta}) \; ,$$

which is orthogonal to $\boldsymbol{\pi}_j$. Since the groups are independent under product-multinomial sampling, the m-th cumulant $\kappa_{\lambda m}(\boldsymbol{\theta})$ of $\psi_\lambda(\boldsymbol{\theta})$ can be obtained as a sum

$$\kappa_{\lambda m}(\boldsymbol{\theta}) = \sum_j \kappa_m(\psi_{\lambda j}(\boldsymbol{\theta})) \; .$$

Introducing for a multinomial $M_K(N, \boldsymbol{\pi})$ vector $\mathbf{Y}$ and a vector $\mathbf{d} \in I\!R^K$ orthogonal to $\boldsymbol{\pi}$ the notation

$$\mu_{\lambda m}(N, \boldsymbol{\pi}, \mathbf{d}) := E_\pi\{[A_\lambda(\mathbf{Y}, N\boldsymbol{\pi}) - e_\lambda(N, \boldsymbol{\pi}) + \mathbf{d}^T\mathbf{Y}]^m\}, \tag{31}$$

we can express the (central) moments of the random variable $\psi_{\lambda j}(\boldsymbol{\theta})$ as

$$E_\theta\{\psi_{\lambda j}^m(\boldsymbol{\theta})\} = \mu_{\lambda m}(N_j, \boldsymbol{\pi}_j, \mathbf{d}_{\lambda j}(\boldsymbol{\theta})) \; . \tag{32}$$

The cumulants are finally obtained from the central moments, using the relations $\kappa_3 = \mu_3$ and $\kappa_4 = \mu_4 - 3\mu_2^2$, as follows

$$\kappa_{\lambda 3}(\boldsymbol{\theta}) = \sum_j \mu_{\lambda 3}(N_j, \boldsymbol{\pi}_j, \mathbf{d}_{\lambda j}(\boldsymbol{\theta})) \; , \tag{33}$$

$$\kappa_{\lambda 4}(\boldsymbol{\theta}) = \sum_j \left[\mu_{\lambda_4}(N_j, \boldsymbol{\pi}_j, \mathbf{d}_{\lambda j}(\boldsymbol{\theta})) - 3\mu_{\lambda 2}^2(N_j, \boldsymbol{\pi}_j, \mathbf{d}_{\lambda j}(\boldsymbol{\theta}))\right] \; . \tag{34}$$

For Pearson's statistic X^2, explicit expressions in terms of multinomial moments are available for the moments (31), which are given in (44) for the binomial case $K = 2$. But for general λ, these moments can only be evaluated using the definition (25).

Bootstrapping the Statistics

A radically different approach to the asymptotic expansions discussed so far is based on a simulation technique, the bootstrap method introduced by Efron (1979) – more recent accounts are given in the book Hall (1992) and the conference proceedings LePage & Billard (1992) and Jöckel et al. (1992). The increasing access of statisticians to powerful and fast computers has established simulation techniques as an important tool to investigate the accuracy of approximations derived from asymptotic results. A simulation study to examine the adequacy of the proposed approximations can only cover particular choices for: (a) the model, given by K and functions G_k, (b) the number J of groups, (c) the sizes N_j and covariates $\mathbf{x}_j$, and (d) the parameter $\boldsymbol{\theta}$. However, in any practical application of a goodness-of-fit test, (a) to (c) are given and a consistent estimate $\hat{\boldsymbol{\theta}} = \boldsymbol{\theta} + O_p(n^{-1/2})$ is available for the unknown parameter. This suggests the following simulation technique known as the *parametric bootstrap*.

A contingency table $\mathbf{Y}^* = (Y^*_{jk})$ is called a *resample* of the original counts $\mathbf{Y} = (Y_{jk})$, if its distribution has the same parametric form as the distribution of $\mathbf{Y}$, with the unknown parameter $\boldsymbol{\theta}$ replaced by its estimate $\hat{\boldsymbol{\theta}}$ based on $\mathbf{Y}$, i.e. all rows $\mathbf{Y}^*_j$ are independent with a $M_K(N_j, \boldsymbol{\pi}_j(\hat{\boldsymbol{\theta}}))$ distribution.

All terms derived from the resample $\mathbf{Y}^*$ are marked by a star, e.g. $\hat{\boldsymbol{\theta}}^*$ is the estimate and $T^* = (SD^* - \mu^*)/\sigma^*$ the standardized power-divergence for $\mathbf{Y}^*$. The distribution of T is now estimated by the distribution of its resample T^*, which leads to the following *bootstrap estimate* for the significance level

$$p := P\{T \geq t\} \approx P\{T^* \geq t\} \; . \tag{BE}$$

Since $\mathbf{T}$ is (asymptotically) pivotal (i.e. its limit distribution does not depend on the parameter $\boldsymbol{\theta}$), the accuracy of this estimate can be obtained as follows (see Hall & Titterington (1989) or Hall (1992, sec. 3.1)). The difference between the simple Edgeworth expansions (using the H_2 term only) for both sides in (BE) has the same order as

$$\rho_3 - \rho_3^* = O_p(J^{-1/2} n^{-1/2}) = AM^{-1/2} \cdot O_p(J^{-1}) \; , \tag{35}$$

since $\rho_3 = O(J^{-1/2})$. Hence the error in (BE) is for a *bounded* arithmetic mean AM of the group sizes not larger than the error $O_p(J^{-1})$ in (EA), but may be considerably less for *unbounded* AM. It must be emphazised, that the order of the error need not be maintained if we bootstrap SD (which is not pivotal) instead of T (Hall and Wilson 1991). Note however, that $\mathbf{T}$ is still pivotal under *fixed-cells* asymptotics, having a standardized χ^2 limit.

The *bootstrap approximation* for p is obtained from a sufficiently large number R of independent resamples $\mathbf{Y}^*_1, ..., \mathbf{Y}^*_R$ as the observed frequency

$$\hat{p} = \#\{r | T_r^* \geq t\}/R \ . \tag{BA}$$

and has variance $\hat{p}(1-\hat{p})/R$. The number R of resamples can be chosen according to the accuracy desired for $\hat{p}$. For less extreme tail probabilities, e.g. $p \geq 1\%$, about $R = 10\ 000$ resamples should be sufficient.

The generation of a resample $\mathbf{Y}^*$ requires – independently for each group – the generation of a $M_K(N, \boldsymbol{\pi})$ random vector for given N and $\boldsymbol{\pi}$, which can be taken as the sum of N independent $M_K(1, \boldsymbol{\pi})$ random vectors. And drawing a random vector $\mathbf{Z}$ from a $M_K(1, \boldsymbol{\pi})$ distribution merely requires a uniform random number U from the unit interval (0,1). The unique category $k = 1, ..., K$ for which $Z_k = 1$ being given by the condition $\gamma_{k-1} < U \leq \gamma_k$, where $\gamma_k = \pi_1 + \cdots + \pi_k$ are the cumulative probabilities.

The bootstrap method is time-consuming, because for each resample $\mathbf{Y}^*$ the estimate $\hat{\boldsymbol{\theta}}^*$ has to be computed and this typically requires an iterative procedure. In this case it is reasonable to reduce the accuracy of the fitting procedure and stop the iteration if the deviance D (being among the statistics of interest) is suitably stable. However, a more radical one-step procedure (starting with the original estimate $\hat{\boldsymbol{\theta}}$) can not be generally recommended, because from our experience the power-divergence statistics thus obtained are not stable enough in sparse data.

The bootstrap estimate of the expectation μ, variance σ^2 and higher cumulants ρ_m of T are the corresponding cumulants of T^*. The error here is $O_p(n^{-1/2}) = AM^{-1/2}\ O_p(J^{-1/2})$ which is, for *bounded AM*, of the same order $O_p(J^{-1/2})$ as the error in the approximations (16), (17) and (29) based on (18), but may again be considerably less for *unbounded AM*.

Binomial Models (Binary Data)

In many situations the response of interest may be classified in two categories, usually termed "success" and "failure". The previous presentation for general K will now be specialized to $K = 2$, leading to some simplifications. For notational simplicity, the row counts $\mathbf{Y}_j = (Y_{j1}, Y_{j2})$ are identified with the first component (success), i.e. $Y_j := Y_{j1}$, the "failures" being determined by $Y_{j2} = N_j - Y_j$. The product-multinomial sampling model now states that all Y_j are independent, each having a binomial $B(N_j, \pi_j)$ distribution, where $(\pi_j, 1-\pi_j)$ corresponds to our previous probability vector $\boldsymbol{\pi}_j$. And the model (1) reduces to a generalized linear model

$$\pi_j = \pi_j(\boldsymbol{\theta}) = G(\mathbf{x}_j^T\boldsymbol{\theta}) \qquad \text{resp.} \qquad \eta_j := g(\pi_j) = \mathbf{x}_j^T\boldsymbol{\theta} \tag{36}$$

with link function $g = G^{-1}$ (cf. McCullagh & Nelder (1989)). The "distance measure" $A_\lambda(p, \pi)$ for $0 \leq p \leq 1$ and $0 < \pi < 1$ is given by

$$A_\lambda(p, \pi) = a_\lambda(p, \pi) + a_\lambda(1-p, 1-\pi)\ . \tag{37}$$

Fixing π, the function $A_\lambda(-,\pi)$ is convex with a minimum $A_\lambda(\pi,\pi)=0$, leading to bounds

$$\begin{array}{rclr} A_\lambda(p,\pi) & \leq & A_\lambda(0,\pi) & \text{for } p \leq \pi \\ A_\lambda(p,\pi) & \leq & A_\lambda(1,\pi) & \text{for } p \geq \pi\ . \end{array} \tag{38}$$

And the power-divergence statistic is

$$SD_\lambda(\hat{\boldsymbol{\theta}}) = \sum_j N_j A_\lambda(\hat{P}_j, \hat{\pi}_j)\ , \tag{39}$$

with $\hat{P}_j := Y_j/N_j$ as the observed proportion of success in group j. The expectation (25) required to compute the moments reduces to

$$E\{f(Y)\} = \sum_{y=0}^{N} b(y) \cdot f(y)\ , \tag{40}$$

where Y is $B(N,\pi)$-distributed with probabilities $b(y) = P\{Y=y\}$. Unless N is small, it is advisable to compute the sum by starting with the "middle term" $y_0 := \mathrm{Int}(N\pi)$ where $b(y)$ attains its maximum, and then working downwards $y_0 - 1, \cdots$, as well as upwards $y_0 + 1, \cdots$ using the recurrence relation between successive values of $b(y)$. For large N (> 100, say) it may not be necessary to complete the sum down to $y = 0$ or up to $y = N$ because the individual contributions become too small to affect the internal representation of the sum in the computer. Since (38) provides bounds for the functions $f(y)$ of interest, an appreciable lower and upper part of the sum may be omitted for large N.

The vector $\mathbf{c}_\lambda$ from (21) is determined by the value

$$c_\lambda(N,\pi) = \mathrm{cov}\{Y, A_\lambda(Y, N\pi)\} \tag{41}$$

for a $B(N,\pi)$-variable Y via $c_{\lambda 1} = c_\lambda$ and $c_{\lambda 2} = -c_\lambda$. And the components of covariance vector (24) reduce to

$$c_{\lambda s}(\boldsymbol{\theta}) = \sum_j x_{js} \cdot c_\lambda(N_j, \pi_j) \cdot G'(\eta_j)/\pi_j(1-\pi_j), \qquad s = 1, ..., S. \tag{42}$$

In order to evaluate the cumulants $\kappa_{\lambda m}(\boldsymbol{\theta})$ of $\psi_\lambda(\boldsymbol{\theta})$ for $m = 3, 4$ we introduce for a $B(N,\pi)$-variable Y and $d \in I\!R$ the following moments corresponding to (31)

$$\mu_{\lambda m}(N,\pi,d) = E_\pi\{[A_\lambda(Y,N\pi) - e_\lambda(N,\pi) + d(Y - N\pi)]^m\}\ . \tag{43}$$

Using this notation, equation (32) remains valid with the vector $\mathbf{d}_{\lambda j}(\boldsymbol{\theta})$ replaced by the scalar

$$d_{\lambda j}(\boldsymbol{\theta}) := -\mathbf{c}_\lambda^T(\boldsymbol{\theta}) \cdot \mathbf{I}^{-1}(\boldsymbol{\theta}) \cdot \mathbf{b}_j(\boldsymbol{\theta}),$$

where the vector $\mathbf{b}_j(\boldsymbol{\theta}) \in I\!R^S$ is defined by

$$b_{js}(\boldsymbol{\theta}) := x_{js} \cdot G'(\eta_j)/\pi_j(1-\pi_j) \ .$$

For Pearson's statistic X^2 we get the familiar quadratic "distance"

$$A_1(p,\pi) = \frac{(p-\pi)^2}{\pi(1-\pi)} \ ,$$

from which the moments (19-20) and (41) are easily obtained as

$$\begin{aligned}
e_1(N,\pi) &= 1 \ , \\
v_1^2(N,\pi) &= 2 + \frac{1}{N}\left[\frac{1}{\pi(1-\pi)} - 6\right] \ , \\
c_1(N,\pi) &= 1 - 2\pi \ .
\end{aligned}$$

Furthermore, the moments (43) can be computed for $\lambda = 1$ using the first eight moments μ_m of the *standardized* binomial $B[N,\pi - N\pi]/s$, with $s^2 = N\pi(1-\pi)$, obtained from a recurrence formulae (cf. Johnson and Kotz 1969, chap. 3):

$$\begin{aligned}
\mu_3 &= (1-2\pi)/s, \\
\mu_4 &= 3(1-2/N) + 1/s^2, \\
\mu_5 &= \mu_3 \cdot [2(5-6/N) + 1/s^2], \\
\mu_6 &= 5(3 - 26/N + 24/N^2) + 5(5-6/N)/s^2 + 1/s^4, \\
\mu_7 &= \mu_3 \cdot [3(35 - 154/N + 120/N^2) + 4(14-15/N)/s^2 + 1/s^4], \\
\mu_8 &= 7[(15 - 340/N + 1044/N^2 - 720/N^3) \\
&\quad +2(35 - 154/N + 120/N^2)/s^2 + (17-18/N)s^4] + 1/s^6,
\end{aligned}$$

thus giving the moments

$$\begin{aligned}
\mu_{12}(N,\pi,d) &= \mu_4 - 1 + 2ds\mu_3 + d^2s^2, \qquad (44) \\
\mu_{13}(N,\pi,d) &= (\mu_6 - 3\mu_4 + 2) + 3ds(\mu_5 - 2\mu_3) \\
&\quad +3d^2s^2(\mu_4 - 1) + d^3s^3\mu_3, \\
\mu_{14}(N,\pi,d) &= (\mu_8 - 4\mu_6 + 6\mu_4 - 3) + 4ds(\mu_7 - 3\mu_5 + 3\mu_3) \\
&\quad +6d^2s^2(\mu_6 - 2\mu_4 + 1) + 4d^3s^3(\mu_5 - \mu_3) + d^4s^4\mu_4.
\end{aligned}$$

Examples

Let us now look at two examples for binary data, one study on infant mortality and one on cancer. Both examples will serve us merely to illustrate the different approximations for the significance level of power-divergence statistics, and no attempt is made here to discuss the studies in detail or to judge the fit using alternative methods.

Example 1: A Study on Infant Mortality

In a larger study on infant mortality Karn and Penrose (1951-52) analyzed data assembled from records of U.C.H. Obstetric Hospital for the years 1935-46, containing information on 13 730 infants (7037 male, 6693 female, no twins) and their mothers. We are interested here only in parts of the data (table 1 from Karn and Penrose), which relates non-survival at 28 days (including stillbirth), regarded as a response, to the following variables:

- birth weight W, recorded in 25 classes: 1.0 (0.5) 13.5 lb.
- gestation time T, recorded in 41 classes: 155 (5) 355 days
- gender G of infant, recorded as a factor: 1=male, 2=female .

Karn and Penrose fitted a linear logistic model (to the survival rate) separately for males and females, using the model

$$1 + W + W^2 + T + T^2 + W.T$$

with $S = 6$ parameters (for the symbolic notation of models see McCullagh & Nelder (1989, sec. 3.4)). We investigate the fit of this model only for the *female* infants in more detail.

The classification of all (female) infants according to W and T yields $J = 345$ groups with a wide range of sizes from 1 to 265 (see table 1) which is typical for larger studies with several observed variables. The relevant information on the power-divergence statistics D, X^2, FT and CR is summarized in table 2. The four statistics differ dramatically, and so do the conclusions based on the classical χ^2-approximation with $df = 339$. However, this approximation is not justified here, due to considerable sparseness of the data. The large fraction of very low group sizes N_j leads to a low harmonic mean $HM = 2.4$ (and a moderate arithmetic mean $AM = 19.4$). This causes extremely small fitted counts for many groups, the overall (female) mortality rate was only 4.1%. Consequently, the asymptotic expectation and variance of these statistics are quite distinct from those of the limiting χ^2-distribution.

Passing to the approximation based on the normal- or the rescaled χ^2-approximation with $df = \nu$ gives comparable results (being quite distinct from the classical χ^2-approximation). The reduced $df = \nu - S$ yields only for X^2 a considerable decrease of the significance level, but the reduction may not be justified here, since ν is not large. Taking skewness and kurtosis into account, which are not negligible for X^2 and CR, the saddlepoint approximation may be more trustworthy than the Edgeworth expansion,

Size	1	2 – 3	4 – 5	6 – 9	10 – 19	20 +
Percent	27	24	10	10	10	19

Table 1: Distribution of the J=345 group sizes N_j for all female infants in example 1.

statistic		FT	D	CR	X^2
parameter λ		$-1/2$	0	2/3	1
power-divergence	SD	442.8	325.7	344.6	402.6
rescaled:	SD/β	550.2	523.3	204.4	22.2
	$df = \nu$	506.3	455.7	171.6	19.0
standardized:	T	1.38	2.24	1.77	0.51
cumulants of SD					
expectation	μ	407.5	283.6	289.4	345.0
variance	σ^2	655.9	353.1	975.9	12526.8
skewness	ρ_3	0.1515	0.1669	2.1663	21.5315
kurtosis	ρ_4	0.0339	0.0439	21.6770	779.4321
P–level approximations					
classical χ^2:	$df = 339$	0.01%	68.81%	40.47%	0.99%
rescaled χ^2:	$df = \nu$	8.68%	1.53%	4.42%	27.58%
	$df = \nu - 6$	6.09%	0.92%	2.17%	5.28%
normal		8.41%	1.25%	3.84%	30.34%
Edgeworth		8.72%	1.61%	5.75%	(< 0)
saddlepoint		8.73%	1.61%	8.37%	8.73%
bootstrap		5.40%	0.77%	2.38%	7.54%
± S.E.		± 0.23%	± 0.09%	± 0.13%	± 0.26%
bootstrapped cumulants of T					
expectation		– 0.2352	– 0.2636	– 0.2232	– 0.1065
variance		0.9673	0.9614	0.8518	0.5457
skewness	ρ_3	0.1712	0.1991	1.1392	9.2451
kurtosis	ρ_4	– 0.0324	0.0137	5.7147	171.0157

Table 2: Power-divergence statistics, cumulants and significance levels for all female infants in example 1 (study on infant mortality). The bootstrap results are based on R = 10 000 resamples.

which leads to a "negative probability" for X^2. Looking finally at the bootstrap estimate for the significance level (based on $R = 10000$ resamples), we find a remarkable agreement with the rescaled χ^2-approximation based on $df = \nu - S$ for D, FT and CR, but less for X^2 (where the latter approximation is questionable, since ν is not large).

Concerning the final decision of the goodness-of-fit test based on the deviance D, the model is rejected on the common 5% level no matter which approximation is used (except the inadequate classical χ^2). But only the bootstrap and the rescaled χ^2 approximation with reduced *df* also reject the model on the 1% level (which may be appropriate in view of the large sample size). However, the test based on X^2, does not reject the model on the 5% level for any approximation, except the inadequate classical χ^2.

Passing to the corresponding analysis of fit for the *males* we only note, that the bootstrapped significance levels ranged from 21% (for X^2) to 35% (for FT) and again were very near the rescaled χ^2 levels (based on the reduced *df*), except for X^2, thus not rejecting the model. For X^2, the Edgeworth and even the saddlepoint approximations gave "negative" P-levels, due to an extremely high kurtosis. But for the remaining statistics these approximation were consistent with the bootstrapped value.

Turning to the cumulants of T, we observe for the females (cf. table 1), that the bootstrap approximation to the expectation is clearly below the nominal value of 0, and this also holds for the males. The bootstrapped variance is also well below the nominal value of 1 for CR and X^2 among the females, (but not among the males).

Example 2: Ille-et-Vilaine Study on Oesophageal Cancer

The data are taken from the Ille-et-Vilaine study on oesophageal cancer as given in Appendix I of Breslow & Day (1980). This is a retrospective case-control study with 975 individuals (200 cases and 775 controls), classified according to the 3 covariables: AGE (6 classes), alcohol (ALC) and tobacco (TOB) consumption (4 classes each). Of the 96 possible combinations only $J = 88$ different groups contained at least one individual at risk, with group sizes from 1 to 60 (see also table 3). Although the sampling scheme of this study is not of the product-binomial type assumed here, we nevertheless apply the approximations for illustration and numerical comparison with significance levels for conditional tests suggested by McCullagh (1985ab).

We only look at the model 1+AGE+ALC+TOB with $S = 12$ parameters, containing the main effects of the covariables viewed as factors. The relevant

Size	1	2 – 3	4 – 5	6 – 9	10 – 19	20 +
Percent	14	17	14	18	22	16

Table 3: Distribution of the J=88 group sizes N_j for example 2.

statistic		FT	D	CR	X^2
parameter λ		$-1/2$	0	2/3	1
power-divergence	SD	112.4	82.3	79.4	86.6
rescaled:	SD/β	95.7	104.9	76.0	32.8
	$df = \nu$	98.5	107.2	76.8	33.3
standardized:	T	-0.20	-0.16	-0.06	-0.07
cumulants of SD					
expectation	μ	115.6	84.2	80.2	88.0
variance	σ^2	271.4	132.2	167.7	464.4
skewness	ρ_3	0.3121	0.2749	0.5298	1.8451
kurtosis	ρ_4	0.1626	0.1192	0.9143	9.9206
P–level approximations					
classical χ^2:	$df = 76$	0.43%	28.98%	37.21%	19.13%
rescaled χ^2:	$df = \nu$	55.98%	54.51%	50.39%	49.42%
	$df = \nu - 12$	23.28%	23.34%	16.08%	5.37%
normal		57.78%	56.27%	52.53%	52.67%
Edgeworth		55.83%	54.51%	49.16%	41.90%
saddlepoint		55.84%	54.51%	49.18%	42.65%
bootstrap		24.19%	20.52%	17.54%	16.92%
± S.E.		± 0.43%	± 0.44%	± 0.38%	± 0.37%
bootstrapped cumulants of T					
expectation		– 0.8204	– 0.9233	– 0.7905	– 0.4634
variance		0.8771	0.8788	0.7081	0.5904
skewness	ρ_3	0.2705	0.2551	0.3426	0.6669
kurtosis	ρ_4	0.0680	0.0771	0.5401	2.4653

Table 4: Power-divergence statistics, cumulants and significance levels for example 2 (Ille-et-Vilaine study on oesophagal cancer). The bootstrap results are based on $R = 10\ 000$ resamples.

information on the four power-divergence statistics D, X^2, FT and CR is summarized in table 4.

The classical χ^2-approximation gives quite different P-levels (the one for FT being extremely low), and again appears unreliable in view of the sparseness of the data (cf. table 3). The normal and rescaled χ^2-approximation with $df = \nu$ gives comparable results for all four statistics, which are more than halved in passing to the reduced $df = \nu - S$, which again may be questionable (for X^2), since ν is not large. For each statistic, the Edgeworth and saddlepoint expansion give almost identical values differing noticably from the normal approximation only for X^2 (which could be expected from the higher cumulants). The significance level based on the bootstrap with $R = 10\ 000$ resamples show moderate variation across the four statistics and are again very close to those from the rescaled χ^2-approximation with reduced $df = \nu - S$ (except for X^2), but are considerably smaller then those based on the other approximations. One possible explanation is that the approximation (16) for the expectation of SD is not accurate enough here, since the bootstrapped expectation of T is clearly below 0 (cf. table 4).

McCullagh (1985ab) computed the *conditional* cumulants of X^2 given $\hat{\boldsymbol{\theta}} = \boldsymbol{\theta}$ as $\kappa_1 = 77.38$, $\kappa_2 = 401.1$ and $\rho_3 = 1.98$. This leads to a conditional significance level of 23.1% (based on an simple Edgeworth expansion, using only the H_2 term), which is quite different from our bootstrapped value.

Hence the model is not rejected by any of the four statistics, no matter which approximation is used (except for FT and the inadequate classical χ^2), in agreement with other considerations by Breslow and Day (1980, Sec.6.5) confirming a satisfactory fit.

Discussion

The following remarks – reflecting our present theoretical knowledge and (limited) practical experience with power-divergence statistics – should only be taken as rough guidelines for applications. It should also be borne in mind that we have assumed a *large sample size* (at least a few hundred, say) and *models with large degrees of freedom* $J - S$ (at least 50, say), and hence a large number J of multinomial groups throughout, with no attempt to pursue the accuracy of the approximations if these assumptions are violated.

1. Although a goodness-of-fit test is usually based on a *single* power-divergence statistic, like the deviance D or Pearson's X^2, the calculation of several statistics SD_λ, at least D and X^2 but preferably also FT and CR, turns out to be a useful diagnostic tool to detect low expected counts.

2. The classical χ^2-approximation for SD_λ with $df = J - S$ is reliable only if all expected counts are not too small, which generally leads to comparable values for the statistics D, X^2, FT and CR, provided the specified model is correct.

3. If considerable variation is observed between D and X^2 (and FT, CR), leading to markedly different P-levels for the classical χ^2-approximation, then the normal- or rescaled χ^2-approximation are preferable, which typically give similar results. The latter is preferable, using the reduced $df = \nu - S$ for large ν, because it provides a smooth transition between the classical χ^2- and the normal approximation (and was very close to the bootstrapped value in our examples, except for X^2).

4. Calculation of the third and fourth cumulant together with the Edgeworth or preferably the saddlepoint approximation is highly desirable to detect skewness and kurtosis. Particulary for sparse data the examples revealed an appreciably skewness and kurtosis of X^2 (but not for D and FT), leading to an insufficient accuracy of the saddlepoint approximation (SA), which may partly be due to extreme discreteness in the upper tail of X^2. In this situation the deviance D and FT may be favoured for a goodness-of-fit test, since they have less skewness and curtosis (leading to comparable results for *all* approximations in the examples). Pearson's X^2, however, has other advantages: its expectation is constant, and explicit expressions for all relevant moments simplifies its computation.

5. The parametric bootstrap is intuitively appealing and at present seems to be the most accurate way to approximate the significance level of any T_λ. Furthermore, the bootstrap remains valid also under fixed-cells asymptotics, thus eliminating the need to chose between the two asymptotics in practice. However, the computing time is now multiplied by the number R of resamples, which should be about 10 000 to obtain a reliable significance level within the range of interest (down to 1%).

6. The additional time to compute the moments and cumulants necessary for all approximations took (in our examples) roughly twice the time as to fit the model. All computations (including the bootstrap) can be performed even on personal computers taking from a few minutes to several hours, depending on the computer's CPU, the sample size and the dimension of the model.

7. Bootstrapping the cumulants of SD_λ in the examples suggests, that the expectation μ_λ tends to be underestimated by (16). An ad-hoc

correction, motivated by (6), is to replace μ_λ by $\mu_\lambda - S$. In our examples, this brings all approximations based on μ_λ much closer to the bootstrapped value, but further investigation concerning the approximation of μ_λ are needed.

In conclusion, for large (and possibly sparse) data and large degrees of freedom, the rescaled χ^2- and saddlepoint approximations (RCA) and (SA) provide a substantial improvement over the often misleading classical χ^2-approximation for the favorite statistics D, X^2 as well for FT and CR, and bootstrapping is even preferable, but much more computer intensive. The unique effort to implement these methods on a computer is not very large (as compared to the total time spent on a single larger study) and enables a state-of-the-art evaluation of significance levels for the power-divergence statistics in all future applications.

References

Barndorff-Nielsen, O.E. & Cox, D.R. (1989). Asymptotic Techniques for Use in Statistics. London: Chapman and Hall.

Breslow, N.E. & Day, N.E. (1980). Statistical methods in cancer research, Volume I: The analysis of case-control studies. International Agency for Research on Cancer, Lyon.

Cochran, W.G. (1952). The χ^2 test of goodness of fit. Annals of Mathematical Statistics 23, 315.

Cressie, N.A.C., & Read, T.R.C. (1984), Multinomial Goodness-of-Fit tests. Journal of the Royal Statistical Society, Ser. B, 46, 440-464.

Dale, J.R. (1986), Asymptotic Normality of Goodness-of-Fit Statistics for Sparse Product Multinomials. Journal of the Royal Statistical Society, Ser. B, 48, 48-59.

Efron, B. (1979). Bootstrap methods: Another look at the jacknife. Annals of Statistics 7, 1-26.

Haberman, S.J. (1974). The analysis of frequency data. The University of Chicago Press, Chicago and London.

Hall, P. (1992). The bootstrap and Edgeworth expansions. New York: Springer.

Hall, P. & Titterington, D.M. (1989). The effect of simulation order on level accuracy and power of Monte Carlo tests. Journal of the Royal Statistical Society, Ser. B, 51, 459-467.

Hall, P. & Wilson, S.R. (1991). Two guidelines for bootstrap hypothesis testing. Biometrics 47, 757-762.

Jöckel, K.-H., Rothe, G. & Sendler, W. (Eds.) (1992). Bootstrapping and related techniques. New York: Springer.

Johnson, N.L. & Kotz, S. (1969). Distributions in statistics: Discrete distributions. Houghton Mifflin Company, Boston.

Karn, M.N. & Penrose, L.S. (1951-52). Birth weight and gestation time in relation to maternal age, parity and infant survival. Annals of Eugenetics (London) 16, 147-164.

Koehler, K.J (1986), Goodness-of-Fit Tests for Log-Linear Models in Sparse Contingency Tables. Journal of the American Statistical Association 81, 483-493.

LePage, R. & Billard, L. (Eds.) (1992). Exploring the limits of bootstrap. New York: Wiley.

McCullagh, P. (1985a), On the Asymptotic Distribution of Pearson's Statistic in Linear Exponential Family Models. International Statistical Review 53, 61-67.

McCullagh, P. (1985b), Sparse Data and Conditional Tests. Bulletin of the International Statistical Institute, Proceedings of the 45th Session of ISI (Amsterdam), Invited Paper 28.3,1-10.

McCullagh, P. (1986), The Conditional Distribution of Goodness-of-Fit Statistics for Discrete Data. Journal of the American Statistical Association 81, 104-107.

McCullagh, P. & Nelder, J.A. (1989), Generalized Linear Models (2nd. Ed.). London: Chapman and Hall.

Osius, G. (1985), Goodness-of-Fit Tests for Binary Data With (Possible) Small Expectations but Large Degrees of Freedom. Statistics & Decision, Suppl. No. 2, 213-224.

Osius, G. & Rojek, D. (1992). Normal goodness-of-fit tests for multinomial models with large degrees of freedom. Journal of the American Statistical Association 87, 1145-1152.

Read, T.R.C., & Cressie, N.A.C. (1988), Goodness-of-Fit Statistics for Discrete Multivariate Data. New York: Springer.

Rojek, D. (1989), Asymptotik für Anpassungstests in Produkt Multinomial Modellen bei wachsendem Freiheitsgrad. Unpublished Ph.D. Thesis, University of Bremen, Germany.

A Multiple Test Procedure for Nested Systems of Hypotheses

Gerhard Hommel, Gudrun Bernhard
Institut für Medizinische Statistik und Dokumentation,
Johannes-Gutenberg-Universität Mainz

Key Words:
multiple testing, nested systems of hypotheses, model search, logical dependence among hypotheses, arrow hypothesis, arrow test strategy (ATS)

Abstract
As an alternative to the classical closure test, we describe a multiple testing procedure called "arrow test strategy" ("ATS"). It is based on "arrow tests", i.e. tests, which are not developed within the full parameter space, but in a restricted one. The ATS may be more powerful than the closure test if the given system of hypotheses is completely or partially nested, which often occurs in model search. The practical performance of the ATS is possible by means of a computer program by Bernhard (1992) using information about logical dependencies among the hypotheses.

Introduction

In model search problems, it is often desirable to select one or more models which are not unnecessarily complex. For this aim, one can apply a multiple test procedure controlling the familywise error rate α in the strong sense (see Hochberg & Tamhane (1987)), which is sometimes referred to as "control of the multiple level α" (Holm (1979)). A common practice for this is the use of the closure test (Marcus et.al (1976)), which is recommended, particularly for model search problems, either directly or in an approximate form, by Edwards & Havranek (1987), Edwards (1992) or Schiller (1988). However, an important assumption for the application of the closure test is that the system of models (= null hypotheses) is closed under intersection which may lead to a substantial inflation of a given system of models. Moreover, the tests used for this strategy are usually tests developed within the saturated model (i.e. the full parameter space), leading to the drawbacks that, e.g. for common multiple regression models, the sums of squares for the nominator of the F statistic may have many degrees of freedom, and, on the other hand, for the error term in the denominator there may remain only few degrees of freedom.

To avoid this disadvantage, Bauer & Hackl (1987) proposed an alternative procedure for completely nested systems of hypotheses. The crucial idea of this procedure (which is denoted as "arrow test strategy" in the sequel) is that the test for a given model M is developed not within the saturated model, but within the closest model which is more complex than M, thus saving d.f. in the nominator of an F statistic, as well as providing more d.f. for the denominator (see also Alt (1988)). The resulting strategy is a sequentially rejective procedure similar to that of Holm (1979), but with "jumps", i.e. when a bound α/m has been used and a corresponding model hypothesis can be rejected, the next bound is not necessarily $\alpha/(m-1)$, but possibly an α/q with $q < m-1$. (It should be noted, that, in contrast to Shaffer's (1986) MRSB procedure, where the sequential bounds $\alpha/t_1, \alpha/t_2, \ldots, \alpha/t_n$ are fixed in advance, the sequential bounds of the Bauer/Hackl strategy depend on the information which model hypotheses already have been rejected - a "dynamic" strategy).

By Hommel & Bernhard (1992) it has been shown that the application of an arrow test strategy is not only possible for completely nested but also for partially nested systems of hypotheses and thus can be applied, in principle, for arbitrary model search problems. It has only to be noticed that for a given model M there can exist more than one closest model which is more complex than M such that M has to be tested by more than one test (called "arrow test" in the following). The practical performance of such a procedure is possible by applying the so-called "Procedure 4" of Bergmann & Hommel (1988), which uses, in its simplest form, Bonferroni global tests as constituents for the strategy based on the idea of the closure test. A computer program performing "Procedure 4" which uses information about the logical structure of the model hypotheses has been written by Bernhard (1992). It should be emphasized that for the application of this procedure it is not necessary that the given system of model hypotheses is closed, a priori.

In this paper, we will first give a description of Bernhard's (1992) rapid algorithm and a formal definition of the arrow test strategy ("ATS"). Then we will show that the resulting multiple test procedure is consistent under quite general assumptions. Since the application of the full ATS may often be too computer-time consuming, we will propose a short-cut procedure which is more conservative than the ATS. Finally, we present the results of a simulation study comparing the closure test and the ATS, showing that the ATS may be superior to the closure test, even if only Bonferroni global tests are used.

Practical performance of "Procedure 4"

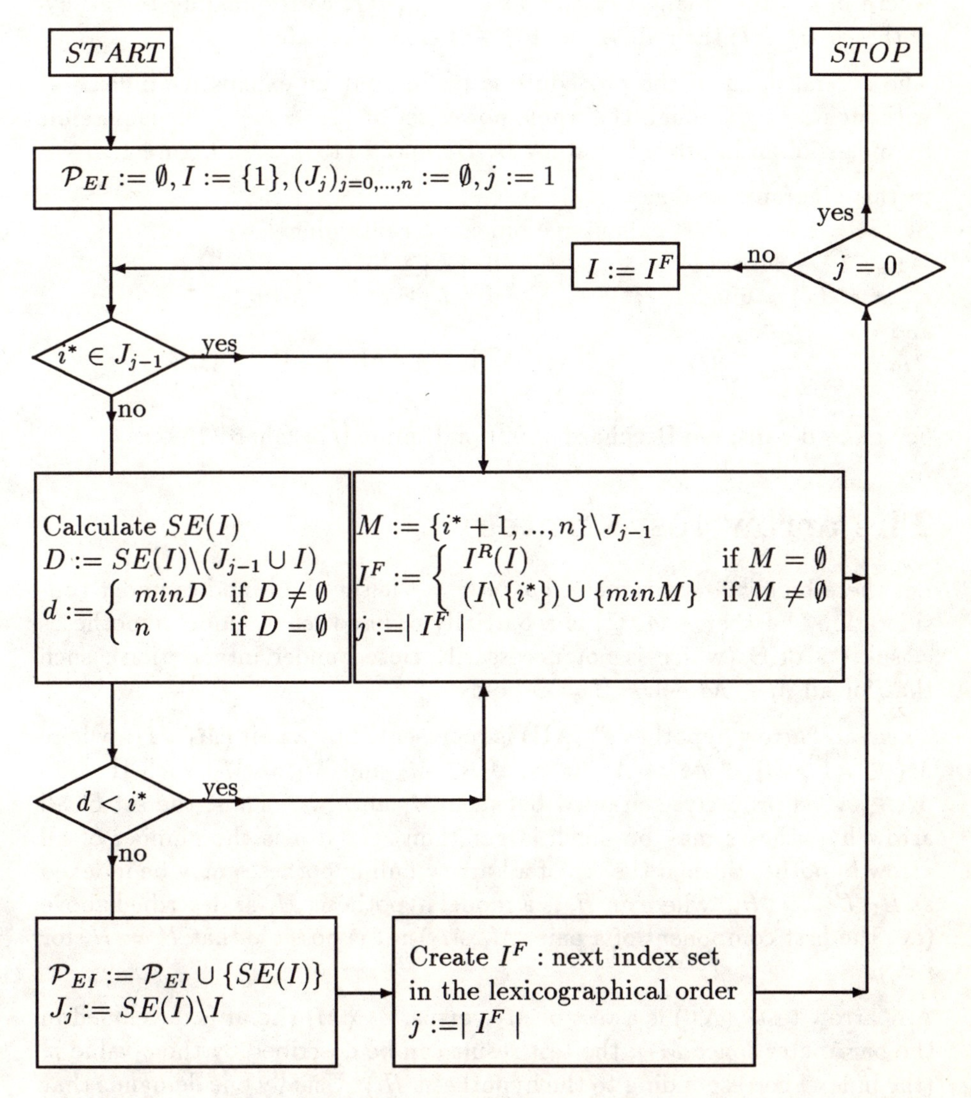

Figure 1: Flow chart of an improved strategy to determine $\mathcal{P}_{EI}$.

Let there be given $n \geq 2$ hypotheses $H_1, \ldots, H_n$. If, for each index set $I \subseteq \{1, \ldots, n\}$, $H_I = \bigcap\{H_i : i \epsilon I\}$, then I is called *exhaustive*, if H_I is non-empty and for all $J \subseteq \{1,\ldots,n\}$ with $H_I = H_J$ it follows that $I \supseteq J$. $\mathcal{P}_{EI}$ is the set of all exhaustive index sets.

The principle of "Procedure 4" is to determine an "acceptance set"

$$A = \cup\{I : I \in \mathcal{P}_{EI}, p_{1:I} > \alpha / \mid I \mid\},$$

where $p_{1:I}$ is the smallest of the p-values p_i, $i \epsilon I$, corresponding to the hypotheses H_i, $i \epsilon I$; then all H_j with $j \notin A$ can be rejected.

The crucial point of the procedure is to find out all exhaustive index sets without passing through the whole power set of $\{1,\ldots,n\}$. This is possible by means of an improved strategy by Bernhard (1992), see Figure 1.

In this diagram, we denote
$SE(I)$ = the smallest exhaustive index set containing I
$i^* := i^*(I) := \max(I)$ if $\mid I \mid \geq 1$
$i^{**} := i^{**}(I) := \max(I \backslash \{i^*\})$ if $\mid I \mid \geq 2$
and

$$I^R := I^R(I) := \begin{cases} (I\backslash\{i^*, i^{**}\}) \cup \{i^{**}+1\} & \text{if } \mid I \mid \geq 2 \\ \emptyset & \text{if } \mid I \mid = 1 \end{cases}$$

For more details, see Bernhard(1992) or Hommel/Bernhard(1992).

The arrow test strategy

Let there be given a parameter space Θ = the set of all models under consideration. Further, $(\mathcal{M}, \subseteq)$ is a partially ordered set of model hypotheses = subsets of Θ (which is not necessarily closed under intersection), such that for all $M \in \mathcal{M}$ $\emptyset \neq M \neq \Theta$ holds.

Then, an "arrow hypothesis" (AH) is represented by a pair (M_1, M_2), where $M_1 \in \mathcal{M}$, $M_2 \in \mathcal{M}$ or $M_2 = \Theta, M_1 \subset M_2$ and $M_1 \neq M_2$, such that no $M_3 \in \mathcal{M}$ is properly contained between M_1 and M_2. Thus, the set of all arrow hypotheses may be much larger than $\mathcal{M}$. If n is the number of all arrow hypotheses, then the set of all arrow null hypotheses may be indexed as $H_1, H_2, \ldots, H_n$, where an H_i is a model hypothesis M_1 as described above (i.e. the first component of a pair (M_1, M_2)); it is possible that $H_i = H_j$ for $i \neq j$.

An "arrow test" (AT) is a test of M_1 versus $M_2 \setminus M_1$ (i.e. it is developed in the parameter space M_2); the test result can be described by the p-value p_i (the index i corresponding to the hypothesis H_i). Usually one demands that an AT is consistent in $M_2 \setminus M_1$ (which means that, under certain regularity conditions, the probability of a true rejection of M_1 tends to 1, if N = the number of independent replications of the experiment tends to ∞). An AT can also be interpreted as a test of M_1 versus $\Theta \setminus M_1$ (which is useful when closure test arguments are applied), but it need not be consistent for all $\vartheta \in \Theta \setminus M_2$.

One possibility to perform a multiple test procedure based on an arrow test strategy ("ATS") controlling the multiple level α is to perform "Procedure 4" of Bergmann & Hommel (1988). In the following, we will use the term "ATS" for this particular multiple test strategy. An important property of it is that the decisions for the arrow hypotheses are identical to those of the closure test for the system of all arrow hypotheses using global Bonferroni tests as constituents (see Bergmann & Hommel (1988)).

Example: Consider all linear regression models with two regressors x_1, x_2 (without interactions)

$$Y_i = \beta_0 + \beta_1 \cdot x_{1i} + \beta_2 \cdot x_{2i} + \epsilon_i, \quad i = 1, \ldots, N,$$

where the $\epsilon_i \sim N(0, \sigma^2)$ are independent.

Then Θ is the set of all three-dimensional real vectors $(\beta_0, \beta_1, \beta_2)$.

We consider the three model hypotheses

$$M_1 : \beta_1 = \beta_2 = 0; \quad M_2 : \beta_1 = 0; \quad M_3 : \beta_2 = 0.$$

We define the three remaining sums of squares (if the respective model is considered as true)

$$RSS_1 = \sum_{i=1}^{N} (y_i - \bar{y})^2, \quad RSS_2 \ , \quad RSS_3 \ ,$$

and SSQ is the sum of squares of the error term, if the saturated model is assumed.

Then the classical closure test may be performed in the following stepwise manner:
Perform a level α test of M_1 using the F test statistic

$$F_1 = \frac{(RSS_1 - SSQ)/2}{SSQ/(N-3)}$$

with $(2; N-3)$ d.f.

If the result is not significant, retain all hypotheses; otherwise reject M_1 and continue as follows:
Perform level α tests of $M_i \quad (i = 2, 3)$ using the F test statistics

$$F_i = \frac{RSS_i - SSQ}{SSQ/(N-3)}$$

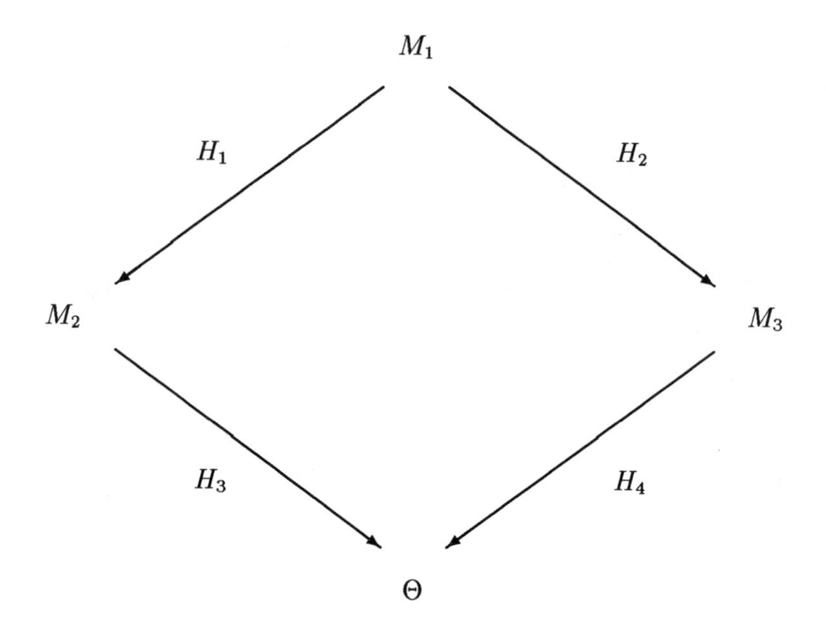

Figure 2: Model hypotheses (M_1, M_2, M_3) and arrow hypotheses (H_1, H_2, H_3, H_4) for the example. For each arrow, the arrow null hypotheses H_i are at the beginning and the corresponding parameter space at its end.

with $(1; N-3)$ d.f. Reject M_i if the corresponding results are significant (at the level α).

There exist four arrow hypotheses (corresponding to the four arrows in Fig.2) in this example, represented by $(M_1, M_2), (M_1, M_3), (M_2, \Theta), (M_3, \Theta)$, so that one can choose $H_1 = H_2 = M_1$, $H_3 = M_2$, $H_4 = M_3$. The arrow tests can be performed by means of the test statistics

$$F_1' = \frac{RSS_1 - RSS_2}{RSS_2/(N-2)}, \quad F_2' = \frac{RSS_1 - RSS_3}{RSS_3/(N-2)},$$

which are, under $H_1 = H_2 = M_1$, F-distributed with $(1; N-2)$ d.f., and by

$$F_3' = F_2, \quad F_4' = F_3,$$

which are, under $H_3 = M_2$ resp. $H_4 = M_3$, F-distributed with $(1; N-3)$ d.f.

It should be emphasized that the tests based on F_1' resp. F_2' are not consistent on the whole of Θ (for example, the F_1' test may be not consistent for $\beta_1 \neq 0$).

The resulting ATS works as follows:
Determine the p-values p'_1, p'_2, p'_3, p'_4 corresponding to F'_i. If $\min\{p'_1, p'_2, p'_3, p'_4\} > \alpha/4$, reject no hypothesis. If $p'_3 \leq \alpha/4$, reject $H_1 = H_2 = M_1$ and $H_3 = M_2$; reject $H_4 = M_3$, in addition, if $p'_4 \leq \alpha$. Analogously, if $p'_4 \leq \alpha/4$, reject $H_1 = H_2 = M_1$ and $H_4 = M_3$; reject $H_3 = M_2$, in addition, iff $p'_3 \leq \alpha$.

The most interesting case occurs, when $p'_1 \leq \alpha/4$ or $p'_2 \leq \alpha/4$. Then one can reject $H_1 = H_2 = M_1$, and the two model hypotheses $H_3 = M_2, H_4 = M_3$ remain. But since both hypotheses cannot be true simultaneously (provided M_1 is really false), the information about logical structures used by "Procedure 4" leads to the consequence that both remaining p-values p'_3, p'_4 need not be compared with $\alpha/2$, but rather with α (i.e. reject, in addition, $H_3 = M_2$ resp. $H_4 = M_3$, iff $p'_3 \leq \alpha$ resp. $p'_4 \leq \alpha$).

Consistency of the ATS

Theorem: Let there be given an ATS (based on "Procedure 4" with global Bonferroni tests).

If every AT (represented by (M_1, M_2), as in the last section) is consistent in $M_2 \setminus M_1$, then the ATS is also consistent, i.e. one has for each $\vartheta \in \Theta$ and for all $M \in \mathcal{M}$ with $\vartheta \notin M$

$$P_\vartheta(M \text{ rejected by the ATS}) \to 1,$$

or equivalently

$$P_\vartheta(H_i \text{ rejected by the ATS}) \to 1,$$

for all H_i with $\vartheta \notin H_i$.

Proof: Because of the remark about the ATS based on "Procedure 4" in the last section, it is sufficient to show that the closure test based on global Bonferroni tests has the asserted consistency property.

The local tests used as constituents for this closure test can be described as follows: Let there be given an intersection hypothesis $H = \bigcap\{H_i : i \in J\}$, where $J \subseteq \{1, \ldots, n\}$. Then one considers the exhaustive index set $I = SE(J)$ (see the second section). The global Bonferroni test of H is defined by the rule: "Reject H if $\min\{p_i : i \in I\} \leq \alpha / \mid I \mid$".

For fixed $\vartheta \in \Theta$, consider a hypothesis H_i with $\vartheta \notin H_i$. Then the corresponding exhaustive index set is $I_i = \{i_1, \ldots, i_m\}$, where $H_{i_1}, \ldots, H_{i_m}$ are *all* hypotheses implied by H_i (i.e. $H_{i_j} \supseteq H_i$). Since $\vartheta \notin H_i$, there exists either a pair (H_{i_j}, H_{i_k}) such that $\vartheta \notin H_{i_j}$, but $\vartheta \in H_{i_k}$, or a pair (H_{i_j}, Θ) such that $\vartheta \notin H_{i_j}$, representing an AH with a consistent AT. Because this AT of H_{i_j} is consistent, the global Bonferroni test of H_i is also consistent.

However, by the closure test H_i is only rejected if all $H = \bigcap\{H_j : j \in J\} \subseteq H_i$, where J is exhaustive, can be rejected. But since $\vartheta \notin H$, there exists an $H_j, j \in J$, with $\vartheta \notin H_j$, for which the global Bonferroni test is consistent (from the same reasons as before). Because one has only a finite number of such intersection hypotheses $H \subseteq H_i$, the probability of a rejection of H_i by the closure test also tends to 1.

Remark: By the theorem it is ensured that asymptotically no false models are selected. However, it should also be desirable to select the most parsimonious correct model. This is possible, under rather weak conditions, if α tends to zero with a suitable speed; see Bauer et al. (1988).

A short-cut procedure

Since the complete ATS, as described before, is often very computer-time consuming, or it is even impossible to obtain the results within an acceptable computing time, we propose the following much simpler, but also more conservative procedure:

START: Compute the p-values $p_1, \ldots, p_n$ for each AH. m is the number of all AH's which have not been rejected at a certain step of the procedure. At the beginning, one has $m = n$.

GENERAL STEP: Consider an AH H_i with $p_i \leq \alpha/m$. Reject all AH's H_j leading by direct implication to H_i (i.e. there exists a path (H_j, H_{i_1}), $(H_{i_1}, H_{i_2}), \ldots, (H_{i_{q-1}}, H_{i_q}), (H_{i_q}, H_i)$ of hypothesis pairs representing AH's). Moreover, reject all AH's H_k, whenever $H_k = H_j$ for such a j. Then, define m as the new number of AH's which have still not been rejected.

STOP, if you cannot find any more H_i as described in the general step.

Theorem: This short-cut procedure("SCP") has the following properties:

1. It rejects only hypotheses/models which are also rejected by the ATS (or the closure test using global Bonferroni tests).

2. It is possible that the ATS rejects a hypothesis/model, but not the SCP.

3. The SCP is coherent and controls the multiple level α.

Proof:

1. To start with, one has to find an H_i with $p_i \leq \alpha/n$. If it is the case, it follows for all intersection hypotheses $H = \bigcap\{H_j : j \in J\}$ implying H_i, with J exhaustive, that $i \in J$. Since $p_i \leq \alpha/ \mid J \mid, H$ is rejected by its Bonferroni global test, and (because this is true for all corresponding intersection hypotheses $\subseteq H$) also by the closure test using global Bonferroni tests.

   In the general step, one assumes that $(n - m)$ hypotheses H_j have already been rejected. Therefore, by the closure test all $H = \bigcap\{H_j : j \in J\}$, with J exhaustive and $\mid J \mid> m$, are rejected. Thus, for a remaining H_i it is sufficient to have $p_i \leq \alpha/m$, and the same arguments as in the introductory step can be used.

2. Assume that one obtains, in the example of Section 3, $p'_1 \leq \alpha/4$, $p'_3 > \alpha/2$, $\alpha/2 < p'_4 \leq \alpha$. Then the ATS rejects $H_1 = H_2 = M_1, M_3 = H_4$ (and $M_2 = H_3$ provided $p'_3 \leq \alpha$). However, the SCP can reject in the first step $H_1 = H_2 = M_1$, but since then $m = 2$ and no remaining $p'_i \leq \alpha/m$ ($i = 3$ or 4), it has to stop.

3. By construction, the SCP is coherent; since the ATS controls the multiple level α, the SCP also does.

Remark: For totally nested systems, the SCP and the ATS coincide and are identical to the Bauer & Hackl (1987) procedure. However, for other systems (e.g. regression models with 3 or more regressors) the SCP may be considerably inferior to the complete ATS.

Simulation results

In order to compare the closure test (CT) and the ATS, two simulation studies were performed for rather simple situations. The first study investigated a completely nested system with only two hypotheses (a polynomial model with maximal degree 2), whereas in the second study the example with two regressor variables of the second section was treated.

Simulation study I:
The model equation was

$$Y_i = \alpha_0 + \alpha_1 \cdot i + \alpha_2 \cdot i^2 + \epsilon_i, \quad i = 1, \ldots, N,$$

where the $\epsilon_i \sim N(0, \sigma^2)$ were independent.

We considered the two model hypotheses

$$M_1 : \alpha_1 = \alpha_2 = 0 \quad \text{(no influence of design points)},$$

$$M_2 : \alpha_2 = 0 \quad \text{(at most degree 1)}.$$

If the remaining sums of squares (under the assumption of the respective model) are defined by

$$RSS_1 = \sum_{i=1}^{N} (y_i - \overline{y})^2, \quad RSS_2 \ ,$$

and SSQ is the sum of squares of the error term for the full model, the following test statistics are used:

$$F_1 = \frac{(RSS_1 - SSQ)/2}{SSQ/(N-3)},$$

which is, under M_1, F-distributed with $(2; N-3)$ d.f., and

$$F_1' = \frac{RSS_1 - RSS_2}{RSS_2/(N-2)}, \quad F_2 = F_2' = \frac{RSS_2 - SSQ}{SSQ/(N-3)},$$

which are, under $H_1 = M_1$ resp. $H_2 = M_2$, F-distributed with $(1; N-2)$ resp. $(1; N-3)$ d.f. .

The CT rejects M_1 if the level α test based on F_1 is significant; M_2 is rejected if M_1 is rejected and the level α test based on F_2 is significant.

The ATS requires for a rejection that one of the two p-values p_1', p_2' (coming from F_1', F_2') is $\leq \alpha/2$. If $p_2' \leq \alpha/2$, M_1 and M_2 are rejected; if $p_1' \leq \alpha/2$, M_1 is rejected, and, in addition, M_2, if $p_2' \leq \alpha$.

For the practical performance, the model equation was expressed in terms of orthogonal polynomials $f_0(i), f_1(i), f_2(i)$ as

$$Y_i = \gamma_0 \cdot f_0(i) + \gamma_1 \cdot f_1(i) + \gamma_2 \cdot f_2(i) + \epsilon_i, \quad i = 1, \ldots, N,$$

where $f_0(i) = 1$, $f_1(i) = i - (N+1)/2$, and $f_2(i) = i^2 - (N+1) \cdot i + (N+1)(N+2)/6$ (see Anderson (1971, p.31 ff.)). So it could be shown that it was sufficient to generate three independent random variables

$$U_1 \sim N(\delta_1, 1), \quad U_2 \sim N(\delta_2, 1), \quad V \sim \chi^2(N-3),$$

with

$$\delta_1 = \Big(\alpha_1 + (N+1) \cdot \alpha_2\Big) \cdot \sqrt{(N-1)N(N+1)/12},$$

$$\delta_2 = \alpha_2 \cdot \sqrt{\sum_{i=1}^{N}\Big(i^2 - (N+1)\cdot i + (N+1)(N+2)/6\Big)}.$$

Then, the three test statistics

$$G_1 = \frac{(U_1^2 + U_2^2)/2}{V/(N-3)}, \quad G_1' = \frac{U_1^2}{(V+U_2^2)/(N-2)}, \quad G_2 = G_2' = \frac{U_2^2}{V/(N-3)}$$

have the same distributions as the corresponding statistics F_1, F_1' and $F_2 = F_2'$.

For $\alpha = .05$, $N = 5$ and $N = 20$, 1000 simulations were performed for different situations. If the global null hypothesis was true ($M_1 : \alpha_1 = \alpha_2 = 0$), the actual type I error rates were about .05 for both procedures.

For the situation that M_2 was true but M_1 not ($\alpha_1 \neq 0, \alpha_2 = 0$) we found that for $N = 5$ the ATS had substantially higher power than the CT; for $N = 20$ the ATS was still slightly superior (Table 1). Nevertheless, the type I error rates (= rejection rates of M_2) were higher for the CT.

N	5					20				
α_1	0.4	0.8	1.2	1.6	2.0	0.06	0.08	0.10	0.12	0.14
M_1 rej. by CT	73	180	331	505	658	244	381	538	730	845
M_1 rej. by ATS	102	271	537	742	895	235	391	558	762	858
M_2 rej. by CT	34	41	45	54	42	38	34	55	43	50
M_2 rej. by ATS	22	27	35	44	38	24	31	46	43	44

Table 1: Simulation study I: Number of rejections for $\alpha_1 \neq 0, \alpha_2 = 0$ (1000 simulations each)

If the model curve was a pure parabola (i.e. $\alpha_2 = c$, $\alpha_1 = -(N+1)\cdot c$ for some $c \neq 0$), we found no substantial differences between both procedures. Finally we considered the situation $\alpha_2 = c, \alpha_1 = -c$ (i.e. the model curve is increasing with a parabolic trend). In this case we found substantial differences (in favour of ATS) only for $N = 5$ and $\alpha_2 < 0.6$ for the rejection rates of M_1; the rejection rates of M_2 were slightly higher for CT than for ATS (Table 2).

Simulation study II:
We considered the example described in Section 3. For sake of simplicity, we assumed further

$$\sum x_{1i} = \sum x_{2i} = 0, \quad \sum x_{1i}^2 = \sum x_{2i}^2, \quad \sigma^2 = 1,$$

N	5					20			
$\alpha_2 = -\alpha_1$	0.2	0.4	0.6	0.8	1.0	0.004	0.006	0.008	0.010
M_1 rej. by CT	272	673	912	985	998	406	741	950	999
M_1 rej. by ATS	352	745	916	981	999	396	734	940	998
M_2 rej. by CT	60	163	256	360	539	72	111	182	241
M_2 rej. by ATS	39	133	237	363	538	60	95	172	240

Table 2: Simulation study I: Number of rejections for $-\alpha_1 = \alpha_2 \neq 0$ (1000 simulations each)

and we defined $q := \sum x_{1i}x_{2i}$ as the formal correlation coefficient of the design points (x_{1i}, x_{2i}).

Also in this case, a simulation step was possible by generating three sufficient independent random variables U_1, U_3, V (where the indices of the U_i correspond to those of the H_i in the third section), where

$$U_i \sim N(\delta_i, 1) \quad (i = 1,3) \text{ with } \delta_1 = q \cdot \beta_1 + \beta_2, \quad \delta_3 = \beta_1 \cdot \sqrt{1-q^2},$$

and $V \sim \chi^2(N-3)$.

Further, if one defines

$$U_2 := q \cdot U_1 + \sqrt{1-q^2} \cdot U_3, \quad U_4 := \sqrt{1-q^2} \cdot U_1 - q \cdot U_3,$$

then

$$U_i \sim N(\delta_i, 1) \quad (i = 2,4) \text{ with } \delta_2 = \beta_1 + q \cdot \beta_2, \quad \delta_4 = \beta_2 \cdot \sqrt{1-q^2},$$

and U_2, U_4 are independent from each other.

Now one can define the test statistics

$$G_0 := \frac{(U_1^2 + U_3^2)/2}{V/(N-3)} = \frac{(U_2^2 + U_4^2)/2}{V/(N-3)},$$

$$G_1 := \frac{U_1^2}{V/(N-2)}, G_2 := \frac{U_2^2}{V/(N-2)}, G_3 := \frac{U_3^2}{V/(N-3)}, G_4 := \frac{U_4^2}{V/(N-3)},$$

and it can be shown by means of linear regression techniques that G_0 has the same distribution as the global test statistic F_1 (from Section 3), and the G_i have the same distributions as the AT statistics F_i', $i = 1,2,3,4$.

In the simulation study, $\alpha = .05$, $N = 5$ and $N = 100$ were chosen. If the global hypothesis was true ($M_1 : \beta_1 = \beta_2 = 0$), the actual rejection rates of the ATS were smaller than those of the CT.

N	5				100			
β_1	5				2.5			
q	0	0.2	0.5	0.8	0	0.2	0.5	0.8
M_1 rej. by CT	502	497	471	500	590	589	608	602
M_1 rej. by ATS	561	570	552	593	500	516	558	558
M_2 rej. by CT	502	497	467	374	574	556	509	303
M_2 rej. by ATS	561	570	514	357	498	504	492	280
M_3 rej. by CT	55	49	51	52	48	57	39	45
M_3 rej. by ATS	54	45	51	51	36	44	38	43

Table 3: Simulation study II: Number of rejections for $N = 5, \beta_1 = 5, \beta_2 = 0$, $N = 100, \beta_1 = 2.5, \beta_2 = 0$ and different values of q
(1000 simulations each)

If M_3 was chosen as true, but M_1 not (i.e. $\beta_1 \neq 0, \beta_2 = 0$), then the ATS had higher power than the CT for $N = 5$, whereas for $N = 100$ the converse result was obtained. Again, the type I error rates were higher for the CT (Table 3).

Finally, the situation was investigated where β_1 and β_2 were different from zero. For the case $\beta_1 = \beta_2 \neq 0$, the CT was by far superior to the ATS (Table 4).

N	5				100			
$\beta_1 = \beta_2$	4				2			
q	0	0.2	0.5	0.8	0	0.2	0.5	0.8
M_1 rej. by CT	555	623	737	810	719	792	871	931
M_1 rej. by ATS	222	239	266	643	518	673	820	917
M_2 rej. by CT	503	534	497	301	467	485	418	221
M_2 rej. by ATS	219	234	238	269	353	443	416	221
M_3 rej. by CT	498	517	501	302	491	480	383	195
M_3 rej. by ATS	219	232	246	271	371	444	380	193

Table 4: Simulation study II: Number of rejections for $N = 5, \beta_1 = \beta_2 = 4$, $N = 100, \beta_1 = \beta_2 = 2$ and different values of q (1000 simulations each)

Concluding remarks

Comparing the CT and the ATS, we found by theoretical considerations and in some practical examples that the ATS might be more powerful than the CT, if the system of models is "largely" completely nested, if only few d.f. remain for the denominator of the test statistics (which may occur, e.g., in the analysis of time series), or if the true model is rather simple (since the tests based on the saturated model are not powerful in this case). The trend of these results was confirmed in our simulation studies. The first simulation study (a completely nested system) showed that the ATS can be substantially superior to the closure test. Also in the second study we found one situation (N small, only one regressor has an influence upon the outcome variable), where the ATS was more powerful.

In many model search problems, the model family is closed under intersection, a priori. Otherwise, for the application of the closure test a larger model family has to be considered. For the ATS, however, one does not need a closed model family, and therefore one can expect a more powerful behaviour of the ATS in such a case. Moreover, it is often possible to improve the Bonferroni adjustment used in the simplest form of the ATS, e.g. by Šidák-style adjustments.

Whereas for each case there may occur specific algorithmic problems when performing the CT, the practical performance of the ATS is possible by means of the computer program written by Bernhard (1992). If the hypotheses can be formulated in a linear way (as for nearly all model search problems), it is particularly easy to use the logical structures among the hypotheses (by considering the ranks of certain matrices). This is also implemented in the computer program.

References

Alt, R. (1988). Hierarchical test problems and the closure principle. In: P. Bauer, G. Hommel, E. Sonnemann (eds.): Multiple Hypothesenprüfung - Multiple Hypotheses Testing, 162-176. Springer, Berlin/ Heidelberg/New York.

Anderson, T.W. (1971). The Statistical Analysis of Time Series. Wiley, New York.

Bauer, P. & Hackl, P. (1987). Multiple testing in a set of nested hypotheses. Statistics *18*, 345-349.

Bauer, P., Pötscher, B. M. & Hackl, P. (1988). Model selection by multiple test procedures. Statistics *19*, 39-44.

Bergmann, B. & Hommel, G. (1988). Improvements of general multiple test procedures for redundant systems of hypotheses. In: P.Bauer, G. Hommel,

E. Sonnemann (eds.): Multiple Hypothesenprüfung - Multiple Hypotheses Testing, 100-115. Springer, Berlin/ Heidelberg/New York.

Bernhard, G. (1992). Computergestützte Durchführung von multiplen Testprozeduren - Algorithmen und Powervergleich. Doctoral thesis, Mainz.

Edwards, D. (1992). Some computational aspects of graphical model selection. In: J. Antoch (ed.): Computational Aspects of Model Choice, 187-210. Physica-Verlag, Heidelberg.

Edwards, D. & Havránek, T. (1987). A fast model selection procedure for large families of models. J. Amer. Statist. Ass. *82*, 205-213.

Hochberg, Y. & Tamhane, A. C. (1987). Multiple Comparison Procedures. Wiley, New York.

Holm, S. (1979). A simple sequentially rejective multiple test procedure. Scand. J. Statist. *6*, 65-70.

Hommel, G. & Bernhard, G. (1992). Multiple hypotheses testing. In: J. Antoch (ed.): Computational Aspects of Model Choice, 211-235. Physica-Verlag, Heidelberg.

Marcus, R., Peritz, E. & Gabriel, K. R. (1976). On closed testing procedures with special reference to ordered analysis of variance. Biometrika *63*, 655-660.

Schiller, K. (1988). Der Abschlußtest zur Unterstützung bei der Modellauswahl loglinearer Modelle. In: P. Bauer, G. Hommel, E. Sonnemann (eds.): Multiple Hypothesenprüfung - Multiple Hypotheses Testing, 177-189. Springer, Berlin/ Heidelberg/New York.

Shaffer, J. P. (1986). Modified sequentially rejective multiple test procedures. J. Amer. Statist. Ass. *81*, 826-831.

Kernel Estimation in the Proportional Hazards Model

Jürgen Kübler
Bayer AG, PH-FE ME MSI
Aprather Weg 18a, D–42096 Wuppertal, Germany

Key Words:
survival analysis, nonparametric regression, generalized additive models

Abstract
In this paper a kernel based estimator for the risk function in the proportional hazards model is discussed and compared with other nonparametric estimators. The concept of generalized additive models is adapted to the proportional hazards model and a method for estimating risk functions based on a generalization of the kernel estimator is proposed. This method has applications in biomedicine if both continuous and categorical data have to be taken into consideration.

Introduction

Regression models for possibly right censored survival time data are widely used in medical statistics. The proportional hazards model (*phm*) which has been introduced by Cox (1972) constitutes an important class of such models where the influence of the covariates can be described as a multiplicative effect on an underlying baseline hazard function, usually assumed to be an arbitrary nuisance parameter. In addition, a parametric risk function describes the effect of the covariates.

Let T_i, $i = 1, \ldots, n$, denote non-negative and continuous random survival times, and $\Delta_i \in \{0, 1\}$ censoring indicators defined as 0 if T_i is censored and as 1 if T_i is not censored. Furthermore, let X_i be vectors of covariates with values in $\mathbb{R}^p$, $p \in \mathbb{N}$. Realizations of these random variables are denoted by t_i, δ_i, and x_i, $i = 1, \ldots, n$, respectively. Under the assumptions of the *phm*, the *hazard function* $\lambda(t \mid x) = \lim_{h \searrow 0} \mathrm{P}_{X=x}(T \leq t + h \mid T > t)$ of the survival time T of an individual is given as

$$\lambda(t \mid x) = \lambda_0(t) R(\eta(x)), t \geq 0, x \in \mathbb{R}^p, \tag{1}$$

where $R : \mathbb{R} \rightarrow \mathbb{R}_+$ is a known and strictly monotone function and $\eta : \mathbb{R}^p \rightarrow \mathbb{R}$ is an unknown regression function. The *baseline hazard function* $\lambda_0(t)$ remains an unknown and unspecified non-negative nuisance function. The hazard function $\lambda(t \mid x)$ can be interpreted as the risk of dying at time t given survival to $t - 0$ and covariate x. The relative risk $\lambda(t \mid x_1)/\lambda(t \mid x_2)$ of two individuals with covariate x_1 and x_2 can be written as $R(\eta(x_1))/R(\eta(x_2))$. If $R(\eta(\cdot))$ is restricted to fulfill $R(\eta(0)) = 1$, it can be regarded as a *relative risk function* where $x = 0$ indicates the absence of additional risk factors.

The main intention of this paper consists in presenting and discussing non-parametric estimators of η. Since these estimators are based on generalizations of techniques used for parametric approaches, we will give a brief summary of the most important tools next.

In a parametric approach R is usually chosen as the exponential function and the regression function as $\eta(x) = x'\beta$, where β is a p-vector of unknown regression coefficients. Cox (1972, 1975) suggests to estimate the regression coefficients β by maximizing the *partial likelihood function* $PL(\beta)$

$$PL(\beta) = \prod_{i=1}^{n} \left\{ \frac{R(\eta(x_i))}{\sum_{j \in \Re_i} R(\eta(x_j))} \right\}^{\delta_i}, \tag{2}$$

where $\Re_i = \{j \in \mathbb{N} \mid t_j \geq t_i\}$ is the set of individuals *under risk* at time $t_i - 0, i = 1, \ldots, n$. Asymptotic properties of the *partial maximum likelihood estimate* $\hat{\beta}_{PL}$ are given in Andersen & Gill (1982).

Note that the representation of the *phm* in (1) is not unique, since a positive factor α can be shifted from the baseline hazard function λ_0 to the risk function R without violating the above assumptions. For $R(\cdot) = \exp(\cdot)$ and a linear regression function $x'\beta$, this fact constitutes no problem, because a unique representation can be achieved by omitting the intercept from the regression function. Thus, it is not at all surprising that the intercept parameter, appearing as a multiplicative constant in the nominator as well as in the denominator of $PL(\beta)$, cannot be estimated from (2).

Besides the estimation of the regression coefficients β, it is also of interest to derive an estimator of the baseline hazard function. Breslow (1972) suggests

$$\hat{\Lambda}_0(t) = \sum_{t_i \leq t} \frac{\delta_i}{\sum\limits_{t_j \in \Re_i} \exp(x_j{}'\hat{\beta}_{PL})}$$

as an estimator for the underlying *cumulative baseline hazard function* $\Lambda_0(t) = \int_{u=0}^{t} \lambda_0(u)du$. In the more general model (1), Breslow's estimator becomes

$$\hat{\Lambda}_0(t) = \sum_{t_i \le t} \frac{\delta_i}{\sum_{t_j \in \Re_i} R(\hat{\eta}(x_j))}, \tag{3}$$

where $\hat{\eta}(\cdot)$ denotes an estimate of $\eta(\cdot)$.

The partial likelihood estimator $\hat{\beta}_{PL}$ is usually calculated by applying the Newton-Raphson algorithm to the derivatives of $\log(PL)$. An alternative approach is based on the full likelihood function

$$L(\eta, \Lambda_0) = \prod_{i=1}^{n} \{\lambda_0(t_i) R(\eta(x_i))\}^{\delta_i} \exp(-\Lambda_0(t_i) R(\eta(x_i))), \tag{4}$$

Clayton & Cuzick (1985) apply the EM algorithm (Dempster et al. (1977)) which leads to a two–step iterative algorithm. In the first step, $\Lambda_0(t_i)$, $i = 1, \ldots, n$, are *estimated* by the Breslow estimator (3) at $t_i, i = 1, \ldots, n$, for an initial or previously estimated value of β. The second step consists in *maximizing* the full likelihood function (4), where the $\Lambda_0(t_i), i = 1, \ldots, n$, are replaced by the estimates obtained in the first step. These two steps are repeated until convergence. It can be shown that the sequence of estimates of β converges to $\hat{\beta}_{PL}$. Note, that replacing Λ_0 in (4) by the Breslow estimator (3) yields the partial likelihood function (2) (c.f. Kalbfleisch & Prentice (1980, p. 78f)).

In practical applications, an investigator is often faced with the problem when and how to include a possible risk factor or covariate in a model and how to check goodness of fit. One possibility is through the analysis of residuals which allows for a graphical inspection of the model. In general, such methods provide no information why the fit is unsatisfactory. Therefore, other techniques have been proposed, which, however, are sensitive only w.r.t. clearly specified alternatives. Thus, both for model selection and for checking goodness of fit, nonparametric approaches can be helpful when estimating the risk function.

The next section provides the theory of nonparametric estimation of the risk function based either on the partial likelihood (2) or on the full likelihood function (4). Different smoothing techniques are applied. Continuous covariates are assumed in many biomedical applications. However, it is necessary to consider also categorical variables such as 'gender' or 'treatment group'. The generalized additive models proposed by Hastie & Tibshirani (1984, 1986) can be used when different types of variables are simultaneously considered. A method for estimating the risk function under this model is proposed in the third section. Finally, the application of this method is illustrated by an example.

Nonparametric risk estimation

Let us first give a survey on nonparametric estimators of the risk function in the *phm*. The techniques discussed here make use of *kernel estimators*, *k-nearest neighbor estimators*, and *spline smoothing*. In the *phm* (1), both the partial likelihood and the full likelihood approach have been combined with smoothing techniques.

Hastie & Tibshirani (1984, 1986) propose a smoothing technique for so–called *generalized additive models*. These models can be interpreted as an extension of the well–known *generalized linear models*, where the linear predictor $x'\beta,\ x,\beta \in \mathbb{R}^p, p \in \mathbb{N}$, is replaced by a sum of smooth functions $\sum_{i=1}^{p} s_i(x_i)$ with $x = (x_1, \ldots, x_p)'$. The authors derive the *local–scoring–algorithm* as smooth extension of Fisher's scoring method to estimate $s_i, i = 1, \ldots, p$, with the *running line smoother*, which refines the k–nearest–neighbor–estimator. The calculation of the k–nearest–neighbor–estimator can be interpreted as fitting a constant line to all the data in a neighborhood of $x_i, i = 1, \ldots, n$. The running line smoother can then be obtained by replacing the local constants with local linear functions, which are again estimated by using data only from the neighborhood of $x_i, i = 1, \ldots, n$. Hastie and Tibshirani apply this method to the partial likelihood function (2) although it has been derived under the assumptions of an ordinary likelihood function. Since the local scoring algorithm is defined as an iterative procedure, it can be interpreted as an algorithm to determine a solution of an underlying fix–point problem. It can be shown, that for this implicitly given fix–point problem a unique solution does not exist (Kübler (1991)), due to the fact, that the identification problem for the *phm* is not taken into account. If $R(\hat{\eta})$ is a solution of the underlying fix–point problem, then each multiple of $R(\hat{\eta})$ is also one.

Another estimator, also based on the partial likelihood function and a running mean smoother, has been proposed later by Tibshirani & Hastie (1987). Before discussing the estimator of the risk-function in the *phm*, let us first fix ideas in the general context of maximum likelihood estimation. Suppose $(y_1, x_1), \ldots, (y_n, x_n)$ are realizations of independent random vectors $(Y_i, X_i) \in \mathbb{R}^2, i = 1, \ldots, n$, with conditional density $f_{Y|X=x}(y) = f_{\theta(x)}(y)$. The *local likelihood estimate* of $\theta(x_i)$ is defined as

$$\hat{\theta}(x_i) = \hat{\beta}_{0i} + x_i\hat{\beta}_{1i},$$

where $\hat{\beta}_{0i}$ and $\hat{\beta}_{1i}$ maximize the *local likelihood function*

$$LL_i(\beta_{0i}, \beta_{1i}) = \prod_{j \in N_i} f_{\beta_{0i}+x_j\beta_{1i}}(y_j), \tag{5}$$

where

$$N_i = \{\max(i - ([\gamma n] - 1)/2, 1), \ldots, \min(i + ([\gamma n] - 1)/2, n)\},$$
$$\gamma \in \mathbb{R}_+ \text{ such that } [\gamma n] \in \{1, 3, 5, \ldots\},\ i = 1, \ldots, n.$$

If the conditional distribution of Y given $X = x$ is the normal distribution with mean function $\theta(x)$ and unknown variance $\sigma^2 > 0$, the local likelihood estimator of $\theta(x)$ yields the ordinary running mean smoother.

Although the partial likelihood function (2) is neither a product of densities nor can it be interpreted as a joint density, Tibshirani & Hastie (1987) construct a *local partial likelihood* function by restricting the product and the sum in (2) as in (5). Choosing $R(\cdot) = \exp(\cdot)$, this leads to

$$\widetilde{PL}_i(\beta_{0i}, \beta_{1i}) = \prod_{j \in N_i} \left\{ \frac{\exp(\beta_{0i} + x_j \beta_{1i})}{\sum_{k \in \Re_j \cap N_i} \exp(\beta_{0i} + x_k \beta_{1i})} \right\}^{\delta_j}, \tag{6}$$

where N_i and γ are defined as in (5). Since the $\exp(\beta_{0i})$ terms cancel out, (6) reduces to

$$PL_i(\beta_i) = \prod_{j \in N_i} \left\{ \frac{\exp(x_j \beta_i)}{\sum_{k \in \Re_j \cap N_i} \exp(x_k \beta_i)} \right\}^{\delta_j}. \tag{7}$$

Only the slopes β_i of the log risk function $log R(\eta(\cdot))$ can be estimated from (7). With $\hat{\beta}_i$ maximizing $PL_i(\beta_i)$ and the trapezoidal rule, the author suggest

$$\hat{\eta}(x_i) = \sum_{j=1}^{i} (x_j - x_{j-1}) \frac{\hat{\beta}_j + \hat{\beta}_{j-1}}{2} \tag{8}$$

as local partial likelihood estimator of η at $x_i, i = 1, \ldots, n$. At a first glance (8) seems to be a natural choice of a smooth estimator of η. However, (8) gives no unique definition of an estimator, because neither x_0 nor $\hat{\beta}_0$ are specified. Moreover, the local partial likelihood approach heavily depends on using nearest neighbor estimators in connection with fitting local linear functions. Therefore, this approach can not be generalized e.g. to a kernel based estimator. The main disadvantage is due to the fact, that the baseline hazard function is allowed to depend on x_i when estimating β_i. From the definition of the *phm* it can be seen, that (1) has to hold only for each neighborhood N_i of x_i. This yields

$$\lambda(t \mid x_i) = \lambda_{0i}(t) \exp(\beta_{0i} + x_i \beta_{1i})), t \geq 0, x \in \mathbb{R}^p,$$

where the baseline hazard functions $\lambda_{0i}(t), i = 1, \ldots, n$, are in general not unique.

An estimator of the log risk function suggested by O'Sullivan (1986, 1988) is based on the penalized partial likelihood approach. Here, in contrast to the kernel or the nearest neighbor method, a *roughness penalty* is substracted from the log partial likelihood function. This yields a cubic spline, when maximization is carried out over the Sobolev space of continuous functions with square integrable second derivative and mean zero to avoid the identification problem. An algorithm to calculate such estimators is described in O'Sullivan (1988).

The running line smoother is also used by Gentleman (1988) to derive an estimator of the log risk function. The proposed algorithm can be interpreted as a smooth extension of Clayton & Cuzick's method (1985) discussed above. The corresponding local full likelihood function is maximized instead of the full likelihood function (4). As for the local scoring algorithm, it can be shown that the underlying fix-point problem has no unique solution (Kübler (1991); Gentleman & Crowley (1991)).

Staniswalis (1989) introduces the concept of weighted likelihood estimation to derive kernel estimates in general likelihood-based models. Using again the notations and assumptions specified for the local likelihood function (5), she defines the *weighted log likelihood function* $wl_\gamma(\theta;x)$ as

$$wl_\gamma(\theta;x) = \sum_{i=1}^{n} K\left(\frac{x-x_i}{\gamma}\right) \log f_\theta(y_i), \tag{9}$$

where $K(\cdot)$ is a *kernel function* and γ is the *bandwidth.* The kernel function is usually chosen as a continuous, bounded, and symmetric real function which integrates to one. For a discussion of different kernel functions and selection of an appropriate bandwidth see e.g. Härdle (1990). The *weighted likelihood estimate* is obtained as the maximizer of (9). Note, that if the conditional distribution of Y given $X = x$ is the normal distribution with mean function $\theta(x)$ and unknown variance $\sigma^2 > 0$, $\hat{\theta}(x)$ is the well known *Nadaraya* (1964)-*Watson* (1964) *estimator*

$$\hat{\theta}(x) = \frac{\sum_{i=1}^{n} K\left(\frac{x-x_i}{\gamma}\right) y_i}{\sum_{i=1}^{n} K\left(\frac{x-x_i}{\gamma}\right)}$$

Staniswalis applies this method also to the *phm.* Similar to Gentleman's (1988) approach, her estimator is based on the full likelihood function, where a spline estimator is used to estimate the *baseline survival function* $S_0(t) = exp(-\Lambda_0(t))$. An initial estimate of $S_0(t)$ is obtained from a *control group* and updated in an iterative algorithm, but it is doubtful, that this approach leads to a unique estimate of the log risk function.

As already mentioned, most of the estimators discussed above disregard that the representation of the *phm* (1) is not unique. With the exception of O'Sullivan's (1986, 1988) they are not uniquely defined. A concept which avoids this problem, when using kernel–based or nearest neighbor estimators has been proposed by Kübler (1991).

The identification problem can be easily evaded, when a parametric regression function η is chosen. Usually the risk function is forced to hold the condition $R(\eta(x)) = 1$. This can be interpreted as selecting a special function out of all admissible risk functions only differing in a positive factor. More general constraints as the one above are introduced by *norming functionals* H as $H : \mathcal{F} \to \mathbb{R}$, $\mathcal{F} = \{f \mid f : \mathbb{R}^p \to \mathbb{R}_+\}$, which satisfy the condition $H(\alpha f) = \alpha H(f)\ \forall\, \alpha \in \mathbb{R}$. Then, if $R(\eta)$ satisfies the *norming condition*

$$H(R(\eta)) = 1 \tag{10}$$

for a certain functional H, a unique represention of the *phm* (1) is obtained. An admissible estimator of η under the assumptions above also has to fulfill condition (10).

The method of Kübler (1991) is based on the full likelihood function (4). For a known baseline distribution the weighted log likelihood function can be written as

$$wl_\gamma(\eta; x) = \sum_{i=1}^{n} w_{\gamma,i}(x) \left\{\delta_i \left[logR(\eta) + log\lambda_0(t_i)\right] - \Lambda_0(t_i)R(\eta)\right\}, \tag{11}$$

where $w_{\gamma,i}(x), i = 1, \ldots, n$, are appropriate weights. Maximization w.r.t. $R(\eta)$ yields

$$R(\hat{\eta}(x)) = \frac{\sum_{i=1}^{n} w_{\gamma,i}(x)\, \delta_i}{\sum_{i=1}^{n} w_{\gamma,i}(x)\, \Lambda_0(t_i)}. \tag{12}$$

Since $\Lambda_0(t_i)$ given $X_i = x_i, i = 1, \ldots, n$, are exponentially distributed with mean $1/R(\eta(x_i))$, it can be seen that (12) is a smooth generalization of the ML estimator of the parameter of the exponential distribution.

In case that $\Lambda_0(t)$ is unknown, the *iterative corrected locally weighted maximum likelihood estimator* (ICME) is calculated iteratively in three steps until convergence. In the m-th iteration:

Step 1: Preliminary estimation of Λ_0

$$\tilde{\Lambda}_0^{(m)}(t) = \hat{\Lambda}_0(t \mid R(\hat{\eta}^{(m-1)})$$

Step 2: Preliminary estimation of $R(\eta)$

$$R(\tilde{\eta}^{(m)}(x)) = \frac{\sum_{i=1}^{n} w_{\gamma,i}(x)\, \delta_i}{\sum_{i=1}^{n} w_{\gamma,i}(x)\, \tilde{\Lambda}_0^{(m)}(t_i)}.$$

Step 3: Correction step

$$\begin{aligned} R(\hat{\eta}^{(m)}(x)) &= R(\tilde{\eta}^{(m)}(x))/H(R(\tilde{\eta}^{(m)})) \\ \hat{\Lambda}_0^{(m)}(t) &= \tilde{\Lambda}_0^{(m)}(t)H(R(\tilde{\eta}^{(m)})). \end{aligned}$$

The final estimator obtained in Step 3 satisfies (10).

Since an arbitrary norming functional H can be used in this algorithm, the question arises how this functional should be chosen. It should be noted that estimators obtained from different choices of the norming functional should again be the same up to a positive factor. Can this condition be fulfilled by an algorithm estimating $R(\eta)$, then the estimator is called *invariant.* Kübler (1991) shows that the ICME is invariant if the estimator $\hat{\Lambda}_0(t \mid r)$ used in Step 1 satisfies

$$\hat{\Lambda}_0(t \mid \alpha r) = \frac{1}{\alpha} \hat{\Lambda}_0(t \mid r) \; \forall \alpha \in \mathbb{R} \; \forall r \in \mathcal{F}. \tag{13}$$

Note, this condition is rather weak, because it says that the product of the risk functions r and αr with the corresponding estimates of the cumulative baseline hazard functions do not depend on α, which is a natural condition under (1). Thus, parametric as well as nonparametric estimators of Λ_0 satisfying (13) can be used for the ICME, but the nonparametric approach seems to be more appropriate in this context.

Following Clayton & Cuzick (1985) it is quite natural to choose Breslow's estimator (3) for Λ_0. In this case the algorithm is invariant. Furthermore, there exists a unique solution of the underlying fix–point problem if $w_{\gamma,i}(x_j) > 0 \; \forall i, j = 1, \ldots, n$ (Kübler (1991)). This condition is obviously fulfilled when choosing the *gaussian kernel*, i.e.

$$w_{i,\gamma}(x) = \varphi(\frac{x - x_i}{\gamma}), \qquad x, x_i \in \mathbb{R}, \; i = 1, \ldots, n \; ,$$

where φ denotes the probability density function of the standard normal distribution.

Generalized additive model

The basic idea of smoothing, namely averaging over neighborhoods, causes some problems in high dimensions. Unfortunately, the number of observations in a neighborhood of an arbitrary point x rapidly decreases with an increase of dimension p of the covariate for a fixed number n of observations. To compensate for this effect, larger bandwidths should be chosen such that a reasonable number of observations is available in each neighborhood. But then local changes in the regression functions cannot be detected and, if the number of covariates p is larger than two, the estimates of the risk functions can no more be visualized. For that reason several proposals for structured models have been made, e.g. *regression trees* (Gordon & Olshen (1980)), *projection pursuit regression* (Friedman & Stützle (1981)), and *generalized additive models* (Hastie & Tibshirani (1986)).

In this paper let us focus on the concept of generalized additive models, which can be interpreted as a smooth extension of the usual parametric model, when the function R in (1) is known. In the parametric approach, usually the regression function $\eta : \mathbb{R}^p \to \mathbb{R}$ is assumed to be $\eta(x) = x'\beta$, where β is a p-vector of unknown regression coefficients. More general, $\eta(x)$ can be considered as a sum of regression functions

$$\eta(x) = \sum_{j=1}^{p} \eta_j(x_j), \; x = (x_1, \ldots, x_p)' \in \mathbb{R}^p, \tag{14}$$

where $\eta_j : \mathbb{R} \to \mathbb{R}, j = 1, \ldots, p$. The basic idea of *generalized additive models* is to consider $\eta_j, j = 1, \ldots, p$, as smooth functions which are to be estimated.

In biomedical applications there is a tendency to incorporate quite a large number of covariates in a model most of them of categorical scale such as 'gender', 'treatment', health categories etc. Thus, mixtures of both nonparametric and parametric approaches are of special interest. Let us assume, that $z \in \mathbb{R}^q$ is an additional vector of covariates. Then (14) is easily extended to

$$\eta^*(x, z) = \eta(x) + z'\beta, \; \beta = (\beta_1, \ldots, \beta_q)' \in \mathbb{R}^q, \tag{15}$$

where $\eta(x) = \sum_{j=1}^{p} \eta_j(x_j)$ (c.f. (14)) denotes the nonparametric and $z'\beta$ denotes the parametric component of the regression function. In the remainder estimates of $\eta_j, j = 1, \ldots, p$, and β are proposed based on a combination of the ICME and the *backfitting algorithm.* The method is described only for R chosen as the exponential function, but can be extented to other choices.

The backfitting algorithm is originally defined as a method of estimating a regression function $s_j(x_j), j = 1, \ldots, p$, in an *additive model* $E(Y \mid X = x) = s_0 + \sum_{j=1}^{p} s_j(x_j)$, but can be extended straightforwardly to a likelihood–based estimation. The algorithm starts with an initial estimate of s_0, say $\hat{s_0}$. The subsequent iterative procedure can be described as follows: Suppose $\hat{s}_j^{(\ell)}, j = 1, \ldots, p$, are the estimates obtained from the ℓ–th iteration. Then, in step $\ell + 1$ the j-th function is estimated conditionally on $s_k = \hat{s}_k^{(\ell)}, k = 1, \ldots, p, \; k \neq j$ and $s_0 = \hat{s_0}$.

Applying this technique to the ICME the second step has to be replaced by a generalized backfitting algorithm to estimate η^*. Thus, Step 2 itself becomes an iterative procedure, where the estimate of Λ_0 plays the role of the initial estimate of s_0. Since (15) is used, each iteration of the backfitting algorithm consists of two parts. Part (a) provides an estimate of β, where a full likelihood approach is used for given functions $\Lambda_0(t)$ and $\eta(x)$. Unfortunately, this estimator cannot be obtained in a closed form. To estimate the j-th component of η in Part (b), a weighted likelihood estimator is used for given functions $\Lambda(t)$, $z'\beta$, and η_k, $k = 1, \dots, p$, $k \neq j$. The correction step is necessary for the functions η_j, but also for the parametric component $z'\beta$ if an intercept parameter is used .

Now, let $z_i \in I\!R^q$ denote the covariates of the parametric part of η^* for the i-th observation with survival time t_i and censoring indicator $\delta_i, i = 1, \dots, n$. The covariates to be smoothed are denoted by $x_i = (x_{i1}, \dots, x_{ip}) \in I\!R^p$, $i = 1, \dots, n$. The weight of the i-th observation for smoothing the j-component of η at $\xi \in I\!R$ with bandwidth γ is given as $w_{\gamma,i;j}(\xi)$, $j = 1, \dots, p$, $i = 1, \dots, n$. Then the ℓ-th loop of the backfitting algorithm embedded in the m-th iteration of the algorithm to calculate $\hat{\eta}^*$ can be described as follows:

Step 1: Preliminary estimation of Λ_0

$$\tilde{\Lambda}_0^{(m)}(t) = \hat{\Lambda}_0(t \mid \exp(\hat{\eta}^{*(m-1)}))$$

Step 2: Backfitting

(a) $\hat{\beta}^{(m,\ell)} = \underset{\beta \in I\!R^q}{\operatorname{argmax}} \, L(\beta \mid \tilde{\Lambda}_0^{(m)}, \hat{\eta}^{(m,\ell-1)})$

(b) $\exp(\tilde{\eta}_j^{(m,\ell)}(\xi)) =$

$$\frac{\sum_{i=1}^{n} w_{\gamma,i;j}(\xi)\, \delta_i}{\sum_{i=1}^{n} w_{\gamma,i;j}(\xi)\, \tilde{\Lambda}_0^{(m)}(t_i) \exp\left(\tilde{\eta}_{-j}^{(m,\ell-1)}(x_i) + z_i'\hat{\beta}^{(m,\ell-1)}\right)}, \ \xi \in I\!R \ ,$$

where

$$\tilde{\eta}_{-j}^{(m,\ell-1)}(x_i) = \sum_{k=1, k\neq j}^{q} \tilde{\eta}_k^{(m,\ell-1)}(x_{ik}), \ j = 1, \dots, p \ .$$

Step 3: Correction step

$$\exp(\hat{\eta}_j^{(m)}(\xi)) = \exp(\tilde{\eta}_j^{(m)}(\xi))/H(\exp(\tilde{\eta}_j^{(m,\ell)})), j = 1, \dots, p \ .$$

Note, that both the bandwidth γ and the norming function H might depend on j.

Although this *generalized ICME* is highly computer intensive, it provides a flexible tool for estimating the risk function in the *phm*. It should be mentioned, that this algorithm yields the ordinary partial likelihood estimator if the nonparametric part is omitted.

Example

The method presented in the above section is in the following illustrated using the Veteran's Administration lung cancer data as reported in Appendix I of Kalbfleisch & Prentice (1980). In this randomized clinical trial, a standard and a test chemotherapy for lung cancer were compared w.r.t. time to death chosen as primary end point. The treated population consisted of 137 males with advanced inoperable lung cancer. In total, only 9 survival times were censored. In addition, six covariates were considered:

1. treatment: 1 = standard, -1 = test
2. prior therapy: 1 = YES, -1 = NO
3. histological type of tumor: squamous, small cell, adeno, large cell
4. Karnofsky scale of the patients' performance status
5. time in months from diagnosis to randomization of the patients
6. age in years.

In the models used for analyzing the data, the categorical covariate "histological type of tumor" is replaced by three variables:

$Cell_1$: 1 = squamous, -1 = large cell, and 0 else,
$Cell_2$: 1 = small cell, -1 = large cell, and 0 else,
$Cell_3$: 1 = adeno, -1 = large cell, and 0 else.

Kalbfleisch & Prentice (1980, p. 89) fit an proportional hazard model to the data. They assume the log risk function to be a linear function of the above 8 variables. Our aim is now to fit a generalized additive model by replacing $\beta_{age} \times age$ with a smooth function $\eta_{age}(age)$. An estimate for $\eta_{age}(\cdot)$ will be obtained by using the generalized ICME.

Comparing the ICME and the partial likelihood estimate problems occur if tied survival times are present. As already mentioned, the generalized ICME reduces to Clayton & Cuzick's method for calculating the partial likelihood estimator in case of no smoothing and no tied survival times. However, in presence of ties, Clayton & Cuzick's method can not handle tied observations in the *phm*. Therefore, randomization has been used for breaking these ties to obtain comparable results. The parameter estimates both for the exact and the randomized partial likelihood model are given in Table 1. Since only minor differences can be observed, randomization seems to be appropriate in this example.

Parameter	PL_{exact}		PL_{rand}		$ICME$
	parameter estimate	*standard error*	*parameter estimate*	*standard error*	*parameter estimate*
β_{treatm}	-0.1473	0.1038	-0.1497	0.1038	-0.1250
β_{prior}	0.0358	0.1162	0.0344	0.1161	0.0283
β_{KPS}	-0.0328	0.0055	-0.0328	0.0055	-0.0315
β_{Cell_1}	-0.6148	0.1768	-0.6117	0.1768	-0.6114
β_{Cell_2}	0.2468	0.1594	0.2490	0.1594	0.2339
β_{Cell_3}	-0.2134	0.1741	-0.2148	0.1742	-0.2023
β_{diag}	0.0000	0.0091	0.0004	0.0092	0.0016
β_{age}	-0.0087	0.0093	0.0087	0.0093	—

Table 1: Parameter estimates

The parameter estimates from the generalized ICME are also given in Table 1. The differences between the randomized partial likelihood estimates and the ICME estimates are again not essential.

The kernel estimate of $\eta_{age}(\cdot)$ and the partial likelihood estimate are shown in Figure 1. Both functions are shifted to have mean zero on $[34, 81]$. These estimates agree with each other for patients between about 60 and 75 years. A marked reduction of risk can be observed for those, which are about 80 years. For young patients the nonparametic estimate has a greater slope than the parametric one.

Since there are differences between the parametric and the nonparametric estimate, it is of interest to have a closer look on the distribution of the risk factor "Age" given in Table 2. The decrease of risk for older patients is due to two patients of 81 years, whereas the next younger patients are only 72. About 60% of patients are between 60 and 72 years old. This large number of patients obviously has a tremendous impact on the parametric estimate and is the reason for the estimates conformity in this range.

Discussion

The calculation of kernel based estimates of the regression function in the proportional hazards model is computer intensive. The effort to calculate such estimates can be reduced taking into account some basic properties of the algorithm. The estimator (3) of the cumulative baseline hazard function

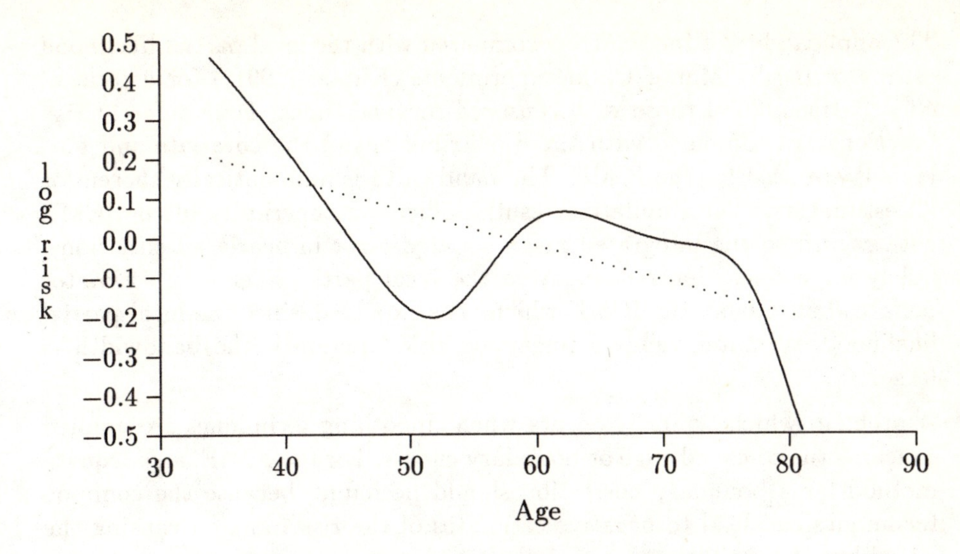

Dotted line: Partial likelihood estimate
Solid line: ICME with Gaussian kernel and bandwidth $\gamma = 4.7$

Figure 1: Estimated Log Risk Function for variable AGE

Age Group	30 - 34	35 - 39	40 - 44	45 - 49	50 - 54
Patients	1	10	10	10	13
Age Group	**55 - 59**	**60 - 64**	**65 - 69**	**70 - 74**	**≥ 75**
Patients	9	40	32	10	2

Table 2: Distribution of Age

proposed by Breslow (1972), which is used in Step 1 of the ICME, does not depend on the whole regression function but only on distinct values $R(\eta(x_i)), i = 1, \ldots, n$, of that function. Thus, the estimator of $R(\eta)$ in Step 2 has to be calculated solely for those values. Moreover, the norming functional to be applied in the correction step can be chosen only w.r.t. its computational simplicity, because the invariance property of the algorithm ensures that the result can be retransformed such that it fulfills the norming condition of choice.

When different estimates are to be compared (c.f. the above example) we recommend using the integrated logarithm of the risk function as norming functional, whereas the condition $R(\eta(x_0)) = 1$ for some fixed value of x_0 is often suitable during the iteration.

The applicability of the ICME is compared with the local partial likelihood estimator (8) by Monte–Carlo experiments (Kübler (1991)) for a total of 48 situations, 36 of those with censored survival times. Four different risk functions are combined with three distributions of the covariate and four kernels are used for the ICME. The bandwidth is automatically chosen for all estimators. The simulation results indicate a superiority of the ICME with regard to the integrated mean squared error in nearly all situations. Solely for a linear log risk function the local partial likelihood estimator behaves better than the ICME which is as expected, since the local partial likelihood estimates yields a linear log risk function if the bandwidth is large.

A problem which typically occurs when smoothing techniques are applied concerns the so-called edge or boundary effects. For the ICME, an adequate method for a boundary correction should be found, because the common techniques can lead to negative estimates of the risk function causing the algorithm to break down.

References

Andersen, P.K. & Gill, R.D. (1982). Cox's regression model for counting processes: a large sample study. Ann. Statist., 10, 1100 - 1120.

Breslow, N.E. (1972). Contribution to the discussion of Cox (1972). J. Roy. Statist. Soc., B 34, 216 - 217.

Clayton, D. & Cuzick, J. (1985). The EM algorithm for Cox's regression model using GLIM. Applied Statistics, 34, 148 - 156.

Cox, D.R. (1972). Regression models and life–tables (with discussion). J. Roy. Statist. Soc., B 34, 187 - 202.

Cox, D.R. (1975). Partial likelihood. Biometrika, 62, 269 - 276.

Dempster, A.P., Laird, N.M. & Rubin, D.B. (1977). Maximium likelihood from incomplete data via the EM algorithm (with discussion). J. Roy. Statist. Soc., B 39, 1 - 38.

Friedman, J.H. & Stützle, W. (1981). Projection pursuit regression. J. Amer. Statist. Assoc., 76 817 - 823.

Gentleman, R. (1988). Non–linear covariates in the proportional hazards model. Technical report series, STAT–88–14, University of Waterloo, Department of Statistics and Acturial Science.

Gentleman, R. & Crowley, J. (1991). Local full likelihood estimation for the proportional hazards model. Biometrics, 47, 1283 - 1296.

Gordon, L. & Olshen, R.A. (1980). Consistent nonparametric regression from recursive partitioning schemes. J. Mult. Anal., 10, 611-627.

Härdle, W. (1990). Applied nonparametric regression, Cambridge University Press, Cambridge, New York.

Hastie, T. & Tibshirani, R. (1984). Generalized additive models. Technical Report No. 2, Stanford University, Department of Statistics.

Hastie, T. & Tibshirani, R. (1986). Generalized additive models (with discussion). Statistical Science, 1, 297 - 318.

Kalbfleisch, J.D. & Prentice, R.L. (1980). The statistical analysis of failure time data, Wiley, New York.

Kübler, J. (1991). Nichtparametrische Schätzer der Risikofunktion im Proportional–Hazards–Modell. Dissertation. Universität Dortmund, Fachbereich Statistik.

Nadaraya, E.A. (1964). On estimating regression. Theor. Probab. Appl., 9, 141 - 142.

O'Sullivan, F. (1986). Nonparametric estimation in the Cox proportional hazards model. Technical Report No. 64, University of California, Department of Statistics.

O'Sullivan, F. (1988). Nonparametric estimation of relative risk functions using splines and cross–validation. SIAM J. Sci. Stat. Comput., 9, 531 - 542.

Staniswalis, J.G. (1989). The kernel estimate of a regression function in likelihood–based models. J. Amer. Statist. Assoc., 84, 276 - 283.

Tibshirani, R. & Hastie, T. (1987). Local likelihood estimation. J. Amer. Statist. Assoc., 82, 559 - 567.

Watson, G.S. (1964). Smooth regression analysis. Sankhyã, A 26, 359 - 372.

Interval Censored Observations in Clinical Trials

Armin Koch
Abteilung Medizinische Biometrie, Universität Heidelberg, Im Neuenheimer Feld 305, D – 69120 Heidelberg

Key Words:
clinical trials, interval censoring, survival analysis

Abstract
Interval censored observations are introduced. Some example situations from clinical trials are mentioned, where in one interval censored observations occur 'by design' and cannot be omitted. A short survey of papers considering this restriction on the observability of the outcome variable is given. For one of the sample situations, the remission duration in a cancer clinical trial, study data are analysed according to a conventional and an interval data approach. Simulation results demonstrate the general need to take the interval structure of the data in the example situation into consideration. The effect of censoring and of larger or smaller follow-up intervals on bias, variance and mean squared error of the two approaches is shown.

Introduction

Time to event is one of the most important target variables in medical research to evaluate the capabilities of a treatment or the value of an intervention. Restrictions on the observability of the target event are common to this sort of investigations and have led to the development of a special methodology, termed survival analysis. Although the interest is, in general, not focused on an individual's total lifetime (i.e. time between birth and death) the terms lifetime and failure-time will be used interchangeable for time to event data in this article.

Methodological investigations were initiated from life testing and carcinogeneity experiments, where not all animals develop a tumor and are sacrificed at the end of the predetermined study period. These experimental designs are leading to observations that are nowadays called singly type I censored. Further developments were concerned with the situation of a clinical trial, where a patient's time under investigation is a random variable, which is determined by the person's entry into the study and the follow-up period of the trial, and even more complex situations.

From a theoretical point of view at least three different directions for methodological developments can be identified: Generalizations of classical methods to censored data situations, to further generalize these methods to apply to more general restrictions on the observability of the outcome variable and to introduce the methodology of stochastic processes into survival analysis.

This article is devoted to so-called "interval censored observations", where instead of the true value of an outcome variable, only a lower and an upper boundary can be specified. This can account for fairly general restrictions on the observability of the response variable: Right censored observations, where only a lower boundary l_i for the true value of the outcome variable is available, can be included as intervals $(l_i, \infty]$. Serial sacrifice experiments are designed to determine e.g. the effect of an agent on the time until tumor onset in a situation, where this event is not directly observable. The presence of a tumor at sacrifice produces a left censored observation, where in contrast an upper boundary r_i is known to restrict the value of the response variable. This can be represented by an interval $(0, r_i]$. Even members of grouped samples can be viewed as interval censored, where the class boundaries provide the interval limits of an observation.

The problem

In the context of clinical trials, at least two situations can be identified where interval censored observations occur in a somehow natural way. In the first example, interval censoring is a consequence of a random process that might be termed patient's compliance, whereas in the second, the interval-valued quality of the response variable is a consequence of the study design, and is in this, inherent to the type of the investigation.

Interval censoring in long term clinical trials

Time to event is usually regarded as a continuous variable. In real world examples it is, however, subjected to measurements with finite precision. Neglecting this might be valid in a situation, where the presence or absence of a condition is monitored every day for several weeks or months. Very often, however, the duration of a clinical trial exceeds the patient's stay in hospital, and the observation of the outcome variable is restricted to discrete follow-up dates. This should lead to grouped, or in the usual case of a limited study duration, to grouped and right censored observations.

A common problem, especially in the investigation of chronical diseases, is, that a patient, who is monitored every three months, misses some of the pre-specified visits and returns under observation after the change in the response state has occurred. Methods for grouped and censored observations are no longer valid as this observation cannot be assigned to one of the intervals.

Methods for interval censored data are appropriate here. Additional problems occur in situations where the observational scheme is not independent from the disease process (informative censoring). For example: the patient misses appointments while feeling healthy and returns because the response state has changed. See Grüger et al. (1991) for a more thorough investigation on likelihood based inference and informative examination schemes in clinical trials.

Remission duration in cancer clinical trials

Remission duration in cancer clinical trials is a meaningful, but nevertheless criticized outcome measure. Deviations from the planned examination scheme are in common: The doctor might postpone the next treatment for some days due to organisational reasons or the patient's bad overall condition; the patient might miss an appointment in the follow-up period of the trial (see Rücker & Messerer (1988)). It is important to notify, that even in a situation, where every patient keeps all visits, interval censored observations for the remission duration will always occur as soon as intervals between two successive clinical appointments differ under treatment conditions and in the follow-up period (equidistant follow-up appointments would impose unacceptable restrictions on most clinical trials):

Recall the common regimen of a treatment with six chemotherapeutic cycles that are administered in one month intervals and a planned two year follow-up, where the patient's remission state is investigated in three month intervals. Both events, the occurence of complete remission, as well as beginning of progression, can only be assigned to one of the respective intervals of observation under treatment or control conditions. The precise limits of these intervals are determined by the date of the last observation of the old tumor, the first date of diagnosis of the complete remission, the last diagnosis of the patient's state in complete remission and the first diagnosis of progress, respectively. A patient who is still in complete remission at the end of the study will contribute a right censored observation.

Clinical events are often recorded when they are actually diagnosed and remission duration is then estimated from the elapsed time between diagnosis of remission and diagnosis of progress or an appropriately right censored observation at the date of the last examination of a patient under study conditions (which will be termed conventional approach below). Let T_R and T_P denote the begin of complete remission and of progression, respectively, limiting the "true" remission duration. Both are only known up to the length of the relevant sub-intervals. As an example, suppose that a patient achieves complete remission between the first and the second cycle, i.e. $T_R \in (1,2]$, and progression happens to fall into the first follow-up interval ($T_P \in (6,9]$). The true remission duration is then bounded with four and eight months. A second example with $T_R \in (3,4]$ and $T_P \in (9,12]$, bounding the true time in remission to the interval between six and eight months, demonstrates, that these data can not be treated as coming from grouped and censored samples.

Short survey of methods

Parametric methods

There exist a number of objections to the approach of estimating model parameters in the situation where only grouped, or even interval censored observations are available, as the justification of the selected model might be difficult.

In case the assumption of a certain parametric failure time model can be justified from external or theoretical reasons, the estimation of the relevant parameters from the interval censored sample is a mere technical problem. Suppose that random variables T_i $(i = 1, \ldots, n)$ represent the survival times of n individuals in a population. $T_1, \ldots, T_n$ are assumed independent and identical distributed with probability density function $f(t)$. As noted earlier the available data are intervals containing the true, but unobserved survival times, i.e. $T_i \in (l_i, r_i]$. Under the above mentioned independence of the examination scheme and the failure time process, the full likelihood is proportional to

$$L = \prod_{i \in L} P(T_i \leq r_i) \prod_{j \in I} P(l_j < T_i \leq r_j) \prod_{k \in R} P(T_k > r_k)$$

where L, I, R denote the index sets for individuals with left-, interval- and right censored observations, respectively. A term for exact observations might be included. It has been, however, omitted as it is difficult to find a practical relevant situation, where an observational plan might lead to exact and interval censored observations at the same time. The asumption of an exponential failure time model (i.e. $T \sim Ex(\lambda)$ and $P(T \leq t) = 1 - \exp(-\lambda t)$) leads to the following log-likelihood function:

$$l = \sum_{i \in L} \log(1 - e^{-\lambda r_i}) + \sum_{j \in I} \log(e^{-\lambda l_j} - e^{-\lambda r_j}) + \lambda \sum_{k \in R} l_k.$$

This is in contrast to the standard situation with exact and right censored observations, where the root of the first derivation of the likelihood function with respect to λ is $\hat{\lambda} = d/\sum t_i$ (d denotes the number of observed failures): Closed form maximum-likelihood estimators are no longer available. Whereas the asymptotic properties of maximum-likelihood estimators are well understood, the effect of interval censoring on bias and mean squared error of these estimators in the above mentioned situations in small or moderate sized samples has not yet been investigated in the literature.

Lognormal distributed failure-times have been investigated in a simulation study by Denby et al. (1975). In that paper the interest was, however, mainly focused on the effect of variations in the experimental conditions between studies (i.e. variations in interval sizes between studies) on the estimation of median survival times, whereas the concern here are varying interval sizes within one study.

In Flygare et al. (1985) the results of a number of simulations on grouped and censored observations from the Weibull distribution are presented. The authors compare the maximum-likelihood estimators from the interval data to the estimators gained from replacing each observation with the midpoint of the censoring interval. The main results are: the latter method, being comparable to the usual approach of a clinical investigation to register an event when it is actually observed (and thus, replacing an interval valued observation with the right interval boundary), is underestimating the true error variance by assuming that the failure times are known with a higher degree of accuracy. If the fraction of right censored observations increases, the difference between the error variances, estimated in the two approaches, becomes smaller: these observations are handled correctly in both approaches. The midpoint method biases the estimation of the scale parameter of the distribution towards an overestimation of the expected survival. On the contrary the estimation of the shape parameter is biased towards unity.

The loss of efficiency in estimating the parameter of the exponential distribution from grouped and singly type I censored observations is investigated in Hamada (1991a) with the aid of theoretical computations and small sample simulations. One major result is, that this loss is small if inspection intervals are small or moderately wide, provided the amount of censoring is not too large.

Nonparametric estimation of the survival function

The Kaplan-Meier-estimator, the nonparametric maximum likelihood estimator of the survivorship function, is based upon the number of persons under risk at a certain time. In a situation with interval censored data, where intervals also might overlap, it is no longer clear, which observations do belong to the risk set for time points in the overlapping area.

One approach to gain a maximum likelihood estimator in this situation is considered by Peto (1973), who proposed a constrained Newton-Raphson method. Turnbull (1976) provides a self-consistent estimator for the survival function from interval censored data. The term is introduced in Efron (1967) in the context of an alternative derivation of the Kaplan-Meier estimator: a recursion formula is applied to an initial estimate. This iterative process is converging to a function that cannot be further improved by its own. Turnbull shows, that the maximum likelihood estimate fulfils the self consistency condition and that the self consistent estimate in the present situation is also maximum likelihood estimate. Estimates for variance and covariance of the parameters are, however, only available from the second derivation of the likelihood function (observed information) which in general is known to approximate the true variance especially in small samples very poorly.

From an isotonic regression approach Groeneboom (see e.g.: Groeneboom (1991) or Groeneboom & Wellner (1992)) derives a necessary and sufficient characterization of the maximum likelihood estimator from interval data, which can also be used to construct an algorithm. Distribution theory is, however, not yet fully developed for the present situation: Groeneboom specifies a working hypothesis under which asymptotic normality and a formula for the asymptotic variance for the nonparametric estimator of the survival function can be derived. Parameter estimates from loglinear models can be used to compute the Turnbull estimator of the survivor function. A description of this is found in Becker & Melbye (1991).

Nonparametric tests

In many occasions, a medical investigation is aimed at a decision, whether two treatments differ with respect to their ability to improve survival. Suppose that a decision has to be based on two samples with sizes n_1, n_2 and survival times for two individuals are censored into intervals $I_{ij} = (l_{ij}, r_{ij}]$ $(i = 1,2, j = 1,\dots,n_i)$ and $I_{kl} = (l_{kl}, r_{kl}]$ $(k = 1,2, l = 1,\dots,n_k)$. In case $l_{kl} \geq r_{ij}$ individual l from sample k reveals the better survival. This shows clearly that ranks can be associated with intervals. Samples with interval censored observations might contain overlapping intervals (the associated true survival times might be tied in this situation), but ties are a known problem in nonparametric statistical methodology. The papers of Gehan (1965a, 1965b) provide the generalization of the Wilcoxon test in the present situation. Mantel (1967) suggests a computationally less expensive formula for the test statistics and the permutational variance estimate:

A general scoring function assesses the comparison of two observations I_{ij} and I_{kl} according to

$$c(I_{ij}, I_{kl}) = \begin{cases} +1, & , \quad \text{if} \quad l_{ij} \geq r_{kl} \\ -1, & , \quad \text{if} \quad r_{ij} \leq l_{kl} \\ 0, & , \quad \text{otherwise} \end{cases}$$

The score of an observation $U_{ij} = \sum_{k=1}^{2} \sum_{l=1}^{n_k} c(I_{ij}, I_{kl})$ is just the difference of the numbers of observations that are definitely less than the current observation and the number of observations that are definitely greater in the pooled samples. The sum of scores for the first sample forms the test statistics $G_M = \sum_{j=1}^{n_1} U_{1j}$ with expectation $E(G_M) = 0$ and variance

$$Var(G_M) = \frac{n_1 n_2}{(n_1 + n_2)(n_1 + n_2 - 2)} \sum_{i=1}^{2} \sum_{j=1}^{n_i} U_{ij}^2$$

under the hypothesis that all rank-permutations have the same probability and $G_M / \sqrt{Var(G_M)}$ can be compared to the standard normal distribution.

This generalized scoring procedure forms the basis for the majority of the methods cited hereafter: Hilgers (1989) supplies the generalization of the U-test for interval censored data, testing the specific hypothesis about a shift-parameter: $H_0 : \delta = 0$. The test statistic is identical to the test of Gehan, but the estimators of the variance differ between the two approaches. Mantel (1967) and Hilgers & Neumann (1989) both note the possibility to perform an exact version of the Gehan test with the above mentioned scores.

In the case of more than two samples, generalizations for almost all non-parametric tests are available. See Schemper (1984b) for the generalization of the Wilcoxon-signed-rank test; Marcuson (1983) and Schemper (1983) independently mention the generalization of the Kruskal-Wallis test; the Jonckheere test for trend is treated in Abel (1986); cf. Schemper (1984a) for a generalised Friedman test; for Kendall's τ and the corresponding test for association refer to Schemper (1982); see Wittkowski (1984) for a generalization of the marginal likelihood principle.

A systematic approach to linear rank tests is found in Self & Grossman (1986), who extend the work of Prentice (1978) on right censored observations for the current situation. From the likelihood based on all rank vectors that are consistent with the observed data the score statistic is derived and shown to have the form of a linear rank statistic. The score of an observation can be written as the weighted average of the uncensored data rank scores. Unfortunately the calculation of the weights necessitates the enumeration of the above mentioned rank vectors, which can be infeasible even for moderate sample sizes due to censoring. The paper presents several proposals for the estimation of optimal weights including simulation approaches.

Regression models

Regression models with interval censored obserations are investigated in three papers by Finkelstein (& Wolfe) (1986, 1985, 1986). In a so-called reversed regression model the joint probability of a response time T and a random covariate X is factorized as $P(X,T) = P(X|T)P(T)$ where the distribution $S(t) = P(T > t)$ is not further constrained. The weakest restriction for the conditional covariate density is the assumption of an isotonic model with a monotonous non-decreasing function $p(t) = P(X = 1|T = t)$ (the proportional hazards model fulfils this condition). The authors specify an EM-algorithm for parameter estimation but state, that convergence of the algorithm cannot be proven via standard approaches.

In Finkelstein & Wolfe (1985) a parametric model for the conditional covariate density is introduced. Parameter estimates can be retrieved via Newton-Raphson methods. A logistic model is further investigated. Besides parameter estimates a score test for the model parameters, which can be used to test a treatment effect, is available.

Finkelstein (1986) is devoted to the development of the likelihood equations for the proportional hazards model. The latter is of greater importance: Let β denote the treatment effect in this model. The partial score test of the hypothesis $H_0 : \beta = 0$ then provides the generalisation of the logrank test for the interval censored sample.

Example

The data from a randomized multicentre clinical trial are investigated for illustrative purposes. Two different therapeutic administrations for the treatment of small cell lung carcinoma with polychemotherapy were compared. In one treatment group, two different drug combinations were administered strictly alternating, whereas in the other group the application of the second drug combination was performed as soon as no further response under the first was observed. For the evaluation of differences in the efficacies of the two regimens, overall survival was stated to be the relevant criterion. Details on this study are published in Wolf, Pritsch, Drings et al. (1991). No differences between the two treatment groups were reported.

Different opinions exist in the medical community on whether time in remission (defined as above), or freedom from progression (where the zero point of the time axis is the start of treatment), is the relevant therapeutic criterion. In a situation, where all patients keep all clinical appointments an analysis of the variable "freedom from progression" will be based on grouped and censored observations. Our interest is therefore focused on the variable remission duration, leading to interval censored obserations "by design". This variable was not included into the former analysis. The examination scheme was planned as follows: 6 chemotherapeutic cycles were given in three-week intervals. Minimum follow-up was 2 years and a patient is seen in this period in three-month intervals.

The main results are: A number of 95 out of 320 randomized patients achieved complete remission under study conditions, 44/159 and 51/161 in the two arms of the study, which is no significant treatment effect in a two-sided test procedure with a type one error of five percent.

For the estimation of remission duration we restrict our consideration on the responders. Two different approaches for the estimation of distributional parameters from this dataset are presented. The first one uses the conventional approach to register an event when it is actually observed. This is leading to 64 exact and 29 right censored observations. The second approach respects the interval structure in the dataset and leads to 7 left-, 57 interval- and 29 right-censored observations (interval data approach): left censored observations do occur in case the upper boundary of the interval that limits the start of complete remission coincides with the lower bound of the interval, in which the begin of progression happens to fall. Two observations were excluded due to right censoring directly after the diagnosis of complete remission. In both approaches they do not contribute to the likelihood function.

Time in remission is assumed to follow an exponential distribution. Whereas for the first approach the maximum likelihood estimator of the parameter is the number of observed relapses divided by the total observed time and can easily be computed by a pocket calculator, the computation of parameter estimates from interval censored observations, based upon iterative solutions of the likelihood equations, is more complicated. SAS Version 6 procedure LIFERREG can be used to analyze the data sets for both approaches.

In the first approach, where observations are considered exact and right censored, the parameter estimates are $\hat{\lambda}_1 = 1.85 \times 10^{-3}$ and $\hat{\lambda}_2 = 3.03 \times 10^{-3}$ for the two treatment groups. From this the estimated median remission times are 539 and 330 days, respectively. For the interval data approach, the parameter estimates are $\hat{\lambda}_1 = 1.95 \times 10^{-3}$ and $\hat{\lambda}_2 = 3.25 \times 10^{-3}$ with median remission times of 512 and 308 days. The conventional approach overestimates the median survial (see also the right part of figure 1).

Standard errors for the estimators are available and can be used to demonstrate that the two different approaches might well lead to different conclusions about a treatment effect in a formal hypothesis testing situation: the P-values are 0.059 and 0.049 for the first and the second approach. In the present situation a treatment comparison should, however, be based on remission rate and remission duration. The "**p**robability of **b**eing in **r**esponse **f**unction" approach (see Temkin (1978)) is capable of integrating the findings of more than one endpoint and is therefore more appropriate for a treatment comparison. The methodology has been developed for grouped and right censored observations.

On the other hand it is more astonishing that the differences between the conventional and the interval data approach are in fact that small. A closer look at the data reveals, that

- the majority of the responders in both treatment arms reaches complete remission after the first or second cycle, so there is only little variation in the first event, that defines remission duration,
- about 30 percent of the observations are right censored and are correctly (and identically) handled in both approaches.

In figure 1 thick lines denote estimates from the data in its interval representation. The left graph contrasts Turnbull- and Kaplan-Meier-estimator. According to the definition the latter curve ends with the largest uncensored observation (at day 488 / 508 in the two treatment groups), whereas Turnbull's algorithm puts a noncero mass on the interval up to the largest censored observation (at day 1047 / 1296, respectively). Gaps between horizontal bars represent regions in which the Turnbull-estimator is undefined (please note that the contribution of a single interval censored observation to the likelihood-function is the difference of the values of the survival function of the two interval end points and does not depend on the detailed behavior within the interval).

A simulation study

For a situation similar to the example mentioned above, a small simulation study was performed to investigate the bias, the mean squared error and the variance of the conventional strategy to record the occurrence of an event when it is actually observed. This is compared with the correct approach to respect the interval nature of this type of data (interval data approach).

Our observational plan, however, assumes that the treatment is ceased as soon as complete remission is diagnosed, and the patient will then be seen on a follow-up schedule with larger time intervals as those usually necessary under treatment. More precisely our model is as follows: Suppose that X represents the time between onset of treatment and the achievement of complete remission, for which we assume either a uniform or an exponential distribution. This is followed by an exponentially distributed time in

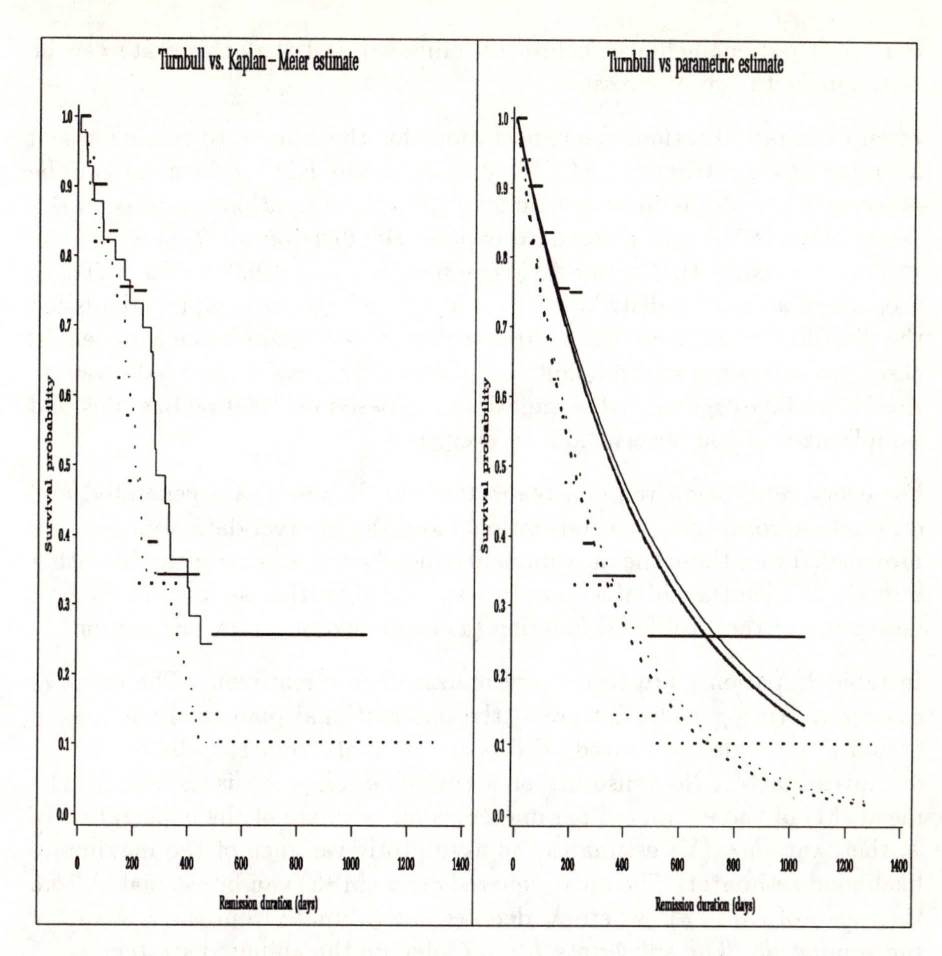

Figure 1: *Remission duration under two therapeutic schemes: thick (thin) lines represent estimates from the interval data (conventional data).*
Left figure: estimates from the Kaplan-Meier estimator are depicted with connected step functions.
Right figure: the smooth plot shows the estimated remission duration in an exponential model.

remission, denoted with a random variable T. The sequential entrance of the patients into the trial, in combination with the usual plan to end the study after a certain amount of time, leads to a mechanism that censors the observability of the outcome variable to the right. Under the assumption of a uniform entry of the patients into the trial, this can be modelled with a uniform random variable $Z \sim U(a, b)$, where a and b denote the minimum and maximum time under investigation, respectively.

The independence of X and T is assumed, i.e. the duration of remission is independent of the time until this state is reached. Further it is assumed, that each patient achieves complete remission and that this state can be distinguished from progression.

In an example situation, the expectations for the time until remission and for remission duration are $E(X) = 3$ months and $E(T) = 24$ months. The exponential model is invariant under scale transformations: remission duration is modelled with a standard exponential distribution $T \sim Ex(1)$, i.e. $P(T > t) = \exp(-t)$. On this scale one month is about 0.042 time units. In the case of a uniform distributed time until the begin of complete remission the distribution is $X \sim U(0, 0.25)$ and it is $X \sim Ex(8)$ for the exponential case. No censoring and amounts of 10% and 20% right censored observations were investigated. All simulations are based on 5000 replications and sample sizes of 100 observations in every run.

For every replication random realizations of X and T are generated and data sets according to the conventional and the interval-data approach are provided. From these the maximum likelihood estimates are available either directly or with the aid of a Newton-Raphson algorithm as roots of the first derivation of the likelihood function (as explained in a previous section).

In table 1, patients are seen every month under treatment. The effect of coarsening (using wider intervals) the observational plan in the follow-up period (FU) is demonstrated. Follow-up intervals from 1 up to 24 months are investigated. No censoring or a complete follow-up is assumed. The mean (M) of the estimated parameters is an estimate of the expectation of $\hat{\lambda}$, their variance (V) estimates the asymptotic variance of the maximum-likelihood estimator. The mean squared error (MSE) can be estimated from the mean of $(\hat{\lambda}_i - \lambda)^2$ where $\hat{\lambda}_i$ denotes the estimate from the i-th run of the simulation. The subscripts I and C denote the affiliated strategies.

For the original data, these values can be computed and used to verify the simulation. The results are $E(\hat{\lambda}) = \lambda n/(n-1) = 1.0101$ (in the simulations, the estimate is denoted with M_V). The asymptotic variance is $AsVar(\hat{\lambda}) = \lambda^2/n = 0.0105$ (denoted with V_V) and the result for the mean squared error is $MSE(\hat{\lambda}) = \lambda^2((n^2/(n^3 - 4n^2 + 5n - 2)) + 1/(n-1)^2) = 0.0105$. Table 2 demonstrates the additional effect of small censoring amounts as it is known, that large amounts of censoring are less problematic in the conventional approach.

Results can be summarized as follows: Enlarging the distance between successive follow-up dates biases the estimation of λ in the conventional approach. Remission duration is overestimated: The correct remission-free five year survival in a situation without censoring and half-year monitoring intervals is 8.2% and would be estimated by 11%. A small amount of right censored observations enlarges this effect. In the interval censoring

FU	M_I	M_C	M_V	V_I	V_C	V_V	MSE_I	MSE_C	MSE_V
1	1.0113	0.9905	1.0108	0.0106	0.0099	0.0106	0.0107	0.0100	0.0107
2	1.0120	0.9705	1.0113	0.0103	0.0088	0.0103	0.0105	0.0097	0.0104
4	1.0111	0.9303	1.0097	0.0105	0.0076	0.0104	0.0106	0.0125	0.0105
6	1.0124	0.8936	1.0105	0.0109	0.0066	0.0108	0.0111	0.0179	0.0109
12	1.0173	0.7922	1.0126	0.0111	0.0039	0.0105	0.0114	0.0471	0.0107
24	1.0204	0.6347	1.0107	0.0121	0.0015	0.0103	0.0125	0.1350	0.0104

Table 1: *Coarsening the observational scheme for different follow-up-periods FU (month). No censoring, patients are seen under treatment in one month intervals. For each strategy (I: interval data, C: conventional approach, V: verification of the simulation) the mean M, the variance V and the mean squared error (MSE) are presented.*

AC	FU	M_I	M_C	M_V	V_I	V_C	V_V	MSE_I	MSE_C	MSE_V
0.0	1	1.0113	0.9905	1.0108	0.0106	0.0099	0.0106	0.0107	0.0100	0.0107
0.1	1	1.0115	0.9882	1.0096	0.0115	0.0107	0.0105	0.0116	0.0109	0.0105
0.2	1	1.0151	0.9890	1.0115	0.0128	0.0120	0.0105	0.0131	0.0122	0.0106
0.0	4	1.0111	0.9303	1.0097	0.0105	0.0076	0.0104	0.0106	0.0125	0.0105
0.1	4	1.0166	0.9321	1.0130	0.0120	0.0086	0.0107	0.0123	0.0132	0.0108
0.2	4	1.0155	0.9289	1.0104	0.0133	0.0097	0.0107	0.0135	0.0147	0.0108
0.0	6	1.0124	0.8936	1.0105	0.0109	0.0066	0.0108	0.0111	0.0179	0.0109
0.1	6	1.0146	0.8931	1.0100	0.0116	0.0071	0.0102	0.0118	0.0185	0.0103
0.2	6	1.0149	0.8908	1.0080	0.0138	0.0084	0.0106	0.0140	0.0203	0.0106

Table 2: *Coarsening the observational scheme for different follow-up periods FU (month). No censoring, 10% and 20% censored observations (AC). Patients are seen under treatment in one month intervals.*

approach, estimation is slightly conservative. Enlarging the follow-up intervals leads correctly to an increase in variance. In the above mentioned sample situation, the five year survival rate is estimated with 7.9%.

Additional simulations with exponential distributed time until remission reinforce the same tendencies. As a result, the conventional approach, to register an event when it is actually observed, cannot be recommended: Anticonservative estimation leads to an overestimation of treatment effects and as the asymptotic variance of this estimator is underestimated, wrong confidence is given to the estimated parameters.

It is to emphasize that the above specified observational plan (i.e. to switch from the treatment schedule to the follow-up schedule as soon as response is observed) is experimental. In comparison with a study, where a fixed number of chemotherapy cycles is performed, this observational plan depicts however, with respect to a coarsening of the plan, the worst situation: in general the patient is monitored in the early phase of the trial more frequently than in later stages.

Programs have been developed in IBM APL2 on an IBM 3090-150 mainframe system under VM/CMS. For generating standard uniform random numbers the function DURAND from the ESSL-Library has been used. Exponential random numbers were generated according to the inverse transform method.

Discussion

Interval censored observations are demonstrated to be a common problem in the practice of clinical trials. The situation, however, has often been neglected due to feasibility problems with respect to the inability of the standard statistical analysis systems to cope with this type of data. In addition the necessary algorithms for estimating parameters of a distribution or the nonparametric survivorship function are complex and do not invite to a quick and dirty self development. Some developments by scientists, working in this field, are not well known.

This situation has changed now: The interrelation between the Turnbull estimator and parameter estimation in the loglinear model makes the survival function from interval censored data readily available in most statistical packages. Besides this, the Turnbull estimator is available in the IMSL subroutine library version 2.0 (see IMSL (1991)).

The SAS procedure LIFEREG (see SAS (1989)) initially computes estimates and standard errors for the parameters of an accelerated failure time model under various distributional assumptions. From these the original estimates can be obtained with the aid of simple computations. In addition the effect of covariables on survival can be investigated. Parametric regression models are also available in the IMSL subroutine library.

Since serious bias and underestimation of the true error variance with the conventional approach can be demonstrated for clinically relevant situations, it is advised to keep the problem of interval censoring in mind and to respect the interval structure in the data.

In the planning of clinical investigations several controllable and uncontrollable factors have to be drawn into consideration. Patient's compliance, seen as a treatment and response-state independent effect, is in many instances hard to influence. Further work has to be done to quantify the effect of lower compliance and an increased amount of interval censored observations on the aim to establish a treatment effect. In addition little is known about the performance of the statistical methods if the assumption of a disease-state independent observational plan is violated. Study design is, in contrast, under the control of the investigator. Two different examination schemes are mentioned above and many others are found in practice (e.g. often the examination intervals are planned smaller in the early phase and less frequent in later phases of a trial). It is therefore mandatory to examine in advance the effect of different examination schemes on the ability to detect a treatment effect in a clinical trial. This can prevent the investigators from a potential loss of efficiency.

Acknowledgements

I wish to thank my colleagues for reading and discussing this paper and M. Pritsch for providing the remission duration dataset. I am especially grateful to the referees for improving the paper with their comments.

References

Abel, U. (1986): A nonparametric test against ordered alternatives for data defined by intervals. *Statistica Neerlandica*, **40**, 87-91.

Becker, N.G. & Melbye, M. (1991): Use of a log-linear model to compute the empirical survival curve from interval-censored data, with application to data on tests for HIV positivity. *Australian Journal of Statistics*, **33**, 125-133.

Denby, L., Fowlkes, E.B. & Roe, R.J. (1975): Estimation from censored interval data of the median breaking point of Polyethylene subjected to stress-cracking tests: A Monte-Carlo study. *Journal of Applied Mechanics*, **42**, 607-612.

Efron, B. (1967): The two sample problem with censored data. In: Le Cam, L.M. & Neyman, J.: Fifth Berkeley Symposium on Mathematical Statistics and Probability. UCLA-Press.

Finkelstein, D.M. (1986): A proportional hazards model for interval-censored failure time data. *Biometrics*, **42**, 845-854.

Finkelstein, D.M. & Wolfe, R.A. (1985): A semiparametric model for regression analysis of interval-censored failure time data. *Biometrics*, **41**, 933-945.

Finkelstein, D.M. & Wolfe, R.A. (1986): Isotonic regression for interval censored survival data using the EM Algorithm. *Communications in Statistics–Theory and Methods*, **15**, 2493-2505.

Flygare, M.E., Austin, J.A. & Buckwalter, R.M. (1985): Maximum likelihood estimation for the 2-parameter Weibull distribution based on interval-data. *IEEE Transactions on Reliability.*, **R-34**, 57-59.

Gehan, E.A. (1965a): A generalized two-sample Wilcoxon test for doubly censored data. *Biometrika*, **52**, 650-653.

Gehan, E.A. (1965b): A generalized Wilcoxon test for comparing arbitrarily singly-censored samples. *Biometrika*, **52**, 203-223.

Groeneboom, P. (1991): Nonparametric maximum likelihood estimators for interval censoring and deconvolution. California, USA: Stanford University; Technical report No. 378; National Science Foundation Grant DM289-05874.

Groeneboom, P. & Wellner, J.A. (1992): Information bounds and nonparametric maximum likelihood estimation. Birkhäuser, Basel.

Grüger, J., Kay, R. & Schumacher, M. (1991): The validity of inferences based on incomplete observations in disease state models. *Biometrics*, **47**, 595-605.

Hamada, M. (1991): The costs of using incomplete exponential data. *Journal of Statistical Planning and Inference*, **27**, 317-324.

Hilgers, R.A. (1989): Nonparametric two-sample tests for general clinical data. *Biometrical Journal*, **31**, 171-185.

Hilgers, R.A. & Neumann, N. (1989): An exact nonparametric randomization test for censored data. *Biometrie und Informatik in Medizin und Biologie*, **20**, 66-73.

IMSL STAT/LIBRARY (1991): User's Manual: FORTRAN subroutines for statistical analysis. Version 2.0. September 1991.

Mantel, N. (1967): Ranking procedures for arbitrarily restricted observation. *Biometrics*, **23**, 65-78.

Marcuson, R. (1983): Simplified computation of the multivariate permutation test for arbitrarily censored survival data. *Statistics in Medicine*, **2**, 327-330.

Peto, R. (1973): Experimental survival curves for intervall censored data. *Journal of the Royal Statistical Society, Series C*, **22**, 86-91.

Prentice, R.L. (1978): Linear rank tests with right censored data. *Biometrika*, **65**, 167-179.

Rücker, G. & Messerer, D. (1988): Remission duration: An example of interval-censored observations. *Statistics in Medicine*, **7**, 1139-1145.

SAS Institute Inc. (1989): SAS/STAT User's Guide, Version 6, Fourth Edition, Volume 2, Cary, NC: SAS Institute Inc.

Schemper, M. (1982): A nonparametric test of association for data defined by intervals. *EDV in Medizin und Biologie*, **13**, 44-46.

Schemper, M. (1983): A nonparametric k-sample test for data defined by intervals. *Statistica Neerlandica*, **37**, 69-71.

Schemper, M. (1984a): A generalized Friedman test for data defined by intervals. *Biometrical Journal*, **26**, 305-308.

Schemper, M. (1984b): A generalized Wilcoxon test for data defined by intervals. *Communications in Statistics, Theory and Methods*, **13**, 681-684.

Self, S.G. & Grossman, E.A. (1986): Linear rank tests for interval censored data with application to PCB levels in adipose tissue of transformer repair workers. *Biometrics*, **42**, 521-530.

Temkin, N.R. (1978): An analysis for transient states with application to tumor shrinkage *Biometrics*, **34**, 571-580.

Turnbull, B.W. (1976): The empirical distribution function with arbitrarily grouped, censored and truncated data. *Journal of the Royal Statistical Society, Series B*, **38**, 290-295.

Wittkowski, K.M. (1984): Semiquantitative Merkmale in der nichtparametrischen Statistik. In: Köhler, C.O., Tautu, P. & Wagner, G.: Der Beitrag der Informationsverarbeitung zum Fortschritt der Medizin. Springer, Berlin.

Wolf, M., Pritsch, M., Drings, P., Hans, K., Schroeder, M., Flechtner, H., Heim, M., Hruska, D., Mende, S., Becker, H., Dannhäuser, J., Lohmüller, R., Gropp, C., Gassel, W.D., Holle, R. & Havemann, K. (1991): Cyclic-alternating versus response-oriented chemotherapy in small-cell lung cancer: A German multicenter randomized trial of 320 patients. *Journal of Clinical Oncology*, **9**, 614-624.

Covariates in Clinical Trials: Effects of Adjustment in Regression Models

Gerd Antes[1], Claudia Schmoor[2]
[1] Institut für Klinische Pharmakologie, Klinik für Tumorbiologie, Breisacher Str.117, D-79106 Freiburg, Germany
[2] Institut für Medizinische Biometrie und Medizinische Informatik, Universität Freiburg, Stefan-Meier-Str. 26, D-79104 Freiburg, Germany

Key Words:
clinical trial, treatment effect, covariate adjustment, misspecification, omission bias

Abstract
Results from classic linear regression regarding the effect of adjusting for covariates upon the properties of the estimator and test of the interesting effect are often assumed to apply more generally to other types of regression models. This assumption is not justified in the logistic and the Cox regression model. The effects of misspecification by omitting or including relevant or nonrelevant variables are discussed with respect to the bias and standard error of estimates and size and power of the test on the treatment effect emphasizing the situation in randomized trials. For the Cox model these effects are investigated by simulations.

Introduction

The aim of most clinical studies is the comparison of different therapies or the evaluation of prognostic factors. In these trials the population under study may be rather heterogeneous with respect to prognosis. Therefore, it is commonplace to adjust for covariates describing the heterogeneity of the patients. Such adjustment is typically based on regression models as the linear, logistic (Cox & Snell (1989)) or the Cox model (Cox (1972)), the model choice given by the response or endpoint under investigation.

However, often factors are not taken into consideration in the analysis because they are not known or because they are not documented. Omission of relevant factors may seriously affect the estimation and tests of the interesting factors. These effects may be described in terms of bias, variance, mean-squared-error and efficiency if estimation is considered, or by size and power comparisons of the corresponding tests. For the classic linear regression model the consequences of mismodelling were investigated several years ago (e.g. Rao (1971)) leading to an extensive literature and specific chapters on this topic in most textbooks on linear regression (e.g. Seber (1977, ch.6)).

The widespread teaching of the linear model, its clear geometric interpretation and easy intuitive grasp may be tempting to transfer its well-known misspecification behaviour to situations where other models are fitted. This approach may lead to serious mistakes if applied to the logistic and the Cox regression model. The differing properties of these two models commonly used for the analysis of clinical trials have attracted considerable attention over the last decade. The effect of omitting covariates was investigated in a series of papers (Gail et al. (1984); Gail (1986); Schumacher et al. (1987); Gail (1988); Gail et al. (1988)) for the investigation of a treatment effect in randomized clinical trials. Lagakos (1988), Lagakos & Schoenfeld (1984), Struthers & Kalbfleisch (1986) and Begg & Lagakos (1993) also examined the omission of covariates and further specific deviations from the model assumptions. Recently, Neuhaus & Jewell (1993) developed an intuitive approach to the effects of omitted covariates by examining the geometric properties of the link function of generalized linear models. Whereas these papers consider various aspects of misspecification from a similar methodological background, the articles by Beach & Meier (1989) and Canner (1991) discuss the magnitude of adjustment in specific situations.

This paper is an extension of the article by Schmoor & Antes (1993). The effects of the adjustment are reviewed and discussed with regard to the bias and the variance of the estimator and the properties of the test on the factor under investigation. The consequences of fitting a wrong model (i.e. violating basic assumptions of the model class) are not addressed. Also the impact on the goodness-of-fit and on the prognostic quality of the model is not considered here.

For the linear model straightforward analytical methods are sufficient whereas for the logistic and for the Cox model asymptotic arguments have to be used and in some cases numerical tools are necessary. Especially the Cox model requires statistical simulations and numerical integration for the distribution of the parameter estimates because of the form of the partial likelihood function. Here simulations are not only used to examine the finite behaviour of asymptotic results or the quality of approximations but in some instances are the only way if analytic approaches are intractable. In this sense this paper shows an example where analytical and numerical tools complement one another yielding results which have important consequences for the model fitting techniques in applied fields as e.g. the analysis of clinical trials.

Conditions for the impact of adjustment on bias, variance and tests

Because of the complexity of the problem the investigation of the effects of covariate adjustment is restricted to the simplest case of two factors. These are described by examining the adjusted and the unadjusted regression model which can be written as

$$E(Y|X_1, X_2) = g^{-1}(\beta_0 + \beta_1 X_1 + \beta_2 X_2) \quad (1)$$
$$E(Y|X_1) = g^{-1}(\beta_0^* + \beta_1^* X_1) , \quad (2)$$

where $g(.)$ denotes the link function (see McCullagh & Nelder (1989)).

The interesting factor is denoted by X_1 and the potential covariate by X_2 with corresponding regression parameters β_1 and β_2. Fitting the full model (1) or the restricted model (2) produces estimates $\hat{\beta}_1$ and $\hat{\beta}_1^*$. The properties of $\hat{\beta}_1^*$, compared to those of $\hat{\beta}_1$, are used to measure the effect of omitting X_2 from the regression. The potentially detrimental effect of neglecting X_2 is governed by the following three conditions.

Condition (a) is whether there is an effect of X_2, i.e. whether $\beta_2 = 0$ or $\beta_2 \neq 0$. If X_2 is adjusted for, the correct model (1) is fitted if $\beta_2 \neq 0$ whereas $\beta_2 = 0$ leads to overspecification of the model. Neglecting X_2 in model (2) is correct if $\beta_2 = 0$ and is equivalent to underspecification of the model if $\beta_2 \neq 0$. These different situations are summarized in table 1.

	adjusting for X_2 (model 1)	not adjusting for X_2 (model 2)
$\beta_2 \neq 0$	correct	underspecification
$\beta_2 = 0$	overspecification	correct

Table 1: Misspecification by omitting or including X_2

Condition (b) is given by the size of the correlation ρ_{12} between X_1 and X_2. In clinical trials comparing different treatments the randomization of patients to the treatment arms is used to achieve a balanced distribution of prognostic factors in the treatment groups as a basis for valid treatment comparison. If X_1 codes the treatment effect and X_2 a prognostic factor randomization guarantees the theoretical correlation ρ_{12} to be 0. In contrast to this situation $\rho_{12} \neq 0$ is found in observational studies where β_1 may be a treatment effect or the effect of a prognostic factor. In this paper we concentrate on the analysis of a randomized treatment effect, i.e. emphasis is put on the situation $\rho_{12} = 0$.

Condition (c) is given by the hypothesized situation defined by the null hypothesis $H_0 : \beta_1 = 0$ or the alternative $H_1 : \beta_1 \neq 0$.

The impact on bias, standard errors and tests may show serious differences in different models and therefore has to be considered separately. The discussion here is restricted to the classic linear model, the logistic model and the Cox model.

Adjustment in the classic linear model

Using common notation consider the classic linear model $Y = X\beta + e$ with $e \sim N(0, \sigma^2)$. The $(n \times 2)$-dimensional covariate matrix X may be written in partitioned form $X = [X_1, X_2]$ (for simplicity of presentation no constant term is considered here).

Fitting the model without X_2 generally introduces bias:

$$E(\hat{\beta}_1^* | X_1, X_2) = \beta_1 + \left(X_1^T X_1\right)^{-1} X_1^T X_2 \beta_2 \tag{3}$$

The bias depends on the size of the omitted effect and on the correlation between X_1 and X_2. The bias is zero if $\beta_2 = 0$ or if X_2 is orthogonal to X_1. Consequently in randomized clinical studies the treatment effect may be estimated without bias.

In the context of variability of the estimator there are some statements in the literature which seem to be inconsistent without careful reading. The reason for these differences lies in a characteristic property of the linear model which distinguishes it from some other generalized linear models. In contrast to those models where the variance is a function of the mean, the classic linear model is described by two independent parameters, the expectation $E(Y)$ and the variance σ^2. Assuming σ^2 known in some cases leads to results which are not valid if an estimate $\hat{\sigma}^2$ is used instead as will be seen below. Comparing the variance formula of the reduced model and the full model it may be shown that

$$var\left(\hat{\beta}_1^* | X_1\right) = \sigma^2 \left(X_1^T X_1\right)^{-1} \leq \sigma^2 \left(X^T X\right)_{11}^{-1} = var\left(\hat{\beta}_1 | X\right) \tag{4}$$

Under both models the variance is correctly described by the expressions in (4). The crucial point here is the assumption that σ^2 is known and equal under both models. Therefore the variance of $\hat{\beta}_1$ is only dependent on X and it may be shown analytically (Rao (1971)) that the omission of a covariate reduces the corresponding diagonal element of $(X^T X)^{-1}$. The "textbook conclusion that omitting a variable reduces variance" (Binkley & Abbott (1987)) may wrongly recommend to drop variates in order to gain precision. This argument is not valid if σ^2 is replaced by $\hat{\sigma}^2$ because the estimation of σ^2 is also influenced under the reduced model. Generally the omission of a relevant variable leads to an upward bias of the residual sum of squares estimator: $E[\hat{\sigma}^2] > \sigma^2$ (Seber (1977, p.141)). The overestimation of σ^2 is in competition with the reduction of the respective term in $(X^T X)^{-1}$ resulting in a change in either direction for the corresponding term in $\hat{\sigma}^2(X^T X)^{-1}$. The principal difference of the two situations σ^2 known and σ^2 unknown is that in the latter the estimated variance $\hat{\sigma}^2$ is a function of the response Y whereas in the first only the covariate spread determines the precision of $\hat{\beta}_1$ and $\hat{\beta}_1^*$.

This point was formalized by Robinson & Jewell (1991). Corresponding to equation (1) and (2) they consider the linear models

$$
\begin{aligned}
E(Y|X_1, X_2) &= g^{-1}(\beta_0 + \beta_1 X_1 + \beta_2 X_2) & var(Y|X_1, X_2) &= \sigma^2_{Y.12} \quad (5)\\
E(Y|X_1) &= g^{-1}(\beta_0^* + \beta_1^* X_1) & var(Y|X_1) &= \sigma^2_{Y.1}\ , (6)
\end{aligned}
$$

where they in contrast to (4) allow for different variances $\sigma^2_{Y.1}$ and $\sigma^2_{Y.12}$.

Estimating β_1 and β_1^* separately under these models yields an expression for the asymptotic relative precision (ARP)

$$
ARP\left(\hat{\beta}_1 \ to \ \hat{\beta}_1^*\right) = \frac{v\hat{a}r\left(\hat{\beta}_1^*|X_1\right)}{v\hat{a}r\left(\hat{\beta}_1|X_1, X_2\right)} = \frac{1-\rho_{12}^2}{1-\rho_{Y2.1}^2} \qquad (7)
$$

ρ_{12} is the simple correlation between X_1 and X_2, $\rho_{Y2.1}$ is the partial correlation between Y and X_2 conditional on fixed X_1, or equivalently the effect of X_2 on Y.

The relationship between the correlation structure of Y, X_1 and X_2 and the precision is:

$$
\begin{aligned}
\rho_{Y2.1} &< \rho_{12} & & & &< 1\\
\rho_{Y2.1} &= \rho_{12} & \Rightarrow & & ARP(\hat{\beta}_1 \ to \ \hat{\beta}_1^*) &= 1\\
\rho_{Y2.1} &> \rho_{12} & & & &> 1
\end{aligned}
$$

Generally adjustment improves precision if the $Y - X_2$ association is stronger than the $X_1 - X_2$ association.

This precision comparison suffers from the fact that $\hat{\beta}_1^*$ and $\hat{\beta}_1$ do not estimate the same population parameter. This drawback is not given in randomized trials because $\rho_{12} = 0$ is equivalent to the estimation of the same parameter under both models. The conclusion is that adjustment is desirable in randomized studies because $ARP(\hat{\beta}_1 \; to \; \hat{\beta}_1^*) > 1$.

The size and the asymptotic relative efficiency of the test of the treatment effect under the misspecified model is also heavily dependent on the assumption of known variance σ^2. Gail et al. (1988) found by simulation an increased size for the score test of $H_0 : \beta_1 = 0$ when σ^2 was taken as unknown, whereas estimating σ^2 from the residual sum of squares keeps the nominal level but shows smaller asymptotic relative efficiency compared to the test based on the correctly specified model.

Adjustment in the logistic model

The logistic model is probably the most frequently fitted model in the analysis of categorical data. Some of its properties are in sharp contrast to other members of the class of generalized linear models, especially if compared to the normal model.

Corresponding to (1) and (2) the two logistic models are

$$log\left[\frac{P(Y=1|X_1,X_2)}{1-P(Y=1|X_1,X_2)}\right] = \beta_0 + \beta_1 X_1 + \beta_2 X_2 \quad (8)$$

$$log\left[\frac{P(Y=1|X_1)}{1-P(Y=1|X_1)}\right] = \beta_0^* + \beta_1^* X_1 \quad (9)$$

The interesting parameters β_1 and β_1^* are estimated by maximum likelihood and denoted by $\hat{\beta}_1$ and $\hat{\beta}_1^*$.

In general the omission of a covariate leads to biased estimation of the interesting effect. One outstanding property of the logistic model in contrast to the classic linear model is the asymptotic bias of $\hat{\beta}_1^*$ due to an omitted covariate even if this covariate is orthogonal to the variable whose effect is estimated ($\rho_{12} = 0$). The estimate is asymptotically biased towards zero, i.e. $\lim |\hat{\beta}_1^*| < |\beta_1|$. The bias only vanishes for $\beta_1 = 0$ or $\beta_2 = 0$. In a geometric approach to the problem of covariate omission in generalized linear models, Neuhaus & Jewell (1993) show that the direction of the bias may be predicted by the curvature of $1/g'(.)$, where $g(.)$ is the link function. In the logistic model where the link function is the logistic function, $1/g'(.)$ is convex leading to a downward bias of $\hat{\beta}_1^*$.

The asymptotic variances of $\hat{\beta}_1$ and $\hat{\beta}_1^*$ also behave in a way not consistent with the results in the linear model. The central point is that in general

$ARP(\hat{\beta}_1 \ to \ \hat{\beta}_1^*) \leq 1$ with equality only if $\beta_2 = 0$ (Robinson & Jewell (1991)). In particular, adjusting for a predictive covariate not correlated with X_1 does not improve precision in contrast to the situation in the classic linear model. Therefore, with respect to the test $H_0 : \beta_1 = 0$ the adjusted and the unadjusted analysis is influenced by two competing effects, the biased but more precise unadjusted estimator versus the unbiased adjusted estimator with larger variability. The score test based on the adjusted analysis using model (8) has greater power than that based on the unadjusted analysis using model (9) (Robinson & Jewell (1991); Begg & Lagakos (1993)). The size of the test is kept even without adjustment.

Therefore in the logistic model the estimate of a randomized treatment effect should also be adjusted with respect to important covariates but for different reasons than in the linear model.

Adjustment in the Cox model

In cancer clinical trials comparing different treatments survival time is usually the main variable for analysis. A multiple regression analysis of possibly censored survival data may be accomplished by the Cox proportional hazards model (Cox (1972)). The Cox model does not fit in the framework of generalized linear models because it is not based on a completely specified parametric density of the exponential family. However, there are some links to generalized linear models which may govern the expectation of what the effects of covariate omission may be. For small risks and a short time interval the logistic model yields similar results as the Cox regression, as shown, among others, by Doksum & Gasko (1990) and Green & Symons (1983). Also there is some formal equivalence of the Cox partial likelihood and the Poisson likelihood (Whitehead (1980)).

In our situation with X_1 denoting the treatment indicator and X_2 denoting the covariate which should be adjusted for the hazard function in the Cox model is

$$\lambda(t) = \lambda_0(t) exp(\beta_1 X_1 + \beta_2 X_2) \tag{10}$$

where $\lambda_0(t)$ denotes the unknown baseline hazard function.

As in the logistic model the omission of a relevant covariate, i.e. the analysis in the Cox model

$$\lambda(t) = \lambda_0(t) exp(\beta_1^* X_1) \tag{11}$$

asymptotically leads to underestimation of the randomized treatment effect in the Cox model, too (Gail et al. (1984); Schumacher et al. (1987)). The effect of β_1 is estimated without bias only if $\beta_1 = 0$. The score test of $H_0 : \beta_1 = 0$ keeps it's nominal size even in the unadjusted case, but the power is increased by the adjustment (Lagakos & Schoenfeld (1984)). Little is known about the behaviour of the variances of the adjusted and unadjusted estimators. Because of the similarity of the logistic and the Cox model mentioned above, the impact of covariate adjustment observed under the logistic model may also be expected under the Cox model.

The properties of the adjusted and the unadjusted estimators were investigated in a small simulation study. Two dichotomous, balanced and uncorrelated covariates X_1 and X_2 with varying effects $exp(\beta_1)$ and $exp(\beta_2)$ were given. Exponentially distributed survival times and uniformly distributed censoring times were generated yielding 20 percent censored observations. The total sample size was $N = 100$ and $N = 500$. For each situation 1000 simulations were generated. The design of the simulation study is summarized in table 2.

The simulation was programmed in SAS using the macro facilities as a full-scale programming language (Antes & Sauerbrei (1991)). This use of SAS is fairly uncommon but convenient if different methods are to be compared because SAS supplies the routines for these methods and subroutine libraries are not needed.

distribution of	survival time: censoring time: covariates:	exponential uniform (20% censoring) Bernoulli, p=0.5 X_1 and X_2 uncorrelated
size of effects	$exp(\beta_1) =$	1; 1.5; 2; 3; 4
	$exp(\beta_2) =$	1; 1.5; 2; 3; 4
sample size	$N =$	100; 500
number of simulations	$n_{sim} =$	1000

Table 2: Design of the simulation study

Each data set was analyzed under model (10) yielding an adjusted estimate $\hat{\beta}_1$ and under model (11) yielding an unadjusted estimate $\hat{\beta}_1^*$. The empirical bias, the estimated asymptotic variance and the power of the test of $H_0 : \beta_1 = 0$ were investigated. The results for $exp(\beta_1) = 2$ and sample size $N = 100$ are presented in the sequel. Runs for different values of $\beta_1 (\neq 0)$ and sample size $N = 500$ produced similar results. Figure 1 shows the mean of the percentage bias of $\hat{\beta}_1$ and $\hat{\beta}_1^*$ from 1000 simulations as function of the effect size of X_2.

As expected the unadjusted estimator is seriously biased downwards for increasing effects of the omitted covariate. The omission of a covariate

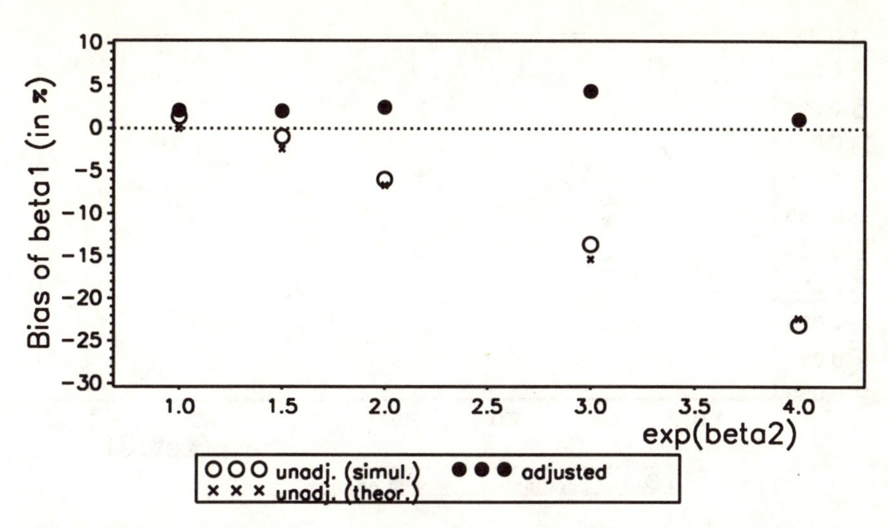

Figure 1: Bias of $\hat{\beta}_1$ (in %) $=100[(1/n_{sim})\sum_{j=1}^{n_{sim}} \hat{\beta}_{1j} - \beta_1]/\beta_1$ as function of $exp(\beta_2)$. $exp(\beta_1) = 2$ and $N = 100$

with a relative risk of $exp(\beta_2) = 4$ leads to a bias of -22%. However, this situation is probably of little practical relevance because a factor with an effect of this size will usually be known in advance and adjusted for. The simulation is well in accordance with theoretically derived asymptotic results (Schumacher et al. (1987)). If X_2 has no effect ($\beta_2 = 0$) both the adjusted and the unadjusted estimator are asymptotically unbiased. This is also true if X_1 has no effect ($\beta_1 = 0$). As observed by some authors before (Johnson et al. (1982); Sayn et al. (1986)) the asymptotically unbiased Cox estimator has slight positive bias for finite sample size.

Figure 2 shows the mean of the estimated asymptotic variance of $\hat{\beta}_1$ and $\hat{\beta}_1^*$ as function of the effect size of X_2. The theoretical results known for the logistic model are reproduced by the simulation study of the Cox model: The adjusted estimator of β_1 shows larger variance than the unadjusted one if X_2 has an effect ($\beta_2 \neq 0$). For $\beta_2 = 0$ they seem to be equal in the Cox model, too.

The arguments from above lead to the "prevent underspecification" recommendation: Also in randomized trials the analysis should be based on the adjustment for important covariates to avoid the underestimation of the treatment effect. This strategy also increases the power of the test of the treatment effect although the adjusted estimator has larger variability than the unadjusted. In contrast to the omission of variables the adjustment for nonrelevant factors (overspecification) asymptotically does not lead to a

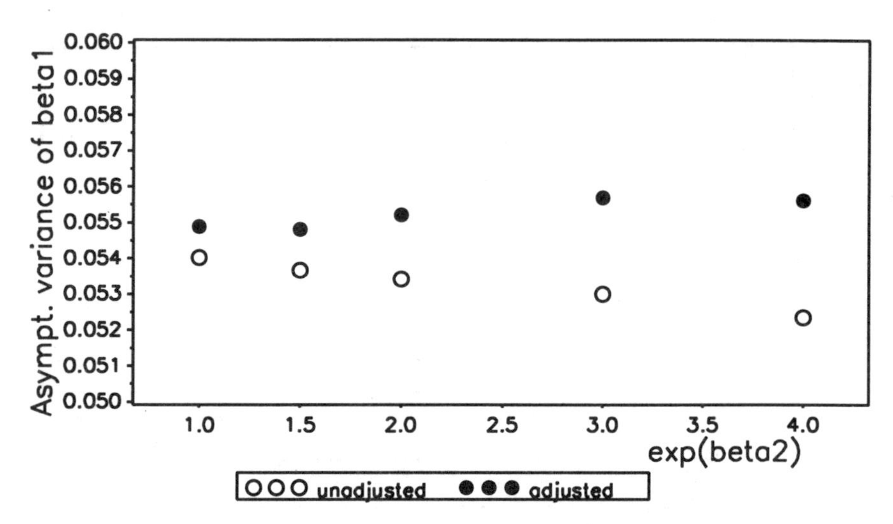

Figure 2: Estimated asymptotic variance of $\hat{\beta}_1$ as function of $exp(\beta_2)$. $exp(\beta_1) = 2$ and $N = 100$

bias and an increased variance of the estimator nor to loss of power of the test of no treatment effect (Andersen et al. (1993)).

This statement may suggest that in practical applications underspecification should be avoided by all means whereas overspecification does not create any problem. A naive conclusion may then be to include all available variables into the model without thinking about variable selection at all. Some further simulations show the effects of this approach. Starting with a factor X_1 with effect $exp(\beta_1) = 2$ covariates $X_2, ..., X_p$ were generated and successively added to the model. These additional variables are dichotomous, balanced, pairwise uncorrelated and uncorrelated with X_1 and have no effect on survival outcome. In order to reduce computing time, simulations were performed only for 1(1)10, 15, 20, 25, 30, 40 , 50 additional variables.

Figure 3 shows the percentage bias of $\hat{\beta}_1$ as function of the number of additionally included covariates. β_1 is increasingly overestimated although the estimator of β_1 is asymptotically unbiased. This outcome may be explained by the increasing number of covariates approaching the fixed sample size not allowing the classic asymptotic arguments. Figure 4 shows the asymptotic variances of $\hat{\beta}_1$ for the same configuration. They also increase as the number of covariates increases. These simulations were repeated for a sample size of $N = 500$. In this situation the inclusion of 50 covariates (10% of the number of observations) corresponded to the inclusion of 10 covariates with $N = 100$.

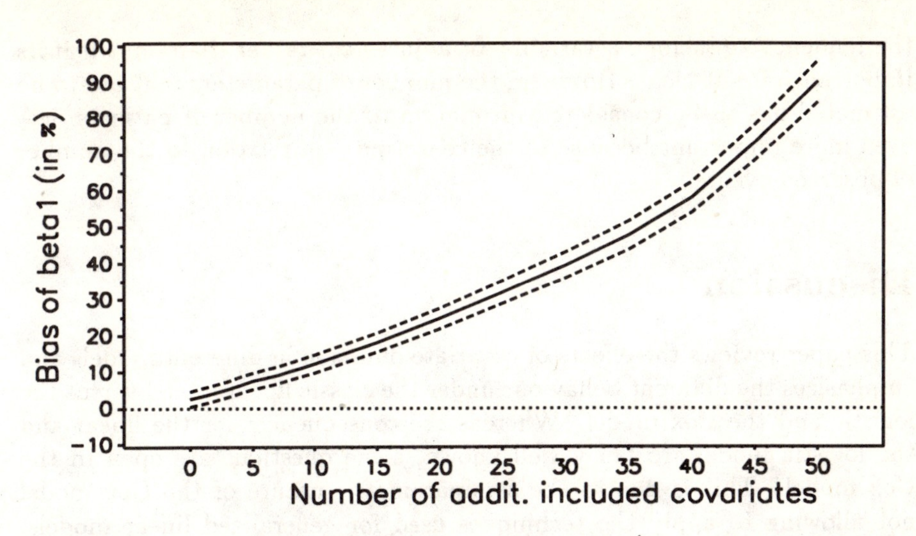

Figure 3: Bias of $\hat{\beta}_1$ (in %) $=100[(1/n_{sim})\sum_{j=1}^{n_{sim}}\hat{\beta}_{1j}-\beta_1]/\beta_1$ with 95%-confidence intervals (derived from the simulation) as function of the number of additionally included unimportant covariates (correlation 0). $exp(\beta_1)=2$ and $N=100$

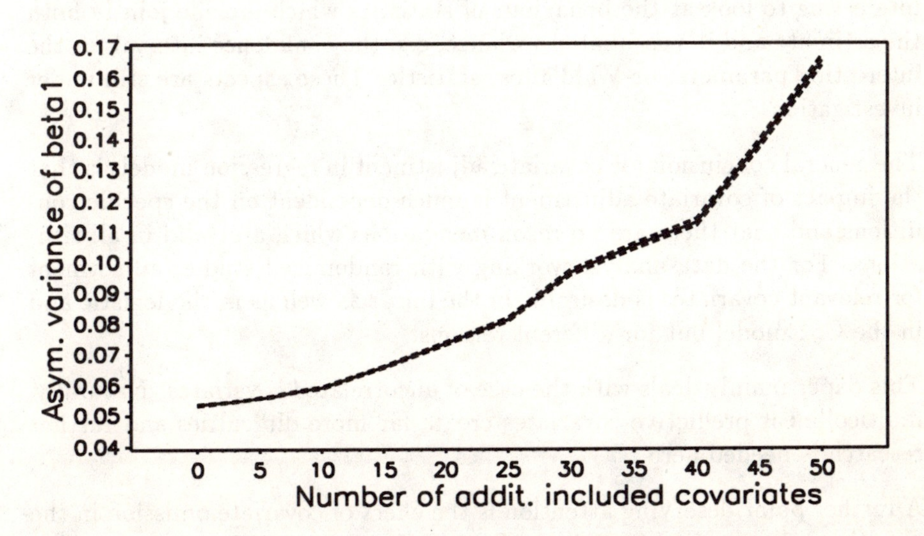

Figure 4: Estimated asymptotic variance of $\hat{\beta}_1$ with 95%-confidence intervals (derived from the simulation) as function of the number of additionally included unimportant covariates (correlation 0). $exp(\beta_1)=2$ and $N=100$

The recommendation for practical applications of the Cox model is to have the tendency to include a variable for adjustment rather than omit it if its importance is not clear. However, the number of parameters that are to be estimated has to be considered in relation to the number of patients or – even more important because of the censoring – in relation to the number of observed events.

Discussion

This paper reviews the effects of covariate omission in different models and emphasizes the different behaviour under the classic linear model versus the logistic and the Cox model. Whereas the consequences for the linear and the logistic model are fairly well known, some questions are open in the Cox model. This is due to the semiparametric nature of the Cox model not allowing to apply the techniques used for generalized linear models. In particular, the behaviour of the asymptotic variance of the estimator of the interesting effect when important covariates are omitted from the model is not fully investigated. Simulation results show similar behaviour as in the logistic model. In this paper bias and variance of the estimates were investigated separately. Especially from a practical point of view, it is interesting to look at the behaviour of statistics which include jointly both the estimate and it's estimated variance, e.g. the confidence interval for the interesting parameter or Wald's test statistic. These aspects are still under investigation.

The general conclusion for covariate adjustment in regression models is that the impact of covariate adjustment is much dependent on the specific conditions and that there are no recommendations which are valid in all situations. For the data analyst working with randomized studies adjustment for relevant covariates is desirable in the linear as well as in the logistic and in the Cox model but for different reasons.

This paper mainly deals with the case of uncorrelated covariates. Naturally, multicollinear predictive covariates create far more difficulties and further research is needed here.

A further point deserving attention is the effect of covariate omission in the situation where covariates are perfectly balanced as arising from e.g. frequency matching or designs with special stratified randomization. This covariate distribution leads to qualitatively different results with respect to the properties of the unadjusted tests as compared to the simple randomization (balanced on average) case (Gail (1988)). However, the basic message of that paper as well is that even in a perfectly balanced situation a covariate used for stratification should be included into the model.

References

Andersen, P.K., Borgan, O., Gill, R.D. & Keiding, N. (1993). Statistical models based on counting processes. Springer Series in Statistics, Springer-Verlag, New York.

Antes, G. & Sauerbrei, W. (1991). Simulations in the statistical system SAS. In: Faulbaum, F. (ed.) Advances in statistical software 3; SoftStat '91, 187-194, Gustav Fischer, Stuttgart

Beach, M.L. & Meier, P. (1989). Choosing covariates in the analysis of clinical trials. Controlled Clinical Trials 10, 161-175.

Begg, M.D. & Lagakos, S. (1993). Loss in efficiency caused by omitting covariates and misspecifying exposure in logistic regression models. Journal of the American Statistical Association, Vol. 88, No. 421, 166-170.

Binkley, J.K. & Abbott, P.C. (1987). The fixed X assumption in econometrics: Can the textbooks be trusted?, The American Statistician 41, 206-214.

Canner, P.L. (1991). Covariate adjustment of treatment effects in clinical trials. Controlled Clinical Trials 12, 359-366.

Cox, D.R. (1972). Regression models and life tables (with discussion). Journal of the Royal Statistical Society B 34, 187-220.

Cox, D.R. & Snell, E.J. (1989). Analysis of binary data. Second Edition. Chapman and Hall. London, New York.

Doksum, K.A. & Gasko, M. (1990). On a correspondence between models in binary regression analysis and in survival analysis. International Statistical Review 58, 243-252.

Gail, M.H. (1986). Adjusting for covariates that have the same distribution in exposed and unexposed cohorts. In Moolgavkar, S.H., Prentice, R.L. (Eds.). Modern statistical methods in chronic disease epidemiology. New York, Wiley, 3-18.

Gail, M.H. (1988). The effect of pooling across strata in perfectly balanced studies. Biometrics 44, 151-162.

Gail, M.H., Tan, W.Y. & Piantadosi, S. (1988). Tests for no treatment effect in randomized clinical trials. Biometrika 75, 1, 57-64.

Gail, M.H., Wieand, S. & Piantadosi, S. (1984). Biased estimates of treatment effect in randomized experiments with nonlinear regressions and omitted covariates. Biometrika 71, 431-444.

Green, M.S. & Symons, M.J. (1983). A comparison of the logistic risk function and the proportional hazards model in prospective epidemiologic studies. Journal of Chronic Diseases 36, 715-724.

Johnson, M.E., Tolley, H.D., Bryson, M.C. & Goldman, A.S. (1982). Covariate analysis of survival data: A small-sample study of Cox's model. Biometrics 38, 685-698.

Lagakos, S.W. & Schoenfeld, D.A. (1984). Properties of proportional-hazard score tests under misspecified regression models. Biometrics 40, 1037-1048.

Lagakos, S.W. (1988). Effects of mismodelling and mismeasuring explanatory variables on tests of their association with a response variable. Statistics in Medicine, Vol. 7, 257-274.

McCullagh, P. & Nelder, J.A. (1989). Generalized linear models. Second Edition. Chapman and Hall. London, New York.

Neuhaus, J.M. & Jewell, N.P. (1993). A geometric approach to assess bias due to omitted covariates in generalized linear models. Technical Report, Division of Biostatistics, Dept of Epidemiology & Biostatistics, University of California, San Francisco.

Rao, P. (1971). Some notes on misspecification in multiple regressions. The American Statistician 25, 5, 37-39.

Robinson, L.D. & Jewell, N.P. (1991). Some surprising results about covariate adjustment in logistic regression models. International Statistical Review 59, 227-240.

Sayn, H., Budde, M. & Schach, S. (1986). Small sample properties of the estimators of the regression coefficients in the Cox model. Statistical Software Newsletter 12, 76-79.

Schmoor, C. & Antes, G. (1993).Effects of covariate adjustment in regression models for the analysis of clinical studies. In: Michaelis, J., Hommel, G. & Wellek, S. (eds.). Europäische Perspektiven der medizinischen Informatik, Biometrie und Epidemiologie. Medizinische Informatik, Biometrie und Epidemiologie 76, MMV Medizin Verlag München, 114-118

Schumacher, M., Olschewski, M. & Schmoor, C. (1987). The impact of heterogeneity on the comparison of survival times. Statistics in Medicine 6, 773-784.

Seber, G.A.F. (1977). Linear regression analysis. John Wiley & Sons, New York.

Struthers, C.A. & Kalbfleisch, J.D. (1986). Misspecified proportional hazard models. Biometrika 73, 2, 363-9.

Whitehead, J. (1980). Fitting Cox's regression model to survival data using GLIM. Applied Statistics 29, 268-275.

Classification and Regression Trees (CART) Used for the Exploration of Prognostic Factors Measured on Different Scales

Berthold Lausen[1], Willi Sauerbrei[2] & Martin Schumacher[2]
[1] Forschungsinstitut für Kinderernährung Dortmund (FKE), Heinstück 11, 44225 Dortmund, Germany
[2] Institut für Medizinische Biometrie und Informatik, Universität Freiburg, Stefan-Meier-Str. 26, 79104 Freiburg, Germany

Key Words:
classification and regression trees (CART), generalized regression, censored data, interactions, adjustment of P-values, prognostic factors, different scales

Abstract
The modelling of the relationship of some response to factors measured on different scales is a common problem in various fields of application. We discuss the regression tree model in the context of general regression models and present the idea of the regression tree algorithm. In this approach all factors under consideration have to be split to binary variables, leading to a high probability of wrongly identifying as influential a variable with many possible splits. We propose a strategy to adjust for such an undesirable effect. Finally, we illustrate our modification of the classification and regression tree method with data from a multicenter randomized clinical trial in patients with brain tumors.

Introduction

The method of classification and regression trees (CART) is one approach for modelling the relationship between a response or dependent variable and factors or independent variables possibly measured on different scales. The book of Breiman et al. (1984) gives a detailed description of various aspects of the CART method. The idea of tree-growing methods was already discussed by Sonquist (1970) or may be seen as adaptation of divisive clustering schemes to regression problems. In recent years there has been renewed interest in this methodology because the approach is very flexible and it does not rely on distributional assumptions as do other regression models. CART allows the analysis of interaction effects between the factors considered and can be seen as a nonparametric analysis of the relationship (Breiman et al. (1984)). The methods are extended for many situations, e.g. for survival data (Segal (1988)), and many applications can be found in the literature. Dirschedl (1991), Sauerbrei et al. (1991) and Schmoor et al. (1993) discuss CART as a tool to find homogeneous prognostic subgroups in a clinical trial. Using CART Schumacher et al. (1993) define a new grading-system for node negative breast cancer patients, and Curran et al. (1993) investigate prognostic factors in three glioma trials.

In the field of computational statistics there is a discussion on CART and similar relatively new and modern methods of data analysis, which estimate or fit graphs to data. For example CART is included in the recent release of the S language (Chambers & Hastie (1992), chapter 9).

We concentrate on regression trees and the problem of analysing factors measured on different scales. In the section "Regression Models and Regression Trees" we introduce the regression tree model. The principles of the regression tree are given in the section "Classification and Regression Tree Algorithm". Using results from maximally selected test statistics (Miller & Siegmund (1982) and Lausen & Schumacher (1992)) we propose a modification of the CART method applied to factors measured on binary, ordinal, quantitative and nominal scales (section "Adjustment for Factors Measured on Different Scales"). We apply this proposal to data from a multicentre clinical trial on patients with brain tumour. Finally we discuss the relevance of the modification and other aspects of CART.

Regression Models and Regression Trees

Being interested in the exploration of the relationship between some response variables and some prognostic factors, we consider the following general situation: $(Y_1, X_1), ..., (Y_n, X_n)$ are stochastically independent observations of a multivariate sample. Y_i denotes the i-th observation of the p-dimensional response variable and X_i denotes the i-th observation of the k factors which may have an influence on Y_i. In the following we consider responses and factors measured on different scales.

As general regression model we assume a functional relationship defined by some characteristic F of the responses conditioned on the factors:

$$F(Y|X) = f(X), \tag{1}$$

where Y denotes the $n \times p$ matrix of the responses, X the $n \times k$ matrix of the factors, F the characteristic of interest and f the function which describes the relation between the characteristic and X.

In simple linear regression the model is given by the expectation of Y given X as characteristic of interest:

$$E(Y|X) = X\beta, \tag{2}$$

where β denotes the unknown parameters and X the design matrix. The design matrix can consist of continuous, ordinal or dummy variables of ordinal or nominal factors. Other special cases with other types of the response variables are generalized linear models (McCullagh & Nelder (1989)) or the Cox model for censored data (Cox (1972)).

Regression tree models (cf. Breiman et al. (1984)) are defined by a functional relationship which consist of a tree structured classification rule on the factors:

$$F(Y|X) = tree(X), \tag{3}$$

where *tree* denotes the unknown tree structured classification rule – the regression tree - and F the characteristic of interest. Examples for F are the conditional expectation of Y given X, the conditional variance of Y given X or the conditional survival function of Y given X.

In principle it is possible to think of the unknown tree structured classification rule as an unknown parameter of such a super model, but the common interpretation (Breiman et al. (1984)) is as a strategy to construct a prediction rule. The prediction rule may be seen as an approximation of the unknown functional relationship or as a method to identify new subgroups in the sample. The regression tree is a rooted binary tree. Each node represents a subset of the sample and each internal node is assigned with a rule depending on the factors, which defines a split or bipartition of the subset of this node. Consequently, two new subsets of the sample are defined in each internal node. The idea of the regression tree is to construct subgroups which are internally as homogenous as possible regarding the characteristic F and externally as separated as possible. Comparing two subsets it is straightforward to use an appropriate two sample statistic T to measure the separation of the two subgroups. If Y is univariate normally distributed the two sample t statistic would be an appropriate choice, or Hotellings T^2 for multivariate normally distributed response variables.

Classification and Regression Tree Algorithm

There exist various definitions of classification and regression tree algorithms. Breiman et al. (1984) define CART as a series of several steps: tree building, tree pruning and amalgamation. We follow Sauerbrei et al. (1991) and define the tree algorithm by the tree building procedure alone.

The tree algorithm can be seen as a divisive clustering scheme. Divisive clustering schemes for multivariate observations are well known in the literature (e.g. Bock (1974)). The classification rules are defined on the factors and the difference between the clusters is measured on the response variables via the characteristic F of interest. The algorithm starts with the set of all observations or objects. In the first step it selects a bipartition which is optimal with respect to a criterion based on the characteristic F of interest. The bipartition is defined by a cutpoint with respect to one quantitative or ordinal factor or by one bipartition of the classes of a nominal factor.

We choose as optimality criterion of the split the maximum of an appropriate two sample statistic with an asymptotic chi square distribution with one degree of freedom for all bipartitions considered. Equivalently we select the bipartition of the observations according to the minimal P-value of this chi square statistic.

In the next step we start with the partition of the set of observations for the subsets, if it consists of at least a minimum number of objects n_{min}. Beside of a predefined fixed number of observations the criterion n_{min} may depend on the size of the study. An often used criterion for n_{min} is $\sqrt{n}$. We apply in each subset the same selection procedure. A further partition of a subset is rejected, if the size of the subset is less than n_{min} or if the minimal P-value is greater than a prespecified value p_{stop}.

More formally the partition procedure is recursive. We start with the set consisting of all observations or objects:

a) The minimal P-value is computed for all k factors and all allowable splits within the factors. An allowable split is given by a possible cutpoint of a quantitative or ordinal factor or some bipartition of the classes of a nominal factor.

b) The set of objects is split into two subsets based on the variable and the corresponding cutpoint with the minimal P-value, if the minimal P-value is smaller or equal to p_{stop}.

c) The partition procedure is stopped if there exists no allowable split, if the minimal P-value is greater than p_{stop} or because of the n_{min} criterion.

d) For each of the two subsets we repeat this procedure.

Running the partition procedure described by steps a), b), c) and d) we get a binary tree with a set of objects, a splitting rule and the minimal P-value at each interior node.

For the observations in a final node we may estimate quantities of interest, like mean values, variances, survival rates etc..

Adjustment for Factors Measured on Different Scales

For a given node the regression tree algorithm computes for each factor the partition which minimizes the P-value of the two sample statistic used. Afterwards the algorithm selects the factor which gives the minimal P-value over all factors. Consequently, the algorithm compares P-values which are minimized over different number of cutpoints or partitions. This definition of the algorithm causes an overestimation of the importance of factors with many allowable splits. The problem is acknowledged by other authors; for example Breiman et al. (1984, p.42) state that *'... variable selection is biased in favour of those variables having more values and thus offering more splits'* and Loh & Vanichsetakul (1988, p.722) observe *'... although the CART method of splitting on categorical variables appears natural, it may favor such variables over ordered ones'.*

Being interested in the exploration of the prognostic factors we are looking for a selection criterion which is not biased for factors measured on different scales and on a different number of partitions. Computing the best split for a factor we compute the maximally selected test statistic as optimality criterion and consequently we have to correct the P-value regarding the maximally selected test statistic. If we do not adjust for the multiplicity of the test statistic, the P-value can be drastically to small. For example Miller & Siegmund (1982) derive the asymptotic distribution of maximally selected chi square statistics in fourfold tables and show that a nominal P-value of 0.05 from the chi square distribution results in a P-value of 0.49 according to the asymptotic distribution of the maximally selected chi square statistic. Lausen & Schumacher (1992) derive the same asymptotic distribution for a wide range of two sample statistics including the t-statistic, rank statistics and the log-rank statistic for censored data.

Miller & Siegmund (1982) and Lausen & Schumacher (1992) consider standardized two sample test statistics with an asymptotic chi square distribution for a fixed cutpoint in a factor measured on a *quantitative or ordinal scale*. They assume that the selection of possible cutpoints is done on a prespecified interval, which guarantees that the subgroups have the minimum size of $[\epsilon n]$ objects, where $[x]$ denotes greatest integer less than x. Consequently, the maximally selected test statistic depends on the parameter ϵ and is denoted by $M(\epsilon)$.

$$M_j(\epsilon) = \max_{\rho \in [x_j(\epsilon), x_j(1-\epsilon)]} T_{j\rho}^2,$$

where $x_j(.)$ denotes the quantile of the j-th factor X_j in the subset considered and $T_{j\rho}$ the standardized two sample statistic which compares the samples defined by the cutpoint ρ.

Allowing for factors measured on different scales we are interested in computing or approximating for each factor P-values of the maximally selected test statistics; i.e. the probability $P(M_j(\epsilon) \geq b^2)$ under the null hypothesis of no influence of the j-th factor, where b^2 denotes the observed value of the random variable $M_j(\epsilon)$.

Miller & Siegmund (1982) give following simple approximation formula for the P-value derived by the asymptotic distribution:

$$P_1(\epsilon) = 4\varphi(b)/b + \varphi(b)(b - 1/b)log_e(\epsilon_2(1-\epsilon_1)/[(1-\epsilon_2)\epsilon_1]), \qquad (4)$$

where $\varphi(b) = exp(-b^2/2)/\sqrt{2\pi}, \epsilon_1 = 1 - \epsilon_2 = \epsilon$. For example we select the maximal value $M_j(0.25) = 6.25$ for $n = 200$. Assuming that we have no ties in the observed prognostic factor X_j, the selection is done on the index set $\{ 50 , ... , 150 \}$ with respect to the order statistic of X_j. Consequently we compute for $b = 2.5$ the P-value $P_1(0.25) = 0.109$ (unadjusted $P_{\chi^2} = 0.012$).

The approximation of the P-value using the asymptotic distribution of the maximally selected test statistic is relatively simple, but it has the disadvantage that it tends to be conservative for a small number of splits considered. Another approximation is based on an improved Bonferroni inequality (Worsley (1982)) and is more appropriate for a smaller number of splits considered. Worsley (1983, eq. 6.4) gives the following approximation:

$$P_2(\epsilon) = 1 - F_1(b^2) + \sum_{i=1}^{k-1} D(l_i, l_{i+1}), \qquad (5)$$

where k denotes the number of different splits, $c_1 < ... < c_k$ denotes the cutpoints which define the splits considered, l_i denotes the size of the subgroup with factor values less or equal to c_i, $D(i,j) = (2/\pi)f_2(b^2)[t_{ij} - (b^2/4 - 1)(t_{ij})^3/6]$, and F_1 is the distribution function of the chi square distribution, $f_2(b^2) = 0.5exp(-b^2/2)$ and $t_{ij} = [1 - i(n-j)/((n-i)j)]^{0.5}$.

Claiming $[\epsilon n] \leq l_1 < ... < l_k \leq n - [\epsilon n]$ the approximation of the improved Bonferroni inequality depends also on ϵ. Looking at the example values above ($n = 200, \epsilon = 0.25, b = 2.5$) and assuming that the cutpoints considered result in $l_1 = 50, l_2 = 60, ..., l_{11} = 150$ we compute the P-value $P_2(0.25) = 0.073$ (cf. Lausen (1990; fig. 5.4.1-4)). In general the approximation above tends to be good if the correlations between sucessive test statistics are high and positive. This is true for the test statistic of the splitting process $M_j(\epsilon)$.

Approximations (4) and (5) tend to be conservative, consequently we use as approximate P-value for $M_j(\epsilon) \geq b^2$ the minimum of P_1 and P_2:

$$P(M_j(\epsilon) \geq b^2) \approx \min(P_1(\epsilon), P_2(\epsilon)). \tag{6}$$

By formula (6) we have an adjustment for factors measured on a quantitative or ordinal scale. For the example values above we get $P(M_j(0.25) \geq 6.25) \approx 0.073$.

Considering factors measured on a *nominal scale* we have no acceptable adjustment procedure in general. Consequently, we propose (cf. Sauerbrei et al. (1991)) to restrict the nominal factors to a small number of categories, e.g. three or maximally four, investigate all possible combinations and adjust the P-values by the improved Bonferroni inequality (5). If the number of categories is much higher, the improved Bonferroni inequality can be very conservative and consequently the effect of the nominal factor can be drastically underestimated.

Brain Tumor Study

Ulm et al. (1989) describe various strategies for the analysis of a study on brain tumors. We use the data of this multicenter, randomized clinical trial to illustrate our regression tree proposals. The study compares two chemotherapeutic schemes. We concentrate on the analysis of the survival time and on the exploration of prognostically homogenous subpopulations defined by twelve potentially prognostic factors (Table 1). One factor is measured on a quantitative scale, three are measured on an ordinal scale and eight are binary.

We ran the regression tree algorithm for the survival times of 447 patients (293 deaths) of the study as described in the previous section without the proposed P-value adjustment. We use as two sample statistic the log-rank statistic. Patients are excluded from the partition process at an inner node, if the observation is missing with respect to the factor of the split criterion. This is only one possible strategy in the case of missing values; other ones are to use 'surrogate splits' (Ulm et al. (1989)) or to restrict the analysis

Factor	Scale (Levels)	number missing
Age	quantitative (in years)	0
Type of resection	ordinal (biopsy, partial, total)	17
Grade of malignancy	ordinal (3, 3-4, 4)	0
Karnofsky index	ordinal (<70, 70-80, >80)	4
Org. psychosyn.	binary (no, yes)	9
Epilepsy	binary (no, yes)	3
Cortison	binary (no, yes)	3
Sex	binary (male, female)	0
Aphasia	binary (no, yes)	4
Patient history	binary ($\leq$ 8 weeks, $>$ 8 weeks)	4
Anticonvulsiva	binary (no, yes)	4
Amnesia	binary (no, yes)	3

Table 1: Twelve potentially prognostic factors considered in the brain tumor study. The measurement scale and the number of missing values are indicated.

to the complete cases in all factors (Schmoor et al. (1993)). The cutpoint selection for the quantitative factor age was performed between the 10 % and 90 % quantiles of the distribution of age in the subsets considered. We stopped the partitioning when the minimal P-value was greater than 0.05 or when the size of the subset was smaller than 22 observations. The latter corresponds to the square root of the total sample size, the so-called $\sqrt{n}$ criterion. For survival time data the $\sqrt{n}$ criterion of the effective sample size, i.e. the number of events, may also be sensible, but that allows even smaller subgroups. Here we use the total sample size.

Figure 1 is a representation of the constructed regression tree. We observe that the quantitative factor age dominates the regression tree. This domination can be seen as at least partially due to an overestimation of the effect of the only quantitative factor by the maximally selected cutpoints and minimal selected P-values for age.

Figure 2 gives the constructed regression tree after adjusting the P-values for factors measured on different scales ($\epsilon = 0.1$). Additionally the nodes are ordered according to the median survival time in the corresponding node. The right branch gives the higher survival time in our representation. Moreover the thickness of the branches shows the relative frequency of the subgroups. A more detailed discussion of the graphical representation is given in Dirschedl (1991). We observe that the split criterion of the root is still determined by the age cutpoint of 55 years and the split criteria

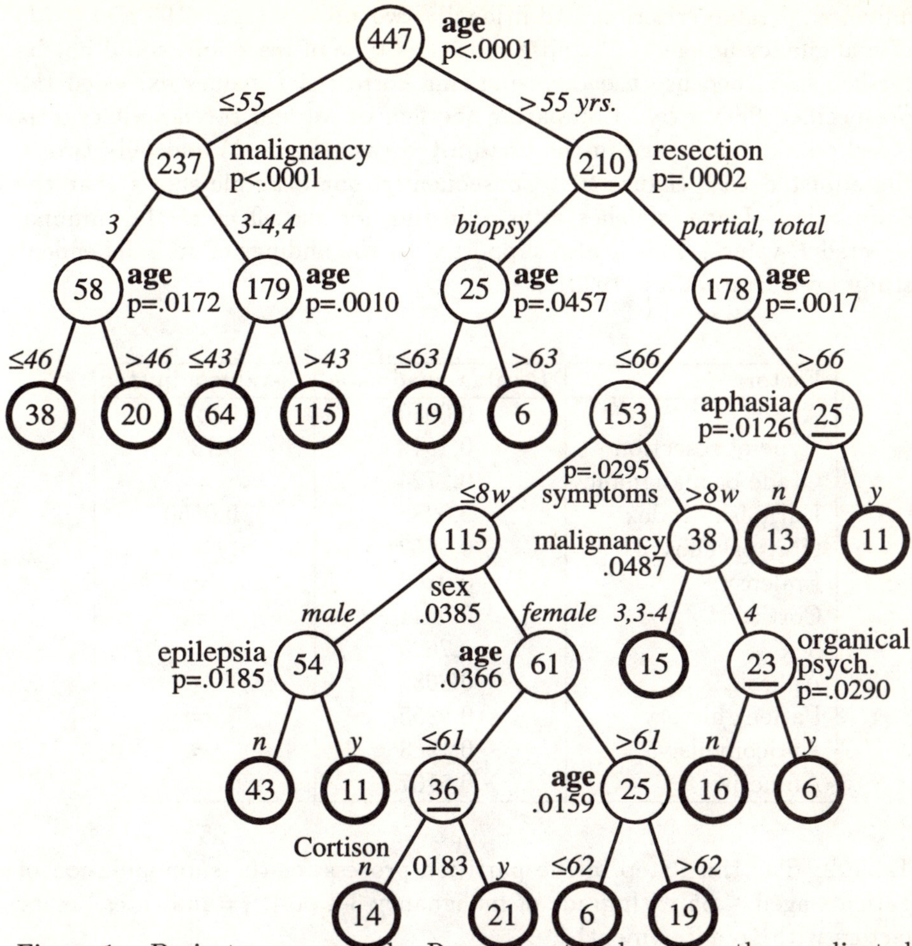

Figure 1: Brain tumor example. Regression tree based on the unadjusted P-values. At each node we give inside the circle the number of patients, outside the circle the factor and P-value of the split. At the branches we state the definition of the split; i.e. the cutpoints selected. The number of patients is underlined, if patients have missing values with respect to the split criterion.

are also not changed in the subsets of the root. But on the next level of the partition we observe that some split critera have changed. For the subset of 179 patients aged 55 or less and grade of malignancy 3-4 or 4 we give the corrected and uncorrected P-values in table 2. In the case of binary variables these values are identical. Whereas age had the smallest uncorrected P-value and was therefore choosen as the split criterion, we had to divide this subgroup by the variable epilepsy when using the the adjusted P-value criterion. Additionally two subsets (age ≤ 55 and grade of malignancy 3; age > 55 and a biopsy as type of resection) could not be further split, because the corresponding corrected P-values exceeded the prespecified 0.05 level. Comparing the figures we find seven split criteria based on the factor age in the unadjusted case (figure 1) and only two in the adjusted case (figure 2). Consequently, our example shows that the domination of age vanishes after adjusting for the effect of the minimal selected P-values. This is also supported by the findings of an independent study by Curran et al. (1993).

Factor	P-value unadjusted	P-value adjusted
Age	0.0010	0.0252
Type of resection	0.4643	0.7307
Grade of malignancy	0.2724	=
Karnofsky index	0.0286	0.0550
Org. psychosyn.	0.0652	=
Epilepsy	0.0162	=
Cortison	0.0641	=
Sex	0.7620	=
Aphasia	0.8089	=
Patient history	0.9955	=
Anticonvulsiva	0.1983	=
Amnesia	0.5167	=

Table 2: The third step in the partition process for the subpopulation of patients aged ≤ 55 with grade of malignancy 3-4 or 4. (= indicates binary factors with no adjustment)

Discussion

So far we have given an adjustment for factors measured on different scales which avoids the overrepresentation of factors with many allowable splits. This adjustment is an important improvement of CART. As in all other tree constructing algorithms we do not control the overall P-value of the 'existence' of the binary tree structure on the possibly prognostic factors. Bonferroni inequalities (Hommel (1988, 1989)) have also been used to obtain an adjustment at each node by Curran el al. (1993). As a further illustrative characteristic one may compute a joint P-value for the multiple test with respect to the compared factors at each node via Bonferroni-type inequalities and include it in the plot. But the conservative character of such

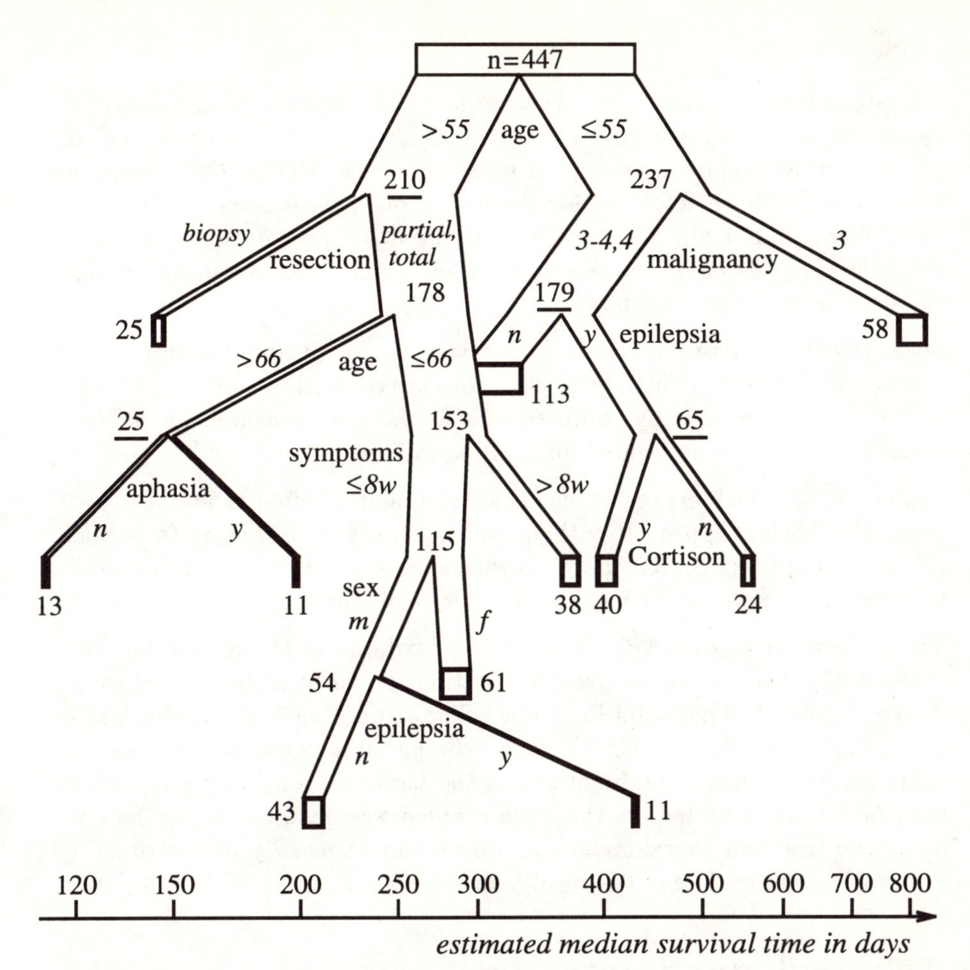

Figure 2: Brain tumor example. Regression tree based on the adjusted P-values. End nodes ordered by estimated median survival time in days. Thickness of the branches gives relative frequency of the subgroups. The number of patients is underlined, if patients have missing values with respect to the split criterion.

inequalities is an obvious limitation of such a proposal and the derivation of the P-value is only justified at the root node.

We propose to use CART as an explorative additional tool for an analysis of prognostically important subgroups. Because CART analyses high-dimensional interactions this approach can help to detect important subgroups which are not found by the more traditional regression techniques. Dirschedl (1991) discusses an example in detail, where it is shown that a subgroup of patients should have been excluded from the analysis of the study.

An interpretation of the tree structure has to be done very cautiously, because the tree obtained has to be seen as an estimator of an underlying structure that might be associated with a large variability. Other methods used for variable selection or for the fitting of many factors to a response suffer from similar difficulties or have to use strict assumptions like linearity. The analysis of interaction is often either not done or it suffers from the drawback of very low power.

Segal (1992) discusses the CART approach for longitudinal data (see also Ciampi et al. (1992)). Such generalizations are covered by our general setup of the second section, but adjustments for factors measured on different scales remain to be developed for multivariate response variables.

Ripley (1993) considers statistical aspects of neural networks and also compares the CART approach with neural networks. Tibshirani & LeBlanc (1992) compare the CART approach with a variant of the multivariate adaptive regression splines (MARS) for logistic regression.

These alternatives to CART increase the flexibility of the model, but they increase also the risk of an overspecification or overfit of the model to the data set at hand. This point has to be acknowledged also for the Breiman et al. (1984) formulation of CART with pruning, amalgamation and possible splits based on linear combinations of the factors. Considering regression trees based on many factors the main problem seems to be the (in-) stability of the tree and therefore this approach should usually be restricted to explorative studies where the results have to be validated with new data.

Acknowledgements

We thank Peter Dirschedl for providing figures 1 and 2. Moreover, we thank an anonymous referee and the editors for valuable comments and support.

References

Bock, H.H. (1974). Automatische Klassifikation. Vandenhoeck & Ruprecht, Göttingen.

Breiman, L., Friedman, J.H., Olshen, R.A. &d Stone, C.J. (1984). Classification and regression trees, Wadsworth, Monterey.

Chambers, J.M. & Hastie, T.J. (1992). Statistical models in S, Wadsworth and Brooks/Cole, Pacific Grove.

Ciampi, A., Hendricks, L. & Lou, Z. (1992). Tree-growing for the multivariate model: the RECPAM approach. In: Dodge, Y. & Whittaker, J. (eds.). Computational Statistics, Vol. 1, Physica-Verlag, Heidelberg.

Cox, D.R. (1972). Regression models and life tables (with Discussion). Journal of the Royal Statistical Society, Series B 74, 187-200.

Curran, W.J. JR., Scott, C.B., Horton, J., Nelson, J.S., Weinstein, A.S., Fischbach, A.J., Chang, C.H., Rotman, M., Asbell, S.O., Krisch, R.E. & Nelson, D.F. (1993). Recursive partitioning analysis of prognostic factors in three radiation therapy oncology group malignant glioma trials, Journal of the National Cancer Institute 85, 9:704-710.

Dirschedl, P. (1991). Klassifikationsbäume - Grundlagen und Neuerungen. In: Fleischer, W., Nagel, M. & Ostermann, R. (eds.). Interaktive Datenanalyse mit ISP, Westarp Verlag, Essen, 15-30.

Hommel, G. (1988). A stagewise rejective multiple test procedure based on a modified Bonferroni test. Biometrika 75, 383-386.

Hommel, G. (1989). A comparison of two modified Bonferroni procedures. Biometrika 76, 624-625.

Lausen, B. (1990). Maximal selektierte Rangstatistiken. PhD dissertation. University Dortmund.

Lausen, B. & Schumacher, M. (1992). Maximally selected rank statistics. Biometrics 48, 73-85.

Loh, W.-Y. & Vanichsetskul, N. (1988). Tree-structured classification via generalized discriminant analysis. Journal of the American Statistical Association 83, 403: 715-728.

McCullagh, P. & Nelder, J. (1989, 2nd ed.). Generalized linear models, Chapman & Hall, London.

Miller, R. & Siegmund, D. (1982). Maximally selected chi-square statistics. Biometrics 38, 1011-1016.

Ripley, B. (1993). Statistical aspects of neural networks. In: Barndorff-Nielsen, O.E., Jensen, J.L., and Kendall, W.S (eds.). Networks and Chaos - Statistical and Probabilistic Aspects, Chapman & Hall, 40-123.

Sauerbrei, W., Zaiss, A., Lausen, B. & Schumacher, M. (1991). Evaluierung prognostisch homogener Subpopulationan mit Hilfe der Methode der Klassifikationsbäume. In: Guggenmoos-Holzmann (ed.). Quantitative Methoden in der Epidemiologie. Springer-Verlag, Berlin Heidelberg, 229-234.

Schmoor, C., Ulm, K. & Schumacher, M. (1993). Comparison of the Cox model and the regression tree procedure in analysing a randomized clinical trial. Statistics in Medicine, 12, 2351-2366.

Schumacher, M., Schmoor, C., Sauerbrei, W., Schauer, A., Ummenhofer, L., Gatzemeier, W. & Rauschecker, H. (1993). The prognostic effect of histological tumor grade in node-negative breast cancer patients. Breast Cancer Research and Treatment, 25, 235-245.

Segal, M.R. (1988). Regression trees for censored data. Biometrics 44, 35-47.

Segal, M.R. (1992). Tree-structured methods for longitudinal data. Journal of the American Statistical Association 87, 418: 407-418.

Sonquist, J.A. (1970). Multivariate model building. Institute for Social Research, University of Michigan.

Tibshirani, R. & LeBlanc, M. (1992). A strategy for binary description and classification. Journal of Computational and Graphical Statistics 1, 1:3-20.

Ulm, K., Schmoor, C., Sauerbrei, W., Kemmler, G., Aydemir, Ü., Müller, B. & Schumacher, M. (1989). Strategien zur Auswertung einer Therapiestudie mit der Überlebenszeit als Zielkriterium. Biometrie und Informatik in Medizin und Biologie 20, 4:171-205.

Worsley, K.J. (1982). An improved Bonferroni inequality and applications. Biometrika 69, 297-302.

Worsley, K.J. (1983). Testing for a two-phase multiple regression. Technometrics 25, 35-42.

Clustering Algorithms and Cluster Validation

Allan D. Gordon, University of St Andrews, Department of Mathematical and Computational Sciences, North Haugh, St.Andrews KY16 9SS, Fife, Scotland

Key Words:
classification, cluster analysis, cluster validation

Abstract
An overview is presented of two topics in cluster analysis: algorithms for obtaining clusterings, and procedures for investigating the validity of the results.

Introduction

Cluster analysis is concerned with the problem of obtaining a valid and parsimonious description of the similarities within a set of objects. The objects are usually described in one of two formats :

1. a *pattern matrix* is an (nxp) matrix $X \equiv (x_{ik})$, where x_{ik} denotes the kth variable describing the ith object ($i = 1,...,n$; $k = 1,...,p$);

2. a *dissimilarity matrix* is an (nxn) matrix $D \equiv (d_{ij})$, where d_{ij} provides a measure of the dissimilarity between the ith and j^{th} objects, satisfying

$$(i)\, d_{ij} \geq 0,\ (ii)\, d_{ii} = 0,\ (iii)\, d_{ji} = d_{ij}\ (i,j = 1,\ldots,n).$$

The two most common types of summary of the data that have been sought are:

1. a partition of the objects into disjoint classes (or clusters) having the property that objects in the same class are similar to one another and dissimilar to objects in other classes;

2. a hierarchical classification, comprising a nested set of partitions of the data, in which information is provided about the resemblances between classes.

A complementary graphical investigation is often undertaken, in which the objects are represented by a set of points in a low-dimensional space, with the property that similar objects are represented by points that are close together.

Many different decisions have to be taken before a cluster analysis can be successfully completed. Some of these decisions are summarized below, although it should be noted that not all of them need be relevant in a particular study.

1. How should the objects for analysis be selected?

2. Which variables should be used to describe the objects?

3. Should any standardization or differential weighting of variables be used in the analysis?

4. If a dissimilarity matrix has to be obtained from a pattern matrix, how should a relevant dissimilarity coefficient be selected?

5. Which clustering and graphical procedures should be used to analyse the data?

6. How should one validate and summarize the results of the investigation?

Introductions to cluster analysis, and partial answers to these questions, are provided by Hartigan (1975), Gordon (1981) and Jain & Dubes (1988). The present paper concentrates on two topics in cluster analysis that may be of particular interest to those involved in computational statistics : clustering algorithms and cluster validation.

Partitioning Algorithms

At the start of a cluster analysis, information is in general lacking on the number (c, say) of homogeneous classes in the data set. It is usual to find an 'optimal' partition of the set of n objects into c classes for a range of values of c, and to compare these partitions with one another, often with the assistance of a graphical representation of the data. One commonly-used strategy has been to obtain a hierarchical classification of the set of objects (see the next section), comprising partitions into c classes for all values of c

between 1 and n, and to select one of these n partitions as the optimal one. Further discussion of these so-called *stopping rules* is postponed to the final section.

Another family of algorithms concentrates on seeking an optimal partition of the data for a specified value of c. These algorithms generally involve the analysis of an (nxp) pattern matrix; since the number of variables, p, is usually much smaller than the number of objects, n, considerable savings in storage requirements result.

The number of different ways, $P(n,c)$, in which n objects can be divided into c non-empty classes satisfies the following equation (see, e.g., Jensen (1969)):

$$P(n,c) = \frac{1}{c!}\sum_{i=1}^{c}(-1)^{c-i}\binom{c}{i} i^n \, .$$

It is clear that, as n and c increase, it soon becomes computationally infeasible to examine all partitions in order to identify an optimal one. This conclusion holds even when — by making use of efficient accounting schemes based on dynamic programming or branch-and-bound methodology (Jensen (1969), Koontz, Narendra & Fukunaga (1975), Diehr (1985)) — some partitions need not be explicitly investigated. It is thus desirable to have algorithms that can identify optimal partitions without examining all, or even a small fraction, of all possible partitions. However, many partitioning problems are NP-hard; see Day (1994) for a discussion of complexity theory with particular reference to clustering problems. For example, if the variability of a class is defined to be either the largest pairwise dissimilarity, or the sum of the pairwise dissimilarities, of its constituent objects, the problems of minimizing either (i) the sum of the class variabilities or (ii) the largest class variability are both NP-hard (Brucker (1978), Křivánek (1986)).

Much attention has therefore been devoted to obtaining efficient approximating algorithms. The most commonly-used approximating partitioning algorithms can be described as iterative relocation algorithms. These comprise three stages.

1. An initial partition of the set of objects into c classes is obtained.

2. A criterion or procedure for measuring the optimality of a partition is defined.

3. The set of allowable transformations of a partition into another partition is defined.

Many different options are available within this general framework.

1.Initial partition: This can be obtained by random (MacQueen (1967)) or systematic (Wishart (1987)) assignments of objects to classes, or from the results of an earlier investigation (e.g., by sectioning a hierarchical classification to provide a partition of the objects into c classes, or by amalgamating a closest pair of classes in the optimal partition that is found of the objects into $(c+1)$ classes (Beale (1969), Wishart (1987))). Alternatively, c seed points can be identified, and the initial partition obtained by assigning each object to the class specified by the seed point to which it is closest. The seed points can be chosen uniformly to cover the range of variation in the data set or to be some of the data points, choosing sequentially objects that are (i) greater than a specified distance from existing seed points, or (ii) farthest from any of the existing seed points (Kennard and Stone (1969)). Strategy (i) has also been used on its own as a *single pass* algorithm for partitioning large data sets. Another variant involves selecting objects from high-density regions of space to be seed points or sub-classes (e.g., Gordon (1986), Nicolau & Brito (1989)). A computationally-demanding global optimization algorithm for finding the configuration of seed points that minimizes the sum of distances of the objects from their nearest seed point is presented by Pintér & Pesti (1991). Fuller details of some initiating rules are given in the review paper by Belbin (1987). It is common for the algorithm to be run several times, using a different initial partition each time. Zhang & Boyle (1991) advocated using a perturbed version of the final partition as the initial partition for a new run of the algorithm.

2.Optimality criterion: One of the earliest iterative relocation clustering algorithms involved modifying a partition by assigning each object to the class to whose centroid it was closest, then recalculating class centroids (e.g., Ball & Hall (1967), MacQueen (1967), Hartigan & Wong (1979)). This *k-means* algorithm does not explicitly involve a measure of the optimality of a partition. It can be shown that (in the absence of ties in the centroid-to-point distances) changes made during an iteration of the k-means algorithm reduce the total within-class sum of squared distances about the class centroid. However, this sum-of-squares criterion would also be reduced by moving the ith point from the jth class to the kth class if (Beale (1969)):

$$D_{ik}^2 < [(n_k+1)n_j/(n_k(n_j-1))]D_{ij}^2$$

where D_{ij} (resp., D_{ik}) denotes the distance of the ith point from the jth (resp., k^{th}) class centroid, and the jth (resp., k^{th}) class contains n_j (resp., n_k) objects.

Other definitions of the variability of a class, which could be summed over all classes to provide a measure of the optimality of a partition, include: (i) the sum of the distances from the class median to each object (Späth

(1976)); (ii) the sum of a scaled version of the squared Mahalanobis distances from the class centroid to each object, where the covariance matrix defining the Mahalanobis distance is defined for each class separately (Diday & Govaert (1977)), thus allowing an adaptive definition of within-class distances. Other possible criteria were suggested by Friedman & Rubin (1967) and Maronna & Jacovkis (1974).

3. Transformations: The k-means algorithm involves considering the simultaneous reassignment of all objects to the class to whose centroids they are closest. Given an explicitly-stated optimality criterion, many algorithms involve moving a single object at a time, if this move would improve the quality of the partition. The object can be moved to the first class for which an improvement results, or to the class which provides the greatest improvement, or a hybrid of these two movement strategies can be adopted (e.g., Ismail & Kamel (1989)). Updating of class properties — such as the coordinates of the class centroid — have been carried out after each move, or only after the class membership of each object in turn has been considered; in the former case, the results could be markedly affected by the order in which the objects are considered. More elaborate transformations have involved the pairwise interchange of the class memberships of two objects (Banfield & Bassill (1977)), and the cumulative assignment of objects to the nearest different class in sorted order of their distances from that class (Friedman & Rubin (1967)). Such algorithms are run until no improvement in the quality of the partition is possible using the current set of transformation options.

Klein & Dubes (1989) and Selim & Asultan (1991) presented simulated annealing (Kirkpatrick, Gelatt & Vecchi (1983)) algorithms, in which at each stage there is a small probability that the transformation that is carried out will degrade the quality of the partition, in the hope of avoiding obtaining an inferior final solution. Simulated annealing algorithms make heavy demands on computing resources, and their success is in general highly dependent on the parameters of the cooling schedule. It remains to be seen if the extra effort required for successful use of such algorithms will be worthwhile for many clustering problems.

Some iterative relocation algorithms allow the possibility of excluding atypical objects from the analysis (Wishart (1987)), or for the value of c to change during the course of the iterations, by amalgamating small or neighbouring classes and dividing large or heterogeneous classes (Ball & Hall (1967), MacQueen (1967)); this activity requires the definition of various threshold values.

Some clustering problems have been reformulated by defining supplementary 0/1 variables, such as :
(i) $y_{ij} = 1$ if object i belongs to class j, otherwise 0;

(ii) $z_{ij} = 1$ if objects i and j belong to the same class, otherwise 0.
This has allowed the reformulated problems to be addressed using mathematical programming algorithms (e.g., Rao (1971), Mulvey & Crowder (1979), Massart, Plastria & Kaufman (1983), Klein & Aronson (1991)). If the (y_{ij}) are not required to be 0/1 variables but only to satisfy

$$y_{ij} \geq 0; \sum_j y_{ij} = 1 \; (i = 1, ..., n),$$

then y_{ij} can be interpreted as a membership function of object i for class j, a formulation that has led to work under the heading of fuzzy clustering (Bezdek (1981,1987)).

In recent years, interest has been increasing in the use of parallel computer hardware (Hwang & Briggs (1984)) to implement clustering algorithms. In such applications, it is common to assign a unique processing element to each data element. Increased efficiency relative to computers that operate serially is possible when an algorithm requires repeated application of the same kind of operation, such as identifying the distance between an object and a class centroid. Published applications include algorithms for seeking the partition with minimum sum of squared distances about class centroids (Ni & Jain (1985), Li & Fang (1989)) and the single pass algorithm (Whaley & Hodes (1991)). It seems likely that the future will see more use of parallel computers for clustering, and that efficient algorithms will exploit particular features of the machine on which they run and thus be highly machine-dependent. There has also been an increased use of neural networks to obtain clusterings of data; a review is provided by Murtagh (1994).

Hierarchical Classification Algorithms

The output produced by these algorithms is a nested set of partitions, usually summarized in a dendrogram. This tree structure associates a height with each class and can be defined by the set of ultrametric distances (h_{ij}), where h_{ij} denotes the lowest value of the height for which objects i and j belong to the same class. The set (h_{ij}) satisfies the ultrametric inequality (Johnson (1967)) : for all i,j,k,

$$h_{ij} \leq max(h_{ik}, h_{jk}).$$

Obtaining a hierarchical classification can thus be viewed as providing a transformation from a set of dissimilarities (d_{ij}) to a set of ultrametric distances (h_{ij}) so as to minimize some measure of the distortion of the transformation, such as

$$\sum_{j<i} |d_{ij} - h_{ij}| \; or \; \sum_{j<i} (d_{ij} - h_{ij})^2.$$

Krivánek & Morávek (1986) and Krivánek (1986) proved that these problems are NP-hard. Chandon, Lemaire & Pouget (1980) described a branch-and-bound algorithm guaranteed to find least-squares ultrametric distances, which was computationally feasible only for small values of n. Approximating algorithms for seeking least-squares ultrametric distances have been proposed by Hartigan (1967), Carroll & Pruzansky (1980) and De Soete (1984).

Most hierarchical classification algorithms that have been proposed belong to the family of agglomerative algorithms. In these, the objects initially belong to n singleton classes, and at each stage a closest pair of classes is amalgamated and 'distances' between this new class and all other classes are recalculated. The most widely-used of all clustering algorithms comprise a subset of the family of agglomerative algorithms, in which the updating formula for calculating the distance between a newly-formed class $C_i \cup C_j$ and another class C_k is given by (Lance & Williams (1967), Jambu (1978)):

$$\begin{aligned} d(C_i \cup C_j, C_k) &= \alpha_i d(C_i, C_k) + \alpha_j d(C_j, C_k) + \beta d(C_i, C_j) \\ &\quad + \gamma |d(C_i, C_k) - d(C_j, C_k)| + \delta_i h(C_i) \\ &\quad + \delta_j h(C_j) + \varepsilon h(C_k), \end{aligned}$$

where initially $d(C_i, C_j) = d_{ij}$; $h(C_i)$ denotes the height of class C_i in the dendrogram; and the parameters (α_i,α_j,β,γ, δ_i,δ_j,ε) define the clustering criterion to be used in the amalgamations. For example, the single link criterion, for which

$$d(C_i \cup C_j, C_k) = min(d(C_i, C_k), d(C_j, C_k))$$

can be obtained by choosing $\alpha_i = \alpha_j = 1/2$, $\gamma = -1/2$, and the other parameters $= 0$. For definitions of other clustering criteria, see, for example, Podani (1989) or Gordon (1994b).

It is convenient to have many different clustering criteria available within this general agglomerative algorithm, but it should be noted that more efficient algorithms exist for particular clustering criteria : for example, see Rohlf (1982) for a discussion of algorithms for obtaining a hierarchical classification using the single link criterion.

Straightforward application of the general agglomerative algorithm would require $O(n^3)$ time and $O(n^2)$ space, but the time requirements can be reduced to $O(n^2 \log n)$ by use of priority queues (Day & Edelsbrunner (1984)).

A useful concept in this context is that of reducibility (Bruynooghe (1978)): for all i,j,k,

$$d(C_i, C_j) \leq min(d(C_i, C_k), d(C_j, C_k))$$

$$\Rightarrow min(d(C_i, C_k), d(C_j, C_k)) \leq d(C_i \cup C_j, C_k).$$

If a clustering criterion satisfies this reducibility condition, all reciprocal nearest neighbours can be amalgamated simultaneously without affecting later stages in the amalgamation; this allows larger data sets to be analysed by examining only dissimilarities less than a threshold value, which is increased during the course of the amalgamations. When this concept is used in conjunction with nearest neighbour chains (Benzécri (1982), Murtagh (1983,1984), Day & Edelsbrunner (1984)), algorithms with time complexity $O(n^2)$ can be constructed.

Divisive algorithms, in which initially all objects belong to a single class and at each stage one of the classes is divided into two, have been used less frequently. The problem of finding an optimal bipartition of a class has been shown to be NP-hard for several clustering criteria (Brucker (1978), Welch (1982)), but there are polynomial-time algorithms for finding a bipartition to minimize the larger of the two class diameters (Guénoche, Hansen & Jaumard (1991)) and the sum of the two class diameters (Hansen & Jaumard (1987)).

Examples of the use of parallel computers to obtain hierarchical classifications are presented by Li & Fang (1989), Rasmussen & Willett (1989) and Li (1990). Further details of hierarchical classification algorithms are given in the review paper by Gordon (1994b).

Cluster Validation

Clustering algorithms can provide misleading summaries of data.

1. A computer program with instructions to obtain a partition (or a nested set of partitions) of a set of objects will provide this type of output, even if the objects do not fall naturally into distinct, homogeneous classes.

2. The analysis of a set of objects by different clustering algorithms can lead to markedly different results, because each clustering criterion implicitly involves a model for the data, and tends to distort the output in an attempt to fit this model. For example, algorithms seeking to partition the data so as to minimize the sum-of-squares criterion tend to produce hyperspherical-shaped clusters, a result that is to be expected given links shown by Scott & Symons (1971) between this clustering criterion and a spherical normal components model.

It is thus important to validate the results of a cluster analysis. This validation has been carried out at various levels of formality. First, the results have been compared with a graphical representation of the data, to assess by eye whether or not each cluster is represented by a homogeneous and isolated set of points. Secondly, the set of objects has been analysed using more than one clustering algorithm and the results synthesized, the rationale being that the parts of the results about which the algorithms are in agreement are unlikely to be artifacts of a clustering criterion and more likely to correspond to genuine class structure (e.g., Gordon (1981,Chapter 6)). Thirdly, and more formally, various tests have been carried out on the results of a cluster analysis.

In statistics, exploratory analyses are often carried out on data, for example to identify appropriate models for them. In general, such studies are followed by a confirmatory analysis on new data. This is seldom the procedure followed in a clustering study : the objects can rarely be summarized by a simple statistical model; further, the investigator is often interested solely in the set of objects being analysed, and not in the properties of a larger population of objects from which this set may be regarded as being a random sample. It is thus usual for validation studies to be carried out on the set of objects that has been clustered.

Reviews of methods of cluster validation have been presented by Perruchet (1983), Jain & Dubes (1988,Chapter 4) and Bock (1989). There are three main classes of models for the complete absence of class structure in the data, the first two of which relate to the pattern matrix and the third to the dissimilarity matrix.

1. *Uniformity or Poisson model* (Bock (1985)) : The objects can be represented by points that are uniformly distributed in some region A of p-dimensional space. Standard definitions of A include a hypercube or hypersphere. Another option is to specify a data-influenced region, such as the convex hull of the points in the data set.

2. *Unimodal model* (Bock (1985)) : The variables describing the objects have a unimodal probability (density) function. Standard definitions include a spherical normal distribution, but again the data could influence the specification of the model.

3. *Random dissimilarity matrix model* (Ling (1973)) : The elements of the (lower triangle of the) dissimilarity matrix are ranked in random order, all $\binom{n}{2}!$ orderings being regarded as equally likely.

There are difficulties with all of these models : the first two require the specification of the region A and the probability (density) function; the random dissimilarity matrix model ignores second- and higher-order relationships

— thus, if objects i and j are close together, one would expect d_{ik} and d_{jk} to have similar ranks for each object k.

Many tests have been proposed of the complete absence of class structure (e.g., Ling (1975), Smith & Jain (1984), Bock (1985), Dubes & Zeng (1987), Hartigan & Mohanty (1992)). Logically, such tests should precede a cluster analysis (although some of them use the results of the clustering), but they have rarely been carried out, possibly because cluster analyses are generally undertaken only when there is a strong presumption that class structure exists.

Jain and Dubes (1988,Chapter 4) categorized validation tests by whether they are concerned with assessing (1) hierarchical classifications, (2) partitions, or (3) individual clusters. Three different types of validation study can be distinguished, involving use of external, internal, or relative indices. External indices compare (an aspect of) a classification with information that was not used in the analysis of the data. Such external information is rarely available in clustering studies, and external indices are not considered further in this account; see Hubert (1987) for relevant methodology. Internal indices are based on the results of the cluster analysis, generally comparing these with the original data. Relative indices compare different classifications of (parts of) the same data set.

(1) Hierarchical classifications : An internal index for assessing a hierarchical classification measures the resemblance between the set of input dissimilarities (d_{ij}) and the set of output ultrametric distances (h_{ij}). One example of such an index is the Goodman-Kruskal γ (Hubert (1974)), defined as $(S_+ - S_-)/(S_+ + S_-)$, where S_+ (resp., S_-) denotes the number of subscript combinations (i, j, k, l) for which $g_{ijkl} \equiv (d_{ij} - d_{kl})(h_{ij} - h_{kl})$ is strictly positive (resp., strictly negative). When $\gamma = 1$, the correspondence is perfect, but it is difficult to assess the import of moderate-sized values. The problem is that even if a null hypothesis of the absence (or only partial presence) of class structure is rejected (this latter hypothesis requiring various decisions to be made before it can be formulated), it does not follow that the complete hierarchy is validated. Further, relative indices that attempt to establish if the hierarchical classification provided by one clustering criterion is superior to that provided by a different clustering criterion encounter the problem that different relative indices may suggest different conclusions : for example, if the relative index is defined to be the correlation between the sets of dissimilarities and ultrametric distances, the hierarchical classification that is deemed optimal is usually the one provided by the group average link criterion (Sokal & Rohlf (1962), Sneath (1969)). In fact, it seems preferable to assess separately the individual clusters and partitions contained in a hierarchical classification.

(2) Partitions : Two questions which are relevant when assessing the results of a partitioning algorithm are as follows.

1. If the set of objects has been partitioned into a specified number (c, say) of classes, does this partition comprise a set of homogeneous, well-separated classes?

2. How many classes are there in the data?

The first question can be addressed by defining a measure of the variability of a partition (e.g., the total within-class sum-of-squares), and assessing whether or not the observed value of the measure is unusually low by comparison with the values to be expected under a null model of the absence of class structure. An appropriate null distribution must be considered: the classes resulting from the application of a clustering algorithm to data are more homogeneous than would be expected of classes belonging to a randomly-generated partition; the relevant distribution is obtained from partitions of randomly-generated data provided by the same clustering procedure. Asymptotic results have been derived for some measures of variability (e.g., Bock (1985)), but their accuracy for finite samples is uncertain, and it has been more common for Monte Carlo tests of significance to be carried out (e.g., Milligan & Sokol (1980)). As indicated earlier, such tests require decisions to be made in specifying relevant null models for the data; it is also unclear whether or not features such as the number of objects in each class should also be taken into account in deriving the null distribution of the measure of variability.

Tests for the existence of c clusters based on such 'internal' indices encounter the problem of the inter-relatedness of class structure at different levels : thus, a partition of the data into c classes might be deemed valid when in fact there are $(c-1)$ or $(c+2)$ classes in the data set. The second question relates to the identification of the most appropriate value of c, and is addressed using relative indices. Uncertainty about the value of c has often led investigators to obtain a complete hierarchical classification of their data, usually using an agglomerative clustering algorithm, and the procedures used to decide on the number of classes have often been referred to as 'stopping rules', as they indicate when amalgamations should cease and the current partition be chosen. Some of the rules are also applicable to sets of non-nested partitions, obtained from the results of using a partitioning algorithm for different values of c; others are based solely on comparing the two classes involved in the latest amalgamation, and so cannot be used for such comparisons.

Many different stopping rules have been proposed in the research literature, but there have been few comparative studies of their performance. The most detailed study of which the author is aware was reported by Milligan & Cooper (1985). These authors examined how thirty stopping rules performed on the results of analysing artificially-generated data by four standard agglomerative clustering algorithms. The data comprised 50 objects described by 4, 6 or 8 numeric variables, assigned to between 2 and 5 well-separated classes, whose sizes varied. The success rates of the rules were accumulated over the four clustering criteria as well as over the different data sets, whereas it might be more relevant to use different stopping rules with different clustering criteria. The main value of Milligan & Cooper's (1985) study is in identifying inferior stopping rules, whose use cannot be recommended since they are unable to detect clear-cut clusters. Although further studies based on other types of data are unlikely to yield the same ranking of the stopping rules, it seems reasonable — until provided with further evidence — to place some confidence in the indices that were rated highly in their study.

Alternative approaches to validating partitions have been based on the idea of the stability or consistency of the results under replication, and have included linking the results of two separate analyses of half of the data (McIntyre & Blashfield (1980), Breckenridge (1989,1990)), and comparing analyses of bootstrap samples of the data (Jain & Moreau (1987))

(3) Individual clusters : It would rarely seem relevant to seek relative indices for investigating whether or not one cluster is more strongly supported than another, but various internal indices have been used to assess the validity of individual clusters. The probability of obtaining a value of the index that is as extreme as the observed one has been derived, bounded or simulated under the null hypothesis of a random dissimilarity matrix (Ling (1973), Bailey & Dubes (1982)) or a random pattern matrix (Gordon (1994a)). As before, care must be taken to specify an appropriate null distribution for the cluster statistic; further, significance levels cannot be interpreted strictly if several different clusters in the same classification are assessed.

References

Bailey, T.A. & Dubes, R.C. (1982). Cluster validity profiles. Pattern Recognition 15, 61-83.

Ball, G.H. & Hall, D.J. (1967). A clustering technique for summarizing multivariate data. Behavioral Science 12, 153-155.

Banfield, C.F. & Bassill, L.C. (1977). Algorithm AS113. A transfer algorithm for non-hierarchical classification. Applied Statistics 26, 206-210.

Beale, E.M.L. (1969). Euclidean cluster analysis. Bulletin of the International Statistical Institute 43(2), 92-94.

Belbin, L. (1987). The use of non-hierarchical allocation methods for clustering large sets of data. Australian Computer Journal 19, 32-41.

Benzécri, J.-P. (1982). Construction d'une classification ascendante hiérarchique par la recherche en chaïne des voisins réciproques. Les Cahiers de l'Analyse des Données 7, 209-218.

Bezdek, J.C. (1981). Pattern Recognition with Fuzzy Objective Function Algorithms. Plenum Press, New York.

Bezdek, J.C. (1987). Some non-standard clustering algorithms. In: Developments in Numerical Ecology (Eds. P. Legendre and L. Legendre), 225-287. Springer-Verlag, Berlin.

Bock, H.H. (1985). On some significance tests in cluster analysis. Journal of Classification 2, 77-108.

Bock, H.H. (1989). Probabilistic aspects in cluster analysis. In: Conceptual and Numerical Analysis of Data (Ed. O. Opitz), 12-44. Springer-Verlag, Berlin.

Breckenridge, J.N. (1989). Replicating cluster analysis : Method, consistency, and validity. Multivariate Behavioral Research 24, 147-161.

Breckenridge, J.N. (1990). Replicated cluster analyses indicate recovery of true structure. Paper presented at CSNA Annual Meeting, Logan, Utah, June 1990.

Brucker, P. (1978). On the complexity of clustering problems. In: Optimization and Operations Research (Eds. R. Henn, B. Korte and W. Oettli), 45-54. Springer-Verlag, Berlin.

Bruynooghe, M. (1978). Classification ascendante hiérarchique des grands ensembles de données : Un algorithme rapide fondé sur la construction des voisinages réductibles. Les Cahiers de l'Analyse des Données 3, 7-33.

Carroll, J.D. & Pruzansky, S. (1980). Discrete and hybrid scaling models. In: Similarity and Choice (Eds. E.D. Lantermann and H. Feger), 108-139. Huber, Bern.

Chandon, J.L., Lemaire, J. & Pouget, J. (1980). Construction de l'ultramétrique la plus proche d'une dissimilarité au sens des moindres carrés. Recherche Operationnelle/Operations Research 14, 157-170.

Day, W.H.E. (1994). Complexity theory : An introduction for practitioners of classification. To appear in: Clustering and Classification (Eds. P. Arabie, L.J. Hubert and G. De Soete). World Scientific, Singapore.

Day, W.H.E. & Edelsbrunner, H. (1984). Efficient algorithms for agglomerative hierarchical clustering methods. Journal of Classification 1, 7-24.

De Soete, G. (1984). A least squares algorithm for fitting an ultrametric tree to a dissimilarity matrix. Pattern Recognition Letters 2, 133-137.

Diday, E. & Govaert, G. (1977). Classification automatique avec distances adaptatives. R.A.I.R.O Informatique/Computer Sciences 11, 329-349.

Diehr, G. (1985). Evaluation of a branch and bound algorithm for clustering. SIAM Journal on Scientific and Statistical Computing 6, 268-284.

Dubes, R.C. & Zeng, G. (1987). A test for spatial homogeneity in cluster analysis. Journal of Classification 4, 33-56.

Friedman, H.P. & Rubin, J. (1967). On some invariant criteria for grouping data. Journal of American Statistical Association 62, 1159-1178.

Gordon, A.D. (1981). Classification : Methods for the Exploratory Analysis of Multivariate Data. Chapman and Hall, London.

Gordon, A.D. (1986). Links between clustering and assignment procedures. In: COMPSTAT 1986 (Eds. F. De Antoni, N. Lauro and A. Rizzi), 149-156. Physica-Verlag, Heidelberg.

Gordon, A.D. (1994a). Identifying genuine clusters in a classification. Computational Statistics & Data Analysis 18 (in press)

Gordon, A.D. (1994b). Hierarchical classification. To appear in: Clustering and Classification (Eds. P. Arabie, L.J. Hubert and G. De Soete). World Scientific, Singapore.

Guénoche, A., Hansen, P. & Jaumard, B. (1991). Efficient algorithms for divisive hierarchical clustering with the diameter criterion. Journal of Classification 8, 5-30.

Hansen, P. & Jaumard, B. (1987). Minimum sum of diameters clustering. Journal of Classification 4, 215-226.

Hartigan, J.A. (1967). Representation of similarity matrices by trees. Journal of American Statistical Association 62, 1140-1158.

Hartigan, J.A. (1975). Clustering Algorithms. Wiley, New York.

Hartigan, J.A. & Mohanty, S. (1992). The runt test for multimodality. Journal of Classification 9, 63-70.

Hartigan, J.A. & Wong, M.A. (1979). Algorithm AS 136. A k-means clustering algorithm. Applied Statistics 28, 100-108.

Hubert, L. (1974). Approximate evaluation techniques for the single-link and complete-link hierarchical clustering procedures. Journal of American Statistical Association 69, 698-704.

Hubert, L.J. (1987). Assignment Methods in Combinatorial Data Analysis Marcel Dekker, New York.

Hwang, K. & Briggs, F.A. (1984). Computer Architecture and Parallel Processing. McGraw-Hill, New York.

Ismail, M.A. & Kamel, M.S. (1989). Multidimensional data clustering utilizing hybrid search strategies. Pattern Recognition 22, 75-89.

Jain, A.K. & Dubes, R.C. (1988). Algorithms for Clustering Data. Prentice Hall, Englewood Cliffs, NJ.

Jain, A.K. & Moreau, J.V. (1987). Bootstrap techniques in cluster analysis. Pattern Recognition 20, 547-568.

Jambu, M. (1978). Classification Automatique pour l'Analyse des Données, Tome 1. Dunod, Paris.

Jensen, R.E. (1969). A dynamic programming algorithm for cluster analysis. Operations Research 17, 1034-1057.

Johnson, S.C. (1967). Hierarchical clustering schemes. Psychometrika 32, 241-254.

Kennard, R.W. & Stone, L.A. (1969). Computer aided design of experiments. Technometrics 11, 137-148.

Kirkpatrick, S., Gelatt, C.D.,Jr. & Vecchi, M.P. (1983). Optimization by simulated annealing. Science 220, 671-680.

Klein, G. & Aronson, J.E. (1991). Optimal clustering : A model and method. Naval Research Logistics 38, 447-461.

Klein, R.W. & Dubes, R.C. (1989). Experiments in projection and clustering by simulated annealing. Pattern Recognition 22, 213-220.

Koontz, W.L.G., Narendra, P.M. & Fukunaga, K. (1975). A branch and bound clustering algorithm. IEEE Transactions on Computers C-24, 908-915.

Krivánek, M. (1986). On the computational complexity of clustering. In: Data Analysis and Informatics 4 (Eds. E.Diday, Y. Escoufier, L. Lebart, J. Pagès, Y. Schektman and R. Tomassone), 89-96. North-Holland, Amsterdam.

Krivánek, M. & Morávek, J. (1986). NP-hard problems in hierarchical-tree clustering. Acta Informatica 23, 311-323.

Lance, G.N. & Williams, W.T. (1967). A general theory of classificatory sorting strategies I. Hierarchical systems. Computer Journal 9, 373-380.

Li, X. (1990). Parallel algorithms for hierarchical clustering and cluster validity. IEEE Transactions on Pattern Analysis and Machine Intelligence 12, 1088-1092.

Li, X. & Fang, Z. (1989). Parallel clustering algorithms. Parallel Computing 11, 275-290.

Ling, R.F. (1973). A probability theory for cluster analysis. Journal of American Statistical Association 68, 159-164.

McIntyre, R.M. & Blashfield, R.K. (1980). A nearest-centroid technique for evaluating the minimum-variance clustering procedure. Multivariate Behavioral Research 15, 225-238.

MacQueen, J. (1967). Some methods for classification and analysis of multivariate observations. Proceedings of 5th Berkeley Symposium 1, 281-297.

Massart, D.L., Plastria, F. & Kaufman, L. (1983). Non-hierarchical clustering with MASLOC. Pattern Recognition 16, 507-516.

Milligan, G.W. & Cooper, M.C. (1985). An examination of procedures for determining the number of clusters in a data set. Psychometrika 50, 159-179.

Milligan, G.W. & Sokol, L.M. (1980). A two-stage clustering algorithm with robust recovery characteristics. Educational and Psychological Measurement 40, 755-759.

Mulvey, J.M. & Crowder, H.P. (1979). Cluster analysis : An application of Lagrangian relaxation. Management Science 25, 329-340.

Murtagh, F. (1983). A survey of recent advances in hierarchical clustering algorithms. Computer Journal 26, 354-359.

Murtagh, F. (1984). Complexities of hierarchic clustering algorithms : State of the art. Computational Statistics Quarterly 1, 101-113.

Murtagh, F. (1994). Neural networks for clustering. To appear in Clustering and Classification (Eds. P. Arabie, L.J. Hubert and G. De Soete). World Scientific, Singapore.

Ni, L.M. & Jain, A.K. (1985). A VLSI systolic architecture for pattern clustering. IEEE Transactions on Pattern Analysis and Machine Intelligence PAMI-7, 80-89.

Nicolau, F.C. & Brito, M.P. (1989). Improvements in NHMEAN method. In Data Analysis, Learning Symbolic and Numeric Knowledge (Ed. E. Diday), 109-116. Nova Science, Commack, NY.

Perruchet, C. (1983). Une analyse bibliographique des épreuves de classifiabilité en analyse des données. Statistiques et Analyse de Données 8, 18-41.

Pintér, J. & Pesti, G. (1991). Set partition by globally optimized cluster seed points. European Journal of Operational Research 51, 127-135.

Podani, J. (1989). New combinatorial clustering methods. Vegetatio 81, 61-77.

Rao, M.R. (1971). Cluster analysis and mathematical programming. Journal of American Statistical Association 66, 622-626.

Rasmussen, E.M. & Willett, P. (1989). Efficiency of hierarchic agglomerative clustering using the ICL distributed array processor. Journal of Documentation 45, 1-24.

Rohlf, F.J. (1982). Single-link clustering algorithms. In Handbook of Statistics Volume 2. Classification, Pattern Recognition and Reduction of Dimensionality (Eds. P.R. Krishnaiah and L.N. Kanal), 267-284. North-Holland, Amsterdam.

Scott, A.J. & Symons, M.J. (1971). Clustering methods based on likelihood ratio criteria. Biometrics 27, 387-397.

Selim, S.Z. & Asultan, K. (1991). A simulated annealing algorithm for the clustering problem. Pattern Recognition 24, 1003-1008.

Smith, S.P. & Jain, A.K. (1984). Testing for uniformity in multidimensional data. IEEE Transactions on Pattern Analysis and Machine Intelligence PAMI-6, 73-81.

Sneath, P.H.A. (1969). Evaluation of clustering methods (with discussion). In Numerical Taxonomy (Ed. A.J. Cole), 257-271. Academic Press, London.

Sokal, R.R. & Rohlf, F.J. (1962). The comparison of dendrograms by objective methods. Taxon 11, 33-40.

Späth, H. (1976). Algorithm 30. L1 cluster analysis. Computing 16, 379-387.

Welch, W.J. (1982). Algorithmic complexity : Three NP-hard problems in computational statistics. Journal of Statistical Computation and Simulation 15, 17-25.

Whaley, R. & Hodes, L. (1991). Clustering a large number of compounds. 2. Using the Connection machine. Journal of Chemical Information and Computer Science 31, 345-347.

Wishart, D. (1987). Clustan User Manual (4th Edition). Computing Laboratory, University of St Andrews.

Zhang, Q. & Boyle, R.D. (1991). A new clustering algorithm with multiple runs of iterative procedures. Pattern Recognition 24, 835-848.

Learning Statistics: Beyond Authoring Systems

Rolf Schulmeister
Interdisziplinäres Zentrum für Hochschuldidaktik,
Universität Hamburg, Sedanstr.19, D — 20148 Hamburg

Key Words:
authoring systems, learning, statistics anxiety, discovery learning

Abstract
Results of an experiment about teaching statistics at the university Hamburg following the concept of discovery learning are reported. This live experiment has since then been transformed into a computer assisted program for learning statistics which does not use the technology of authoring systems but utilizes a hypertext approach instead and attempts to design interactive exercises which follow the didactical concept of discovery learning.

Introduction

When I first learned statistics it was offered in lectures, sometimes brilliant lectures, some of them well structured lectures, but lectures. It was obvious that the learning success of the students was directly correlated with the transmission method chosen. Moreover students in psychology demonstrated an open disinclination to learn statistics, a motivational construct which after several evaluations was clearly identified as something similar to number anxiety and which from then on we used to call "statistics anxiety" (Schulmeister (1983)). Others were driven by an ideal conception of psychology as therapy and were as unlikely as the others to feel empathy for statistics. Thus only about 20% of the students really learned statistics by sitting in lectures, another 40% had to prepare themselves at home or underwent extreme drill in study groups. About 40% of them never made it.

This situation was our reason to change the setting[2]: We abolished the lectures, introduced small group teaching in seminars, gave them written exercises intended to stimulate problem solving, and tried to handle the feedback sessions as good as we could. We evaluated learning processes, made surveys, investigated loud thinking (Schulmeister & Birkhan (1983), Birkhan & Schulmeister (1983)), and learned how to improve the teaching material (Bogun et.al. (1983)). As the exercises improved, so did the students' achievement: About 60% of them learned the basics in the classroom, 20% still the hard way. Only 20% never made it or dropped out early.

What we called "statistics anxiety" was a conglomeration of several related factors:

- lack of faith in empirical methods
- number anxiety
- disinclination against systematical activity
- discrepancy between a humanistic motivation and the formal character of methods.

The following figure illustrates the interaction of these constructs:

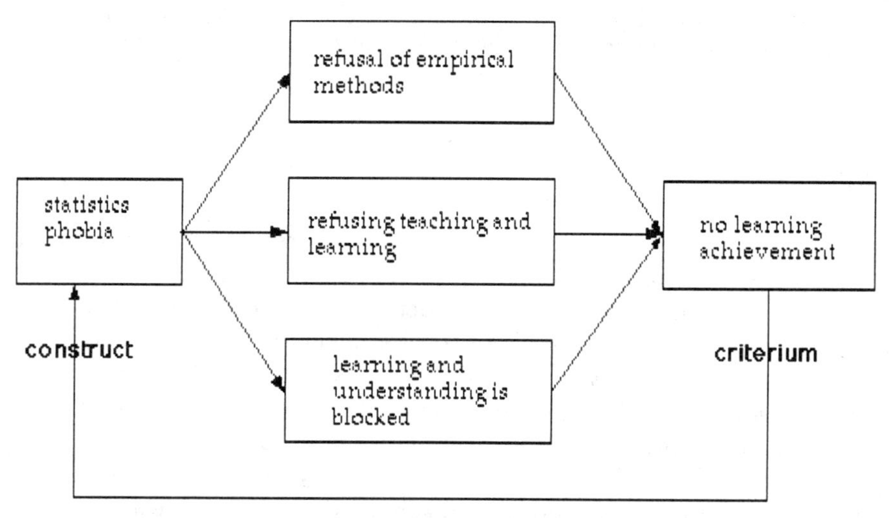

Figure 1: statictics phobia

[2]The research project "E.L.M.A." (Evaluation von Lernprozessen in der Methoden-Ausbildung) was sponsored by a grant from the German Research Foundation (DFG). Its aim was to investigate cognitive problems of students learning statistics.

Our hypothesis: Teaching and learning using another didactical concept might influence the achievement of the students. The location for an intervention is illustrated in Figure 2:

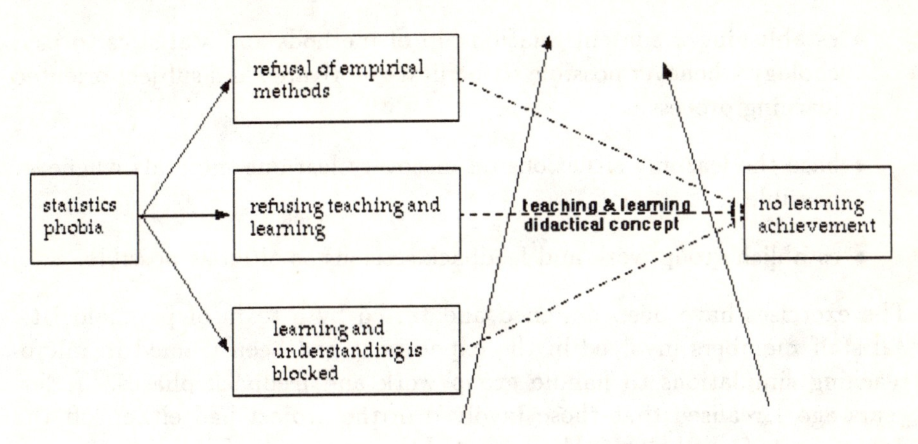

Figure 2: didactic intervention

The following two tables illustrate the changes in the distribution of the constructs "statistic phobia" and "criticism of statistics" after three terms of learning statistics:

t	disclination	neutral	no desclination
1.term	73%	12%	15%
2.term	25%	48%	27%
3.term	31%	26%	53%

Table 1: Statistics phobia (N = 123)

t	criticism	neutral	no criticism
1.term	66%	6%	8%
2.term	31%	42%	27%
3.term	34%	13%	47%

Table 2: Criticism towards statistics (N = 123)

The original critical attitude passed a temporary indifferent phase and reached a bimodal stage. The correlation between emotional disclination and critical attitude towards statistics as research method decreased from $r = 0.71$ to $r = 0.39$ within three terms.

The didactical concept used in E.L.M.A. stressed four principles:

- establishing a content relationship of methods and statistics to psychology whenever possible to facilitate motivated and subject oriented learning processes
- base the learning situations on discovery learning methods whenever possible
- establish group work and feedback sessions as often as possible.

The exercises have been designed and tested by a team of psychologists. All staff members involved in the experiment had been trained in micro-teaching simulations to handle group work and feedback phases. A few years ago I realised that those involved in the project had either left the department of psychology or had devoted themselves to other teaching areas. The knowledge of the concept and the experience gained in the trainings had been lost. I had the choice of starting all over again or try another approach this time.

This is the reason why two years ago I started to design the paper-and-pencil exercises developed in E.L.M.A. on a computcr. For selecting an easy programmable development tool I principally had two alternatives: An authoring system or a hypertext environment.

Authoring systems are nowadays often picked when people decide to develop a computer assisted learning application. It seems to be an obvious choice because the authoring system provides the author with a number of practical tools thus facilitating the task of structuring and programming the tests, the sequences, and the evaluation of the learning process. It might even be an even more natural choice for people in medical education in Germany where the central examinations are presented as multiple-choice tests.

Hypertext systems, however, present an open environment in which the author himself/herself has to find his or her own concept, provide orientation for students, take care for the navigation of the students, give feedback to the students, and find a way to evaluate of the learning achievement etc. The hypertext features of the system have not been used in this case. The environment of the hypertext system was chosen because of its openness.

"LearnStat" was developed exclusively with HyperCard version 2.1 on the Apple Macintosh.The programming environment is responsible as well for the advantages (rapid prototyping, design flexibility, comfortable browsing facilities, easily manageable multimedia features, and external commands functionality) as well as for the disadvantages (no spreadsheet windows, no fields with multiple cells, no records or arrays in the programming language HyperTalk). HyperCard is an objectoriented programming environment making available for the developer quite a number of sophisticated and efficient program structures and features: It saves its files as piles of database cards, so-called "stacks", offers hypertext features, an easy iconic form of navigation, message parsing etc. But before I start to explain what "LearnStat" is — I would like to include an explanation about authoring systems.

Authoring Systems

The relatively low degree of difficulty in using authoring systems is a resulting function of the fact that the lecturer does not have to learn a programming language — at least for using a basic version of an authoring system. This of course only makes sense (or is possible) if the software produces applications according to a very strict regular procedure and always generates the same structures. The result of this form of easiness is a simplicity (more powerful authoring systems therefore use some kind of programming language of their own). The basic concept of authoring systems is derived from the behavioristic model of learning, which in the earlier times of computing was incorporated in the learning machines of programmed instruction (cf. Schulmeister (1981); concerning the decline of programmed instruction in Germany cf. Gorny (1982), 346ff.). Lecturers who generally are not experts in computing expect that authoring systems enable easy prototyping. Their expectation obviously is not directed towards a software which is original, presents a diversity of methods, and allows creative didactical concepts for teaching. They tend to neglect that the consequence of easy prototyping yields a restricted and schematic product for the learner.

The behavioristic concept of authoring systems forces the author to divide the subject matter into small steps and to form out of these small pieces a learning sequence which does not confront the learner with gaps. The author has to be careful to provide the right clues and feed-back mechanisms (reinforcement) to keep up the motivation of the student. These didactic aspects are non-trivial, but they are usually not known to the teacher who decides to use authoring systems. The amount of time needed to produce a consistent learning application is relatively high. Ottmann (1987, p.65) reports "that for a lesson of one hour about 80 to 120 hours of programming time have to be invested". This estimatation I believe is a lower limit, and regarding the shortage of personnel and auxiliary help and their poor training in didactics, his suggestion "that a greater part of this time [...] might be spent by auxiliary personnel" does not seem to be very acceptable to me.

Let's first have a look into authoring systems (for a more complete description and criticism of authoring systems cf. Schulmeister (1989)). They typically consist of two programs:

A. the authoring tool used to define tasks and questions and the evaluation, the program which automatically generates the learning application (B).

B. the learning application used by the student in an individual learning situation which introduces the student to a certain subject matter, presents multiple choice questions-and-answers, and gives some sort of feedback to the learners' responses.

A. The Authoring Part of Authoring Systems

Authoring systems demand from the lecturer certain preparations and requirements:

1. Description of a task (statement or question or movement)

2. Definition of possible correct or false responses (reactions) to the task (1)

3. Reactions of the application in regard to the learner's responses (evaluation of input, branching if the answer is correct, repetition if the answer is wrong, help, hints).

In judging the quality of authoring systems it is essential to know which modalities are provided by the system with respect to selection mechanisms, question-and-answer types, types of evaluation (analysis of answers), and forms of branching (see Fankhänel et.al. (1988)). Selection mechanisms may be serial, random or mixed forms. Question-and-answer types may be the following:

Yes/No (true/false)

Multiple choice

Fill-in forms

Full text answer and derivatives of these forms.

In some cases the names of authoring programs suggest what they deal with: e.g. Storyboard, Choicemaster, Testmaster, Gapmaster, Matchmaster, Pinpoint, Wordstore, and Vocab are modules of the computer aided language learning package CALL (Eurocentres Learning Service). Well known and far spread packages in this field are PLATO (Control Data), CALOP (Utrecht), TOPIC (ECOL).

A critical point with authoring programs is the evaluation of full text answers. Full text input generally is only evaluated according to sameness or synonyms. A semantic analysis of texts is not possible, only a tolerance with respect to orthography. This is a very decisive aspect for the methodological and learning theoretical evaluation of authoring systems. Because learning progress in these systems is dependent from correct responses of the learner, it really does not matter at all how 'socratic' or liberal questions are put to the student — the evaluation of the answer has to follow the restricted evaluation rules.

B. The Student Part of Authoring Systems

The exercise appearing on the screen of the student consists mainly of a sequence of consecutively arranged inputs, questions, and a number of prearranged answers, out of which the student might choose one or more. Then the predetermined mechanisms are activated: e.g. reinforcement of a correct response, presentation of the next question, or repeating the last question, maybe paired with a comment.

The original behaviorism (i.e. Pawlow's classical conditioning) stated that certain stimuli (=S) were followed by certain reactions (Response=R), and that these S-R-connections could be chained (paired associations) and habitualised, especially, if desired (correct) reactions are reinforced, while undesired reactions do not receive feed-back and are consequently tilt. Skinner (1953) assumes that behavior is not only reactive but also appears spontaneous. Such behavior he calls "operant". Operant behavior according to his theory appears more frequently if it is reinforced. Skinner (1954, 1958) has developed a behavioristic learning model which as "Programmed Instruction" in the 60s and 70s seemed to meet the demands of schools and universities. Computer Assisted Instruction (CAI) basically is derived from this model.

The description of a task has to be defined precisely in order to prepare the answer. But task and question should not be suggestive else the question would be superfluous. The question has to be formulated exactly without being trivial. The cognitive steps demanded from the learner should not be too big, otherwise gaps would develop during the learning process. Thus the learning stuff has to be fragmented into small pieces and continuously is endangered to be suggestive and trivial. This especially is the case if predefined answers are arranged in multiple-choice manner. The result very often is some kind of "Taylorization" of contents and learning processes. Applications are not adjustable to natural learning processes. Learning is cut into bits and pieces which have to be learned in strict sequential order. Students very often have a distinct feeling to be confined in a very narrow costume, because loose thinking, associative ideas, or hypothetical inferences and modelling is not part of the hidden curriculum of authoring systems.

Having this characterization of authoring systems in mind, I do not hesitate to criticize Ottmann's (1987, p.64) statement: "One should not mistake CAI with Programmmed Instruction. Poeple thinking of using a computer for teaching at a university do not intend to use the computer machine for drill exercises. They indeed hope to convey subject knowledge with the help of a computer as a thinking laboratory which can only be taught with a computer or which at least can be taught by experiencing more understandably." Ottmann does not see that the intention of a teacher using authorware is irrelevant in this case where the structure of the program dictates the result. Underneath the attractive interface resides the behavioristic principle of taylorizing the subject matter and forcing everything through the narrow road of a right-wrong answer testing mechanism (cf. Streitz (1985, 57)). Anyway, the efficiency of these learning systems is not as great as expected (overview in Kulik et al. (1980)): "Interestingly the CAI is known to have only short-time effects which moreover seem to be overlaid by the Hawthorne effect of experiencing something 'new': After using CAI for a longer time teachers complain about boredom and lessening student motivation." (Fischer (1985, 69)).

Studies pointing to evidence about different learner types (e.g. Goldman (1972); Pask & Scott (1972)) might be able to explain why after some time of curiosity a certain unwillingness originates to use more of these programs: This type of learning might be suitable for students who predominantly learn in a sequential manner, but they are totally unable to motivate students who apply holistic thinking and cognitive problem solving. Moreover the cognitive processes are embedded in motivational conditions and personality factors:

Rheinberg (1985, 98) states the hypothesis that a possible reason for the attraction computers present to youth might be the fact that their achievement motivation is assisted. On the other hand also students motivated by fear of failure are not terrified by computers, while applications which exercise strict control and guidance are not accepted by both groups. The explanation might be that these programs do not accentuate the achievement motif, and that by fear of failure motivated students do not feel free of control and judgement. Rheinberg who did a study on youth and motivation understands this result not as a fundamental objection against authoring systems. He proposes a number of techniques to restructure programs in such a way that students might be better motivated (99ff.). This reminds me very much of psychological techniques of manipulation. I would not suggest to enhance the functionality this way but to restrict authoring programs to the purpose they were invented for: memorising, drill – tasks which demand a lot of facts to be learned by heart (cf. Computer Board (1983, p.11)).

Recently developers of authoring systems try to refine the rather simple structure by incorporating routines for simulations and problem solving (cf. Ottmann (1987, p.64)). They even boast to develop authoring systems on top of expert shells which claim to have a natural language interface, provide tutoring, more sophisticated evaluation instruments, and intelligent branching methods. A quick overview over expert systems teaches us that this is not the case. And even a very pragmatic implementation of some of these methods raises the development time enormously (cf. Anderson (1984)). Anyway, insofar authoring systems still stick to testing the responses of the learners, even more elaborate mechanisms of routing the program flow cannot change their behavioristic bias.

Principially, authoring systems may only then overcome their built-in limitations if they are successfully incorporated into "intelligent" environments which have a knowledge base and tutorial functions, and which moreover permit a "natural language" access and dialogue (from a psychological point of view cf. Spada & Opwis (1985, 14/15); examples p. 15). But then they will no longer be authoring systems.

The Application LearnStat

In order to avoid misconceptions, I would like to stress the following statement: "LearnStat" is not 'just another' statistics program, it is **not** a statistics program **at all**. In writing "LearnStat" I did not intend to compete with commercial statistics software like SPSS or SYSTAT. What then is "LearnStat"? "LearnStat" is a program for **teaching and learning** statistics in psychology, social sciences, medicine and education.

"LearnStat"[3] covers the whole range of descriptive statistics at the moment (further plans include probability and hypotheses testing, inference statistics): population, distribution, arithmetic mean, variance, correlations, factor analysis and regression. For each of these topics "LearnStat" offers a variety of interactive exercises and an online help.

The exercises in "LearnStat" are in many cases not comparable to traditional statistics exercises, but represent specific didactic material for tutorial

[3] The program is made available by the author as public domain.

experiments in the classroom. Most exercises have been designed to stimulate the understanding of statistical concepts from the grass roots by way of discovery learning, a concept developed by Bruner (1964) in accordance with Piaget's theory of cognitive development.

Some exercises had been designed as pencil-and-paper tests before. On a computer these exercises may be repeated easily and as often as you like. In most cases the computer calculates the formula or functions so that the student saves time and does not have to repeat calculations when varying the data. "LearnStat" may be used in the classroom, but students may also use it for self-paced studies at home.

Components of "LearnStat"

Figure 3: The navigator palette

The navigator palette is present throughout the application. Buttons in the palette enable the user to jump to the next exercise, to open the help file with explanations for the actual exercise or definitions of the inherent statistical concepts, and to open a formulary enabling the user to trace statistical formulae.

The help file is available throughout the program. It consists of a stack with a hierarchically structured index and explanations of statistical concepts, covering almost a complete statistics manual. The help window may be kept open all the time. If the user selects the help button, the help window opens exactly on that page which contains information related to the actual exercise.

The Formulary lists formulae and offers the user a chance to fill in rows of data and let the program calculate the actual formula with the actual data. In a second window appears a tracer (similar to the "observe window" used in Pascal or C compilers) which dissects the formula and calculates the data for the different components of the formula.

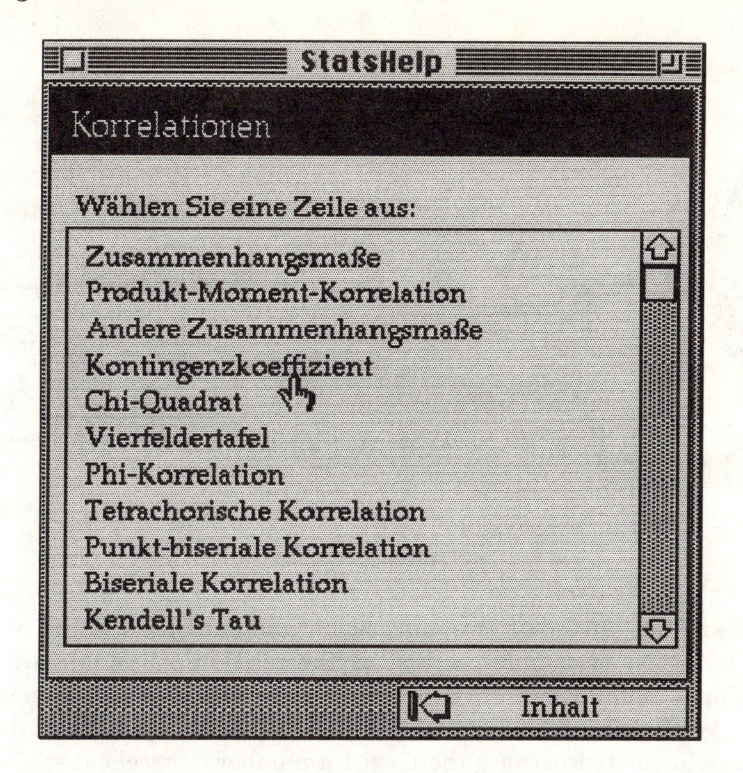

Figure 4: The help file

The protocol: "LearnStat" registers the actions of the user in a kind of protocol, it writes information about the topics, the exercises, and the results obtained in the exercises into a file. Students may look up their protocol, may copy it for a lecturer to evaluate it or they may decide to delete it. The purpose of protocolling is not to control the students, but to give them a chance of receiving expert advice and guidance.

The exercises attempt to implement certain didactic principles essential for motivated cognitive learning: discovery learning, visualisation, animation and learning by doing.

Discovery learning: Some exercises in "LearnStat" present the challange to discover statistical concepts independently through puzzles, riddles, and cognitive dissonances, starting from the students' previously acquired knowledge and giving them a chance to discover the next step on their own. It is not demanded that the solution is a complete one or that it is very precise, but it is intended that students grasp a concept, explore an explanation and thus train their problem-solving behaviour.

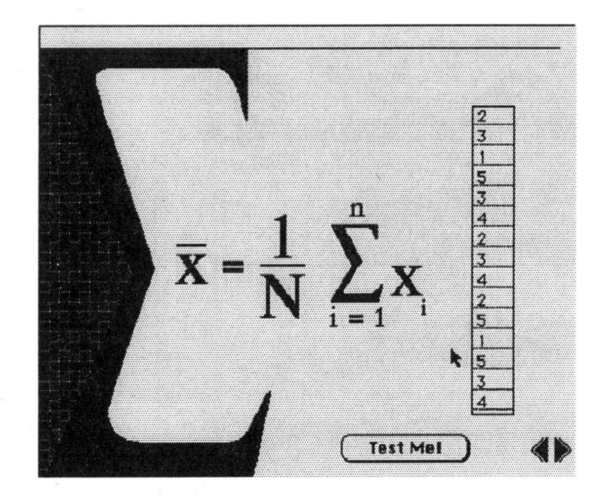

Figure 5: Formulary and formula tracer

Visualisation: Different two-dimensional methods are used to visualize the distribution of samples as well as the relations of variables. Three-dimensional methods are used to explore the concept of multivariate relations in factor analysis. The methods of visualisation are technically crude and unsophisticated because there exist a number of excellent applications exactly for that purpose like SYSTAT, JMP, DataDesk. Sample sets of data are visualised with the help of MacSpin and incorporated in "LearnStats" as screen movies.

Animated data illustrate the functional mechanisms of formulae by successively highlighting the data used for calculating the formulae, and the effect of relations is demonstrated by animating the spread of different parameters if students vary the data. The idea of factor analysis is demonstrated by varying the relations and subsets in a rotating three-dimensional scatterplot.

Learning by doing: From a cognitive perspective, an exercise has to offer a chance to transform an inactive and experiential mode of learning into an active acquisition and a more formal, abstract understanding of concepts. Nearly all exercises in "LearnStat" offer students a chance to manipulate data and relations of data rows and to repeat this process as often as necessary. Students may observe the results of their manipulations while forming and testing their own cognitive hypotheses.

Complex concepts are split up into series of steps, e.g. the calculation of factor loadings has been designed as a stepwise matrix calculation prompted by 13 buttons.

Some Examples of Exercises from "LearnStat"

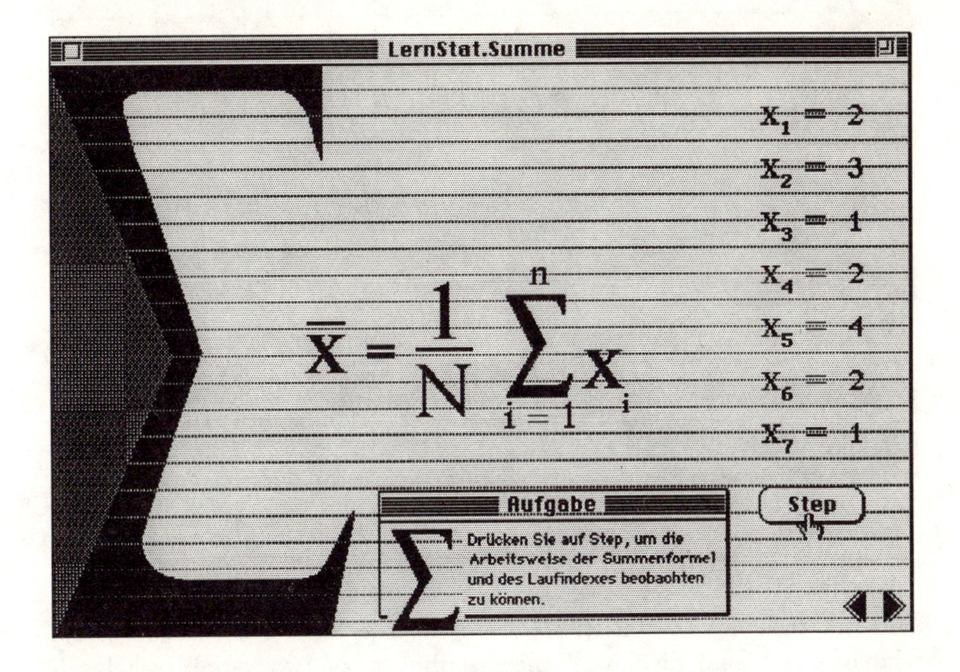

Figure 6: Sigma

Many scientists involved in teaching statistics believe the sum to be the simplest element of statistics and tend to neglect this part of statistics, but the sum is the major initial source of cognitive misconceptions for students in psychology. Although from a mathematical point of view rather simple, students show difficulties to solve the formula correctly especially when they encounter the sum enclosed in more complex formulae. Sigma is the basic form of statistics, recurring in nearly every other formula. That is the reason why "LearnStat" presents several animated exercises demonstrating that the sum is a calculation mechanism.

The concept of dispersion in statistics needs to be motivated in a special way. The concepts of central values or parameters are immediately understood by the students, the concept of dispersion is not. Variance cannot be read by following the scale as is the case with standard deviation, and it is difficult to understand why it is squared. The natural intuitive cognitive concept is a "mean deviation". Students start with an exercise which asks them to develop a descriptive concept for a deviation problem.

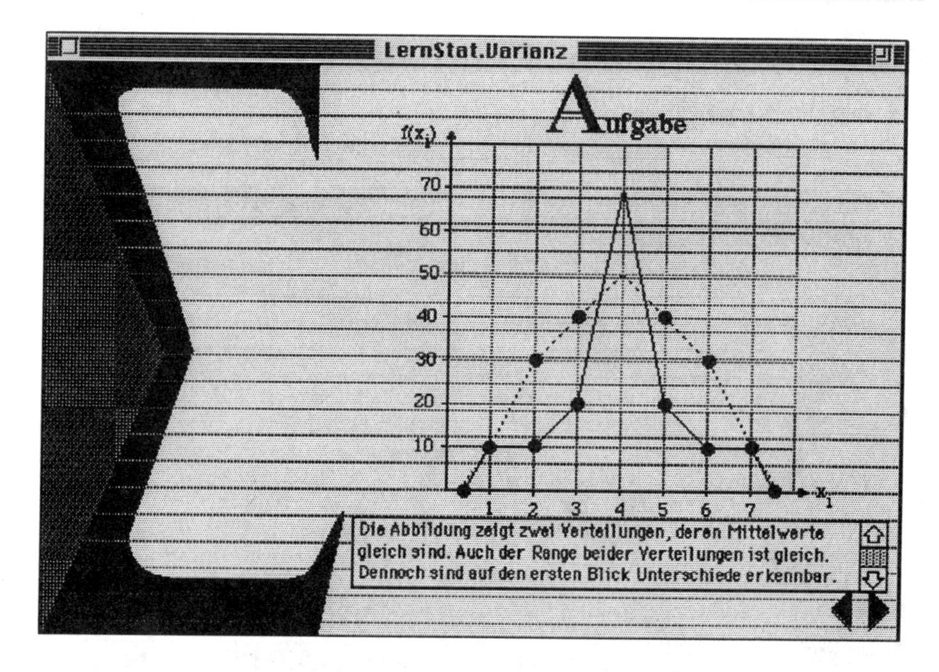

Figure 7: Dispersion or deviation

The next difficulty in statistics is to grasp the idea of a relation of two variables. A visual display of the distribution of two variables helps to facilitate the understanding of the relation of two variables. Students may enter and manipulate the values, observe how parameters change in the distribution, observe the effect of extreme values on the mean etc.

Correlation exercises try to stress the concept of covarying of pairs of variables. The basic correlation exercise presents two data rows. Students may change the values. They are then asked to predict if the calculation will result in a big or low, a positive or a negative correlation. The manipulation demystifies correlation by relating it to data rows. This enactive cognitive level is for most students a necessary condition for developing a more formal concept of correlation.

A table is a relational concept of the next higher order, the concept of amount becomes frequency, the concept of relation is one of cells in tables. For non-parametric correlations the exercises follow a similar principle as the exercise above. As shown in the next example, an exercise teaching Chi-square, students may vary the data in the table, observe the changes in the expected values and in Chi-square. "Learning by variation and repetition" may help students to grasp a notion of the kind of relations present in the

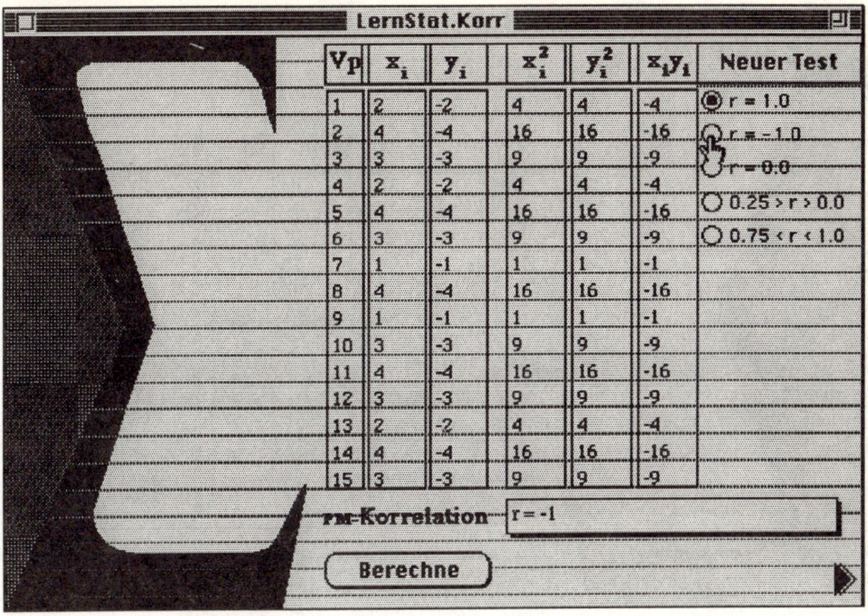

LernStat.Korr

Vp	x_i	y_i	x_i^2	y_i^2	$x_i y_i$
1	2	-2	4	4	-4
2	4	-4	16	16	-16
3	3	-3	9	9	-9
4	2	-2	4	4	-4
5	4	-4	16	16	-16
6	3	-3	9	9	-9
7	1	-1	1	1	-1
8	4	-4	16	16	-16
9	1	-1	1	1	-1
10	3	-3	9	9	-9
11	4	-4	16	16	-16
12	3	-3	9	9	-9
13	2	-2	4	4	-4
14	4	-4	16	16	-16
15	3	-3	9	9	-9

Neuer Test
- r = 1.0
- r = -1.0
- r = 0.0
- 0.25 > r > 0.0
- 0.75 < r < 1.0

rxy-Korrelation r = -1

Berechne

Figure 8: Correlation: Togetherness as a case of covariance

table.

The logical problem of selecting the correct type of correlation to compute is conceptualised in "LearnStat" as a decision matrix containing the dependencies of scales and correlations as rules. The two dimensions of the matrix are formed by the quality of measurement scales of two variables. The student is asked to select a certain type of correlation for his problem and is then prompted to find out if his decision was correct. This is done in a way similar to a mini-expert system.

Exercises in the tutorial for factor analysis include an example of rotating data in a three-dimensional scatterplot. The three-dimensional case provides a better understanding of the motivation underlying the method of factor analysis, the "Why" of the factor analysis problem. Only if factor analysis is motivated reasonably enough then there is any use in proceeding to the technical aspects of the factor analysis. Figure 11 is a still from a screen movie which demonstrates a three-dimensional data distribution and rotation. The movie offers an insight into the dimensional relation of data viewed from different angles. The scatterplot is in the movie differentiated by viewpoints, by subsets, and by using time as a fourth dimension to focus on the data.

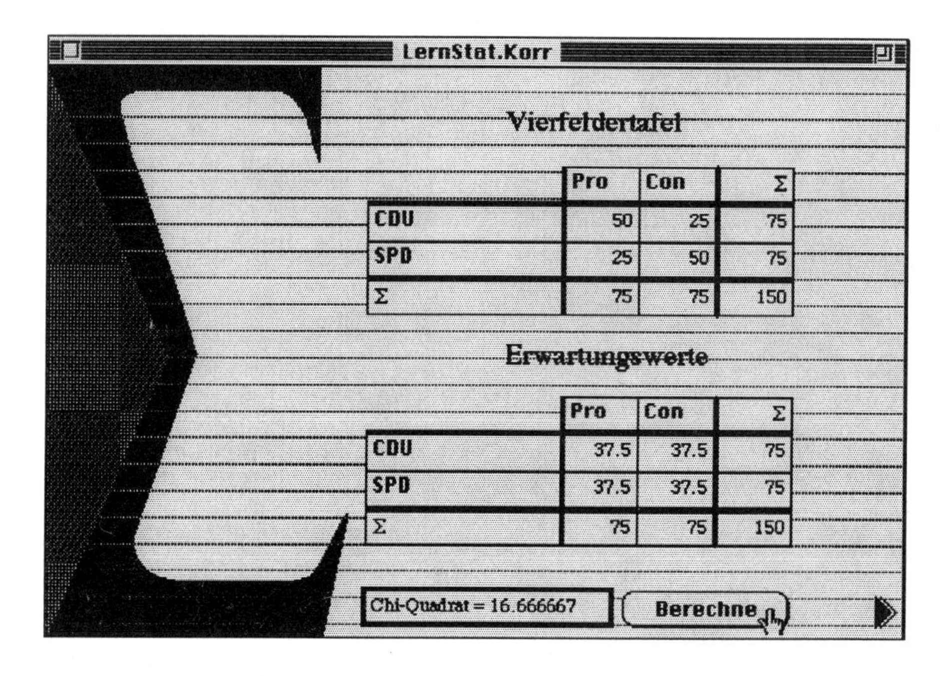

Figure 9: Chi-square exercise

The technical aspects, the "How" of factor analysis is dealt with by offering a complete factor analysis, split up into 13 steps which demonstrate how matrix calculation solves the problem. Nowadays it is no longer demanded that students actually learn how to calculate a factor analysis. But nevertheless a certain degree of insight into matrix calculation seems to form a sound basis for the understanding of the concept of "factors".

Interpretations and Conclusions

So far we did not have the chance to gather much experience with this new approach. The program was tested once with a group of 25 students. We made the mistake of accepting students for the seminar who had never seen a computer before. Thus we had to deal with two difficulties at once: The inexperience to use a computer and the problem of learning statistics. However, after a somewhat turbulent start, students were able to use "LearnStat" on their own, in the classroom as well as for self-paced studies during the week. It was often observed that they repeated the exercises more often than they would have done with paper-and-pencils tests, thereby varying the respective parameters of the exercises. A formal evaluation was

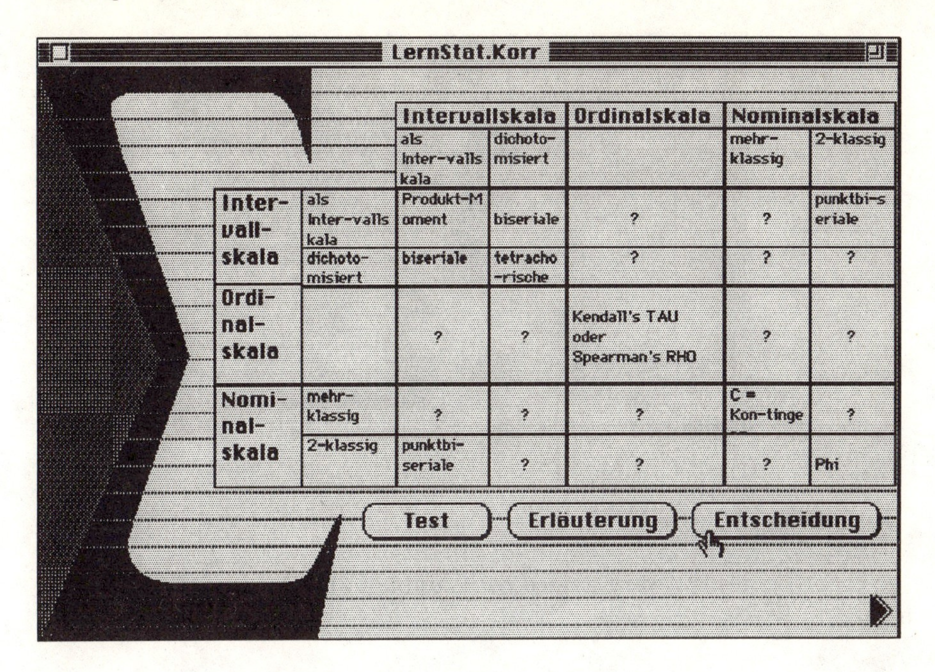

Figure 10: The concept of dependencies

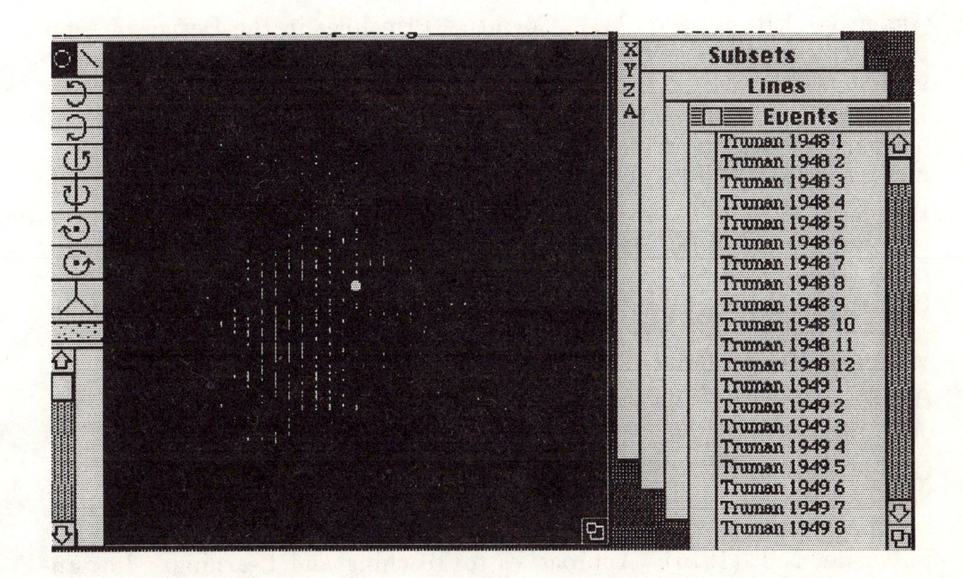

Figure 11: Motivating the notion of multivariate relations

not yet performed for this first trial. But my informal impression was that

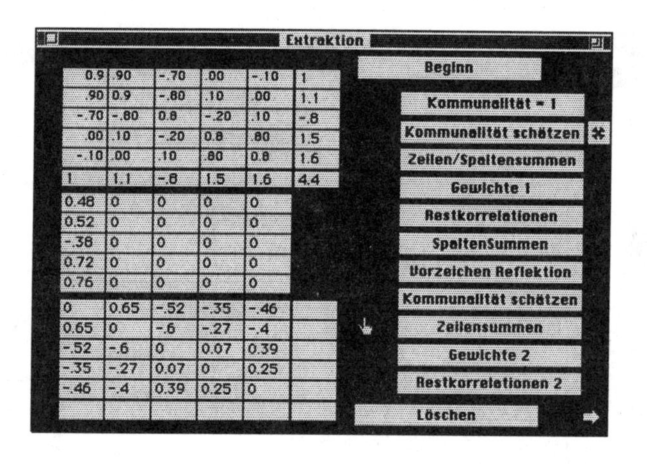

Figure 12: Facilitating complex concepts by splitting complexity

the cognitive results were indeed comparable to those of our former experiments.

References

Anderson, J.R. et al. (1984): Cognitive Principles in the Design of Computer Tutors. Pittsburgh 1984. Carnegie-Mellon-University, Technical Report ONR-84-1.

Birkhan, G. & Schulmeister, R. (1983): Untersuchung kognitiver Probleme beim Lernen der Statistik: Kognitive Operationen und Lernstile. -In: Schulmeister (1983), 44-84.

Bogun, M., Erben, C. & Schulmeister, R. (1983): Einführung in die Statistik. Ein Lernbuch für Psychologen und Sozialwissenschaftler. Weinheim/ Basel: Beltz 1983.

Bogun, M., Erben, C. & Schulmeister, R. (1983): Begleitbuch zur Einführung in die Statistik. Didaktischer Kommentar und Unterrichtsübungen. Weinheim/Basel: Beltz 1983.

Bruner, J.S. (1964): The Course of Cognitive Growth. American Psychologist 19 (1964) 1ff.

Computer Board for Universities and Research Councils (ed.) (1983): Report of a Working Party on Computer facilities for teaching in universities. London 1983.

Entwistle, N.J. (1976): Approaches to Teaching and Learning. -In: Entwistle, N.J. (ed.): Strategies for Reasearch and Development in Higher Education. Amsterdam 1976, 9ff.

Fankhänel, K., Schlageter, G. & Stern, W. (1988): Lehrsysteme für Personal Computer. Autoren- und Tutorsysteme. Fernuniversität Hagen 1988.

Fischer, P.M. (1985): Wissenserwerb mit interaktiven Feedbacksystemen. -In: Mandl & Fischer (1985), 68ff.

Goldman, R.D. (1972): Effects of Logical Versus a Mnemonic Learning Strategy on Performance in Two Undergraduate Psychology Classes. J. of Educational Psychology 63 (1972) 347ff.

Gorny, P. (1982): New Information Technologies in Education in the Federal Republic of Germany. European Journal of Education 17 (1982) 339ff.

Kulik, J.A., Kulik, C.L.C. & Cohen, P.A. (1980): Effectiveness of computer-based college teaching: A meta-analysis of findings. Review of Educational Research 50 (1980) 525ff.

Leiblum, M.D. (1981): Factors overlooked and underestimated in the selection and success of CAL as an instructional Medium. Computers in Education 1981, 277ff.

Mandl, H. & Fischer, P.M. (eds.) (1985): Lernen im Dialog mit dem Computer. München/Wien/Baltimore 1985.

Ottmann, Th.: Entwicklung und Einsatz computergestützter Unterrichtslektionen für den Informatikunterricht an der Hochschule. CAK 4 (1987) 63ff. (Computer Anwendungen Universität Karlsruhe).

Ottmann, Th. & Widmayer, P. (1986): Modellversuch computergestützter Informatikunterricht: Algorithmen und Datenstrukturen. -In: GI-Fachtagung: Informatikgrundbildung in Schule und Beruf. Kaiserslautern 1986. Springer-Verlag (=Informatik Fachberichte 129).

Pask, G. & Scott, B.C.E. (1972): Learning Strategies and Individual Competence. International Journal of Man-Machine Studies 4 (1972) 217ff.

Piaget, J. (1972): Intellectual Evolution from Adolescence to Adulthood, Human Development, 15, 3ff

Rheinberg, F. (1985): Motivationsanalysen zur Interaktion mit Computern. -In: Mandl & Fischer (1985), 83ff.

Schulmeister, R. (ed.) (1983): Angst vor Statistik. Empirische Untersuchungen zum Problem des Statistik-Lehrens und -Lernens. Hamburg: AHD 1983.

Schulmeister, R. & Birkhan, G. (1983): Untersuchung kognitiver Probleme beim Erlernen der Statistik: Denkniveaus und kognitive Komplexität. -In: Schulmeister (1983).

Schulmeister, R.(1989): Autorensysteme und Alternativen. Hochschulausbildung. Zeitschrift für Hochschuldidaktik und Hochschulforschung 1989. Also: Eberle (ed.): Informationstechnik in der Juristenausbildung. München: Beck 1989. Also: CAK 5/6 (= Computer Anwendungen - Computer Applications) Universität Karlsruhe 1989.

Schulmeister, R.: Authoring Systems and Alternatives. Wheels for the Mind Europe 1989.

Skinner, B.F. (1953): Science and Human Behavior. Macmillan Co. 1953.

Skinner, B.F. (1954): The science of learning and the art of teaching. Harvard Educational Review 24 (1954) 86ff.

Skinner, B.F. (1958): Teaching machines. Science 128 (1958) 969ff.

Spada, H. & Opwis, K. (1985): Intelligente tutorielle Systeme aus psychologischer Sicht. -In: Mandl & Fischer (1985), 13ff.

Streitz, N.A. (1985): Kognitionspsychologische Grundlagen der Gestaltung von Dialogstrukturen bei interaktiven Lehr-Lern-Systemen. -In: Mandl & Fischer (1985), 54ff.

Remarks on Protecting Patient Data Against Misuse and on its Consequences Concerning their Statistical Data Analysis

Reinhold Haux
University of Heidelberg, Institute for Medical Biometry and Informatics,
Im Neuenheimer Feld 400, D-69120 Heidelberg, Germany

Key Words:
statistical analysis systems, data protection, medical biometry.

Abstract
Besides the need for easily collecting, storing and statistically analyzing data there is also the need for protecting data against misuse. For medical data, especially for patient data in hospitals, special care has to be taken concerning data protection. Storage and statistical analysis of such data for medical research usually is only allowed by law and with informed consent of the patient. Online networks of statistical analysis systems in multicenter clinical trials and registers should be avoided. Often this can be done by simple organizational procedures and without reducing the semantic integrity of the data and the quality of statistical data analysis.

Introduction

This paper discusses in a simplified and somewhat superficial manner aspects of data protection with respect to storing and analyzing *medical* (esp. patient) data for research purposes, one of the most sensitive areas of statistical data analysis. Here (multicenter) clinical trials as well as clinical registers have to be considered. As data protection law differs for each country (and as it even may differ within each state of a country) I will focus my remarks on data protection aspects with respect to Germany under consideration of the federal German law, including state laws. This will be done due to my (former) experiences in designing, conducting and analyzing clinical studies, and due to my (present) work on hospital information systems. Answers will be proposed for the following questions of interest:

1. How can data protection be achieved in the context of statistical data analysis?

2. What are the consequences for the organization of clinical (especially multicenter) trials and registers?

3. What are the consequences for the construction of statistical analysis systems?

It is assumed that the reader is familiar with clinical trials and registers as well as with its general ethical rules e.g. the declaration of Helsinki and the rules for good clinical practice. For the sake of simplicity databases for storing data of clinical trials, clinical registers, etc. are all regarded as databases of statistical analysis systems. Obviously they could also be stored separately in database systems or in other application systems. In addition it is also assumed that for a (multicenter) clinical trial or register there exists one 'central' database on the logical level (independent of its kind of distribution on the physical level of the database).

Within this paper only some aspects of data protection will be discussed. Data protection and encoding techniques within computer networks e.g. will not be mentioned. The argumentation will focus on (prolective and prospective) multicenter clinical trials and registers. Similar arguments for patient records could be given.

References on the topic of this paper are contained in Haag et.al. (1992) and Mann & Haux (1993).

Laws for analyzing medical data

In order to analyze data of patients admitted to hospital, there exist in Germany a variety of federal and state laws. Some of them will be mentioned here:

- The state data protection laws (Landesdatenschutzgesetze) in combination with the federal law on data protection (Bundesdatenschutzgesetz) specifies in general data protection aspects for data stored on computer, e.g. which data are allowed to be stored in a machine readable way, for which technical and organizational procedures one has to take care in order to prevent personal data against misuse, and how an informed consent of a person has to be achieved with respect to storing her or his data.

- The state hospital laws (Landeskrankenhausgesetze) among others specify how and to which extent data may be stored, collected and used with respect to in- and outpatients in hospitals. In the context of this paper they can be regarded as refinements of the data protection laws.

- The federal social law, esp. its 5th volume (Sozialgesetzbuch, 5.Buch), demands that (multicenter) quality assurance of hospitals using patient data should be done. In its new 1993 revision (Gesundheitsstrukturgesetz) a variety of new specifications for documenting diagnoses and medical procedures, and for the general exchange of basic, esp. administrative data on chip card (Krankenversichertenkarte) are given.

- The state archiving laws (Landesarchivgesetze) specify for universities and for their hospitals how documents have to be stored in archives.

- There exist different laws on how to obtain comparative or nationwide statistics, e.g. on diagnoses of patients (Krankenhausstatistikgesetz, Bundespflegesatzverordnung, Pflegepersonalregelung,...).

- There exist different laws on registers for certain diseases, mainly for cancer registers (Krebsregistergesetze).

The intention of all laws may be summarized in the following, simplified way: Analyzing patient data has explicitely to be allowed by law, otherwise it is prohibited. Information processing in hospitals is mainly concerned with the collection, storage and analysis of data for patient care and for administrative purposes (billing, controlling). For quality assurance and for medical research data can also be stored and analyzed. It should be done

- (with exceptions) in an anonymized way, so that a certain patient can not be reidentified by data, stored for research purposes, and

- (especially for clinical registers) inside a hospital.

Networking of statistical analysis systems

Especially for multicenter clinical trials and registers as well as for 'distributed computing' there is a demand of having computer networks. Nationwide and international computer networks facilitate the exchange of data and knowledge as well as, with respect to distributed computing, the efficient use of computing resources.

Because of the above mentioned data protection aspects for patient data we should be careful in networking statistical analysis systems between different hospitals. Let us consider the semantic data integrity in multicenter clinical trials and registers. In order to obtain high quality data the semantic integrity has to be checked as early as possible. Without any online communication between hospitals this can be achieved on the attribute level and on the level of 'cases' admitted within a hospital. The most important aspect here is, that data input has to be done as early as possible after asking for, measuring or observing the data, so that corrections are possible.

Only for the case of referential integrity aspects on the database of a trial or register as a whole, access is needed to all data. E.g. questions like "Has a certain patient of a new inpatient case in a certain hospital, relevant for a certain clinical multicenter register, already been admitted in another hospital contributing data to this register?". In such a case it has usually to be checked whether some basic data such as the birth date of the patient is identical. However there may also exist consequences e.g. on (stratified) randomization. Usually integrity constraints for referential integrity do not necessarily have to be checked immediately online. Often a telephone call may suffice here or data exchange on disc. The same holds usually also for having online access to the central database of a trial or a register.

The need for distributed computing in medical biometry concerning the statistical analysis of patient data is in general also not of great importance (if at all), as here sample sizes usually are relative small and the algorithms for statistical data analysis (for data description, for confirmatory and exploratory data analysis) usually are 'not much' time consuming. It can even be observed that biometrical centers for clinical trials and for clinical registers no longer see the need for storing and analyzing patient data on mainframe computers and that statistical analysis systems on workstations are not only powerful enough but also have better user interfaces.

In any way attributes that enable a patient identification have to be omitted as far as possible. However we know, that data reidentification procedures exist, even if the patients name and other attributes like birth date are not or at least only partially stored.

Hence, at least in my opinion, storage and analysis of patient data in statistical analysis systems, which are not online interfaced between different hospitals, is usually sufficient for clinical trials and registers and also supports data protection.

Constructing statistical analysis systems

As data storage and analysis is usually done in one (biometrical) center and/or within one hospital, and considering the arguments mentioned above, there is in my opinion no strong need to consider strategies for distributed database management or for secure communication protocols for communicaton in national or international networks. It is of more importance to keep the workstation or the local network of workstations on which the data is stored separate from (inter)national computer networks.

In addition it has to be taken care especially for organizational restrictions such as preventing access for not authorized persons to the room the computer is in. Technical restrictions are mainly user access control by password either to the workstation or to the database containing the trial data or the register.

Hence, at least in my opinion, there is no need in considering special aspects for constructing statistical analysis systems with respect to data protection that go beyond considerations that are well known for database systems.

Final remarks

As noted in the first section, only some aspects of data protection could be discussed here. So, e.g., appropriate data protection and encoding techniques within computer networks have not been discussed. In addition, clinical trials and registers within hospitals have been the basis of my argumentation. Taking epidemiological studies, the argumentation would have to be different.

As quintessence let me propose three (may be subjectively biased) answers on the questions mentioned above.

Answer to question 1: "How can data protection be achieved in the context of statistical data analysis?". It can be achieved by considering the existing laws, mainly those dealing with data storage for medical research and for data protection.

Answer to question 2: "What are the consequences for the organization of clinical (especially multicenter) trials and registers?". Computers (independent if they are mainframes, workstations or personal computers), that store data of clinical trials and registers should not be interfaced to (inter)national computer networks. Well known technical and organizational procedures for data protection should be considered.

Answer to question 3: "What are the consequences for the construction of statistical analysis systems?". Although there is a considerable need to improve the architecture of statistical analysis systems in a variety of aspects, there is no special need with respect to data protection, that goes beyond considerations that are well known for the architecture of database systems.

References

Note: References may serve as 'pointers' especially to references on constructing statistical analysis systems, and on database design aspects.

Haag, U., Haux, R. & Kieser, M. (1992). Statistische Auswertungssysteme. Stuttgart: Fischer.

Mann, G., & Haux, R. (1993). Database Scheme Design for Clinical Studies Based on a Semantic Data Model. Comp. Statist. & Data Analysis 15, 81-108.

Index

χ^2-metric, 243

accuracy, 111
additive model
    generalized, 327, 338, 381, 435, 438
adjustment
    covariate, 469
adjustment of p-values, 483
agglomerative algorithms, 503
algebra
    incidence, 355, 365
    rank, 366
algorithm
    backfitting, 443
    CFA, 360
    EM, 345, 349, 437
allocation
    optimum, 191
alpha design, 196
alternagraphics, 290
analysis
    cluster, 497, 498
    complete case, 345, 346
    configural frequency, 355
    correspondence, 285
    data, 37, 237, 277
    environmental, 295
    exploratory, 317
    exploratory data, 259, 276, 316
    factor, 527
    principal component, 301
    procrustes, 291
    prospecting data, 276
    sensitivity, 177, 183
    survival, 435, 451
analysis of variance, 191, 192
analysis system, 8
Andrews plot, 281
animated data, 524
animation, 266
anxiety
    statistics, 513, 514
approximation
    saddlepoint, 395
arrow hypothesis, 419
arrow test, 422
arrow test strategy, 419
assurance
    quality, 111
asymptotic properties, 159, 190
asymptotic relative precision, 473
authoring systems, 513, 516, 517
autocorrelation
    spatial, 307
average derivative estimation, 327, 337

backfitting algorithm, 443
balanced design, 197
bandwidth, 162, 246, 440
base
    data, 73
behaviorism, 519
behavioristic bias, 521
behaviour
    tail, 240
Beyer ratio, 131
bias
    behavioristic, 521
    omission, 469
    selection, 345, 347

binary data, 407
binomial coefficients, 355
binomial data, 395, 407
biplot, 284
block design, 187, 197
Bonferroni global test, 420
Bonferroni inequality
    improved, 488
bootstrap, 160, 183, 287, 395, 406
    iterated, 143, 154
    two-step, 143, 154
    world, 144
box & whisker-plot, 251
boxplot, 318, 319
branch-and-bound algorithms, 499

CADEMO, 223, 225, 226
CAI, 519
CART, 483
categorical data, 159
censored data, 239, 483
censored survival data, 475
censored survival time data, 435
censoring
    interval, 451
CFA, 355, 360
CFA algorithm, 360
choropleth-map, 305
chronic, 17
classification, 295, 497, 498
    visual, 298
classification and regression trees, 483
clinical trial, 451, 469
closure test, 419
cluster analysis, 497, 498
cluster validation, 497, 498
coefficient
    binomial, 355
communication protocols, 537
comparison of means, 194
complete case analysis, 345, 346
computational statistics, 37
    history, 3
computer, 37
computer assisted instruction, 519
computer science, 177, 185
computing
    statistical, 37
computing language, 56
conditional scatterplot, 279
confidence intervals, 327, 334
confidence sets, 145
configural frequency analysis, 355
configuration, 357
confounding, 197, 199
contingency table, 159, 355–357, 397
control
    quality, 181
convergence
    rates-of-, 150
correlation, 526, 527
    spatial auto-, 307
correspondence analysis, 285
    joint, 287
    multiple, 287
covariate adjustment, 469
coverage probability, 143, 150
Cox model, 475
    grouped, 392
cross-validation, 160

data
    animated, 524
    binary, 407
    binomial, 395, 407
    categorical, 159
    censored, 239, 483
    censored survival, 475
    discrete, 379
    grouped survival, 379
    longitudinal, 379
    Longley, 111
    missing, 239
    multinomial, 163, 395, 397

sparse, 395, 397
spatial, 295, 298, 315, 318, 319, 324
survival, 475
data analysis, 37, 237, 277
exploratory, 259, 276, 316
languages, 295, 296
prospecting, 276
data base, 73
data base management, 78
Data Desk, 319, 325
data graphics, 61
data protection, 533
clinical trials, 533
laws, 534
data sets
large, 5, 295, 313
dendrogram, 502
densitogram, 252
density
smoothed, 246
density estimator, 237
histogram, 244
histogram density, 244
kernel, 246, 327, 328
dependence among hypotheses
logical, 419
design
alpha, 196
balanced, 197
block, 187, 197
experimental, 187, 205
factorial, 187, 197, 198
incomplete, 197, 198
mixture, 187, 197
optimal, 187, 191, 195, 201
determination
sample size, 187, 192, 203
development
technological, 37
deviance, 395, 396
deviation, 525
diagnostic plots, 237
diagnostics
regression, 270
dialogue system, 203
different scales, 483
discovery learning, 513, 522, 523
discrete data, 379
discrete time series, 381
discretization, 240
disease mapping, 295
dispersion, 525
display
multiwindow, 279
dissimilarity matrix, 497
distribution
exact, 355, 356
hypergeometric, 355
hypergeometric iterated, 358, 359
permutation, 357
distribution function, 248
divisive algorithms, 510
dynamic graphics, 259, 290, 298
dynamic model, 379
dynamic programing algorithms, 499

EDA, 295
edgeworth expansion, 395, 403
effect
treatment, 469
EM algorithm, 345, 349, 437
environment
programming, 71
environmental analysis, 295
epidemiology, 177
error resistence, 67, 68
estimates
kernel, 438
estimation
average derivative, 327, 337
kernel density, 246
maximum likelihood, 348
nonparametric, 327
estimator

density, 237
kernel density, 327, 328
evaluation
program, 6, 10
software, 6, 10, 12–14
exact distribution, 355, 356
excess mass, 237, 253
expansion
edgeworth, 395, 403
saddlepoint, 403
expectation maximization algorithm, 349
experimental design, 187, 205
experiments
Monte Carlo, 159
Monte-Carlo, 165
expert system, 203, 211, 212, 222
statistical, 211, 217
exploratory analysis, 317
exploratory data analysis, 259, 276, 316
external clustering indices, 506

factor analysis, 527
factorial design, 187, 197, 198
factors
prognostic, 483
filtering, 379
finite-sample properties, 159
Freeman-Tukey's statistic, 396
frequency polygon, 245
function
distribution, 248
hazard, 436
risk, 435
fuzzy clustering, 502

generalized additive model, 327, 338, 381, 435, 438, 442
generalized regression, 483
generator
inversive congruential, 125
linear congruential, 125
matrix, 125
multiple-recursive, 138
pseudorandom number, 125
geographic information system, 315
Gibbs sampling, 389
GIS, 315, 318
global test
Bonferroni, 420
goodness of fit, 243, 247, 395
grand tour, 268, 290
graphics, 77
data, 61
dynamic, 259, 290, 298
interactive, 295, 296, 315, 316, 318
interactive statistical, 325
multivariate, 277
presentation, 77
grouped survival data, 379

h-plot, 285
hazard function, 436
heuristic strategy, 211, 214, 217, 227, 229
hierarchical classification, 498
high-dimensional parameter, 143
histogram, 242
histogram density estimator, 244
history, 37
of the working groups, 17
history of computational statistics, 3
HOMALS, 289
hypergeometric distribution, 355
iterated, 358, 359
hypergeometric matrix, 358
hypertext systems, 516
hypotheses
logical dependence among, 419
nested systems of, 419
hypothesis
arrow, 419

ICD, 305

icon, 57, 278
Ille-et-Vilaine study, 412
implementation of kernel estimates, 327, 333
improved Bonferroni inequality, 488
incidence algebra, 355, 365
incomplete design, 197, 198
inequality
 improved Bonferroni, 488
infant mortality, 410
inference
 Monte Carlo statistical, 185
inference mechanism, 211
information service, 6, 13
information system
 geographic, 315
instruction
 computer assisted, 519
integrated mean square error, 244
interactions, 483
interactive graphics, 295, 296, 315, 316, 318
interactive statistical graphics, 325
interactivity, 54
interface issue, 64
internal clustering indices, 506
interpolation
 plot, 266
interval censoring, 451
intervals
 confidence, 327, 334
invariant, 442
inversive congruential generators, 125
isosurfaces, 303
ISP, 55, 73, 319
issue
 interface, 64
iterated bootstrap, 143, 154
iterated hypergeometric distribution, 358, 359
iterative relocation algorithms, 499

jackknife, 159, 160
JCA, 287
joint correspondence analysis, 287

k-means algorithms, 510
k-nearest neighbor, 438
Kalman filtering, 386
kernel density estimation, 327
kernel density estimator, 246, 328
kernel estimates, 438
 implementation of, 327, 333
kernel regression, 327, 334
knowledge representation, 211, 227
knowledge-based system, 5, 7, 211–217, 223, 225, 227, 229, 230
Kolmogorov-Smirnov, 248
Kuiper metric, 254

language
 computing, 56
 natural, 56
languages
 data analysis, 295, 296
large data sets, 5, 295, 313
lattice structure, 125
learning, 513
 discovery, 513, 522, 523
learning by doing, 524
likelihood
 partial, 436
 penalized, 380, 385
 weighted, 440
likelihood ratio statistic, 395
linear congruential generators, 125
linear model, 472
linking, 265
Lisp-Stat, 259
log-rank statistic, 487, 489
logical dependence among hypotheses, 419
logistic model, 474
longitudinal data, 379
Longley data, 111

LOWESS, 281

Möbius matrix, 367
management
    data base, 78
map, 312
    choropleth, 305
    thematical, 295–297
mass
    excess, 237, 253
matrix
    hypergeometric, 358
    Möbius, 367
matrix generators, 125
maximal period, 129
maximally selected test statistic, 484, 488
maximum likelihood
    estimation, 348
    principle, 345
MCA, 287
MDS, 285
mean square error
    integrated, 244
means
    comparison of, 194
measurement, 289
mechanism
    inference, 211
medical biometry, 533
metric
    χ^2, 243
    Kuiper, 254
Minkowski reduced bases, 125
missing at random, 349
missing data, 239
missing values, 345
misspecification, 469
mixture design, 197
model
    Cox, 475
    dynamic, 379
    generalized additive, 327, 338, 435, 442
    linear, 472
    logistic, 179, 474
    regression, 350, 469
    search, 419
    selection, 188, 238
    single index, 327, 337
    special - on lattices, 295
    state space, 379
Monte Carlo, 37, 143, 148
    experiments, 159
    statistical inference, 185
    tests, 177, 183, 507
Monte-Carlo
    experiments, 165
mortality
    infant, 410
multicategorical time series, 383
multidimensional scaling, 285
multimodality, 240, 252
multinomial data, 163, 395, 397
multiple correspondence analysis, 287
multiple testing, 419
multiple-recursive generator, 138
multivariate graphics, 277
multivariate regression, 327, 336
multiwindow display, 279

natural language, 56
nested systems of hypotheses, 419
networking, 535
neural networks, 502
Newton Raphson, 349
NLM, 285
nonlinear mapping, 285
nonlinear principal component analysis, 289
nonlinear regression, 191
nonparametric estimation, 327
nonparametric regression, 435
NPCA, 289

object oriented programming, 60, 259
odds ratio, 159
omission bias, 469
optimal design, 187, 191, 195, 201
optimum allocation, 191
overspecification, 471

P-P-plot, 250
p-values
  adjustment of, 483
parallel computers, 502
parallel coordinates, 281
parameter
  high-dimensional, 143
partial likelihood, 436
partition, 498
pattern matrix, 497
Pearson's statistic, 396
penalized likelihood, 380, 385
percentage-percentage-plot, 250
perception, 240
period
  maximal, 129
permutation distribution, 357
plot
  Andrews, 281
  box & whisker, 251
  h-, 285
  P-P, 250
  percentage-percentage, 250
  principle component, 284
  projection, 285
  Q-Q, 249
  quantile-quantile, 249
  rotating, 318
  shorth, 237, 254
plot interpolation, 266
plots
  diagnostic, 237
Poisson model, 505
polygon
  frequency, 245
portability, 6, 8–11, 13
posets, 355
posterior mode smoothing, 385
power-divergence statistic, 395
precesion, 188
precision, 190
  asymptotic relative, 473
presentation graphics, 77
prevalence, 305
principal component analysis, 301
  nonlinear, 289
principle component plot, 284
probability
  coverage, 143, 150
problem
  selection, 203
Procedure 4, 420
procrustes analysis, 291
prognostic factors, 483
prognostic subgroup, 484
program evaluation, 6, 10, 13
program packages, 6–8
programming
  object oriented, 60, 259
programming environment, 71
projection plot, 285
projection pursuit, 290, 327, 330, 337
Prolog, 221, 222
properties
  asymptotic, 159, 190
  finite-sample, 159
prospecting data analysis, 276
prototyping
  rapid, 77, 312
pseudorandom number generators, 125

Q-Q-plot, 249
quality assurance, 111
quality control, 181
quantile-quantile-plot, 249

random dissimilarity matrix model, 500
rank algebra, 366
rank preserving, 367
rapid prototyping, 312
rates-of-convergence, 150
ratio
  Beyer, 131
  likelihood – statistic, 395
reciprocal nearest neighbours, 504
record keeping, 69, 70, 75
reducibility, 503
referential integrity, 536
regression
  generalized, 483
  kernel, 327, 334
  multivariate, 327, 336
  nonlinear, 191
  nonparametric, 435
  sliced inverse, 283
regression diagnostics, 270
regression model, 350, 469
Reisensburg Symposia, 4, 10, 12, 14
relative clustering indices, 506
relative risk, 477
representation
  knowledge, 211
residual, 277
resistence
  error, 67, 68
risk
  relative, 477
risk function, 435
rotating plot, 318
rotation, 264

S-Plus, 319
saddlepoint approximation, 395
saddlepoint expansion, 403
sample size, 190
sample size determination, 187, 192, 203
sampling
  Gibbs, 389
scales
  different, 483
scatterplot, 245, 318, 320, 324
  conditional, 279
  matrix, 280
scatterplot brushing, 290
scatterplot-matrix, 301, 306
scatterplotsmoother, 281
search
  model, 419
selection
  model, 188, 238
selection bias, 345, 347
selection problem, 203
sensitivity analysis, 177, 183
service
  information, 13
set
  confidence, 145
shorth plot, 237
shorth-plot, 254
significance test, 276
  controversy, 276
silhouette, 252
simulated annealing algorithms, 501
simulation study, 476
single index model, 327, 337
single pass algorithm, 500
singular value decomposition, 283
size
  sample, 190
skewness, 240
sliced inverse regression, 283
small tour, 289
smoothed density, 246
smoothing, 379
  posterior mode, 385
  splines, 438
software
  statistical, 37
software evaluation, 6, 10, 12–14

software information, 9, 10
software testing, 111
sparse data, 395, 397
spatial autocorrelation, 307
spatial data, 295, 298, 315, 318, 319, 324
special model on lattices, 295
spline smoothing, 438
split plot, 198
stability, 508
statistic
  Freeman-Tukey's, 396
  likelihood ratio, 395
  log-rank, 487, 489
  Pearson's, 396
  power-divergence, 395
statistical analysis systems, 37, 185, 533
statistical computing, 37
statistical expert system, 211, 217
statistical graphics
  interactive, 325
statistical inference
  Monte Carlo, 185
statistical software, 37
Statistical Software Newsletter, 12
statistical system, 8–10, 12–14
statistics, 37, 177, 182
  computational, 37
statistics anxiety, 513, 514
stopping rules, 499
strategy
  arrow test, 419
  heuristic, 211, 214, 217, 227, 229
study
  Ille-et-Vilaine, 412
  simulation, 476
study types, 178
subgroup
  prognostic, 484
survival analysis, 435, 451
survival data, 475
  censored, 475
  grouped, 379
survival time data
  censored, 435
symmetry, 240, 246
system
  analysis, 8
  authoring, 513, 516, 517
  dialogue, 203
  expert, 203, 211, 212, 222
  geographic information, 315
  hypertext, 516
  knowledge-based, 5, 7, 211–217, 223, 225, 227, 229, 230
  statistical, 8–10, 12–14
  statistical expert, 211, 217
systems
  statistical analysis, 37
systems of hypotheses
  nested, 419

table, 526
  contingency, 159, 355–357, 397
tail behaviour, 240
technological development, 37
test
  arrow, 422
  Bonferroni global, 420
  closure, 419
  Monte Carlo, 177, 183
  significance, 276
test statistic
  maximally selected, 484, 488
testing
  multiple, 419
thematical maps, 295–297
time series, 379
  discrete, 381
  multicategorical, 383
tour
  grand, 268, 290
  small, 289
transformation, 239

treatment effect, 469
trees
    classification and regression, 483
trial
    clinical, 451, 469
    variety, 187
two-step bootstrap, 143, 154

ultrametric, 502
underspecification, 471
uniformity model, 512
unimodal model, 505

validation
    cluster, 497, 498
validity, 177, 181
variance
    analysis of, 191, 192
visual classification, 298
visualisation, 523

WAMASTEX, 223, 226–229
weighted likelihood, 440
working groups history, 17

Alphabetical List of Authors, Reviewers and Editors

Klaus Abt	Abteilung für Biomathematik, Universität Frankfurt, Germany
Gerd Antes	Klinik für Tumorbiologie, Freiburg, Germany
Peter Bauer	Institut für Medizinische Statistik, Universität Wien, Austria
Richard Becker	AT&T Bell Laboratories, Murray Hill NJ, USA
Axel Benner	Deutsches Krebsforschungszentrum, Heidelberg, Germany
Rudolf Beran	Department of Statistics, University of California, Berkeley CA, USA
Gudrun Bernhard	PHARMA - Beratungsgesellschaft, Kronberg, Germany
Hans Hermann Bock	Institut für Statistik und Wirtschaftsmathematik, RWTH Aachen, Germany
Peter Darius	Laboratory of Statistics and Experimental Design, Katholieke Universiteit Leuven, Belgium
Peter Dirschedl	Institut für Medizinische Informationsverarbeitung, Biometrie und Epidemiologie, Universität München, Germany
Lutz Edler	Deutsches Krebsforschungszentrum, Heidelberg, Germany
Friedhelm Eicker	Fachbereich Statistik, Universität Dortmund, Germany

Ludwig Fahrmeir	Seminar für Statistik, Universität München, Germany
Michael Falk	Mathematisch-Geographische Fakultät, Katholische Universität Eichstätt, Germany
Uwe Feldmann	Institut für Mediz. Biometrie, Epidemiologie und Informatik, Homburg, Germany
Karl Fröschl	Institut für Statistik und Informatik, Universität Wien, Austria
Allan Gordon	Department of Mathematical and Computational Sciences, University of St.Andrews, Scotland
Uwe Haag	Institut für Medizinische Biometrie und Informatik, Universität Heidelberg, Germany
Wolfgang Härdle	Institut für Statistik und Ökonometrie, Humboldt-Universität zu Berlin, Germany
Reinholg Haux	Institut für Medizinische Biometrie und Informatik, Universität Heidelberg, Germany
Siegfried Heiler	Fakultät für Wirtschaftswissenschaften und Statistik, Universität Konstanz, Germany
Allmut Hörmann	medis Institut, GSF, Oberschleißheim, Germany
Gerhard Hommel	Institut für Medizinische Statistik und Dokumentation, Universität Mainz, Germany
Peter J. Huber	Lehrstuhl für Angewandte Mathematik, Universität Bayreuth, Germany
Karl-Heinz Jöckel	BIPS Bremer Institut für Präventionsforschung und Sozialmedizin, Bremen; Germany
Michael Kaplan	Institut für Mathematik, Technische Universität München, Germany
Meinhard Kieser	Schwabe Arzneimittel, Karlsruhe
Sigbert Klinke	C.O.R.E & Institute de Statistique, Université Catholique de Louvain, Lovain-La-Neuve, Belgium
Armin Koch	Institut für Medizinische Biometrie und Informatik, Universität Heidelberg

Wolfgang Köpcke	Institut für Mediz. Informatik u. Biomathematik, Universität Münster, Germany
Andreas Krause	Institut für Statistik und Ökonometrie, Universität Basel, Switzerland
Lothar Kreienbrock	Fachbereich Arbeitssicherheit und Umweltmedizin, Bergische Universität-GH Wuppertal, Germany
Jürgen Kübler	Pharmaforschungszentrum Bayer, Wuppertal, Germany
Joachim Kunert	Fachbereich Statistik, Universität Dortmund, Germany
Berthold Lausen	Forschungsinstitut für Kinderernährung, Dortmund, Germany
Walter Lehmacher	Institut für Biometrie und Epidemiologie, Tierärztliche Hochschule Hannover, Germany
Jürgen Lehn	Institut für Mathematik, Technische Hochschule Darmstadt, Germany
Enno Mammen	Institut für Stochastik, Fachbereich Mathematik, Humboldt-Universität zu Berlin, Germany
Herbert Matschinger	Zentralinstitut für Seelische Gesundheit, Mannheim, Germany
Marlene Müller	Institut für Statistik und Ökonometrie, Humboldt-Universität zu Berlin
Matthias Nagel	Zentrum für Epidemiologie und Gesundheitsforschung, Zwickau, Germany
Gerdard Osius	Institut für Statistik, Fachbereich Mathematik und Informatik, Universität Bremen
Rüdiger Ostermann	Hochschulrechenzentrum, Universität-GH Siegen, Germany
Iris Pigeot	Fachbereich Statistik, Universität Dortmund, Germany
Dieter Rasch	Deparment of Mathematics, Wageningen Agricultural University, The Netherlands

Stefan Rettig	AG Stochastik und Operation Research, Technische Hochschule Darmstadt, Germany
Joachim Röhmel	Institut für Arzneimittel, Bundesgesundheitsamt, Berlin, Germany
Wilhelm Sauerbrei	Institut für Medizinische Biometrie und Informatik, Universität Freiburg, Germany
Günther Sawitzki	Institut für Angewandte Mathematik, Universität Heidelberg, Germany
Claudia Schmoor	Institut für Medizinische Biometrie und Informatik, Universität Freiburg
Rainer Schnell	Lehrstuhl für Soziologie und Wissenschaftslehre, Universität Mannheim, Germany
Rolf Schulmeister	Interdisziplinäres Zentrum für Hochschuldidaktik, Universität Hamburg, Germany
Martin Schumacher	Institut für Medizinische Biometrie und Informatik, Universität Freiburg
Peter P. Sint	Österreichische Akademie der Wissenschaften, Wien, Austria
Bernd Streitberg	†
Christian Tismer	AFB-Parexel, Berlin, Germany
Gerhard Tutz	Institut für Statistik und Wirtschaftsmathematik, Technische Universität Berlin, Germany
Antony Unwin	Department of Statistics, Trinity College, University of Dublin, Ireland
Werner Vach	Institut für Medizinische Biometrie und Informatik, Universität Freiburg, Germany
Norbert Victor	Institut für Medizinische Biometrie und Informatik, Universität Heidelberg, Germany
Leland Wilkinson	SYSTAT Inc., Evanston IL, USA
Knut Wittkowski	Institut für Medizinische Biometrie, Universität Tübingen, Germany

Acknowledgements

Cordially we would like to thank the following sponsors who supplied the necessary financial backing

APPLE Computer GmbH, Ismaning, Germany
bias GmbH, München, Germany
BMDP Statistical Software Inc., Cork, Ireland
Boehringer Mannheim GmbH, Mannheim, Germany
Majih Software (DataDesk), Dublin, Ireland
Dr. Karl Thomae GmbH, Biberach, Germany
GraS GmbH (S-Plus), München, Germany
Hewlett-Packard GmbH, Böblingen, Germany
IDV Datenanalyse und Versuchsplanung, Gauting, Germany
SAS Institute, Heidelberg, Germany
SPSS GmbH Software, München, Germany
SUN Microsystems GmbH, Grasbrunn, Germany
StatCon gbR (SYSTAT), Witzenhausen, Germany
VGSPS mbH, Bonn, Germany